The Oxford Book of Essays

JOHN GROSS is the author of *The Rise and Fall of the Man of Letters* (1969) and editor of *The Oxford Book of Aphorisms* (1983), among other publications. He was editor of the *Times Literary Supplement* from 1974 to 1981, and is currently theatre critic of the *Sunday Telegraph*.

The Oxford Book of
Essays

Chosen and Edited by John Gross

Oxford   New York

OXFORD UNIVERSITY PRESS

Oxford University Press, Walton Street, Oxford OX2 6DP

Oxford New York
Athens Auckland Bangkok Bombay
Calcutta Cape Town Dar es Salaam Delhi
Florence Hong Kong Istanbul Karachi
Kuala Lumpur Madras Madrid Melbourne
Mexico City Nairobi Paris Singapore
Taipei Tokyo Toronto

and associated companies in
Berlin Ibadan

Oxford is a trade mark of Oxford University Press

Introduction and selection © John Gross 1991

First published 1991
First issued as an Oxford University Press paperback 1992

British Library Cataloguing in Publication Data

Data available

Library of Congress Cataloging in Publication Data
Data available
ISBN 0-19-282970-X

9 10

Printed in Great Britain by
Biddles Ltd
Guildford and King's Lynn

Contents

Acknowledgements

The editor and publisher are grateful for permission to include the following copyright material in this volume.

James Baldwin, 'Stranger in the Village', reprinted from *The Price of the Ticket*, © 1985 by James Baldwin, by permission of Michael Joseph Ltd., and Richard Curtis for and on behalf of the author.

Reyner Banham, 'The Crisp at the Crossroads', reprinted from *The Arts in Society*, ed. Paul Barker (Fontana/Collins, 1977), by permission of Collins Publishers.

Jacques Barzun, 'What If—? English Versus German and French', © 1986 by Jacques Barzun, reprinted from *A Word or Two Before You Go.... Brief Essays On Language*, by permission of University Press of New England.

Max Beerbohm, 'A Clergyman' reprinted from *And Even Now* (Heinemann, 1920), by permission of Mrs Eva Reichmann.

Hilaire Belloc, 'On the Departure of a Guest' reprinted from *On Nothing* (Methuen & Co), by permission of the Peters Fraser & Dunlop Group Ltd.

Sir Isaiah Berlin, 'Churchill and Roosevelt' reprinted from the essay entitled 'Winston Churchill' in *Personal Impressions* copyright 1949, 1964 by Isaiah Berlin, by kind permission of the author, Chatto & Windus Ltd., and of Viking Penguin, a division of Penguin Books USA Inc.

John Betjeman, 'A New Westminster', reprinted from *Points of View: A Selection from The Spectator*, edited by Brian Inglis (Longman, 1962), © John Betjeman, by permission of Curtis Brown Group Ltd.

John Jay Chapman, 'William James' reprinted from *The Selected Writings of John Jay Chapman*, edited by Jacques Barzun.

Winston Churchill, 'The Dream', reprinted from Michael Gilbert, *Winston Spencer Churchill*, vol. viii (Heinemann, 1987), by permission of Curtis Brown Ltd. on behalf of the Estate of Sir Winston Churchill. Copyright the Estate of Sir Winston Churchill.

Richard Cobb, 'The Homburg Hat', reprinted from *People and Places*, © Oxford University Press 1985, by permission of Oxford University Press.

Cyril Connolly, 'The Ant-Lion', reprinted from *The Condemned Playground* (Routledge, 1945), by permission of Rogers Coleridge & White Ltd.

Joan Didion, 'At the Dam', reprinted from *The White Album*, © 1970,

1979, 1989 by Joan Didion, by permission of Weidenfeld & Nicolson Ltd. and Farrar, Straus & Giroux Inc.

Loren Eiseley, 'The Snout', copyright 1950 by Loren Eiseley, reprinted from *The Immense Journey* by permission of Random House Inc.

T. S. Eliot, 'Marie Lloyd', reprinted from *Selected Essays*, copyright 1950 by Harcourt Brace Jovanovich, Inc., and renewed 1978 by Esme Valerie Eliot by permission of Harcourt Brace Jovanovich, Inc., and Faber & Faber Ltd.

William Empson, 'The Faces of Buddha', reprinted from *Argufying* (Chatto, 1987), by permission of Chatto & Windus Ltd. on behalf of Lady Empson.

D. J. Enright, 'The Marquis and the Madame', reprinted from *Conspirators and Poets* (Chatto, 1966), by permission of Watson, Little Ltd.

Joseph Epstein, reprinted from 'About Face' from *The Middle of My Tether. Familiar Essays* by Joseph Epstein, by permission of the author and the publisher, W. W. Norton & Co. Inc., © 1983 by Joseph Epstein.

M. F. K. Fisher, 'Young Hunger', reprinted from *As They Were* © 1982 by M. F. K. Fisher, by permission of A. M. Heath & Co. Ltd., Authors' Agents and Alfred A. Knopf Inc.

E. M. Forster, 'My Own Centenary', reprinted from *Abinger Harvest* (Edward Arnold, 1936), reprinted by permission of King's College, Cambridge and The Society of Authors as the literary representatives of the E. M. Forster Estate.

Robert Graves, 'The Case for Xanthippe', reprinted from *The Crane Bag* (Cassell, 1969), by permission of A. P. Watt Ltd. on behalf of the Executors of the Estate of Robert Graves.

Graham Greene, 'The Lost Childhood', reprinted from *Collected Essays* (Bodley Head Ltd), by permission of Laurence Pollinger Ltd. and International Creative Management.

J. B. S. Haldane, 'On Being the Right Size', reprinted from *Possible Worlds*, copyright 1928 by Harper & Row, Publishers, Inc., © renewed 1965 by J. B. S. Haldane, by permission of Chatto & Windus on behalf of the Estate of J. B. S. Haldane and Harper & Row, Publishers Inc.

Elizabeth Hardwick, 'The Apotheosis of Martin Luther King', reprinted from *Bartleby in Manhattan and Other Essays*, © 1983 by Elizabeth Hardwick, by permission of Weidenfeld & Nicolson Ltd. and Random House Inc.

Aldous Huxley, 'Meditation on the Moon', reprinted from *Music at Night and Other Essays*, copyright 1931 by Aldous Huxley, renewed © 1959 by Aldous Huxley, by permission of Chatto & Windus on behalf of Mrs Laura Huxley and The Hogarth Press as UK publisher and of Harper & Row, Publishers Inc.

Dan Jacobson, 'A Visit from Royalty', reprinted from *The Time of Arrival* (Weidenfeld, 1963), by permission of A. M. Heath.

Clive James, 'A Blizzard of Tiny Kisses', reprinted from *From the Land of Shadows* (Cape, 1982), by permission of the Peters Fraser & Dunlop Group Ltd.

Randall Jarrell, 'Bad Poets', reprinted from 'A Verse Chronicle' in *Poetry and the Age* (Knopf, 1955). Copyright Mrs Mary Jarrell and used with her permission.

Pauline Kael, 'Movies on Television', reprinted from *Kiss Kiss Bang Bang,* © 1968 by Pauline Kael, by permission of Marion Boyars Publishers Ltd. and Curtis Brown Ltd.

P. J. Kavanagh, 'Is It Alas, Yorick?', reprinted from *People and Places* (1988), by permission of Carcanet Press Ltd.

Joseph Wood Krutch, 'The Colloid and the Crystal', reprinted from *The Best of Two Worlds* (New York, 1950). Copyright the Estate of Joseph Wood Krutch.

Philip Larkin, 'The Savage Seventh', reprinted from *Required Writing*, © 1982, 1983 by Philip Larkin, by permission of Faber & Faber Ltd. and Farrar Straus & Giroux Inc.

Rose Macaulay, 'Into Evening Parties', reprinted from *A Casual Commentary* (Methuen & Co.), by permission of the Peters Fraser & Dunlop Group Ltd.

Desmond McCarthy, 'Invective', reprinted from *Experience* (Putnam, 1935), by permission of Hugh Cecil.

H. L. Mencken, 'The Libido for the Ugly' and 'Funeral March', copyright 1927 by Alfred A. Knopf Inc., and renewed 1955 by H. L. Mencken, reprinted from *A Mencken Chrestomathy*, by permission of Alfred A. Knopf Inc. and Jonathan Cape Ltd. on behalf of the Estate of H. L. Mencken.

Marianne Moore, 'What There is to See at the Zoo', reprinted from *Complete Prose*, by permission of Faber & Faber Ltd. Copyright in the US by Viking Penguin, a division of Penguin USA.

Jan Morris, 'La Paz', reprinted from *Cities* (Faber, 1963), by permission of A. P. Watt Ltd. on behalf of Jan Morris.

V. S. Naipaul, 'Columbus and Crusoe', reprinted from *The Overcrowded Barracoon* (Deutsch, 1972), by permission of Aitken & Stone Ltd.

Lewis Namier, 'Symmetry and Repetition', reprinted from *Conflicts* (1942), by Professor L. B. Namier, by permission of Macmillan, London and Basingstoke.

Conor Cruise O'Brien, 'The People's Victor', reprinted from *Writers and Politics* (Chatto, 1965/Penguin Books, 1976), by permission of Elaine Greene Ltd. First appeared in *The Spectator*, April 1956.

George Orwell, 'Reflections on Ghandhi' reprinted from *Shooting an Elephant*, copyright 1949 by Sonia Brownell Orwell and renewed

1976 by Sonia Orwell, by permission of A. M. Heath & Co. Ltd. for the Estate of the late Sonia Brownell Orwell and the publisher Secker & Warburg, and Harcourt Brace Jovanovich

Katherine Anne Porter, 'The Necessary Enemy', reprinted from *The Collected Essays and Occasional Writings of Katherine Anne Porter*, © 1970 by Katherine Anne Porter, by permission of Houghton Mifflin Co.

J. B. Priestley, 'The Toy Farm', reprinted from *Open House* (Heinemann, 1929), by permission of the Peters Fraser & Dunlop Group Ltd.

V. S. Pritchett, 'Our Half-Hogarth', reprinted from *The Living Novel* (Chatto, 1947), by permission of Chatto & Windus Ltd. on behalf of the author and the Peters Fraser & Dunlop Group Ltd.

Maurice Richardson, 'In Search of Nib-Joy', reprinted from *Fits and Starts* (Michael Joseph, 1979), by permission of David Higham Associates Ltd.

Bertrand Russell, 'On Being Modern-Minded', reprinted from *Unpopular Essays*, copyright 1950 by Bertrand Russell, by permission of Unwin Hyman Ltd. and Simon & Schuster Inc.

George Santayana, 'Intellectual Ambition' and 'Intuitive Morality', both reprinted from *Little Essays* (1920) by permission of Constable & Co. Ltd. and the MIT Press.

Bernard Shaw, 'Sir George Grove and Beethoven', reprinted from *Pen Portraits and Reviews* (Constable, 1932), by permission of The Society of Authors on behalf of The Bernard Shaw Estate.

James Stephens, 'Finnegans Wake', reprinted from *James, Seumas and Jacques* (Macmillan, 1964), by permission of The Society of Authors on behalf of the copyright owner, Mrs Iris Wise.

Lytton Strachey, 'Creighton', reprinted from *Portraits in Miniature* (Chatto & Windus Ltd., 1931).

Arthur Symons, 'Cordova', reprinted from *Cities and Sea-Coasts* (Collins, 1918), by permission of Mr B. Read.

Lewis Thomas, 'To Err is Human', reprinted from *The Medusa and the Snail,* © 1976 by Lewis Thomas, by permission of Viking Penguin, a division of Penguin, USA.

James Thurber, 'My Own Ten Rules for a Happy Marriage', from *Thurber Country*, © 1953 James Thurber, © 1981 Helen Thurber and Rosemary A. Thurber, published in America by Simon & Schuster, reprinted by permission of Lucy Kroll Agency and Hamish Hamilton Ltd.

H. R. Trevor-Roper, 'Thomas Hobbes', reprinted from *Historical Essays* (Macmillan, 1957), by permission of the Peters Fraser & Dunlop Group Ltd.

Lionel Trilling, 'Adams at Ease' reprinted from *A Gathering of Fugitives*, © 1956 by Lionel Trilling and renewed 1984 by Diana Trilling, by permission of Harcourt Brace Jovanovich Inc.

John Updike, 'The Bankrupt Man', © 1976 by John Updike, reprinted from *Hugging the Shore* (1983), by permission of Alfred A. Knopf Inc. and André Deutsch Ltd.

Gore Vidal, 'Robert Graves and the Twelve Caesars', reprinted from *Rocking the Boat* (Heinemann, 1963).

Robert Warshow, 'The Gangster as Tragic Hero', copyright 1948 by The Partisan Review, reprinted from *The Immediate Experience* by Robert Warshow, by permission of Doubleday, a division of Bantam, Doubleday, Dell Publishing Group Inc.

Evelyn Waugh, 'Well-Informed Circles and How to Move In Them', reprinted from *The Essays, Articles and Reviews of Evelyn Waugh*, ed. Donat Gallagher, © 1983 by the Estate of Laura Waugh, by permission of Little Brown & Co. and the Peters Fraser & Dunlop Group Ltd.

Rebecca West, 'The Sterner Sex', reprinted from *The Young Rebecca* (Macmillan, 1982), by permission of the Peters Fraser & Dunlop Group Ltd.

E. B. White, 'About Myself', reprinted from *The Second Tree From the Corner*, copyright 1945 by E. B. White, by permission of Harper & Row Publishers Inc.

Edmund Wilson, 'A Preface to Persius . . .', reprinted from *The Shores of Light*, copyright 1952 by Edmund Wilson, renewal © 1980 by Helen Wilson, by permission of Farrar, Straus & Giroux Inc.

Virginia Woolf, 'Harriette Wilson', reprinted from *The Moment and Other Essays*, coypright 1948 by Harcourt Brace Jovanovich Inc., and renewed 1976 by Harcourt Brace Jovanovich Inc. and Marjorie T. Parsons; 'The Death of the Moth', reprinted from *The Death of the Moth and Other Essays*, copyright 1942 by Harcourt Brace Jovanovich Inc. and renewed 1970 by Marjorie T. Parsons, Executrix. By permission of the Hogarth Press on behalf of the Executors of the Virginia Woolf Estate and Harcourt Brace Jovanovich Inc.

G. M. Young, 'The Greatest Victorian', reprinted from *Today and Yesterday* (1948), by permission of Grafton Books, a division of the Collins Publishing Group.

Introduction

ESSAYS come in all shapes and many sizes. There are essays on Human Understanding, and essays on What I Did in the Holidays; essays on Truth, and essays on potato crisps (see page 617); essays that start out as book reviews, and essays that end up as sermons. Even more than most literary forms, the essay defies strict definition. It can shade into the character sketch, the travel sketch, the memoir, the *jeu d'esprit*.

Yet amid all this variety there is, or was until recent times, a central tradition of essay-writing—a tradition that looks back to the first and greatest of essayists, Montaigne. Montaigne in his turn could summon up classical precedents (the 'moral essays' of Plutarch, for instance), but for most modern readers he represents a clear point of departure.

No matter how large its subject, the distinguishing marks of an essay by Montaigne are intimacy and informality. In the words of Hazlitt, who was in many ways his nearest English equivalent, 'he did not set up for a philosopher, wit, orator or moralist, but he became all these by merely daring to tell us whatever passed through his mind'. His watchword was 'Que sais-je?'—'What do I know?', not 'What am I supposed to know?'—and in setting down his thoughts, he refused to be hampered by preconceived notions of order and regularity. For Dr Johnson, two hundred years later, the 'irregular' nature of the essay was still its most obvious characteristic: the first definition he gives in his dictionary is 'a loose sally of the mind'.

Not that there was anything notably loose or self-revealing about the first major English essayist, Francis Bacon. Bacon borrowed the term 'essay' from Montaigne, but his own essays aspire to a measured impersonality. They are masterpieces of rhetoric; their glowing commonplaces have never been surpassed. But they have no real literary progeny. It is the character-writers and the more homely moralists of the seventeenth century who point the way forward to the future.

The true familiar essay made a tentative appearance at the Restoration, in the work of Cowley and others, but it needed journalistic outlets and a journal-reading public before it could come into its own. In due course Addison and Steele's *Tatler* (1709–11) and *Spectator* (1711–13) ushered in a century of 'periodical essayists', and if any one man has a claim to be the father of the English essay, it is surely Addison.

It is a claim that many of his successors would be eager to deny. Judged by the highest or the fiercest standards he seems too worldly, too complacent, too preoccupied with minor social amenities. But his achievements were great. He transformed the essay into a civilizing force, an engine against coarseness and pedantry. What he lacked in depth, he made up for in range of interests and keenness of observation. He taught his readers to appreciate the middle ground of human nature, and fashioned the perfect prose style for the purpose.

The *Tatler* and *Spectator*, bound up as books, were eventually joined on every self-respecting gentleman's shelves by the *Rambler*, the *Adventurer*, the *World*, the *Connoisseur*, the *Mirror*, the *Lounger*. Chalmers' standard collection of *British Essayists*, published in 1808, ran to forty-five volumes. Most of this material has naturally died with the social demand it was designed to serve, but two of the later eighteenth-century essayists survive along with Addison as undisputed classics. Johnson may not have been at home with the lighter kinds of satire, but his finest essays, the ones given over to moral reflection, bear the stamp of his mature wisdom. Goldsmith kept closer to the *Spectator* model, but he wrote with greater freedom than Addison and an easier humour.

It might seem inevitable that the revolutionary upheavals that marked the end of the eighteenth century would produce a new kind of essay. And so they did—but it took time. It was not until the closing years of the Napoleonic wars that the foremost essayists of the Romantic era, Hazlitt and Lamb, began to find their feet.

Both men were better suited to writing essays than anything else. Both of them were masters of the art of talking on paper—unconstrained, independent in their tastes, determined to keep close to the weave and texture of their own experience. But there were big differences between them, too, differences that in the present century have increasingly told in Hazlitt's favour. Hazlitt is forthright and direct; for all his egoism, he has the ability to lose himself in his subject. Lamb, on the other hand, trades much too heavily on his charm: many of the idiosyncrasies that once seemed endearing now merely irritate. But he was capable of subtle insights, and if you can learn to live with the affectations his most characteristic pieces still have a rare tenderness.

For the Victorians and their American counterparts the essay offered (though not all at once) a pulpit, an extension of the novel, a lecture-platform, a diversion. It also offered space—enough space to accommodate Carlyle's brilliant harangues and Macaulay's incomparable history lessons. Too much space, on occasion. Emerson, with his genius for the aphorism and the pregnant observation, was betrayed into impossible prolixity; lesser men were encouraged to stretch material that was thin to start with even thinner. Yet on the whole, nineteenth-century essays, like the journals in which most of them first appeared, testify to a remarkably rich cultural life. An anthology that was limited to a single decade of the Victorian age, perhaps even a single year, would still be able to draw on work of outstanding scope and quality.

Towards the end of the century a change sets in. As you move forward from the mid-Victorians, you become aware of more and more essays being written for their own sake, rather than for the sake of the subject; there is a shift from matter to manner, from discussion to conversation. The essays of Robert Louis Stevenson are one symptom of the new climate—ethical studies conducted in a vagabond mood; and there were plenty of lesser portents among Stevenson's contemporaries. Augustine Birrell, for example, whose good-humored, whimsical musings, principally among books, can be sampled in *Obiter Dicta* and similar collections. They were popular enough in their day to give rise to a new literary term—'birrelling'.

After Birrell came E. V. Lucas, Maurice Hewlett, 'Alpha of the Plough', Robert Lynd, Christopher Morley in America, and a hundred others. Long rows of little books bear witness to a continuing cult of the familiar essay (with Charles Lamb—who else? —as patron saint). It lasted down to the 1930s, in a few cases even longer.

The essayists who birrelled and whiffled their way through this silver age ultimately helped to give the essay a bad name—not least in the schoolroom, where they were all too often held up as models. (Oh, the horrors of being told to write a light-hearted essay all about nothing.) Today, on the other hand, they are so hopelessly out of fashion, and out of print, that it is tempting to make out a case for their virtues. They were humane, they thought of literature as a living thing, they wrote a good deal more readably than some of their critics. But they are past reviving, even so: their cosiness and bookishness tell fatally against them.

The real strengths of the twentieth-century essay have been immensely more varied. What generalization can usefully cover Beerbohm and James Baldwin, J. B. S. Haldane and D. H. Lawrence, Mencken and Virginia Woolf? None, except that they all share the old Montaigne virtues of informality and independence. Any of them might have taken as his or her motto the heading under which George Orwell used to contribute his weekly essay to *Tribune*—'I Write as I Please.'

Today good writers continue to write as they please, although it is true that they are less likely to talk about essays than 'pieces'. 'Essay' has come to sound a little too leisurely: 'piece' strikes the required note of journalistic toughness. And the demands of journalism have in fact pushed writers who might once have set up as essayists further and further in the direction of reportage, travel writing, profiles, instant comment. There is less time and scope for the essay proper. But essays still get written, and still force themselves forward when they can (in the guise of book reviews, for example, as they always have). Certainly there is nothing to suggest that the essay is dying; but then perhaps a form that has already led so many lives is virtually unkillable.

When I began putting this anthology together, I decided to exclude essays on literary themes. I had a romantic notion that they belonged in an anthology of criticism, and that this was going to be a collection of essays about Life. But it was a decision that I very soon abandoned. It became clear that I was depriving myself of far too much valuable material: too many indispensable essayists were at their best writing about other writers.

The other self-imposed rule that I started out with seems to me more reasonable. There was to be no cutting: in the phrase that used to adorn Penguin books in their early days, every essay included was to be 'complete and unabridged'. But this too was a policy that was eventually abandoned, or at any rate modified. There were a number of major writers—nineteenth-century writers, on the whole—who could only be adequately represented by essays that would have swamped the book if they had been reproduced in their entirety. It became a question of printing either extracts or nothing, and extracts seemed preferable.

Space in general proved a constant problem. Favourite essays had a way of always turning out to be longer than I recalled, never shorter. There were authors who only wrote at a length that would have made them loom disproportionately large, and above all there were far too many well-qualified candidates clamouring for admission. Sacrificing the essayists whom I had originally hoped to include was a painful business; there were many times when I envied Chalmers his forty-five volumes.

In making my final selection I have tried to follow three principles. Some essays are included because they seem to me the best of their kind, some because they particularly appeal to me, a few because they are historically representative (though none, I hope, because they are *merely* representative). In a better world all three categories would no doubt coincide, but in the world as it is this means that there are some inevitable inconsistencies. Most authors are represented by a single item, for instance, but it isn't necessarily meant to sum up their most imposing qualities. In some cases, in order to preserve the balance and variety of the book as a whole, I have simply chosen something I find unusually interesting or entertaining.

The date at the end of an essay is that of its first appearance in print, in either periodical or book form, except in those cases where it seemed more appropriate to give the approximate date of composition instead. In the seventeenth-century essays, most notably Bacon's, spelling and punctuation have been modernized.

JOHN GROSS

Of Truth

'What is Truth?' said jesting Pilate; and would not stay for an answer. Certainly there be that delight in giddiness, and count it a bondage to fix a belief; affecting free-will in thinking, as well as in acting. And though the sects of philosophers of that kind be gone, yet there remain certain discoursing wits which are of the same veins, though there be not so much blood in them as was in those of the ancients. But it is not only the difficulty and labour which men take in finding out of truth; nor again that when it is found it imposeth upon men's thoughts; that doth bring lies in favour; but a natural though corrupt love of the lie itself. One of the later school of the Grecians examineth the matter, and is at a stand to think what should be in it, that men should love lies, where neither they make for pleasure, as with poets, nor for advantage, as with the merchant; but for the lie's sake. But I cannot tell: this same truth is a naked and open daylight, that doth not shew the masks and mummeries and triumphs of the world, half so stately and daintily as candlelights.

Truth may perhaps come to the price of a pearl, that sheweth best by day; but it will not rise to the price of a diamond or carbuncle, that sheweth best in varied lights. A mixture of a lie doth ever add pleasure. Doth any man doubt, that if there were taken out of men's minds vain opinions, flattering hopes, false valuations, imaginations as one would, and the like, but it would leave the minds of a number of men poor shrunken things, full of melancholy and indisposition, and unpleasing to themselves? One of the Fathers, in great severity, called poesy *vinum daemonum*, because it filleth the imagination; and yet it is but with the shadow of a lie. But it is not the lie that passeth through the mind, but the lie that sinketh in and settleth in it, that doth the hurt; such as we spake of before. But howsoever these things are thus in men's depraved judgments and affections, yet truth, which only doth judge itself, teacheth that the inquiry of truth, which is the love-making or wooing of it, the knowledge of truth, which is the presence of it, and the belief of truth, which is the enjoying of it, is the sovereign good of human nature.

The first creature of God, in the works of the days, was the light of the sense; the last was the light of reason; and his sabbath work ever since, is the illumination of his Spirit. First he breathed light upon the face of the matter or chaos; then he breathed light into the face of man; and still he breatheth and inspireth light into the face of his chosen. The poet that beautified the sect that was otherwise inferior to the rest, saith yet excellently well: 'It is a pleasure to stand upon the shore, and to see ships tossed upon the sea; a pleasure to stand in the window of a castle, and to see a battle and the adventures thereof below: but no pleasure is comparable to the standing upon the vantage ground of Truth, (a hill not to be commanded, and where the air is always clear and serene,) and to see the errors, and wanderings, and mists, and tempests, in the vale below'; so always that this prospect be with pity, and not with swelling or pride. Certainly, it is heaven upon earth, to have a man's mind move in charity, rest in providence, and turn upon the poles of truth.

To pass from theological and philosophical truth, to the truth of civil business; it will be acknowledged even by those that practise it not, that clear and round dealing is the honour of man's nature; and that mixture of falsehood is like allay in coin of gold and silver, which may make the metal work the better, but it embaseth it. For these winding and crooked courses are the goings of the serpent; which goeth basely upon the belly, and not upon the feet. There is no vice that doth so cover a man with shame as to be found false and perfidious. And therefore Montaigne saith prettily, when he inquired the reason, why the word of the lie should be such a disgrace and such an odious charge? Saith he, 'If it be well weighed, to say that a man lieth, is as much to say, as that he is brave towards God and a coward towards men.' For a lie faces God, and shrinks from man. Surely the wickedness of falsehood and breach of faith cannot possibly be so highly expressed, as in that it shall be the last peal to call the judgments of God upon the generations of men; it being foretold, that when Christ cometh, 'he shall not find faith upon the earth'.

1625

Of Revenge

REVENGE is a kind of wild justice; which the more man's nature runs to, the more ought law to weed it out. For as for the first wrong, it doth but offend the law; but the revenge of that wrong putteth the law out of office. Certainly, in taking revenge, a man is but even with his enemy; but in passing it over, he is superior; for it is a prince's part to pardon. And Solomon, I am sure, saith, 'It is the glory of a man to pass by an offence.' That which is past is gone, and irrevocable; and wise men have enough to do with things present and to come; therefore they do but trifle with themselves, that labour in past matters. There is no man doth a wrong for the wrong's sake; but thereby to purchase himself profit, or pleasure, or honour, or the like. Therefore why should I be angry with a man for loving himself better than me? And if any man should do wrong merely out of illnature, why, yet it is but like the thorn or briar, which prick and scratch, because they can do no other. The most tolerable sort of revenge is for those wrongs which there is no law to remedy; but then let a man take heed the revenge be such as there is no law to punish; else a man's enemy is still before hand, and it is two for one. Some, when they take revenge, are desirous the party should know whence it cometh. This the more generous. For the delight seemeth to be not so much in doing the hurt as in making the party repent. But base and crafty cowards are like the arrow that flieth in the dark. Cosmus, duke of Florence, had a desperate saying against perfidious or neglecting friends, as if those wrongs were unpardonable; 'You shall read (saith he) that we are commanded to forgive our enemies; but you never read that we are commanded to forgive our friends.' But yet the spirit of Job was in a better tune: 'Shall we (saith he) take good at God's hands, and not be content to take evil also?' And so of

friends in a proportion. This is certain, that a man that studieth
revenge keeps his own wounds green, which otherwise would
heal and do well. Public revenges are for the most part fortunate;
as that for the death of Caesar; for the death of Pertinax; for the
death of Henry the Third of France; and many more. But in private revenges it is not so. Nay rather, vindictive persons live the
life of witches; who, as they are mischievous, so end they infor-
tunate.

1625

Of Boldness

It is a trivial grammar-school text, but yet worthy a wise man's
consideration. Question was asked of Demosthenes, 'what was
the chief part of an orator?' he answered, 'action': what next?
'action': what next again? 'action'. He said it that knew it best,
and had by nature himself no advantage in that he commended.

A strange thing, that that part of an orator which is but
superficial, and rather the virtue of a player, should be placed so
high, above those other noble parts of invention, elocution, and
the rest; nay almost alone, as if it were all in all. But the reason is
plain. There is in human nature generally more of the fool than
of the wise; and therefore those faculties by which the foolish
part of men's minds is taken are most potent.

Wonderful like is the case of Boldness, in civil business; what
first? Boldness: what second and third? Boldness. And yet boldness is a child of ignorance and baseness, far inferior to other
parts. But nevertheless it doth fascinate and bind hand and foot
those that are either shallow in judgment or weak in courage,
which are the greatest part; yea and prevaileth with wise men at
weak times. Therefore we see it hath done wonders in popular
states; but with senates and princes less; and more ever upon
the first entrance of bold persons into action than soon after;
for boldness is an ill keeper of promise. Surely as there are
mountebanks for the natural body, so are there mountebanks for
the politic body; men that undertake great cures, and perhaps
have been lucky in two or three experiments, but want the

grounds of science, and therefore cannot hold out. Nay you shall see a bold fellow many times do Mahomet's miracle. Mahomet made the people believe that he would call an hill to him, and from the top of it offer up his prayers for the observers of his law. The people assembled; Mahomet called the hill to come to him, again and again; and when the hill stood still, he was never a whit abashed, but said, 'If the hill will not come to Mahomet, Mahomet will go to the hill.' So these men, when they have promised great matters and failed most shamefully, yet (if they have the perfection of boldness) they will but slight it over, and make a turn, and no more ado. Certainly to men of great judgment, bold persons are a sport to behold; nay and to the vulgar also, boldness hath somewhat of the ridiculous. For if absurdity be the subject of laughter, doubt you not but great boldness is seldom without some absurdity. Especially it is a sport to see, when a bold fellow is out of countenance; for that puts his face into a most shrunken and wooden posture; as needs it must; for in bashfulness the spirits do a little go and come; but with bold men, upon like occasion, they stand at a stay; like a stale at chess, where it is no mate, but yet the game cannot stir. But this last were fitter for a satire than for a serious observation. This is well to be weighed; that boldness is ever blind; for it seeth not dangers and inconveniences. Therefore it is ill in counsel, good in execution; so that the right use of bold persons is, that they never command in chief, but be seconds, and under the direction of others. For in counsel it is good to see dangers; and in execution not to see them, except they be very great.

1625

Of Innovations

As the births of living creatures at first are ill-shapen, so are all Innovations, which are the births of time. Yet notwithstanding, as those that first bring honour into their family are commonly more worthy than most that succeed, so the first precedent (if it be good) is seldom attained by imitation. For Ill, to man's nature as it stands perverted, hath a natural motion, strongest in continuance; but Good, as a forced motion, strongest at first. Surely every medicine is an innovation; and he that will not apply new remedies must expect new evils; for time is the greatest innovator; and if time of course alter things to the worse, and wisdom and counsel shall not alter them to the better, what shall be the end? It is true, that what is settled by custom, though it be not good, yet at least it is fit: and those things which have long gone together, are as it were confederate within themselves; whereas new things piece not so well; but though they help by their utility, yet they trouble by their inconformity. Besides, they are like strangers; more admired and less favoured. All this is true, if time stood still; which contrariwise moveth so round, that a froward retention of custom is as turbulent a thing as an innovation; and they that reverence too much old times, are but a scorn to the new.

It were good therefore that men in their innovations would follow the example of time itself; which indeed innovateth greatly, but quietly, and by degrees scarce to be perceived. For otherwise, whatsoever is new is unlooked for; and ever it mends some, and pairs other; and he that is holpen takes it for a fortune, and thanks the time; and he that is hurt, for a wrong, and imputeth it to the author. It is good also not to try experiments in states, except the necessity be urgent, or the utility evident; and well to beware that it be the reformation that draweth on the change, and not the desire of change that pretendeth the reformation. And lastly, that the novelty, though it be not rejected, yet be held for a suspect; and, as the Scripture saith, 'that we make a stand upon the ancient way, and then look about us, and discover what is the straight and right way, and so to walk in it'.

1625

Of Masques and Triumphs

THESE things are but toys, to come amongst such serious observations. But yet, since princes will have such things, it is better they should be graced with elegancy than daubed with cost. Dancing to song, is a thing of great state and pleasure. I understand it, that the song be in quire, placed aloft, and accompanied with some broken music; and ditty fitted to the device. Acting in song, especially in dialogues, hath an extreme good grace; I say acting, not dancing (for that is a mean and vulgar thing); and the voices of the dialogue would be strong and manly, (a base and a tenor; no treble;) and the ditty high and tragical; not nice or dainty. Several quires, placed one over against another, and taking the voice by catches, anthem-wise, give great pleasure. Turning dances into figure is a childish curiosity. And generally let it be noted, that those things which I here set down are such as do naturally take the sense, and not respect petty wonderments.

It is true, the alterations of scenes, so it be quietly and without noise, are things of great beauty and pleasure; for they feed and relieve the eye, before it be full of the same object. Let the scenes abound with light, specially coloured and varied; and let the masquers, or any other, that are to come down from the scene, have some motions upon the scene itself, before their coming down; for it draws the eye strangely, and makes it with great pleasure to desire to see that it cannot perfectly discern. Let the songs be loud and cheerful, and not chirpings or pulings. Let the music likewise be sharp and loud, and well placed. The colours that shew best by candle-light, are white, carnation, and a kind of sea-water-green; and oes, or spangs, as they are of no great cost, so they are of most glory. As for rich embroidery, it is lost and not discerned. Let the suits of the masquers be graceful, and such as become the person when the vizards are off; not after examples of known attires; Turks, soldiers, mariners, and the like. Let anti-masques not be long; they have been commonly of fools, satyrs, baboons, wild-men, antics, beasts, sprites, witches, Ethiops, pigmies, turquets, nymphs, rustics, Cupids, statua's moving, and the like. As for angels, it is not comical enough to put them in anti-masques; and any thing that is hideous, as devils, giants, is on the other side as unfit. But chiefly, let the music of them be recreative, and with some strange changes. Some sweet odours suddenly coming forth, without any drops falling, are, in such a company as there is steam and heat, things of great pleasure and refreshment. Double masques, one of men, another of ladies, addeth state and variety. But all is nothing except the room be kept clear and neat.

For justs, and tourneys, and barriers; the glories of them are chiefly in the chariots, wherein the challengers make their entry; especially if they be drawn with strange beasts: as lions, bears, camels, and the like; or in the devices of their entrance; or in the bravery of their liveries; or in the goodly furniture of their horses and armour. But enough of these toys.

1625

A Chamber-maid

SHE is her mistress's she-secretary, and keeps the box of her teeth, her hair, and her painting very private. Her industry is upstairs, and downstairs like a drawer: and by her dry hand you may know she is a sore starcher. If she lie at her master's bed's feet, she is quit of her green sickness forever; for she hath terrible dreams when she's awake, as if she were troubled with the nightmare. She hath a good liking to dwell in the country, but she holds London the goodliest forest in England, to shelter a great belly. She reads Greene's works over and over, but is so carried away with the Mirror of Knighthood, she is many times resolved to run out of herself, and become a lady-errant. If she catch a clap, she divides it so equally between the master and the serving-man, as if she had cut out the getting of it by a thread.... The pedant of the house, though he promise her marriage, cannot grow further inward with her, for she hath paid for her credulity often, and now grows weary. She likes the form of our marriage very well, in that a woman is not tied to answer any articles concerning questions of virginity: her mind, her body, and clothes, are parcels loosely packed together, and for want of good utterance, she perpetually laughs out her meaning. Her mistress and she help to make away time, to the idlest purpose that can be, either for love or money. In brief, these chamber-maids, are like lotteries: you may draw twenty, ere one worth anything.

Published posthumously, 1615

A Fair and Happy Milkmaid*

Is a country wench, that is so far from making herself beautiful by art, that one look of hers is able to put all face-physic out of countenance. She knows a fair look is but a dumb orator to commend virtue, therefore minds it not. All her excellencies stand in her so silently, as if they had stolen upon her without her knowledge. The lining of her apparel (which is herself) is far better than the outsides of tissue: for though she be not arrayed in the spoil of the silk-worm, she is decked in innocency, a far better wearing. She doth not, with lying long abed, spoil both her complexion and conditions; nature hath taught her, too immoderate sleep is rust to the soul: she rises therefore with chanticleer, her dame's cock, and at night makes the lamb her curfew. In milking a cow, and straining the teats through her fingers, it seems that so sweet a milk-press makes the milk the whiter or sweeter; for never came almond glove or aromatic ointment on her palm to taint it. The golden ears of corn fall and kiss her feet when she reaps them, as if they wished to be bound and led prisoners by the same hand that felled them. Her breath is her own, which scents all the year long of June, like a new made haycock. She makes her hand hard with labour, and her heart soft with pity: and when winter evenings fall early (sitting at her merry wheel), she sings a defiance to the giddy wheel of fortune. She doth all things with so sweet a grace, it seems ignorance will not suffer her to do ill, being her mind is to do well. She bestows her year's wages at next fair; and in choosing her garments, counts no bravery in the world, like decency. The garden and the bee-hive are all her physic and chirurgery, and she lives the longer for it. She dares go alone, and unfold sheep in the night, and fears no manner of ill, because she means none: yet to say truth, she is never alone, for she is still accompanied with old songs, honest thoughts, and prayers, but short ones; yet they have their efficacy, in that they are not palled with ensuing idle cogitations. Lastly, her dreams are so chaste, that she dare tell them; only a

* This sketch, though published as Overbury's, was probably written by John Webster, the dramatist.

Friday's dream is all her superstition: that she conceals for fear of anger. Thus lives she, and all her care is that she may die in the spring-time, to have store of flowers stuck upon her winding sheet.

1615

JOHN EARLE

An Antiquary

H<small>E</small> is a man strangely thrifty of time past, and an enemy indeed to his maw, whence he fetches out many things when they are now all rotten and stinking. He is one that hath that unnatural disease to be enamoured of old age and wrinkles, and loves all things (as Dutchmen do cheese) the better for being mouldy and worm-eaten. He is of our Religion because we say it is most ancient; and yet a broken statue would almost make him an idolater. A great admirer he is of the rust of old monuments, and reads only those characters where time hath eaten out the letters. He will go you forty miles to see a Saint's Well or ruined Abbey: and if there be but a cross or stone foot-stool in the way, he'll be considering it so long, till he forget his journey. His estate consists much in shekels and Roman coins, and he hath more pictures of Caesar than James or Elizabeth. Beggars cozen him with musty things which they have raked from dunghills, and he preserves their rags for precious relics. He loves no library but where there are more spiders' volumes than authors', and looks with great admiration on the antique work of cobwebs. Printed books he contemns, as a novelty of this latter age; but a Manuscript he pores on everlastingly, especially if the cover be all moth-eaten, and the dust make a parenthesis between every syllable. He would give all the books in his study (which are rarities all) for one of the old Roman binding or six lines of Tully in his own hand. His chamber is hung commonly with strange beasts' skins, and is a kind of charnel-house of bones extraordinary; and his discourse upon them, if you will hear him, shall last longer. His very attire is that which is the eldest out of fashion, and you may pick a criticism out of his breeches. He never looks upon himself till he is gray-haired, and then he is pleased with his own antiquity. His grave does not

fright him for he has been used to sepulchres, and he likes death
the better because it gathers him to his fathers.

1628

A Good Old Man

Is the best antiquity, and which we may with least vanity
admire. One whom time hath been thus long a-working, and like
winter fruit ripened when others are shaken down. He hath
taken out as many lessons of the world as days, and learnt the
best thing in it, the vanity of it. He looks over his former life as a
danger well past, and would not hazard himself to begin again.
His lust was long broken before his body, yet he is glad this
temptation is broke too, and that he is fortified from it by this
weakness. The next door of death sads him not, but he expects it
calmly as his turn in nature; and fears more his recoiling back to
childishness than dust. All men look on him as a common father,
and on old age for his sake as a reverent thing. His very presence
and face puts vice out of countenance, and makes it an indec-
orum in a vicious man. He practises his experience on youth
without the harshness of reproof, and in his counsel is good
company. He has some old stories still of his own seeing to
confirm what he says, and makes them better in the telling; yet is
not troublesome neither with the same tale again, but remembers
with them how oft he has told them. His old sayings and morals
seem proper to his beard; and the poetry of Cato does well out of
his mouth, and he speaks it as if he were the author. He is not
apt to put the boy on a younger man, nor the fool on a boy; but
can distinguish gravity from a sour look, and the less testy he is,
the more regarded. You must pardon him if he like his own
times better than these, because those things are follies to him
now that were wisdom then; yet he makes us of that opinion too,
when we see him and conjecture those times by so good a relic.
He is a man capable of a dearness with the youngest men; yet he
not youthfuller for them, but they older for him; and no man
credits more his acquaintance. He goes away at last, too soon

whensoever, with all men's sorrow but his own; and his memory is fresh when it is twice as old.

1628

A Pot-Poet

Is the dregs of wit; yet mingled with good drink may have some relish. His inspirations are more real than others'; for they do but feign a God, but he has his by him. His verses run like the tap, and his invention as the barrel ebbs and flows at the mercy of the spigot. In thin drink he aspires not above a ballad, but a cup of sack inflames him and sets his muse and nose afire together. The Press is his Mint, and stamps him now and then a sixpence or two in reward of the baser coin his pamphlet. His works would scarce sell for three half-pence, though they are given oft for three shillings, but for the pretty title that allures the country Gentleman: and for which the printer maintains him in ale a fortnight. His verses are like his clothes, miserable centos and patches, yet their pace is not altogether so hobbling as an Almanac's. The death of a great man or the burning of a house furnish him with an argument, and the Nine Muses are out straight in mourning gown, and Melpomene cries 'Fire, Fire.' His other poems are but briefs in rhyme, and like the poor Greeks' collection to redeem from captivity. He is a man now much employed in commendations of our Navy, and a bitter inveigher against the Spaniard. His frequentest works go out in single sheets, and are chanted from market to market, to a vile tune and a worse throat; whilst the poor country wench melts like her butter to hear them. And these are the stories of some men of Tyburne, or a strange monster out of Germany: or sitting in a bawdy-house, he writes God's judgements. He ends at last in some obscure painted cloth, to which himself made the verses, and his life like a can too full spills upon the bench. He leaves twenty shillings on the score, which my Hostess loses.

1628

How the Distempers of these Times should affect wise Men

THE distempers of these times would make a wise man both merry and mad. Merry, to see how vice flourishes but a while, and, being at last frustrate of all her fair hopes, dies in a dejected scorn; which meets with nothing, in the end, but beggary, baseness, and contempt. To see how the world is mistaken in opinion, to suppose those best that are wealthiest. To see how the world thinks to appal the mind of nobleness with misery; while true resolution laughs at their poor impotency, and slights even the utmost spite of tyranny. To see how men buy offices at high rates, which, when they have, prove gins to catch their souls in, and snare their estates and reputations. To see how foolishly men cozen themselves of their souls, while they think they gain, by their cunning defrauding another. To see how the projectors of the world, like the spoke of the wheel of Sesostris' chariot, are tumbled up and down, from beggary to worship, from worship to honour, from honour to baseness again. To see what idle compliments are current among some that affect the fantastic garb: as if friendship were nothing but an apish salute, glossed over with nothing but the varnish of a smooth tongue. To see a strutting prodigal overlook a region, with his waving plume; as if he could as easily shake that, as his feather; yet in private will creep, like a crouching spaniel, to his base muddy prostitute. To see how pot-valour thunders in a tavern, and appoints a duel; but goes away, and gives money to have the quarrel taken up underhand. Mad, on the other side, to see how vice goes trapped with rich furniture, while poor virtue hath nothing but a bridle and saddle, which only serve to increase her bondage. To see Machiavel's tenets held as oracles; honesty reputed shallowness; justice bought and sold; as if the world went about to disprove

Zorobabel, and would make him confess money to be stronger than truth. To see how flattery creeps into favour with greatness, while plain dealing is thought the enemy of state and honour. To see how the papists (for promotion of their own religion) invent lies, and print them; that they may not only cozen the present age, but gull posterity, with forged actions. To see how well-meaning simplicity is footballed. To see how religion is made a politician's visor; which, having helped him to his purpose, he casts by, like Sunday apparel, not thought of all the week after. And, which would mad a man more than all, to know all this, yet not know how to help it.

c.1620

On Dreams

Half our days we pass in the shadow of the earth; and the brother of death exacteth a third part of our lives. A good part of our sleep is peered out with visions and fantastical objects, wherein we are confessedly deceived. The day supplieth us with truths; the night with fictions and falsehoods, which uncomfortably divide the natural account of our beings. And, therefore, having passed the day in sober labours and rational enquiries of truth, we are fain to betake ourselves unto such a state of being, wherein the soberest heads have acted all the monstrosities of melancholy, and which unto open eyes are no better than folly and madness.

Happy are they that go to bed with grand music, like Pythagoras, or have ways to compose the fantastical spirit, whose unruly wanderings take off inward sleep, filling our heads with St Anthony's visions, and the dreams of Lipara in the sober chambers of rest.

Virtuous thoughts of the day lay up good treasures for the night; whereby the impressions of imaginary forms arise into sober similitudes, acceptable unto our slumbering selves and preparatory unto divine impressions. Hereby Solomon's sleep was happy. Thus prepared, Jacob might well dream of angels upon a pillow of stone. And the best sleep of Adam might be the best of any after.

That there should be divine dreams seems unreasonably doubted by Aristotle. That there are demoniacal dreams we have little reason to doubt. Why may there not be angelical? If there be guardian spirits, they may not be inactively about us in sleep; but may sometimes order our dreams: and many strange hints, instigations, or discourses, which are so amazing unto us, may arise from such foundations.

But the phantasms of sleep do commonly walk in the great

road of natural and animal dreams, wherein the thoughts or actions of the day are acted over and echoed in the night. Who can therefore wonder that Chrysostom should dream of St Paul, who daily read his epistles; or that Cardan, whose head was so taken up about the stars, should dream that his soul was in the moon! Pious persons, whose thoughts are daily busied about heaven, and the blessed state thereof, can hardly escape the nightly phantasms of it, which though sometimes taken for illuminations, or divine dreams, yet rightly perpended may prove but animal visions, and natural night-scenes of their awaking contemplations.

Many dreams are made out by sagacious exposition, and from the signature of their subjects; carrying their interpretation in their fundamental sense and mystery of similitude, whereby, he that understands upon what natural fundamental every notion dependeth, may, by symbolical adaptation, hold a ready way to read the characters of Morpheus. In dreams of such a nature, Artemidorus, Achmet, and Astrampsichus, from Greek, Egyptian, and Arabian oneirocriticism, may hint some interpretation: who, while we read of a ladder in Jacob's dream, will tell us that ladders and scalary ascents signify preferment; and while we consider the dream of Pharaoh, do teach us that rivers overflowing speak plenty, lean oxen, famine and scarcity; and therefore it was but reasonable in Pharaoh to demand the interpretation from his magicians, who, being Egyptians, should have been well versed in symbols and the hieroglyphical notions of things. The greatest tyrant in such divinations was Nabuchodonosor, while, besides the interpretation, he demanded the dream itself; which being probably determined by divine immission, might escape the common road of phantasms, that might have been traced by Satan.

When Alexander, going to besiege Tyre, dreamt of a Satyr, it was no hard exposition for a Grecian to say, 'Tyre will be thine.' He that dreamed that he saw his father washed by Jupiter and anointed by the sun, had cause to fear that he might be crucified, whereby his body would be washed by the rain, and drop by the heat of the sun. The dream of Vespasian was of harder exposition; as also that of the emperor Mauritius, concerning his successor Phocas. And a man might have been hard put to it, to interpret the language of Aesculapius, when to a consumptive person he held forth his fingers; implying thereby that his cure lay in dates, from the homonomy of the Greek, which signifies dates and fingers.

We owe unto dreams that Galen was a physician, Dion an historian, and that the world hath seen some notable pieces of Cardan; yet, he that should order his affairs by dreams, or make the night a rule unto the day, might be ridiculously deluded; wherein Cicero is much to be pitied, who having excellently discoursed of the vanity of dreams, was yet undone by the flattery of his own, which urged him to apply himself unto Augustus.

However dreams may be fallacious concerning outward events, yet may they be truly significant at home; and whereby we may more sensibly understand ourselves. Men act in sleep with some conformity unto their awaked senses; and consolations or discouragements may be drawn from dreams which intimately tell us ourselves. Luther was not like to fear a spirit in the night, when such an apparition would not terrify him in the day. Alexander would hardly have run away in the sharpest combats of sleep, nor Demosthenes have stood stoutly to it, who was scarce able to do it in his prepared senses. Persons of radical integrity will not easily be perverted in their dreams, nor noble minds do pitiful things in sleep. Crassus would have hardly been bountiful in a dream, whose fist was so close awake. But a man might have lived all his life upon the sleeping hand of Antonius.

There is an art to make dreams, as well as their interpretation; and physicians will tell us that some food makes turbulent, some gives quiet, dreams. Cato, who doated upon cabbage, might find the crude effects thereof in his sleep; wherein the Egyptians might find some advantage by their superstitious abstinence from onions. Pythagoras might have calmer sleeps, if he totally abstained from beans. Even Daniel, the great interpreter of dreams, in his leguminous diet, seems to have chosen no advantageous food for quiet sleeps, according to Grecian physic.

To add unto the delusion of dreams, the fantastical objects seem greater than they are; and being beheld in the vaporous state of sleep, enlarge their diameters unto us; whereby it may prove more easy to dream of giants than pigmies. Democritus might seldom dream of atoms, who so often thought of them. He almost might dream himself a bubble extending unto the eighth sphere. A little water makes a sea; a small puff of wind a tempest. A grain of sulphur kindled in the blood may make a flame like Aetna; and a small spark in the bowels of Olympias a lightning over all the chamber.

But, beside these innocent delusions, there is a sinful state of dreams. Death alone, not sleep, is able to put an end unto sin; and there may be a night-book of our iniquities; for beside the transgressions of the day, casuists will tell us of mortal sins in dreams, arising from evil precogitations; meanwhile human law regards not noctambulos; and if a night-walker should break his neck, or kill a man, takes no notice of it.

Dionysius was absurdly tyrannical to kill a man for dreaming that he had killed him; and really to take away his life, who had but fantastically taken away his. Lamia was ridiculously unjust to sue a young man for a reward, who had confessed that pleasure from her in a dream which she had denied unto his awaking senses: conceiving that she had merited somewhat from his fantastical fruition and shadow of herself. If there be such debts, we owe deeply unto sympathies; but the common spirit of the world must be ready in such arrearages.

If some have swooned, they may also have died in dreams, since death is but a confirmed swooning. Whether Plato died in a dream, as some deliver, he must rise again to inform us. That some have never dreamed, is as improbable as that some have never laughed. That children dream not the first half-year; that men dream not in some countries, with many more, are unto me sick men's dreams; dreams out of the ivory gate, and visions before midnight.

*c.*1650

Of Anger

ANGER is one of the sinews of the soul; he that wants it hath a maimed mind, and with Jacob, sinew-shrunk in the hollow of his thigh, must needs halt. Nor is it good to converse with such as cannot be angry, and with the Caspian sea never ebb nor flow. This anger is either heavenly, when one is offended for God; or hellish, when offended with God and goodness; or earthly, in temporal matters. Which earthly anger, whereof we treat, may also be hellish, if for no cause, no great cause, too hot, or too long.

1. *Be not angry with any without a cause.* If thou beest, thou must not only, as the proverb saith, be appeased without amends, having neither cost nor damage given thee, but, as our Saviour saith, be in danger of the judgment.

2. *Be not mortally angry with any for a venial fault.* He will make a strange combustion in the state of his soul, who at the landing of every cockboat sets the beacons on fire. To be angry for every toy debases the worth of thy anger; for he will be angry for anything, who will be angry for nothing.

3. *Let not thy anger be so hot, but that the most torrid zone thereof may be habitable.* Fright not people from thy presence with the terror of thy intolerable impatience. Some men, like a tiled house, are long before they take fire, but once on flame there is no coming near to quench them.

4. *Take heed of doing irrevocable acts in thy passion.* As the revealing of secrets, which makes thee a bankrupt for society ever after: neither do such things which done once are done for ever, so that no bemoaning can amend them. Samson's hair grew again, but not his eyes: time may restore some losses, others are never to be repaired. Wherefore in thy rage make no Persian decree which cannot be reversed or repealed; but rather Polonian laws, which, they say, last but three days: do not in an instant what an age cannot recompense.

5. *Anger kept till the next morning, with manna, doth putrefy and corrupt;* save that manna corrupted not at all, and anger most of all, kept the next sabbath. St Paul saith, *Let not the sun go down on your wrath;* to carry news to the antipodes in another world of thy revengeful nature. Yet let us take the apostle's meaning rather than his words, with all possible speed to depose our passion, not understanding him so literally that we may take leave to be angry till sunset: then might our wrath lengthen with the days; and men in Greenland, where day lasts above a quarter of a year, have plentiful scope of revenge. And as the English, by command from William the Conqueror, always raked up their fire and put out their candles when the curfew bell was rung, let us then also quench all sparks of anger and heat of passion.

6. *He that keeps anger long in his bosom, giveth place to the devil.* And why should we make room for him, who will crowd in too fast of himself? Heat of passion makes our souls to chap, and the devil creeps in at the crannies; yea, a furious man in his fits may seem possessed with a devil, foams, fumes, tears himself, is deaf and dumb in effect, to hear or speak reason: sometimes swallows, stares, stamps, with fiery eyes and flaming cheeks. Had Narcissus himself seen his own face when he had been angry, he could never have fallen in love with himself.

1642

A Degenerate Noble: or One That is Proud of his Birth

Is like a turnip, there is nothing good of him, but that which is underground, or rhubarb, a contemptible shrub, that springs from a noble root. He has no more title to the worth and virtue of his ancestors, than the worms that were engendered in their dead bodies, and yet he believes he has enough to exempt himself and his posterity from all things of that nature forever. This makes him glory in the antiquity of his family, as if his nobility were the better, the further off it is in time, as well as desert, from that of his predecessors. He believes the honour that was left him, as well as the estate, is sufficient to support his quality, without troubling himself to purchase any more of his own; and he meddles as little with the management of the one as the other, but trusts both to the government of his servants, by whom he is equally cheated in both. He supposes the empty title of honour sufficient to serve his turn, though he has spent the substance and reality of it, like the fellow that sold his ass, but would not part with the shadow of it; or Apicius, that sold his house, and kept only the balcony, to see and be seen in. And because he is privileged from being arrested for his debts, supposes he has the same freedom from all obligations he owes humanity and his country, because he is not punishable for his ignorance and want of honour, no more than poverty or unskilfulness is in other professions, which the law supposes to be punishment enough to itself. He is like a fanatic, that contents himself with the mere title of a saint, and makes that his privilege to act all manner of wickedness; or the ruins of a noble structure, of which there is nothing left but the foundation, and that obscured and buried under the rubbish of the superstructure. The living honour of his ancestors is long ago departed, dead and gone, and his is but the ghost and shadow of it, that haunts the house with horror and disquiet, where once it lived. His nobility is truly descended from the glory of his forefathers, and may be rightly said to fall to him; for it will never rise again to the height it was in them, by his means; and he succeeds then as candles do the office of the sun. The confidence of nobility has rendered him ignoble, as the opinion of wealth makes some men poor; and as those that are born to estates neglect industry, and have no business, but to spend; so he being born to honour believes he is no further concerned, than to consume and waste it. He is but a copy, and so ill done, that there is no line of the original in him, but the sin only. He is like a word that by ill custom and mistake has utterly lost the sense of that from which it was derived, and now signifies quite contrary: for the glory of noble ancestors will not permit the good or bad of their posterity to be obscure. He values himself only upon his title, which being only verbal gives him a wrong account of his natural capacity; for the same words signify more or less according as they are applied to things as ordinary and extraordinary do at court; and sometimes the greater sound has the less sense, as in accounts though four be more than three, yet a third in proportion is more than a fourth.

c.1668

Of Charity, or the Love of God

LOVE is the greatest thing that God can give us; for himself is love: and it is the greatest thing we can give to God; for it will also give ourselves, and carry with it all that is ours. The apostle calls it the band of perfection: it is the old, and it is the new, and it is the great commandment, and it is all the commandments; for it is the fulfilling of the law. It does the work of all other graces without any instrument but its own immediate virtue. For, as the love to sin makes a man sin against all his own reason, and all the discourses of wisdom, and all the advices of his friends, and without temptation, and without opportunity, so does the love of God; it makes a man chaste without the laborious arts of fasting and exterior disciplines, temperate in the midst of feasts, and is active enough to choose it without any intermedial appetites, and reaches at glory through the very heart of grace, without any other arms but those of love. It is a grace that loves God for himself, and our neighbours for God. The consideration of God's goodness and bounty, the experience of those profitable and excellent emanations from him, may be, and most commonly are, the first motive of our love; but when we are once entered, and have tasted the goodness of God, we love the spring for its own excellency, passing from passion to reason, from thanking to adoring, from sense to spirit, from considering ourselves to an union with God: and this is the image and little representation of heaven; it is beatitude in picture, or rather the infancy and beginnings of glory.

We need no incentives by way of special enumeration to move us to the love of God; for we cannot love any thing for any reason, real or imaginary, but that excellence is infinitely more eminent in God. There can but two things create love — perfection and usefulness: to which answer on our part, 1. Admiration; and, 2. Desire; and both these are centred in love. For the entertainment of the first, there is in God an infinite nature, immensity or vastness without extension or limit, immutability, eternity, omnipotence, omniscience, holiness, dominion, providence, bounty, mercy, justice, perfection in himself, and the end to which all things and all actions must be directed, and will at last arrive. The consideration of which may be heightened, if we consider our distance from all these glories; our smallness and limited nature, our nothing, our inconstancy, our age like a span, our weakness and ignorance, our poverty, our inadvertency and inconsideration, our disabilities and disaffections to do good, our harsh natures and unmerciful inclinations, our universal iniquity, and our necessities and dependencies, not only on God originally and essentially, but even our need of the meanest of God's creatures, and our being obnoxious to the weakest and most contemptible. But, for the entertainment of the second, we may consider that in him is a torrent of pleasure for the voluptuous; he is the fountain of honour for the ambitious, an inexhaustible treasure for the covetous. Our vices are in love with fantastic pleasures and images of perfection, which are truly and really to be found no where but in God. And therefore our virtues have such proper objects that it is but reasonable they should all turn into love; for certain it is that this love will turn all into virtue. For in the scrutinies for righteousness and judgment, when it is inquired whether such a person be a good man or no, the meaning is not, What does he believe? or what does he hope? but what he loves.

1650

Of Avarice

THERE are two sorts of avarice: the one is but of a bastard kind, and that is, the rapacious appetite of gain; not for its own sake, but for the pleasure of refunding it immediately through all the channels of pride and luxury: the other is the true kind, and properly so called; which is a restless and unsatiable desire of riches, not for any farther end or use, but only to hoard, and preserve, and perpetually increase them. The covetous man, of the first kind, is like a greedy ostrich, which devours any metal; but it is with an intent to feed upon it, and in effect it makes a shift to digest and excern it. The second is like the foolish chough, which loves to steal money only to hide it. The first does much harm to mankind; and a little good too, to some few: the second does good to none; no, not to himself. The first can make no excuse to God, or angels, or rational men, for his actions: the second can give no reason or colour, not to the devil himself, for what he does; he is a slave to Mammon, without wages. The first makes a shift to be beloved; ay, and envied, too, by some people: the second is the universal object of hatred and contempt. There is no vice has been so pelted with good sentences, and especially by the poets, who have pursued it with stories and fables, and allegories, and allusions; and moved, as we say, every stone to fling at it: among all which, I do not remember a more fine and gentleman-like correction than that which was given it by one line of Ovid:

> Desunt luxuriæ multa, avaritiæ omnia.
> Much is wanting to luxury, all to avarice.

To which saying, I have a mind to add one member, and tender it thus;

> Poverty wants some, luxury many, avarice all things.

Somebody says of a virtuous and wise man, 'that having nothing, he has all:' this is just his antipode, who, having all things, yet has nothing. He is a guardian eunuch to his beloved gold: 'audivi eos amatores esse maximos, sed nil potesse.' They are the fondest lovers but impotent to enjoy.

> And, oh, what man's condition can be worse
> Than his, whom plenty starves, and blessings curse;
> The beggars but a common fate deplore,
> The rich poor man's emphatically poor.

I wonder how it comes to pass, that there has never been any law made against him: against him, do I say? I mean, for him: as there are public provisions made for all other mad-men: it is very reasonable that the king should appoint some persons (and I think the courtiers would not be against this proposition) to manage his estate during his life (for his heirs commonly need not that care): and out of it to make it their business to see, that he should not want alimony befitting his condition, which he could never get out of his own cruel fingers. We relieve idle vagrants, and counterfeit beggars; but have no care at all of these really poor men, who are (methinks) to be respectfully treated, in regard of their quality. I might be endless against them, but I am almost choked with the super-abundance of the matter; too much plenty impoverishes me, as it does them.

*c.*1665

'Chaucer'

(*From* Preface to the Fables)

IN the first place, as he is the father of English poetry, so I hold him in the same degree of veneration as the Grecians held Homer, or the Romans Virgil. He is a perpetual fountain of good sense; learn'd in all sciences; and, therefore, speaks properly on all subjects. As he knew what to say, so he knows also when to leave off; a continence which is practised by few writers, and scarcely by any of the ancients, excepting Virgil and Horace. One of our late great poets* is sunk in his reputation, because he could never forgive any conceit which came in his way; but swept like a drag-net, great and small. There was plenty enough, but the dishes were ill sorted; whole pyramids of sweetmeats for boys and women but little of solid meat for men. All this proceeded not from any want of knowledge, but of judgment. Neither did he want that in discerning the beauties and faults of other poets, but only indulged himself in the luxury of writing; and perhaps knew it was a fault, but hoped the reader would not find it. For this reason, though he must always be thought a great poet, he is no longer esteemed a good writer; and for ten impressions, which his works have had in so many successive years, yet at present a hundred books are scarcely purchased once a twelvemonth; for, as my last Lord Rochester said, though somewhat profanely, *Not being of God, he could not stand.*

Chaucer followed Nature everywhere, but was never so bold to go beyond her; and there is a great difference of being *poeta* and *nimis poeta,* if we may believe Catullus, as much as betwixt a modest behaviour and affectation. The verse of Chaucer, I confess, is not harmonious to us; but 'tis like the eloquence of one whom Tacitus commends, it was *auribus istius temporis accom-*

* *Editor's note*: the 'late great poet' was Abraham Cowley.

modata: they who lived with him, and some time after him, thought it musical; and it continues so, even in our judgment, if compared with the numbers of Lidgate and Gower, his contemporaries: there is the rude sweetness of a Scotch tune in it, which is natural and pleasing, though not perfect. 'Tis true, I cannot go so far as he who published the last edition of him; for he would make us believe the fault is in our ears, and that there were really ten syllables in a verse where we find but nine: but this opinion is not worth confuting; 'tis so gross and obvious an error, that common sense (which is a rule in everything but matters of Faith and Revelation) must convince the reader that equality of numbers, in every verse which we call *heroic,* was either not known, or not always practised, in Chaucer's age. It were an easy matter to produce some thousands of his verses, which are lame for want of half a foot, and sometimes a whole one, and which no pronunciation can make otherwise. We can only say, that he lived in the infancy of our poetry, and that nothing is brought to perfection at the first. We must be children before we grow men. There was an Ennius, and in process of time a Lucilius, and a Lucretius, before Virgil and Horace; even after Chaucer there was a Spenser, a Harrington, a Fairfax, before Waller and Denham were in being; and our numbers were in their nonage till these last appeared. I need say little of his parentage, life, and fortunes; they are to be found at large in all the editions of his works. He was employed abroad, and favoured, by Edward the Third, Richard the Second, and Henry the Fourth, and was poet, as I suppose, to all three of them. In Richard's time, I doubt, he was a little dipt in the rebellion of the Commons; and being brother-in-law to John of Ghant, it was no wonder if he followed the fortunes of that family; and was well with Henry the Fourth when he had deposed his predecessor. Neither is it to be admired, that Henry, who was a wise as well as a valiant prince, who claimed by succession, and was sensible that his title was not sound, but was rightfully in Mortimer, who had married the heir of York; it was not to be admired, I say, if that great politician should be pleased to have the greatest wit of those times in his interests, and to be the trumpet of his praises. Augustus had given him the example, by the advice of Maecenas, who recommended Virgil and Horace to him; whose praises helped to make him popular while he was alive, and after his death have made him precious to posterity. As for the religion of our poet, he seems to have some little bias towards the opinions of Wicliffe, after John of Ghant his patron; somewhat of which appears in the tale of *Piers Plowman*: yet I cannot blame him for inveighing so sharply against the vices of the clergy in his age: their pride, their ambition, their pomp, their avarice, their worldly interest, deserved the lashes which he gave them, both in that, and in most of his *Canterbury Tales*. Neither has his contemporary Boccace spared them: yet both those poets lived in much esteem with good and holy men in orders; for the scandal which is given by particular priests reflects not on the sacred function. Chaucer's *Monk*, his *Canon*, and his *Friar*, took not from the character of his *Good Parson*. A satirical poet is the check of the laymen on bad priests. We are only to take care that we involve not the innocent with the guilty in the same condemnation. The good cannot be too much honoured, nor the bad too coarsely used, for the corruption of the best becomes the worst. When a clergyman is whipped, his gown is first taken off, by which the dignity of his order is secured. If he be wrongfully accused, he has his action of slander; and 'tis at the poet's peril if he transgress the law. But they will tell us that all kind of satire, though never so well deserved by particular priests, yet brings the whole order into contempt. Is then the peerage of England anything dishonoured when a peer suffers for his treason? If he be libelled, or any way defamed, he has his *scandalum magnatum* to punish the offender. They who use this kind of argument seem to be conscious to themselves of somewhat which has deserved the poet's lash, and are less concerned for their public capacity than for their private; at least there is pride at the bottom of their reasoning. If the faults of men in orders are only to be judged among themselves, they are all in some sort parties; for, since they say the honour of their order is concerned in every member of it, how can we be sure that they will be impartial judges? How far I may be allowed to speak my opinion in this case, I know not; but I am sure a dispute of this nature caused mischief in abundance betwixt a King of England and an Archbishop of Canterbury; one standing up for the laws of his land, and the other for the honour (as he called it) of God's Church; which ended in the murder of the prelate, and in the whipping of his Majesty from post to pillar for his penance. The learned and

ingenious Dr Drake has saved me the labour of inquiring into
the esteem and reverence which the priests have had of old; and
I would rather extend than diminish any part of it: yet I must
needs say that when a priest provokes me without any occasion
given him, I have no reason, unless it be the charity of a
Christian, to forgive him: *prior laesit* is justification sufficient in
the civil law. If I answer him in his own language, self-defence, I
am sure must be allowed me; and if I carry it further, even to
a sharp recrimination, somewhat may be indulged to human
frailty. Yet my resentment has not wrought so far but that I have
followed Chaucer in his character of a holy man, and have
enlarged on that subject with some pleasure; reserving to myself
the right, if I shall think fit hereafter, to describe another sort of
priests, such as are more easily to be found than the Good Parson; such as have given the last blow to Christianity in this age,
by a practice so contrary to their doctrine. But this will keep cold
till another time. In the meanwhile, I take up Chaucer where I
left him.

He must have been a man of a most wonderful comprehensive
nature, because, as it has been truly observed of him, he has
taken into the compass of his *Canterbury Tales* the various
manners and humours (as we now call them) of the whole
English nation in his age. Not a single character has escaped
him. All his pilgrims are severally distinguished from each other;
and not only in their inclinations, but in their very physi-
ognomies and persons. Baptista Porta could not have described
their natures better, than by the marks which the poet gives
them. The matter and manner of their tales, and of their telling,
are so suited to their different educations, humours, and callings,
that each of them would be improper in any other mouth. Even
the grave and serious characters are distinguished by their several sorts of gravity: their discourses are such as belong to their
age, their calling, and their breeding; such as are becoming of
them, and of them only. Some of his persons are vicious, and
some virtuous; some are unlearn'd, or (as Chaucer calls them)
lewd, and some are learn'd. Even the ribaldry of the low characters is different: the Reeve, the Miller, and the Cook, are several men, and distinguished from each other as much as the
mincing Lady-Prioress and the broad-speaking, gap-toothed
Wife of Bath. But enough of this; there is such a variety of game

springing up before me that I am distracted in my choice, and
know not which to follow. 'Tis sufficient to say, according to the
proverb, that *here is God's plenty*. We have our forefathers and
great-grand-dames all before us, as they were in Chaucer's days:
their general characters are still remaining in mankind, and even
in England, though they are called by other names than those of
Monks, and Friars, and Canons, and Lady Abbesses, and Nuns;
for mankind is ever the same, and nothing lost out of Nature,
though everything is altered.

1700

A Treatise on Good Manners and Good Breeding

Good manners is the art of making those people easy with whom we converse.

Whoever makes the fewest persons uneasy is the best bred in the company.

As the best law is founded upon reason, so are the best manners. And as some lawyers have introduced unreasonable things into common law, so likewise many teachers have introduced absurd things into common good manners.

One principal point of this art is to suit our behaviour to the three several degrees of men; our superiors, our equals, and those below us.

For instance, to press either of the two former to eat or drink is a breach of manners; but a farmer or a tradesman must be thus treated, or else it will be difficult to persuade them that they are welcome.

Pride, ill nature, and want of sense, are the three great sources of ill manners; without some one of these defects, no man will behave himself ill for want of experience; or of what, in the language of fools, is called knowing the world.

I defy any one to assign an incident wherein reason will not direct us what we are to say or do in company, if we are not misled by pride or ill nature.

Therefore I insist that good sense is the principal foundation of good manners; but because the former is a gift which very few among mankind are possessed of, therefore all the civilized nations of the world have agreed upon fixing some rules for common behaviour, best suited to their general customs, or fancies, as a kind of artificial good sense, to supply the defects of reason. Without which the gentlemanly part of dunces would be perpetually at cuffs, as they seldom fail when they happen to be drunk, or engaged in squabbles about women or play. And, God be thanked, there hardly happens a duel in a year, which may not be imputed to one of those three motives. Upon which account, I should be exceedingly sorry to find the legislature make any new laws against the practice of duelling; because the methods are easy and many for a wise man to avoid a quarrel with honour, or engage in it with innocence. And I can discover no political evil in suffering bullies, sharpers, and rakes, to rid the world of each other by a method of their own; where the law hath not been able to find an expedient.

As the common forms of good manners were intended for regulating the conduct of those who have weak understandings; so they have been corrupted by the persons for whose use they were contrived. For these people have fallen into a needless and endless way of multiplying ceremonies, which have been extremely troublesome to those who practise them, and insupportable to everybody else: insomuch that wise men are often more uneasy at the over civility of these refiners, than they could possibly be in the conversations of peasants or mechanics.

The impertinencies of this ceremonial behaviour are nowhere better seen than at those tables where ladies preside, who value themselves upon account of their good breeding; where a man must reckon upon passing an hour without doing any one thing he has a mind to; unless he will be so hardy to break through all the settled decorum of the family. She determines what he loves best, and how much he shall eat; and if the master of the house happens to be of the same disposition, he proceeds in the same tyrannical manner to prescribe in the drinking part: at the same time, you are under the necessity of answering a thousand apologies for your entertainment. And although a good deal of this humour is pretty well worn off among many people of the best fashion, yet too much of it still remains, especially in the country; where an honest gentleman assured me, that having been kept four days, against his will, at a friend's house, with all the circumstances of hiding his boots, locking up the stable, and other contrivances of the like nature, he could not remember, from the moment he came into the house to the moment he left it, any one thing, wherein his inclination was not directly contradicted; as if the whole family had entered into a combination to torment him.

But, besides all this, it would be endless to recount the many foolish and ridiculous accidents I have observed among these unfortunate proselytes to ceremony. I have seen a duchess fairly knocked down, by the precipitancy of an officious coxcomb running to save her the trouble of opening a door. I remember, upon a birthday at court, a great lady was utterly desperate by a dish of sauce let fall by a page directly upon her head-dress and brocade, while she gave a sudden turn to her elbow upon some point of ceremony with the person who sat next her. Monsieur Buys, the Dutch envoy, whose politics and manners were much of a size, brought a son with him, about thirteen years old, to a great table at court. The boy and his father, whatever they put on their plates, they first offered round in order, to every person in the company; so that we could not get a minute's quiet during the whole dinner. At last their two plates happened to encounter, and with so much violence, that, being china, they broke in twenty pieces, and stained half the company with wet sweetmeats and cream.

There is a pedantry in manners, as in all arts and sciences; and sometimes in trades. Pedantry is properly the overrating any kind of knowledge we pretend to. And if that kind of knowledge be a trifle in itself, the pedantry is the greater. For which reason I look upon fiddlers, dancing-masters, heralds, masters of the ceremony, etc to be greater pedants than Lipsius, or the elder Scaliger. With these kind of pedants, the court, while I knew it, was always plentifully stocked; I mean from the gentleman usher (at least) inclusive, downward to the gentleman porter; who are, generally speaking, the most insignificant race of people that this island can afford, and with the smallest tincture of good manners, which is the only trade they profess. For being wholly illiterate, and conversing chiefly with each other, they reduce the whole system of breeding within the forms and circles of their several offices; and as they are below the notice of ministers, they live and die in court under all revolutions, with great obsequiousness to those who are in any degree of favour or credit, and with rudeness or insolence to everybody else. Whence I have long concluded, that good manners are not a plant of the court growth: for if they were, those people who have understandings directly of a level for such acquirements, and who have served such long apprenticeships to nothing else, would certainly have picked them up. For as to the great officers, who attend the prince's person or councils, or preside in his family, they are a transient body, who have no better a title to good manners than their neighbours, nor will probably have recourse to gentlemen ushers for instruction. So that I know little to be learnt at court upon this head, except in the material circumstance of dress; wherein the authority of the maids of honour must indeed be allowed to be almost equal to that of a favourite actress.

I remember a passage my Lord Bolingbroke told me, that going to receive Prince Eugene of Savoy at his landing, in order to conduct him immediately to the Queen, the Prince said, he was much concerned that he could not see her Majesty that night; for Monsieur Hoffman (who was then by) had assured his Highness that he could not be admitted into her presence with a tied-up periwig; that his equipage was not arrived; and that he had endeavoured in vain to borrow a long one among all his valets and pages. My lord turned the matter into a jest, and brought the Prince to her Majesty; for which he was highly censured by the whole tribe of gentlemen ushers; among whom Monsieur Hoffman, an old dull resident of the Emperor's, had picked up this material point of ceremony; and which, I believe, was the best lesson he had learned in five-and-twenty years' residence.

I make a difference between good manners and good breeding; although, in order to vary my expression, I am sometimes forced to confound them. By the first, I only understand the art of remembering and applying certain settled forms of general behaviour. But good breeding is of a much larger extent; for besides an uncommon degree of literature sufficient to qualify a gentleman for reading a play, or a political pamphlet, it takes in a great compass of knowledge; no less than that of dancing, fighting, gaming, making the circle of Italy, riding the great horse, and speaking French; not to mention some other secondary, or subaltern accomplishments, which are more easily acquired. So that the difference between good breeding and good manners lies in this, that the former cannot be attained to by the best understandings, without study and labour; whereas a tolerable degree of reason will instruct us in every part of good manners, without other assistance.

I can think of nothing more useful upon this subject, than to point out some particulars, wherein the very essentials of good manners are concerned, the neglect or perverting of which doth very much disturb the good commerce of the world, by introducing a traffic of mutual uneasiness in most companies.

First, a necessary part of good manners, is a punctual observance of time at our own dwellings, or those of others, or at third places; whether upon matter of civility, business, or diversion; which rule, though it be a plain dictate of common reason, yet the greatest minister I ever knew was the greatest trespasser against it; by which all his business doubled upon him, and placed him in a continual arrear. Upon which I often used to rally him, as deficient in point of good manners. I have known more than one ambassador, and secretary of state with a very moderate portion of intellectuals, execute their offices with good success and applause, by the mere force of exactness and regularity. If you duly observe time for the service of another, it doubles the obligation; if upon your own account, it would be manifest folly, as well as ingratitude, to neglect it. If both are concerned, to make your equal or inferior attend on you, to his own disadvantage, is pride and injustice.

Ignorance of forms cannot properly be styled ill manners; because forms are subject to frequent changes; and consequently, being not founded upon reason, are beneath a wise man's regard. Besides, they vary in every country; and after a short period of time, very frequently in the same; so that a man who travels, must needs be at first a stranger to them in every court through which he passes; and perhaps at his return, as much a stranger in his own; and after all, they are easier to be remembered or forgotten than faces or names.

Indeed, among the many impertinencies that superficial young men bring with them from abroad, this bigotry of forms is one of the principal, and more prominent than the rest; who look upon them not only as if they were matters capable of admitting of choice, but even as points of importance; and are therefore zealous on all occasions to introduce and propagate the new forms and fashions they have brought back with them. So that, usually speaking, the worst bred person in the company is a young traveller just returned from abroad.

Published posthumously, 1754

A Meditation Upon a Broom-Stick

ACCORDING TO THE STYLE AND MANNER OF THE
HONOURABLE ROBERT BOYLE'S MEDITATIONS

THIS single stick, which you now behold ingloriously lying in
that neglected corner, I once knew in a flourishing state in a forest; it was full of sap, full of leaves, and full of boughs; but now,
in vain does the busy art of man pretend to vie with nature, by
tying that withered bundle of twigs to its sapless trunk; 'tis now
at best but the reverse of what it was, a tree turned upside down,
the branches on the earth, and the root in the air; 'tis now
handled by every dirty wench, condemned to do her drudgery,
and, by a capricious kind of fate, destined to make other things
clean, and be nasty itself: at length, worn to the stumps in the
service of the maids, 'tis either thrown out of doors, or condemned to its last use, of kindling a fire. When I beheld this I
sighed, and said within myself, *Surely mortal man is a Broomstick!*
Nature sent him into the world strong and lusty, in a thriving
condition, wearing his own hair on his head, the proper branches
of this reasoning vegetable, till the axe of intemperance has
lopped off his green boughs, and left him a withered trunk: he
then flies to art, and puts on a periwig, valuing himself upon an
unnatural bundle of hairs, all covered with powder, that never
grew on his head; but now should this our broomstick pretend
to enter the scene, proud of those birchen spoils it never bore,
and all covered with dust, though the sweepings of the finest
lady's chamber, we should be apt to ridicule and despise its vanity. Partial judges that we are of our own excellencies, and other
men's defaults!

But a broomstick, perhaps you will say, is an emblem of a tree
standing on its head; and pray what is man, but a topsyturvy
creature, his animal faculties perpetually mounted on his rational,
his head where his heels should be, grovelling on the earth! And
yet with all his faults, he sets up to be an universal reformer and
corrector of abuses, a remover of grievances, rakes into every
slut's corner of Nature, bringing hidden corruptions to the light,
and raises a mighty dust where there was none before; sharing

deeply all the while in the very same pollutions he pretends to sweep away. His last days are spent in slavery to women, and generally the least deserving, till, worn out to the stumps, like his brother besom, he is either kicked out of doors, or made use of to kindle flames for others to warm themselves by.

1701

JOSEPH ADDISON

Thoughts in Westminster Abbey

W HEN I am in a serious humour, I very often walk by myself in Westminster Abbey; where the gloominess of the place, and the use to which it is applied, with the solemnity of the building, and the condition of the people who lie in it, are apt to fill the mind with a kind of melancholy, or rather thoughtfulness, that is not disagreeable. I yesterday passed a whole afternoon in the churchyard, the cloisters, and the church, amusing myself with the tombstones and inscriptions that I met with in those several regions of the dead. Most of them recorded nothing else of the buried person, but that he was born upon one day, and died upon another: the whole history of his life being comprehended in those two circumstances, that are common to all mankind. I could not but look upon these registers of existence, whether of brass or marble, as a kind of satire upon the departed persons; who had left no other memorial of them, but that they were born and that they died. They put me in mind of several persons mentioned in the battles of heroic poems, who have sounding names given them, for no other reason but that they may be killed, and are celebrated for nothing but being knocked on the head. The life of these men is finely described in holy writ by 'the path of an arrow', which is immediately closed up and lost.

Upon my going into the church, I entertained myself with the digging of a grave; and saw in every shovelful of it that was thrown up, the fragment of a bone or skull intermixt with a kind of fresh mouldering earth, that some time or other had a place in the composition of a human body. Upon this I began to consider with myself what innumerable multitudes of people lay confused together under the pavement of that ancient cathedral; how men and women, friends and enemies, priests and soldiers, monks and prebendaries, were crumbled amongst one another, and blended together in the same common mass; how beauty, strength, and

youth, with old age, weakness, and deformity, lay undistinguished in the same promiscuous heap of matter.

After having thus surveyed this great magazine of mortality,
as it were, in the lump; I examined it more particularly by the
accounts which I found on several of the monuments which are
raised in every quarter of that ancient fabric. Some of them were
covered with such extravagant epitaphs, that, if it were possible
for the dead person to be acquainted with them, he would blush
at the praises which his friends have bestowed upon him. There
are others so excessively modest, that they deliver the character
of the person departed in Greek of Hebrew, and by that means
are not understood once in a twelvemonth. In the poetical quarter, I found there were poets who had no monuments, and
monuments which had no poets. I observed, indeed, that the
present war had filled the church with many of these uninhabited
monuments, which had been erected to the memory of persons
whose bodies were perhaps buried in the plains of Blenheim, or
in the bosom of the ocean.

I could not but be very much delighted with several modern
epitaphs, which are written with great elegance of expression and
justness of thought, and therefore do honour to the living as
well as to the dead. As a foreigner is very apt to conceive an idea
of the ignorance or politeness of a nation, from the turn of their
public monuments and inscriptions, they should be submitted to
the perusal of men of learning and genius, before they are put
in execution. Sir Cloudesly Shovel's monument has very often
given me great offence: instead of the brave rough English
Admiral, which was the distinguishing character of that plain
gallant man, he is represented on his tomb by the figure of a
beau, dressed in a long periwig, and reposing himself upon velvet cushions under a canopy of state. The inscription is answer-
able to the monument; for instead of celebrating the many
remarkable actions he had performed in the service of his
country, it acquaints us only with the manner of his death, in
which it was impossible for him to reap any honour. The Dutch,
whom we are apt to despise for want of genius, show an infinitely greater taste of antiquity and politeness in their buildings
and works of this nature, than what we meet with in those of our
own country. The monuments of their admirals, which have
been erected at the public expense, represent them like them-

selves; and are adorned with rostral crowns and naval ornaments, with beautiful festoons of seaweed, shells, and coral.

But to return to our subject. I have left the repository of our
English kings for the contemplation of another day, when I shall
find my mind disposed for so serious an amusement. I know that
entertainments of this nature are apt to raise dark and dismal
thoughts in timorous minds and gloomy imaginations; but for
my own part, though I am always serious, I do not know what it
is to be melancholy; and can therefore take a view of nature in
her deep and solemn scenes, with the same pleasure as in her
most gay and delightful ones. By this means I can improve
myself with those objects which others consider with terror.
When I look upon the tombs of the great, every emotion of envy
dies in me; when I read the epitaphs of the beautiful, every in-
ordinate desire goes out; when I meet with the grief of parents
upon a tombstone, my heart melts with compassion; when I see
the tomb of the parents themselves, I consider the vanity of
grieving for those whom we must quickly follow: when I see
kings lying by those who deposed them, when I consider rival
wits placed side by side, or the holy men that divided the world
with their contests and disputes, I reflect with sorrow and astonishment on the little competitions, factions, and debates of mankind. When I read the several dates of the tombs, of some that
died yesterday, and some six hundred years ago, I consider that
great day when we shall all of us be contemporaries, and make
our appearance together.

1711

The Royal Exchange

THERE is no place in the town which I so much love to frequent as the Royal Exchange. It gives me a secret satisfaction,
and, in some measure, gratifies my vanity, as I am an Englishman, to see so rich an assembly of countrymen and foreigners
consulting together upon the private business of mankind, and
making this metropolis a kind of emporium for the whole earth.
I must confess I look upon high-change to be a great council, in

which all considerable nations have their representatives. Factors
in the trading world are what ambassadors are in the politic
world; they negotiate affairs, conclude treaties, and maintain a
good correspondence between those wealthy societies of men
that are divided from one another by seas and oceans, or live on
the different extremities of a continent. I have often been pleased
to hear disputes adjusted between an inhabitant of Japan and an
alderman of London, or to see a subject of the Great Mogul
entering into a league with one of the Czar of Muscovy. I am
infinitely delighted in mixing with these several ministers of
commerce, as they are distinguished by their different walks and
different languages: sometimes I am justled among a body of
Armenians; sometimes I am lost in a crowd of Jews; and sometimes make one in a group of Dutchmen. I am a Dane, Swede, or
Frenchman at different times; or rather fancy myself like the old
philosopher, who upon being asked what countryman he was,
replied, that he was a citizen of the world.

Though I very frequently visit this busy multitude of people,
I am known to nobody there but my friend Sir Andrew, who
often smiles upon me as he sees me bustling in the crowd, but
at the same time connives at my presence without taking any
further notice of me. There is indeed a merchant of Egypt, who
just knows me by sight, having formerly remitted me some
money to Grand Cairo; but as I am not versed in the modern
Coptic, our conferences go no further than a bow and a grimace.

This grand scene of business gives me an infinite variety of
solid and substantial entertainments. As I am a great lover of
mankind, my heart naturally overflows with pleasure at the sight
of a prosperous and happy multitude, insomuch, that at many
public solemnities I cannot forbear expressing my joy with tears
that have stolen down my cheeks. For this reason I am wonderfully delighted to see such a body of men thriving in their
own private fortunes, and at the same time promoting the public
stock; or, in other words, raising estates for their own families,
by bringing into their country whatever is wanting, and carrying
out of it whatever is superfluous.

Nature seems to have taken a peculiar care to disseminate the
blessings among the different regions of the world, with an eye
to this mutual intercourse and traffic among mankind, that the
natives of the several parts of the globe might have a kind of

dependence upon one another, and be united together by this common interest. Almost every degree produces something peculiar to it. The food often grows in one country, and the sauce in another. The fruits of Portugal are corrected by the products of Barbadoes; the infusion of a China plant sweetened with the pith of an Indian cane. The Philippine Islands give a flavour to our European bowls. The single dress of a woman of quality is often the product of a hundred climates. The muff and the fan come together from the different ends of the earth. The scarf is sent from the torrid zone, and the tippet from beneath the pole. The brocade petticoat rises out of the mines of Peru, and the diamond necklace out of the bowels of Indostan.

If we consider our own country in its natural prospect, without any of the benefits and advantages of commerce, what a barren, uncomfortable spot of earth falls to our share! Natural historians tell us, that no fruit grows originally among us besides hips and haws, acorns and pig-nuts, with other delicacies of the like nature; that our climate of itself, and without the assistance of art, can make no further advances towards a plum than to a sloe, and carries an apple to no greater a perfection than a crab: that our melons, our peaches, our figs, our apricots, and cherries, are strangers among us, imported in different ages, and naturalized in our English gardens; and that they would all degenerate and fall away into the trash of our own country, if they were wholly neglected by the planter, and left to the mercy of our sun and soil. Nor has traffic more enriched our vegetable world, than it has improved the whole face of nature among us. Our ships are laden with the harvest of every climate: our tables are stored with spices, and oils, and wines; our rooms are filled with pyramids of China, and adorned with the workmanship of Japan: our morning's draught comes to us from the remotest corners of the earth; we repair our bodies by the drugs of America, and repose ourselves under Indian canopies. My friend Sir Andrew calls the vineyards of France our gardens; the spice-islands our hot-beds; the Persians our silk-weavers, and the Chinese our potters. Nature indeed furnishes us with the bare necessaries of life, but traffic gives us a great variety of what is useful, and at the same time supplies us with everything that is convenient and ornamental. Nor is it the least part of this our happiness, that while we enjoy the remotest products of the north and south, we are free from those extremities of weather which give them birth; that our eyes are refreshed with the green fields of Britain, at the same time that our palates are feasted with fruits that rise between the tropics.

For these reasons there are not more useful members in a commonwealth than merchants. They knit mankind together in a mutual intercourse of good offices, distribute the gifts of nature, find work for the poor, and wealth to the rich, and magnificence to the great. Our English merchant converts the tin of his own country into gold, and exchanges his wool for rubies. The Mahometans are clothed in our British manufacture, and the inhabitants of the frozen zone warmed with the fleeces of our sheep.

When I have been upon the Change, I have often fancied one of our old kings standing in person, where he is represented in effigy, and looking down upon the wealthy concourse of people with which that place is every day filled. In this case, how would he be surprised to hear all the languages of Europe spoken in this little spot of his former dominions, and to see so many private men, who in his time would have been the vassals of some powerful baron, negotiating like princes for greater sums of money than were formerly to be met with in the royal treasury! Trade, without enlarging the British territories, has given us a kind of additional empire: it has multiplied the number of the rich, made our landed estates infinitely more valuable than they were formerly, and added to them an accession of other estates as valuable as the lands themselves.

1711

Sir Roger in Westminster Abbey

My friend Sir Roger de Coverley told me the other night, that he had been reading my paper upon Westminster Abbey, in which, says he, there are a great many ingenious fancies. He told me at the same time, that he observed I had promised another paper upon the tombs, and that he should be glad to go and see them with me, not having visited them since he had read history. I could not at first imagine how this came into the knight's head, till I recollected that he had been very busy all last summer upon Baker's Chronicle, which he has quoted several times in his dispute with Sir Andrew Freeport, since his last coming to town. Accordingly I called upon him the next morning, that we might go together to the Abbey.

I found the knight under his butler's hands, who always shaves him. He was no sooner dressed, than he called for a glass of the widow Trueby's water, which he told me he always drank before he went abroad. He recommended to me a dram of it at the same time, with so much heartiness, that I could not forbear drinking it. As soon as I had got it down, I found it very unpalatable; upon which the knight observing that I had made several wry faces, told me that he knew I should not like it at first, but that it was the best thing in the world against the stone or gravel.

I could have wished, indeed, that he had acquainted me with the virtues of it sooner; but it was too late to complain, and I knew what he had done was out of good-will. Sir Roger told me further, that he looked upon it to be very good for a man whilst he staid in town, to keep off infection, and that he got together a quantity of it upon the first news of the sickness being at Dantzic: when of a sudden turning short to one of his servants, who stood behind him, he bid him call a hackney coach, and take care it was an elderly man that drove it.

He then resumed his discourse upon Mrs Trueby's water, telling me that the widow Trueby was one who did more good than all the doctors and apothecaries in the county: that she distilled every poppy that grew within five miles of her, that she distributed her water gratis among all sorts of people; to which the knight added that she had a very great jointure, and that the whole country would fain have it a match between him and her; 'and truly,' says Sir Roger, 'if I had not been engaged, perhaps I could not have done better.'

His discourse was broken off by his man's telling him he had called a coach. Upon our going to it, after having cast his eye upon the wheels, he asked the coachman if his axletree was good; upon the fellow's telling him he would warrant it, the knight turned to me, told me he looked like an honest man, and went in without further ceremony.

We had not gone far, when Sir Roger, popping out his head, called the coachman down from his box, and upon his presenting himself at the window, asked him if he smoked; as I was considering what this would end in, he bid him stop by the way at any good tobacconist's, and take in a roll of their best Virginia. Nothing material happened in the remaining part of our journey, till we were set down at the west end of the Abbey.

As we went up the body of the church the knight pointed at the trophies upon one of the new monuments, and cried out, 'A brave man I warrant him!' Passing afterwards by Sir Cloudesly Shovel, he flung his hand that way, and cried, 'Sir Cloudesly Shovel! a very gallant man!' As we stood before Busby's tomb, the knight uttered himself again after the same manner, 'Dr Busby, a great man! he whipped my grandfather; a very great man! I should have gone to him myself, if I had not been a blockhead; a very great man!'

We were immediately conducted into the little chapel on the right hand. Sir Roger planting himself at our historian's elbow, was very attentive to everything he said, particularly to the account he gave us of the lord who had cut off the king of Morocco's head. Among several other figures, he was very well pleased to see the statesman Cecil upon his knees; and, concluding them all to be great men, was conducted to the figure which represents that martyr to good housewifery, who died by the prick of a needle. Upon our interpreter's telling us, that she was a maid of honour to Queen Elizabeth, the knight was very inquisitive into her name and family; and after having regarded her finger for some time, 'I wonder, (says he,) that Sir Richard Baker has said nothing of her in his Chronicle.'

We were then conveyed to the two coronation-chairs, where my old friend, after having heard that the stone underneath the most ancient of them, which was brought from Scotland, was called Jacob's Pillow, sat himself down in the chair; and looking like the figure of an old Gothic king, asked our interpreter, what authority they had to say that Jacob had ever been in Scotland? The fellow, instead of returning him an answer, told him, that he hoped his Honour would pay his forfeit. I could observe Sir Roger a little ruffled upon being thus trepanned; but our guide not insisting upon his demand, the knight soon recovered his good humour, and whispered in my ear, that if Will. Wimble were with us, and saw those two chairs, it would go hard but he would get a tobacco-stopper out of one or t'other of them.

Sir Roger, in the next place, laid his hand upon Edward the Third's sword, and leaning upon the pummel of it, gave us the whole history of the Black Prince; concluding, that in Sir Richard Baker's opinion, Edward the Third was one of the greatest princes that ever sat upon the English throne.

We were then shown Edward the Confessor's tomb; upon which Sir Roger acquainted us, that he was the first that touched for the Evil; and afterwards Henry the Fourth's, upon which he shook his head, and told us, there was fine reading of the casualties of that reign.

Our conductor then pointed to that monument where there is the figure of one of our English kings without an head; and upon giving us to know, that the head, which was of beaten silver, had been stolen away several years since: 'Some Whig, I'll warrant you, (says Sir Roger;) you ought to lock up your kings better; they will carry off the body too, if you do not take care.'

The glorious names of Henry the Fifth and Queen Elizabeth gave the knight great opportunities of shining, and of doing justice to Sir Richard Baker, who, as our knight observed with some surprise, had a great many kings in him, whose monuments he had not seen in the Abbey.

For my own part, I could not but be pleased to see the knight show such an honest passion for the glory of his country, and such a respectful gratitude to the memory of its princes.

I must not omit, that the benevolence of my good old friend, which flows out towards every one he converses with, made him very kind to our interpreter, whom he looked upon as an extraordinary man; for which reason he shook him by the hand at parting, telling him, that he should be very glad to see him at his lodgings in Norfolk-buildings, and talk over these matters with him more at leisure.

1712

Sir Roger at Vauxhall

As I was sitting in my chamber, and thinking on a subject for my next Spectator, I heard two or three irregular bounces at my landlady's door, and upon the opening of it, a loud cheerful voice inquiring whether the philosopher was at home. The child who went to the door answered very innocently, that he did not lodge there. I immediately recollected that it was my good friend Sir Roger's voice; and that I had promised to go with him on the water to Spring-Garden, in case it proved a good evening. The knight put me in mind of my promise from the staircase, but told me that if I was speculating, he would stay below till I had done. Upon my coming down, I found all the children of the family got about my old friend, and my landlady herself, who is a notable prating gossip, engaged in a conference with him; being mightily pleased with his stroking her little boy upon the head, and bidding him be a good child, and mind his book.

We were no sooner come to the Temple-stairs, but we were surrounded with a crowd of watermen, offering their respective services. Sir Roger, after having looked about him very attentively, spied one with a wooden leg, and immediately gave him orders to get his boat ready. As we were walking towards it, 'You must know (says Sir Roger), I never make use of anybody to row me that has not either lost a leg or an arm. I would rather bate him a few strokes of his oar, than not employ an honest man that has been wounded in the Queen's service. If I was a lord or a bishop, and kept a barge, I would not put a fellow in my livery that had not a wooden leg.'

My old friend, after having seated himself, and trimmed the boat with his coachman, who, being a very sober man, always serves for ballast on these occasions, we made the best of our way for Fox-hall. Sir Roger obliged the waterman to give us the history of his right leg, and hearing that he had left it at La Hogue, with many particulars which passed in that glorious action, the knight in the triumph of his heart made several reflections on the greatness of the British nation; as, that one Englishman could beat three Frenchmen; that we could never be in danger of Popery so long as we took care of our fleet; that the

Thames was the noblest river in Europe; that London bridge was a greater piece of work than any other of the seven wonders of the world; with many other honest prejudices which naturally cleave to the heart of a true Englishman.

After some short pause, the old knight, turning about his head twice or thrice to take a survey of this great metropolis, bid me observe how thick the city was set with churches, and that there was scarce a single steeple on this side Temple-bar. 'A most heathenish sight! (says Sir Roger): There is no religion at this end of the town. The fifty new churches will very much mend the prospect; but church-work is slow, church-work is slow!'

I do not remember I have anywhere mentioned in Sir Roger's character, his custom of saluting everybody that passes by him with a good-morrow or a good-night. This the old man does out of the overflowings of humanity, though at the same time it renders him so popular among all his country neighbours, that it is thought to have gone a good way in making him once or twice knight of the shire. He cannot forbear this exercise of benevolence even in town, when he meets with any one in his morning or evening walk. It broke from him to several boats that passed by us upon the water; but to the knight's great surprise, as he gave the good-night to two or three young fellows a little before our landing, one of them, instead of returning the civility, asked us what queer old put we had in the boat, and whether he was not ashamed to go a wenching at his years? with a great deal of the like Thames ribaldry. Sir Roger seemed a little shocked at first, but at length assuming a face of magistracy, told us, 'that if he were a Middlesex justice, he would make such vagrants know that her Majesty's subjects were no more to be abused by water than by land.'

We were now arrived at Spring-Garden, which is exquisitely pleasant at this time of year. When I considered the fragrancy of the walks and bowers, with the choirs of birds that sung upon the trees, and the loose tribe of people that walked under their shades, I could not but look upon the place as a kind of Mahometan paradise. Sir Roger told me it put him in mind of a little coppice by his house in the country, which his chaplain used to call an aviary of nightingales. 'You must understand (says the knight), there is nothing in the world that pleases a man in love so much as your nightingale. Ah, Mr Spectator! the many moonlight nights that I have walked by myself, and thought on the widow by the music of the nightingale!' He here fetched a deep sigh, and was falling into a fit of musing, when a mask, who came behind him, gave him a gentle tap upon the shoulder, and asked him if he would drink a bottle of mead with her? But the knight being startled at so unexpected a familiarity, and displeased to be interrupted in his thoughts of the widow, told her, 'She was a wanton baggage,' and bid her go about her business.

We concluded our walk with a glass of Burton ale, and a slice of hung-beef. When we had done eating ourselves, the knight called a waiter to him, and bid him carry the remainder to a waterman that had but one leg. I perceived the fellow stared upon him at the oddness of the message, and was going to be saucy; upon which I ratified the knight's commands with a peremptory look.

As we were going out of the garden my old friend, thinking himself obliged, as a member of the Quorum, to animadvert upon the morals of the place, told the mistress of the house, who sat at the bar, 'that he should be a better customer to her garden, if there were more nightingales and fewer strumpets.'

1712

On Recollections of Childhood;
Death of Parents; First Love

THERE are those among mankind, who can enjoy no relish of their being, except the world is made acquainted with all that relates to them, and think every thing lost that passes unobserved; but others find a solid delight in stealing by the crowd, and modelling their life after such a manner, as is as much above the approbation as the practice of the vulgar. Life being too short to give instances great enough of true friendship or good will, some sages have thought it pious to preserve a certain reverence for the manes of their deceased friends; and have withdrawn themselves from the rest of the world at certain seasons, to commemorate in their own thoughts such of their acquaintance who have gone before them out of this life. And indeed, when we are advanced in years, there is not a more pleasing entertainment, than to recollect in a gloomy moment the many we have parted with, that have been dear and agreeable to us, and to cast a melancholy thought or two after those, with whom, perhaps, we have indulged ourselves in whole nights of mirth and jollity. With such inclinations in my heart I went to my closet yesterday in the evening, and resolved to be sorrowful; upon which occasion I could not but look with disdain upon myself, that though all the reasons which I had to lament the loss of many of my friends are now as forcible as at the moment of their departure, yet did not my heart swell with the same sorrow which I felt at the time; but I could, without tears, reflect upon many pleasing adventures I have had with some, who have long been blended with common earth. Though it is by the benefit of nature, that length of time thus blots out the violence of afflictions; yet, with tempers too much given to pleasure, it is almost necessary to revive the old places of grief in our memory;

and ponder step by step on past life, to lead the mind into that sobriety of thought which poises the heart, and makes it beat with due time, without being quickened with desire, or retarded with despair, from its proper and equal motion. When we wind up a clock that is out of order, to make it go well for the future, we do not immediately set the hand to the present instant, but we make it strike the round of all its hours, before it can recover the regularity of its time. Such, thought I, shall be my method this evening; and since it is that day of the year which I dedicate to the memory of such in another life as I much delighted in when living, an hour or two shall be sacred to sorrow and their memory, while I run over all the melancholy circumstances of this kind which have occurred to me in my whole life.

The first sense of sorrow I ever knew was upon the death of my father, at which time I was not quite five years of age; but was rather amazed at what all the house meant, than possessed with a real understanding why nobody was willing to play with me. I remember I went into the room where his body lay, and my mother sat weeping alone by it. I had my battledore in my hand, and fell a beating the coffin, and calling Papa; for, I know not how, I had some slight idea that he was locked up there. My mother catched me in her arms, and, transported beyond all patience of the silent grief she was before in, she almost smothered me in her embraces; and told me in a flood of tears, 'Papa could not hear me, and would play with me no more, for they were going to put him under ground, whence he could never come to us again.' She was a very beautiful woman, of a noble spirit, and there was a dignity in her grief amidst all the wildness of her transport; which, methought, struck me with an instinct of sorrow, that, before I was sensible of what it was to grieve, seized my very soul, and has made pity the weakness of my heart ever since. The mind in infancy is, methinks, like the body in embryo; and receives impressions so forcible, that they are as hard to be removed by reason, as any mark with which a child is born is to be taken away by any future application. Hence it is, that good-nature in me is no merit; but having been so frequently overwhelmed with her tears before I knew the cause of any affliction, or could draw defences from my own judgment, I imbibed commiseration, remorse, and an unmanly gentleness of mind, which has since insnared me into ten thousand calamities; and from whence I can reap no advantage, except it be, that, in such a humour as I am now in, I can the better indulge myself in the softnesses of humanity, and enjoy that sweet anxiety which arises from the memory of past afflictions.

We, that are very old, are better able to remember things which befell us in our distant youth, than the passages of later days. For this reason it is, that the companions of my strong and vigorous years present themselves more immediately to me in this office of sorrow. Untimely and unhappy deaths are what we are most apt to lament; so little are we able to make it indifferent when a thing happens, though we know it must happen. Thus we groan under life, and bewail those who are relieved from it. Every object that returns to our imagination raises different passions, according to the circumstance of their departure. Who can have lived in an army, and in a serious hour reflect upon the many gay and agreeable men that might long have flourished in the arts of peace, and not join with the imprecations of the fatherless and widow on the tyrant to whose ambition they fell sacrifices? But gallant men, who are cut off by the sword, move rather our veneration than our pity; and we gather relief enough from their own contempt of death, to make that no evil, which was approached with so much cheerfulness, and attended with so much honour. But when we turn our thoughts from the great parts of life on such occasions, and instead of lamenting those who stood ready to give death to those from whom they had the fortune to receive it; I say, when we let our thoughts wander from such noble objects, and consider the havock which is made among the tender and the innocent, pity enters with an unmixed softness, and possesses all our souls at once.

Here (were there words to express such sentiments with proper tenderness) I should record the beauty, innocence, and untimely death, of the first object my eyes ever beheld with love. The beauteous virgin! how ignorantly did she charm, how carelessly excel! Oh death! thou hast right to the bold, to the ambitious, to the high, and to the haughty; but why this cruelty to the humble, to the meek, to the undiscerning, to the thoughtless? Nor age, nor business, nor distress, can erase the dear image from my imagination. In the same week, I saw her dressed for a ball, and in a shroud. How ill did the habit of death become the pretty trifler! I still behold the smiling earth—A large train of disasters were coming on to my memory, when my servant knocked at my closet-door, and interrupted me with a letter, attended with a hamper of wine, of the same sort with that which is to be put to sale on Thursday next, at Garraway's coffee-house. Upon the receipt of it, I sent for three of my friends. We are so intimate, that we can be company in whatever state of mind we meet, and can entertain each other without expecting always to rejoice. The wine we found to be generous and warming, but with such a heat as moved us rather to be cheerful than frolicksome. It revived the spirits, without firing the blood. We commended it until two of the clock this morning; and having to-day met a little before dinner, we found, that though we drank two bottles a man, we had much more reason to recollect than forget what had passed the night before.

1710

Upon Affectation

MOST people complain of fortune, few of nature; and the kinder they think the latter has been to them, the more they murmur at what they call the injustice of the former.

Why have not I the riches, the rank, the power, of such and such, is the common expostulation with fortune; but why have not I the merit, the talents, the wit, or the beauty, of such and such others, is a reproach rarely or never made to nature.

The truth is, that nature, seldom profuse, and seldom niggardly, has distributed her gifts more equally than she is generally supposed to have done. Education and situation make the great difference. Culture improves, and occasions elicit, natural talents. I make no doubt but that there are potentially, if I may use that pedantic word, many Bacons, Lockes, Newtons, Caesars, Cromwells, and Marlboroughs at the plough-tail, behind counters, and, perhaps, even among the nobility; but the soil must be cultivated, and the season favourable, for the fruit to have all its spirit and flavour.

If sometimes our common parent has been a little partial, and not kept the scales quite even; if one preponderates too much, we throw into the lighter a due counterpoise of vanity, which never fails to set all right. Hence it happens, that hardly any one man would, without reserve, and in every particular, change with any other.

Though all are thus satisfied with the dispensations of nature, how few listen to her voice! how few follow her as a guide! In vain she points out to us the plain and direct way to truth; vanity, fancy, affectation, and fashion assume her shape, and wind us through fairy-ground to folly and error.

These deviations from nature are often attended by serious consequences, and always by ridiculous ones; for there is nothing truer than the trite observation, 'that people are never ridiculous for being what they really are, but for affecting what they really are not.' Affectation is the only source, and at the same time the only justifiable object, of ridicule. No man whatsoever, be his pretensions what they will, has a natural right to be ridiculous; it is an acquired right, and not to be acquired without some industry, which perhaps is the reason why so many people are so jealous and tenacious of it. Even some people's vices are not their own, but affected and adopted, though at the same time unenjoyed, in hopes of shining in those fashionable societies where the reputation of certain vices gives lustre. In these cases, the execution is commonly as awkward as the design is absurd, and the ridicule equals the guilt.

This calls to my mind a thing that really happened not many years ago. A young fellow of some rank and fortune, just let loose from the university, resolved, in order to make a figure in the world, to assume the shining character of what he called a rake. By way of learning the rudiments of his intended profession, he frequented the theatres, where he was often drunk, and always noisy. Being one night at the representation of that most absurd play, the *Libertine destroyed*, he was so charmed with the profligacy of the hero of the piece that, to the edification of the audience, he swore many oaths that he would be the libertine *destroyed*. A discreet friend of his who sat by him, kindly represented to him that to be the *libertine* was a laudable design, which he greatly approved of; but that to be the libertine *destroyed*, seemed to him an unnecessary part of his plan, and rather rash. He persisted, however, in his first resolution, and insisted upon being the libertine, and *destroyed*. Probably he was so; at least the presumption is in his favour. There are, I am persuaded, so many cases of this nature, that for my own part I would desire no greater step towards the reformation of manners for the next twenty years, than that our people should have no vices but *their own*.

The blockhead who affects wisdom, because nature has given him dulness, becomes ridiculous only by his adopted character; whereas he might have stagnated unobserved in his native mud, or perhaps have engrossed deeds, collected shells, and studied heraldry, or logic, with some success.

The shining coxcomb aims at all, and decides finally upon everything, because nature has given him pertness. The degree of parts and animal spirits, necessary to constitute that character, if properly applied, might have made him useful in many parts of life; but his affectation and presumption make him useless in most, and ridiculous in all.

The septuagenary fine gentleman might probably, from his long experience and knowledge of the world, be esteemed and respected in the several relations of domestic life, which, at his age, nature points out to him: he will most ridiculously spin out the rotten thread of his former gallantries. He dresses, languishes, ogles, as he did at five-and-twenty; and modestly intimates that he is not without a *bonne fortune*; which *bonne fortune* at last appears to be the prostitute he had long kept not to himself, whom he marries and owns, because *the poor girl was so fond of him and so desirous to be made an honest woman.*

The sexagenary widow remembers that she was handsome, but forgets that it was thirty years ago, and thinks herself so, or at least, very *likeable*, still. The pardonable affectations of her youth and beauty unpardonably continue, increase even with her years, and are doubly exerted in hopes of concealing the number. All the gaudy glittering parts of dress, which rather degraded than adorned her beauty in its bloom, now expose to the highest and justest ridicule her shrivelled or her overgrown carcass. She totters or sweats under the load of her jewels, embroideries, and brocades, which, like so many Egyptian hieroglyphics, serve only to authenticate the venerable antiquity of her august mummy. Her eyes dimly twinkle tenderness, or leer desire: their language, however inelegant, is intelligible, and the half-pay captain understands it. He addresses his vows to her vanity, which assures her they are sincere. She pities him, and prefers him to credit, decency, and every social duty. He tenderly prefers her, though not without some hesitation, to a jail.

Self-love, kept within due bounds, is a natural and useful sentiment. It is, in truth, social love too, as Mr Pope has very justly observed: it is the spring of many good actions, and of no ridiculous ones. But self-flattery is only the ape, or caricature of self-love, and resembles it no more than to heighten the ridicule. Like other flattery, it is the most profusely bestowed and greedily swallowed, where it is the least deserved. I will conclude this subject with the substance of a fable of the ingenious Monsieur de la Motte, which seems not unapplicable to it.

Jupiter made a lottery in heaven, in which mortals, as well as gods, were allowed to have tickets. The prize was WISDOM; and Minerva got it. The mortals murmured, and accused the gods of foul play. Jupiter, to wipe off this aspersion, declared another lottery, for mortals singly and exclusively of the gods. The prize was FOLLY. They got it and shared it among themselves. All were satisfied. The loss of WISDOM was neither regretted nor remembered; FOLLY supplied its place, and those who had the largest share of it, thought themselves the wisest.

1755

The Levee

IN the first chapter of Job we have an account of a transaction said to have arisen in the court, or at the *levee*, of the best of all possible princes, or of governments by a single person, viz. that of God himself.

At this *levee*, in which the sons of God were assembled, Satan also appeared.

It is probable the writer of that ancient book took his idea of this *levee* from those of the eastern monarchs of the age he lived in.

It is to this day usual, at the *levees* of princes, to have persons assembled who are enemies to each other, who seek to obtain favor by whispering calumny and detraction, and thereby ruining those that distinguish themselves by their virtue and merit. And kings frequently ask a familiar question or two, of every one in the circle, merely to show their benignity. These circumstances are particularly exemplified in this relation.

If a modern king, for instance, finds a person in the circle, who has not lately been there, he naturally asks him how he has passed his time since he last had the pleasure of seeing him. The gentleman perhaps replies, that he has been in the country to view his estates, and visit some friends. Thus Satan, being asked whence he cometh, answers, 'From going to and fro in the earth, and walking up and down in it.' And being further asked, whether he had considered the uprightness and fidelity of the prince's servant Job, he immediately displays all the malignance of the designing courtier, by answering with another question; 'Doth Job serve God for naught? Hast thou not given him immense wealth, and protected him in the possession of it? Deprive him of that, and he will curse thee to thy face.' In modern phrase, 'Take away his places and his pensions, and your Majesty will soon find him in the opposition.'

This whisper against Job had its effect. He was delivered into the power of his adversary, who deprived him of his fortune, destroyed his family, and completely ruined him.

The Book of Job is called by divines a sacred poem, and, with the rest of the Holy Scriptures, is understood to be written for our instruction.

What then is the instruction to be gathered from this supposed transaction?

Trust not a single person with the government of your state. For if the Deity himself, being the monarch, may for a time give way to calumny, and suffer it to operate the destruction of the best of subjects; what mischief may you not expect from such power in a mere man, though the best of men, from whom the truth is often industriously hidden, and to whom falsehood is often presented in its place, by artful, interested, and malicious courtiers?

And be cautious in trusting him even with limited powers, lest sooner or later he sap and destroy those limits, and render himself absolute.

For by the disposal of places, he attaches to himself all the placeholders, with their numerous connexions, and also all the expecters and hopers of places, which will form a strong party in promoting his views. By various political engagements for the interest of neighbouring states or princes, he procures their aid in establishing his own personal power. So that, through the hopes of emolument in one part of his subjects, and the fear of his resentment in the other, all opposition falls before him.

1779

The Poor and their Betters

OF all the oppressions which the rich are guilty of, there seems to be none more impudent and unjust than their endeavour to rob the poor of a title which is most clearly the property of the latter. Not contented with all the honourables, worshipfuls, reverends, and a thousand other proud epithets which they exact of the poor, and for which they give in return nothing but dirt, scrub, mob, and such like, they have laid violent hands on a word to which they have not the least pretence or shadow of any title.

The word I mean is the comparative of the adjective good, namely *better*, or as it is usually expressed in the plural number *betters*. An appellation which all the rich usurp to themselves, and most shamefully use when they speak of, or to the poor: for do we not every day hear such phrases as these: Do not be saucy to your *betters*. Learn to behave yourself before your *betters*. Pray know your *betters*, etc.

It is possible that the rich have been so long in possession of this, that they now lay a kind of prescriptive claim to the property; but however that be, I doubt not but to make it appear, that if the word better is to be understood as the comparative of good, and is meant to convey an idea of superior goodness, it is with the highest impropriety applied to the rich, in comparison with the poor.

And this I the rather undertake, as the usurpation which I would obviate, hath produced a very great mischief in society; for the poor having been deceived into an opinion (for monstrous as it is, such an opinion hath prevailed) that the rich are their betters, have been taught to honour, and of consequence to imitate the examples of those whom they ought to have despised; while the rich on the contrary are misled into a false contempt of what they ought to respect, and by this means lose all the advantage which they might draw from contemplating the exemplary lives of these their real betters.

First then let us imagine to ourselves, a person wallowing in wealth, and lolling in his chariot, his mind torn with ambition, avarice, envy, and every other bad passion, and his brain distracted with schemes to deceive and supplant some other man, to cheat his neighbour or perhaps the public, what a glorious use might such a person derive to himself, as he is rolled through the outskirts of the town, by due meditations, on the lives of those who dwell in stalls and cellars! What a noble lesson of true Christian patience and contentment may such a person learn from his betters, who enjoy the highest cheerfulness in their poor condition; their minds being disturbed by no unruly passion, nor their heads by any racking cares!

Where again shall we look for an example of temperance? In the stinking kitchens of the rich, or under the humble roofs of the poor? Where for prudence but among those who have the fewest desires? Where for fortitude, but among those who have every natural evil to struggle with?

In modesty, I think, there will be little difficulty in knowing where we are to find our betters: for to this virtue there can be nothing more diametrically opposite than pride. Whenever therefore we observe persons stretching up their heads, and looking with an air of a contempt on all around them, we may be well assured there is no modesty there. Indeed I never yet heard it enumerated among all the bad qualities of an oyster-woman or a cider-wench, that she had a great deal of pride, and consequently there is at least a possibility that such may have a great deal of modesty, whereas it is absolutely impossible that those to whom much pride belongs, should have any tincture of its opposite virtue.

Nor are the pretensions of these same betters less strongly supported in that most exalted virtue of justice, witness the daily examples which they give of it in their own persons. When a man was punished for his crimes the Greeks said that he gave justice. Now this is a gift almost totally confined to the poor, and it is a gift which they very seldom fail of making as often as there is any very pressing occasion. Who can remember to have seen a rich man whipt at the cart's tail! And how seldom (I am sorry to say it) are such exalted to the pillory, or sentenced to transportation! And as for the most reputable, namely the capital punishments, how rarely do we see them executed on the rich! Whereas their betters, to their great honour be it spoken, do very constantly make all these gifts of justice to the society, which the other part have it much more in their power to serve by showing the same regard to this virtue.

As for chastity, it is a matter which I shall handle with great delicacy and tenderness, as it principally concerns that lovely part of the creation, for whom I have the sincerest regard. On this head therefore, I shall only whisper, that if our ladies of fashion were sometimes for variety only to take a ride through St Giles's, they might find something in the air there as wholesome as in that of Hanover or Grosvenor Square.

It may perhaps be objected to what I have hitherto advanced that I have only mentioned the cardinal virtues, which (possibly from the popish epithet assigned to them) are at present held in so little repute, that no man is conceived to be the better for possessing them, or the worse for wanting them. I will now therefore proceed to a matter so necessary to the genteel character, that a superior degree of excellence in it hath been universally allowed by all gentlemen, in the most essential manner, to constitute our betters.

My sagacious reader, I make no doubt, already perceives I am going to mention decency, the characteristic, as it is commonly thought, of a gentleman; and perhaps it formerly was so; but at present it is so far otherwise, that, if our people of fashion will examine the matter fairly and without prejudice, they cannot have the least decency left, if they refuse to allow that, in this instance, the mob are most manifestly their betters.

Who that hath observed the behaviour of an audience at the playhouse, can doubt a moment to what part he should give the preference in decency! Here indeed I must be forced, however against my inclination, to prefer the upper ladies (I mean those who sit in the upper regions of the house), to the lower. Some, perhaps, may think the pit an exception to this rule; but I am sorry to say, that I have received information by some of my spies, that the example of the boxes hath of late corrupted the manners of their betters in the pit; and that several shopkeepers' wives and daughters have begun to interrupt the performance, by laughing, tittering, giggling, chattering, and such like behaviour, highly unbecoming all persons who have any regard to decency: whereas nothing of this kind hath been imputed, as I have yet heard, to the ladies in either gallery, who may be truly said to be above all these irregularities.

I readily allow, that on certain occasions the gentlemen at the top of the house are rather more vociferous than those at the bottom. But to this I shall give three answers: first, that the voice of men is stronger than that of beaux. Secondly, that on these occasions, as at the first night of a new play, the entertainment is to be considered as among the audience, all of whom are actors in such scenes. Lastly, as these entertainments all begin below-stairs, the concurrence of the galleries is to be attributed to the politeness of our betters who sit there, and to that decent condescension which they show in concurring with the manners of their inferiors.

Nor do these, our betters, give us examples of decency in their own persons only; they take the utmost care to preserve decency in their inferiors, and are a kind of deputies to the censor in all public places. Who is it that prevents the stage being crowded with grotesque figures, a mixture of the human with the baboon species? Who (I say) but the mob? The gentlemen in the boxes observe always the profoundest tranquillity on all such occasions; but no sooner doth one of these apparitions present its frightful figure before the scenes, than the mob, from their profound regard to decency, are sure to command him off.

And should any person of fashion in the boxes expose themselves to public notice by any indecent particularities of behaviour; from whom would they receive immediate correction and admonishment, but from the mob who are (for this purpose perhaps) placed over them?

Was it not for this tender care of decency in the mob, who knows what spectacles the desire of novelty and distinction would often exhibit in our streets? For let persons be guilty of the highest enormities of this kind, they may meet a hundred people of fashion without receiving a single rebuke. But the mob never fail to express their indignation on all indecencies of this kind: and it is, perhaps, the awe of the mob alone which prevents people of condition, as they call themselves, from becoming more egregious apes than they are, of all the extravagant modes and follies of Europe.

Thus, I think, I have fully proved what I undertook to prove. I do not pretend to say, that the mob have no faults; perhaps they have many. I assert no more than this, that they are in all laudable qualities very greatly superior to those who have hitherto, with much injustice, pretended to look down upon them.

In this attempt, I may perhaps have given offence to some of the inferior sort, but I am contented with the assurance of having espoused the cause of truth; and in so doing, I am well convinced I shall please all who are really my betters.

1752

Dignity and Uses of Biography

ALL joy or sorrow for the happiness or calamities of others is produced by an act of the imagination that realizes the event, however fictitious, or approximates it, however remote, by placing us for a time in the condition of him whose fortune we contemplate. So that we feel, while the deception lasts, whatever motions would be excited by the same good or evil happening to ourselves.

Our passions are therefore more strongly moved in proportion as we can more readily adopt the pains or pleasure proposed to our minds by recognizing them at once our own or considering them as naturally incident to our state of life. It is not easy for the most artful writer to give us an interest in happiness or misery which we think ourselves never likely to feel and with which we have never yet been made acquainted. Histories of the downfall of kingdoms and revolutions of empires are read with great tranquillity. The imperial tragedy pleases common auditors only by its pomp of ornament and grandeur of ideas; and the man whose faculties have been engrossed by business, and whose heart never fluttered but at the rise or fall of stocks, wonders how the attention can be seized or the affection agitated by a tale of love.

Those parallel circumstances and kindred images to which we readily conform our minds are, above all other writings, to be found in narratives of the lives of particular persons; and therefore no species of writing seems more worthy of cultivation than biography, since none can be more delightful or more useful, none can more certainly enchain the heart by irresistible interest, or more widely diffuse instruction to every diversity of condition.

The general and rapid narratives of history, which involve a thousand fortunes in the business of a day and complicate innumerable incidents in one great transaction, afford few lessons applicable to private life, which derives its comforts and its wretchedness from the right or wrong management of things which nothing but their frequency makes considerable—'Parva si non fiunt quotidie,' says Pliny—and which can have no place in those relations which never descend below the consultation of senates, the motions of armies, and the schemes of conspirators.

I have often thought that there has rarely passed a life of which a judicious and faithful narrative would not be useful. For not only every man has, in the mighty mass of the world, great numbers in the same condition with himself, to whom his mistakes and miscarriages, escapes and expedients, would be of immediate and apparent use, but there is such an uniformity in the state of man, considered apart from adventitious and separable decorations and disguises, that there is scarce any possibility of good or ill but is common to human kind. A great part of the time of those who are placed at the greatest distance by fortune or by temper must unavoidably pass in the same manner, and though, when the claims of nature are satisfied, caprice and vanity and accident begin to produce discriminations and peculiarities, yet the eye is not very heedful or quick which cannot discover the same causes still terminating their influence in the same effects, though sometimes accelerated, sometimes retarded, or perplexed by multiplied combinations. We are all prompted by the same motives, all deceived by the same fallacies, all animated by hope, obstructed by danger, entangled by desire, and seduced by pleasure.

It is frequently objected to relations of particular lives that they are not distinguished by any striking or wonderful vicissitudes. The scholar who passed his life among his books, the merchant who conducted only his own affairs, the priest whose sphere of action was not extended beyond that of his duty, are considered as no proper objects of public regard, however they might have excelled in their several stations, whatever might have been their learning, integrity, and piety. But this notion arises from false measures of excellence and dignity, and must be eradicated by considering that in the esteem of uncorrupted reason what is of most use is of most value.

It is, indeed, not improper to take honest advantages of prejudice and to gain attention by a celebrated name; but the business of the biographer is often to pass slightly over those performances and incidents which produce vulgar greatness, to lead the thoughts into domestic privacies, and display the minute details of daily life where exterior appendages are cast aside and men excel each other only by prudence and by virtue. The account of Thuanus is, with great propriety, said by its author to have been written that it might lay open to posterity the private and familiar character of that man, *cujus ingenium et candorum ex ipsius scriptis sunt olim semper miraturi*, whose candour and genius will to the end of time be by his writings preserved in admiration.

There are many invisible circumstances which, whether we read as inquirers after natural or moral knowledge, whether we intend to enlarge our science or increase our virtue, are more important than public occurrences. Thus Sallust, the great master of nature, has not forgot in his account of Cataline to remark that his walk was now quick and again slow, as an indication of a mind revolving something with violent commotion. Thus the story of Melanchthon affords a striking lecture on the value of time by informing us that when he made an appointment he expected not only the hour but the minute to be fixed, that the day might not run out in the idleness of suspense. And all the plans and enterprises of De Wit are now of less importance to the world than that part of his personal character which represents him as careful of his health and negligent of his life.

But biography has often been allotted to writers who seem very little acquainted with the nature of their task or very negligent about the performance. They rarely afford any other account than might be collected from public papers, but imagine themselves writing a life when they exhibit a chronological series of actions or preferments, and so little regard the manners or behaviour of their heroes, that more knowledge may be gained of a man's real character by a short conversation with one of his servants than from a formal and studied narrative begun with his pedigree and ended with his funeral.

If now and then they condescend to inform the world of particular facts, they are not always so happy as to select the most important. I know not well what advantage posterity can receive from the only circumstance by which Tickell has distinguished Addison from the rest of mankind, the irregularity of his pulse.

Nor can I think myself overpaid for the time spent in reading the life of Malherbe by being enabled to relate, after the learned biographer, that Malherbe had two predominant opinions; one that the looseness of a single woman might destroy all her boast of ancient descent, the other that the French beggars made use very improperly and barbarously of the phrase *noble gentleman*, because either word included the sense of both.

There are, indeed, some natural reasons why these narratives are often written by such as were not likely to give much instruction or delight, and why most accounts of particular persons are barren and useless. If a life be delayed till interest and envy are at an end, we may hope for impartiality but must expect little intelligence. For the incidents which give excellence to biography are of a volatile and evanescent kind, such as soon escape the memory and are rarely transmitted by tradition. We know how few can portray a living acquaintance except by his most prominent and observable particularities and the grosser features of his mind; and it may be easily imagined how much of this little knowledge may be lost in imparting it, and how soon a succession of copies will lose all resemblance of the original.

If the biographer writes from personal knowledge and makes haste to gratify the public curiosity, there is danger lest his interest, his fear, his gratitude, or his tenderness overpower his fidelity and tempt him to conceal if not to invent. There are many who think it an act of piety to hide the faults or failings of their friends, even when they can no longer suffer by their detection. We therefore see whole ranks of characters adorned with uniform panegyric, and not to be known from one another but by extrinsic and casual circumstances. 'Let me remember,' says Hale, 'when I find myself inclined to pity a criminal, that there is likewise a pity due to the country.' If we owe regard to the memory of the dead, there is yet more respect to be paid to knowledge, to virtue, and to truth.

1750

Conversation

NONE of the desires dictated by vanity is more general, or less blamable, than that of being distinguished for the arts of conversation. Other accomplishments may be possessed without opportunity of exerting them, or wanted without danger that the defect can often be remarked; but as no man can live, otherwise than in an hermitage, without hourly pleasure or vexation, from the fondness or neglect of those about him, the faculty of giving pleasure is of continual use. Few are more frequently envied than those who have the power of forcing attention wherever they come, whose entrance is considered as a promise of felicity, and whose departure is lamented, like the recess of the sun from northern climates, as a privation of all that enlivens fancy, or inspirits gaiety.

It is apparent, that to excellence in this valuable art, some peculiar qualifications are necessary: for every one's experience will inform him, that the pleasure which men are able to give in conversation, holds no stated proportion to their knowledge or their virtue. Many find their way to the tables and the parties of those who never consider them as of the least importance in any other place: we have all, at one time or other, been content to love those whom we could not esteem, and been persuaded to try the dangerous experiment of admitting him for a companion, whom we knew to be too ignorant for a counsellor, and too treacherous for a friend.

I question whether some abatement of character is not necessary to general acceptance. Few spend their time with much satisfaction under the eye of uncontestable superiority; and therefore, among those whose presence is courted at assemblies of jollity, there are seldom found men eminently distinguished for powers or acquisitions. The wit whose vivacity condemns slower tongues to silence, the scholar whose knowledge allows no man to fancy that he instructs him, the critick who suffers no fallacy to pass undetected, and the reasoner who condemns the idle to thought, and the negligent to attention, are generally praised and feared, reverenced and avoided.

He that would please must rarely aim at such excellence as

depresses his hearers in their own opinion, or debars them from the hope of contributing reciprocally to the entertainment of the company. Merriment, extorted by sallies of imagination, sprightliness of remark, or quickness of reply, is too often what the Latins call, the Sardinian laughter, a distortion of the face without gladness of heart.

For this reason, no style of conversation is more extensively acceptable than the narrative. He who has stored his memory with slight anecdotes, private incidents, and personal peculiarities, seldom fails to find his audience favourable. Almost every man listens with eagerness to contemporary history; for almost every man has some real or imaginary connexion with a celebrated character, some desire to advance or oppose a rising name. Vanity often co-operates with curiosity. He that is a hearer in one place, qualifies himself to become a speaker in another; for though he cannot comprehend a series of argument, or transport the volatile spirit of wit without evaporation, he yet thinks himself able to treasure up the various incidents of a story, and please his hopes with the information which he shall give to some inferior society.

Narratives are for the most part heard without envy, because they are not supposed to imply any intellectual qualities above the common rate. To be acquainted with facts not yet echoed by plebeian mouths, may happen to one man as well as to another; and to relate them when they are known, has in appearance so little difficulty, that every one concludes himself equal to the task.

But it is not easy, and in some situations of life not possible, to accumulate such a stock of materials as may support the expense of continual narration; and it frequently happens, that they who attempt this method of ingratiating themselves, please only at the first interview; and, for want of new supplies of intelligence, wear out their stories by continual repetition.

There would be, therefore, little hope of obtaining the praise of a good companion, were it not to be gained by more compendious methods; but such is the kindness of mankind to all, except those who aspire to real merit and rational dignity, that every understanding may find some way to excite benevolence; and whoever is not envied may learn the art of procuring love. We are willing to be pleased, but are not willing to admire: we favour the mirth or officiousness that solicits our regard, but oppose the worth or spirit that enforces it.

The first place among those that please, because they desire only to please, is due to the *merry fellow*, whose laugh is loud, and whose voice is strong; who is ready to echo every jest with obstreperous approbation, and countenance every frolick with vociferations of applause. It is not necessary to a merry fellow to have in himself any fund of jocularity, or force of conception; it is sufficient that he always appears in the highest exaltation of gladness, for the greater part of mankind are gay or serious by infection, and follow without resistance the attraction of example.

Next to the merry fellow is the *good-natured man*, a being generally without benevolence, or any other virtue, than such as indolence and insensibility confer. The characteristick of a good-natured man is to bear a joke; to sit unmoved and unaffected amidst noise and turbulence, profaneness and obscenity; to hear every tale without contradiction; to endure insult without reply; and to follow the stream of folly, whatever course it shall happen to take. The good-natured man is commonly the darling of the petty wits, with whom they exercise themselves in the rudiments of raillery; for he never takes advantage of failings, nor disconcerts a puny satirist with unexpected sarcasms; but while the glass continues to circulate, contentedly bears the expense of an uninterrupted laughter, and retires rejoicing at his own importance.

The *modest man* is a companion of a yet lower rank, whose only power of giving pleasure is not to interrupt it. The modest man satisfies himself with peaceful silence, which all his companions are candid enough to consider as proceeding not from inability to speak, but willingness to hear.

Many, without being able to attain any general character of excellence, have some single art of entertainment which serves them as a passport through the world. One I have known for fifteen years the darling of a weekly club, because every night, precisely at eleven, he begins his favourite song, and during the vocal performance, by corresponding motions of his hand, chalks out a giant upon the wall. Another has endeared himself to a long succession of acquaintances by sitting among them with his wig reversed; another by contriving to smut the nose of any stranger who was to be initiated in the club; another by purring like a cat, and then pretending to be frighted; and another by yelping like a hound, and calling to the drawers to drive out the dog.

Such are the arts by which cheerfulness is promoted, and sometimes friendship established; arts, which those who despise them should not rigorously blame, except when they are practised at the expense of innocence; for it is always necessary to be loved, but not always necessary to be reverenced.

1751

Debtors' Prisons (1)

TO THE IDLER

Sɪʀ,

As I was passing lately under one of the gates of this city, I was struck with horror by a rueful cry, which summoned me 'to remember the poor debtors'.

The wisdom and justice of the English laws are, by Englishmen at least, loudly celebrated; but scarcely the most zealous admirers of our institutions can think that law wise which, when men are capable of work, obliges them to beg; or just which exposes the liberty of one to the passions of another.

The prosperity of a people is proportionate to the number of hands and minds usefully employed. To the community sedition is a fever, corruption is a gangrene, and idleness an atrophy. Whatever body, and whatever society, wastes more than it acquires must gradually decay; and every being that continues to be fed, and ceases to labour, takes away something from the public stock.

The confinement, therefore, of any man in the sloth and darkness of a prison is a loss to the nation, and no gain to the creditor. For of the multitudes who are pining in those cells of misery, a very small part is suspected of any fraudulent act by which they retain what belongs to others. The rest are imprisoned by the wantonness of pride, the malignity of revenge, or the acrimony of disappointed expectation.

If those who thus rigorously exercise the power which the law has put into their hands be asked why they continue to imprison those whom they know to be unable to pay them, one will answer that his debtor once lived better than himself; another, that his wife looked above her neighbours, and his children went in silk clothes to the dancing school; and another, that he pretended to be a joker and a wit. Some will reply that if they were in debt they should meet with the same treatment; some, that they owe no more than they can pay, and need therefore give no account of their actions. Some will confess their resolution that their debtors shall rot in jail; and some will discover that they hope, by cruelty, to wring the payment from their friends.

The end of all civil regulations is to secure private happiness from private malignity; to keep individuals from the power of one another; but this end is apparently neglected when a man, irritated with loss, is allowed to be the judge of his own cause, and to assign the punishment of his own pain; when the distinction between guilt and unhappiness, between casualty and design, is entrusted to eyes blind with interest, to understandings depraved by resentment.

Since poverty is punished among us as a crime, it ought at least to be treated with the same lenity as other crimes; the offender ought not to languish at the will of him whom he has offended, but to be allowed some appeal to the justice of his country. There can be no reason why any debtor should be imprisoned, but that he may be compelled to payment; and a term should therefore be fixed in which the creditor should exhibit his accusation of concealed property. If such property can be discovered, let it be given to the creditor; if the charge is not offered, or cannot be proved, let the prisoner be dismissed.

Those who made the laws have apparently supposed that every deficiency of payment is the crime of the debtor. But the truth is that the creditor always shares the act, and often more than shares the guilt of improper trust. It seldom happens that any man imprisons another but for debts which he suffered to be contracted in hope of advantage to himself, and for bargains in which he proportioned his profit to his own opinion of the hazard; and there is no reason why one should punish the other for a contract in which both concurred.

Many of the inhabitants of prisons may justly complain of

harder treatment. He that once owes more than he can pay is often obliged to bribe his creditor to patience, by increasing his debt. Worse and worse commodities, at a higher and higher price, are forced upon him; he is impoverished by compulsive traffic, and at last overwhelmed, in the common receptacles of misery, by debts which, without his own consent, were accumulated on his head. To the relief of this distress, no other objection can be made but that by an easy dissolution of debts, fraud will be left without punishment, and imprudence without awe, and that when insolvency shall be no longer punishable, credit will cease.

The motive to credit is the hope of advantage. Commerce can never be at a stop while one man wants what another can supply; and credit will never be denied while it is likely to be repaid with profit. He that trusts one whom he designs to sue is criminal by the act of trust; the cessation of such insidious traffic is to be desired, and no reason can be given why a change of the law should impair any other.

We see nation trade with nation, where no payment can be compelled. Mutual convenience produces mutual confidence, and the merchants continue to satisfy the demands of each other, though they have nothing to dread but the loss of trade.

It is vain to continue an institution which experience shows to be ineffectual. We have now imprisoned one generation of debtors after another, but we do not find that their numbers lessen. We have now learned that rashness and imprudence will not be deterred from taking credit; let us try whether fraud and avarice may be more easily restrained from giving it.

I am, Sir, etc.

1758

Debtors' Prisons (2)

SINCE the publication of the letter concerning the condition of those who are confined in gaols by their creditors, an enquiry is said to have been made by which it appears that more than twenty thousand are at this time prisoners for debt.

We often look with indifference on the successive parts of that which, if the whole were seen together, would shake us with emotion. A debtor is dragged to prison, pitied for a moment, and then forgotten; another follows him, and is lost alike in the caverns of oblivion; but when the whole mass of calamity rises up at once, when twenty thousand reasonable beings are heard all groaning in unnecessary misery, not by the infirmity of nature, but the mistake or negligence of policy, who can forbear to pity and lament, to wonder and abhor?

There is here no need of declamatory vehemence; we live in an age of commerce and computation; let us therefore coolly enquire what is the sum of evil which the imprisonment of debtors brings upon our country.

It seems to be the opinion of the later computists that the inhabitants of England do not exceed six millions, of which twenty thousand is the three-hundredth part. What shall we say of the humanity or the wisdom of a nation that voluntarily sacrifices one in every three hundred to lingering destruction!

The misfortunes of an individual do not extend their influence to many; yet, if we consider the effects of consanguinity and friendship, and the general reciprocation of wants and benefits, which make one man dear or necessary to another, it may reasonably be supposed that every man languishing in prison gives trouble of some kind to two others who love or need him. By this multiplication of misery we see distress extended to the hundredth part of the whole society.

If we estimate at a shilling a day what is lost by the inaction and consumed in the support of each man thus chained down to involuntary idleness, the public loss will rise in one year to three hundred thousand pounds; in ten years to more than a sixth part of our circulating coin.

I am afraid that those who are best acquainted with the state of our prisons will confess that my conjecture is too near the truth when I suppose that the corrosion of resentment, the heaviness of sorrow, the corruption of confined air, the want of exercise, and sometimes of food, the contagion of diseases from which there is no retreat, and the severity of tyrants against whom there can be no resistance, and all the complicated horrors of a prison put an end every year to the life of one in four of those that are shut up from the common comforts of human life.

Thus perish yearly five thousand men, overborne with sorrow, consumed by famine, or putrified by filth; many of them in the most vigorous and useful part of life; for the thoughtless and imprudent are commonly young, and the active and busy are seldom old.

According to the rule generally received, which supposes that one in thirty dies yearly, the race of man may be said to be renewed at the end of thirty years. Who would have believed till now that of every English generation a hundred and fifty thousand perish in our gaols! That in every century, a nation eminent for science, studious of commerce, ambitious of empire, should willingly lose, in noisome dungeons, five hundred thousand of its inhabitants: a number greater than has ever been destroyed in the same time by the pestilence and sword!

A very late occurrence may show us the value of the number which we thus condemn to be useless; in the re-establishment of the trained bands, thirty thousand are considered as a force sufficient against all exigencies: while, therefore, we detain twenty thousand in prison, we shut up in darkness and uselessness two thirds of an army which ourselves judge equal to the defence of our country.

The monastic institutions have been often blamed as tending to retard the increase of mankind. And perhaps retirement ought rarely to be permitted, except to those whose employment is consistent with abstraction, and who, though solitary, will not be idle; to those whom infirmity makes useless to the commonwealth, or to those who have paid their due proportion to society, and who, having lived for others, may be honourably dismissed to live for themselves. But whatever be the evil or the folly of these retreats, those have no right to censure them whose prisons contain greater numbers than the monasteries of other countries. It is, surely, less foolish and less criminal to permit inaction than compel it; to comply with doubtful opinions of happiness than condemn to certain and apparent misery; to indulge the extravagancies of erroneous piety than to multiply and enforce temptations to wickedness.

The misery of gaols is not half their evil; they are filled with every corruption which poverty and wickedness can generate between them; with all the shameless and profligate enormities that can be produced by the impudence of ignominy, the rage of want, and the malignity of despair. In a prison the awe of the public eye is lost, and the power of the law is spent; there are few fears, there are no blushes. The lewd inflame the lewd, the audacious harden the audacious. Every one fortifies himself as he can against his own sensibility, endeavours to practise on others the arts which are practised on himself; and gains the kindness of his associates by similitude of manners.

Thus some sink amidst their misery, and others survive only to propagate villainy. It may be hoped that our lawgivers will at length take away from us this power of starving and depraving one another: but, if there be any reason why this inveterate evil should not be removed in our age, which true policy has enlightened beyond any former time, let those whose writings form the opinions and the practices of their contemporaries endeavour to transfer the reproach of such imprisonment from the debtor to the creditor, till universal infamy shall pursue the wretch whose wantonness of power, or revenge of disappointment, condemns another to torture and to ruin; till he shall be hunted through the world as an enemy to man, and find in riches no shelter from contempt.

Surely, he whose debtor has perished in prison, though he may acquit himself of deliberate murder, must at least have his mind clouded with discontent when he considers how much another has suffered from him; when he thinks on the wife bewailing her husband, or the children begging the bread which their father would have earned. If there are any made so obdurate by avarice or cruelty as to revolve these consequences without dread or pity, I must leave them to be awakened by some other power, for I write only to human beings.

1758

Of the Dignity or Meanness of Human Nature

THERE are certain sects, which secretly form themselves in the learned world, as well as factions in the political; and though sometimes they come not to an open rupture, they give a different turn to the ways of thinking of those who have taken part on either side. The most remarkable of this kind are the sects, founded on the different sentiments with regard to the *dignity of human nature*; which is a point that seems to have divided philosophers and poets, as well as divines, from the beginning of the world to this day. Some exalt our species to the skies, and represent man as a kind of human demigod, who derives his origin from heaven, and retains evident marks of his lineage and descent. Others insist upon the blind sides of human nature, and can discover nothing, except vanity, in which man surpasses the other animals, whom he affects so much to despise. If an author possess the talent of rhetoric and declamation, he commonly takes part with the former: If his turn lie towards irony and ridicule, he naturally throws himself into the other extreme.

I am far from thinking, that all those, who have depreciated our species, have been enemies to virtue, and have exposed the frailties of their fellow creatures with any bad intention. On the contrary, I am sensible that a delicate sense of morals, especially when attended with a splenetic temper, is apt to give a man a disgust of the world, and to make him consider the common course of human affairs with too much indignation. I must, however, be of opinion, that the sentiments of those, who are inclined to think favourably of mankind, are more advantageous to virtue, than the contrary principles, which give us a mean opinion of our nature. When a man is prepossessed with a high notion of his rank and character in the creation, he will naturally endeavour to act up to it, and will scorn to do a base or vicious action, which might sink him below that figure which he makes in his own imagination. Accordingly we find, that all our polite and fashionable moralists insist upon this topic, and endeavour to represent vice as unworthy of man, as well as odious in itself.

We find few disputes, that are not founded on some ambiguity in the expression; and I am persuaded, that the present dispute, concerning the dignity or meanness of human nature, is not more exempt from it than any other. It may, therefore, be worth while to consider, what is real, and what is only verbal, in this controversy.

That there is a natural difference between merit and demerit, virtue and vice, wisdom and folly, no reasonable man will deny: Yet is it evident, that in affixing the term, which denotes either our approbation or blame, we are commonly more influenced by comparison than by any fixed unalterable standard in the nature of things. In like manner, quantity, and extension, and bulk, are by everyone acknowledged to be real things: But when we call any animal *great* or *little*, we always form a secret comparison between that animal and others of the same species; and it is that comparison which regulates our judgment concerning its greatness. A dog and a horse may be of the very same size, while the one is admired for the greatness of its bulk, and the other for the smallness. When I am present, therefore, at any dispute, I always consider with myself, whether it be a question of comparison or not that is the subject of the controversy; and if it be, whether the disputants compare the same objects together, or talk of things that are widely different.

In forming our notions of human nature, we are apt to make a comparison between men and animals, the only creatures endowed with thought that fall under our senses. Certainly this comparison is favourable to mankind. On the one hand, we see a creature, whose thoughts are not limited by any narrow bounds, either of place or time; who carries his researches into the most distant regions of this globe, and beyond this globe, to the planets and heavenly bodies; looks backward to consider the first origin, at least, the history of the human race; casts his eye forward to see the influence of his actions upon posterity, and the judgments which will be formed of his character a thousand years hence; a creature, who traces causes and effects to a great length and intricacy; extracts general principles from particular appearances; improves upon his discoveries; corrects his mistakes; and makes his very errors profitable. On the other hand, we are presented with a creature the very reverse of this; limited in its observations and reasonings to a few sensible objects which surround it; without curiosity, without foresight; blindly conducted by instinct, and attaining, in a short time, its utmost perfection, beyond which it is never able to advance a single step. What a wide difference is there between these creatures! And how exalted a notion must we entertain of the former, in comparison of the latter!

There are two means commonly employed to destroy this conclusion: *First*, by making an unfair representation of the case, and insisting only upon the weaknesses of human nature. And *secondly*, by forming a new and secret comparison between man and beings of the most perfect wisdom. Among the other excellencies of man, this is one, that he can form an idea of perfections much beyond what he has experience of in himself; and is not limited in his conception of wisdom and virtue. He can easily exalt his notions and conceive a degree of knowledge, which, when compared to his own, will make the latter appear very contemptible, and will cause the difference between that and the sagacity of animals, in a manner, to disappear and vanish. Now this being a point, in which all the world is agreed, that human understanding falls infinitely short of perfect wisdom; it is proper we should know when this comparison takes place, that we may not dispute where there is no real difference in our sentiments. Man falls much more short of perfect wisdom, and even of his own ideas of perfect wisdom, than animals do of man; yet the latter difference is so considerable, that nothing but a comparison with the former can make it appear of little moment.

It is also usual to *compare* one man with another; and finding very few whom we can call *wise* or *virtuous*, we are apt to entertain a contemptible notion of our species in general. That we may be sensible of the fallacy of this way of reasoning, we may observe, that the honourable appellations of wise and virtuous, are not annexed to any particular degree of those qualities of *wisdom* and *virtue*; but arise altogether from the comparison we make between one man and another. When we find a man, who arrives at such a pitch of wisdom as is very uncommon, we pronounce him a wise man: So that to say, there are few wise men in the world, is really to say nothing; since it is only by their scarcity, that they merit that appellation. Were the lowest of our species as wise as Tully, or lord Bacon, we should still have reason to say, that there are few wise men. For in that case we should exalt our notions of wisdom, and should not pay a singular honour to any one, who was not singularly distinguished by his talents. In like manner, I have heard it observed by thoughtless people, that there are few women possessed of beauty, in comparison of those who want it; not considering, that we bestow the epithet of *beautiful* only on such as possess a degree of beauty, that is common to them with a few. The same degree of beauty in a woman is called deformity, which is treated as real beauty in one of our sex.

As it is usual, in forming a notion of our species, to *compare* it with the other species above or below it, or to compare the individuals of the species among themselves; so we often compare together the different motives or actuating principles of human nature, in order to regulate our judgment concerning it. And, indeed, this is the only kind of comparison, which is worth our attention, or decides anything in the present question. Were our selfish and vicious principles so much predominant above our social and virtuous, as is asserted by some philosophers, we ought undoubtedly to entertain a contemptible notion of human nature.

There is much of a dispute of words in all this controversy. When a man denies the sincerity of all public spirit or affection to a country and community, I am at a loss what to think of him. Perhaps he never felt this in so clear and distinct a manner as to remove all his doubts concerning its force and reality. But when he proceeds afterwards to reject all private friendship, if no interest or self-love intermix itself; I am then confident that he abuses terms, and confounds the ideas of things; since it is impossible for any one to be so selfish, or rather so stupid, as to make no difference between one man and another, and give no preference to qualities, which engage his approbation and esteem. Is he also, say I, as insensible to anger as he pretends to be to friendship? And does injury and wrong no more affect him than kindness or benefits? Impossible: He does not know himself: He has forgotten the movements of his heart; or rather he makes use of a different language from the rest of his countrymen, and calls not things by their proper names. What say you of natural affection? (I subjoin) Is that also a species of self-love? Yes: All is self-love. *Your* children are loved only because they are yours: *Your* friend for a like reason: And *your* country engages you only so far as it has a connexion with *yourself*: Were the idea of self removed, nothing would affect you: You would be altogether unactive and insensible: Or, if you ever gave yourself any movement, it would only be from vanity, and a desire of fame and reputation to this same self. I am willing, reply I, to receive your interpretation of human actions, provided you admit the facts. That species of self-love, which displays itself in kindness to others, you must allow to have great influence over human actions, and even greater, on many occasions, than that which remains in its original shape and form. For how few are there, who, having a family, children, and relations, do not spend more on the maintenance and education of these than on their own pleasures? This, indeed, you justly observe, may proceed from their self-love, since the prosperity of their family and friends is one, or the chief of their pleasures, as well as their chief honour. Be you also one of these men, and you are sure of everyone's good opinion and good will; or not to shock your ears with these expressions, the self-love of everyone, and mine among the rest, will then incline us to serve you, and speak well of you.

In my opinion, there are two things which have led astray those philosophers, that have insisted so much on the selfishness of man. In the *first* place, they found, that every act of virtue or friendship was attended with a secret pleasure; whence they concluded, that friendship and virtue could not be disinterested. But the fallacy of this is obvious. The virtuous sentiment or passion produces the pleasure, and does not arise from it. I feel a pleasure in doing good to my friend, because I love him; but do not love him for the sake of that pleasure.

In the *second* place, it has always been found, that the virtuous are far from being indifferent to praise; and therefore they have been represented as a set of vain-glorious men, who had nothing in view but the applauses of others. But this also is a fallacy. It is very unjust in the world, when they find any tincture of vanity in a laudable action, to depreciate it upon that account, or ascribe it entirely to that motive. The case is not the same with vanity, as with other passions. Where avarice or revenge enters into any seemingly virtuous action, it is difficult for us to determine how far it enters, and it is natural to suppose it the sole actuating principle. But vanity is so closely allied to virtue, and to love the fame of laudable actions approaches so near the love of laudable actions for their own sake, that these passions are more capable of mixture, than any other kinds of affection; and it is almost impossible to have the latter without some degree of the former. Accordingly, we find, that this passion for glory is always warped and varied according to the particular taste or disposition of the mind on which it falls. Nero had the same vanity in driving a chariot, that Trajan had in governing the empire with justice and ability. To love the glory of virtuous deeds is a sure proof of the love of virtue.

1741

On Dress

FOREIGNERS observe, that there are no ladies in the world more beautiful, or more ill-dressed than those of England. Our countrywomen have been compared to those pictures, where the face is the work of a Raphael; but the draperies thrown out by some empty pretender, destitute of taste, and entirely unacquainted with design.

If I were a poet, I might observe, on this occasion, that so much beauty set off with all the advantages of dress would be too powerful an antagonist for the opposite sex and therefore it was wisely ordered, that our ladies should want taste, lest their admirers should entirely want reason.

But to confess a truth, I do not find they have a greater aversion to fine clothes than the women of any other country whatsoever. I cannot fancy that a shopkeeper's wife in Cheapside has a greater tenderness for the fortune of her husband than a citizen's wife in Paris; or that miss in a boarding-school is more an economist in dress than mademoiselle in a nunnery.

Although Paris may be accounted the soil in which almost every fashion takes its rise, its influence is never so general there as with us. They study there the happy method of uniting grace and fashion, and never excuse a woman for being awkwardly dressed, by saying her clothes are made in the mode. A French woman is a perfect architect in dress; she never, with Gothic ignorance, mixes the orders; she never tricks out a squabby Doric shape with Corinthian finery; or, to speak without metaphor, she conforms to general fashion, only when it happens not to be repugnant to private beauty.

Our ladies, on the contrary, seem to have no other standard for grace but the run of the town. If fashion gives the word, every distinction of beauty, complexion, or stature ceases. Sweeping trains, Prussian bonnets, and trollopees, as like each other, as if cut from the same piece, level all to one standard. The Mall, the gardens, and the playhouses are filled with ladies in uniform, and their whole appearance shows as little variety or taste as if their clothes were bespoke by the colonel of a marching regiment, or fancied by the same artist who dresses the three battalions of guards.

But not only ladies of every shape and complexion, but of every age too, are possessed of this unaccountable passion of dressing in the same manner. A lady of no quality can be distinguished from a lady of some quality only by the redness of her hands; and a woman of sixty, masked, might easily pass for her grand-daughter. I remember, a few days ago, to have walked behind a damsel, tossed out in all the gaiety of fifteen; her dress was loose, unstudied, and seemed the result of conscious beauty. I called up all my poetry on this occasion, and fancied twenty Cupids prepared for execution in every folding of her white negligee. I had prepared my imagination for an angel's face; but what was my mortification to find that the imaginary goddess was no other than my cousin Hannah, four years older than myself, and I shall be sixty-two the twelfth of next November.

After the transports of our first salute were over, I could not avoid running my eye over her whole appearance. Her gown was of cambric, cut short before, in order to discover a high-heeled shoe, which was buckled almost at the toe. Her cap, if cap it might be called that cap was none, consisted of a few bits of cambric, and flowers of painted paper stuck on one side of her head. Her bosom, that had felt no hand but the hand of time, these twenty years, rose suing, but in vain, to be pressed. I could, indeed, have wished her more than an handkerchief of Paris-net to shade her beauties; for, as Tasso says of the rose-bud, *Quanto si mostra men, tanto è più bella*, I should think her's most pleasing when least discovered.

As my cousin had not put on all this finery for nothing, she was at that time sallying out to the Park, when I had overtaken her. Perceiving, however, that I had on my best wig, she offered, if I would 'squire her there, to send home the footman. Though I trembled for our reception in public, yet I could not, with any civility, refuse; so, to be as gallant as possible, I took her hand in my arm, and thus we marched on together.

When we made our entry at the Park, two antiquated figures,

so polite and so tender as we seemed to be, soon attracted the eyes of the company. As we made our way among crowds who were out to show their finery as well as we, wherever we came I perceived we brought good humour in our train. The polite could not forbear smiling, and the vulgar burst out into a horse laugh at our grotesque figures. Cousin Hannah, who was perfectly conscious of the rectitude of her own appearance, attributed all this mirth to the oddity of mine; while I as cordially placed the whole to her account. Thus, from being two of the best-natured creatures alive, before we got half-way up the Mall, we both began to grow peevish, and, like two mice on a string, endeavoured to revenge the impertinence of others upon ourselves. 'I am amazed, cousin Jeffrey,' says miss, 'that I can never get you to dress like a Christian. I knew we should have the eyes of the Park upon us, with your great wig so frizzed, and yet so beggarly, and your monstrous muff. I hate those odious muffs.' I could have patiently borne a criticism on all the rest of my equipage; but, as I had always a peculiar veneration for my muff, I could not forbear being piqued a little; and throwing my eyes with a spiteful air on her bosom, 'I could heartily wish, madam,' replied I, 'that for your sake, my muff was cut into a tippet.'

As my cousin, by this time, was grown heartily ashamed of her gentleman usher, and as I was never very fond of any kind of exhibition myself, it was mutually agreed to retire for a while to one of the seats, and from that retreat remark on others as freely as they had remarked on us.

When seated, we continued silent for some time, employed in very different speculations. I regarded the whole company, now passing in review before me, as drawn out merely for my amusement. For my entertainment the beauty had all that morning been improving her charms, the beau had put on lace, and the young doctor a big wig, merely to please me. But quite different were the sentiments of cousin Hannah; she regarded every well-dressed woman as a victorious rival, hated every face that seemed dressed in good humour, or wore the appearance of greater happiness than her own. I perceived her uneasiness, and attempted to lessen it, by observing, that there was no company in the Park to-day. To this she readily assented; 'and yet,' says she, 'it is full enough of scrubs of one kind or another.' My smiling at this observation gave her spirits to pursue the bent of her inclination, and now she began to exhibit her skill in secret history, as she found me disposed to listen. 'Observe,' says she to me, 'that old woman in tawdry silk, and dressed out even beyond the fashion. That is Miss Biddy Evergreen. Miss Biddy, it seems, has money, and as she considers that money was never so scarce as it is now, she seems resolved to keep what she has to herself. She is ugly enough you see; yet, I assure you, she has refused several offers to my own knowledge, within this twelvemonth. Let me see, three gentlemen from Ireland who study the law, two waiting captains, her doctor, and a Scotch preacher, who had like to have carried her off. All her time is passed between sickness and finery. Thus she spends the whole week in a close chamber, with no other company but her monkey, her apothecary, and cat, and comes dressed out to the Park every Sunday, to show her airs, to get new lovers, to catch a new cold, and to make new work for the doctor.

'There goes Mrs Roundabout, I mean the fat lady in the lutestring trollopee. Between you and I, she is but a cutler's wife. See how she's dressed, as fine as hands and pins can make her, while her two marriageable daughters, like bunters, in stuff gowns, are now taking six pennyworth of tea at the White Conduit-house. Odious puss! how she waddles along, with her train two yards behind her. She puts me in mind of my Lord Bantam's Indian sheep, which are obliged to have their monstrous tails trundled along in a go-cart. For all her airs, it goes to her husband's heart to see four yards of good lutestring wearing against the ground, like one of his knives on a grindstone. To speak my mind, cousin Jeffrey, I never liked tails; for suppose a young fellow should be rude, and the lady should offer to step back in a fright, instead of retiring she treads upon her train, and falls fairly on her back; and then you know, cousin—her clothes may be spoiled.

'Ah! Miss Mazzard! I knew we should not miss her in the Park; she in the monstrous Prussian bonnet. Miss, though so very fine, was bred a milliner, and might have had some custom if she had minded her business; but the girl was fond of finery, and instead of dressing her customers, laid out all her goods in adorning herself. Every new gown she put on impaired her credit; she still, however, went on improving her appearance and lessening her little fortune, and is now, you see, become a belle and a bankrupt.'

My cousin was proceeding in her remarks, which were interrupted by the approach of the very lady she had been so freely describing. Miss had perceived her at a distance, and approached to salute her. I found, by the warmth of the two ladies' protestations, that they had been long intimate esteemed friends and acquaintance. Both were so pleased at this happy rencounter, that they were resolved not to part for the day. So we all crossed the Park together, and I saw them into a hackney coach at the gate of St James's. I could not, however, help observing, 'That they are generally most ridiculous themselves, who are apt to see most ridicule in others.'

1759

A Little Great Man

FROM LIEN CHI ALTANGI, TO FUM HOAM,
FIRST PRESIDENT OF THE CEREMONIAL ACADEMY,
AT PEKIN, IN CHINA

IN reading the newspapers here, I have reckoned up not less than twenty-five great men, seventeen very great men, and nine very extraordinary men in less than the compass of half a year. These, say the gazettes, are the men that posterity are to gaze at with admiration; these the names that fame will be employed in holding up for the astonishment of succeeding ages. Let me see—forty-six great men in half a year, amounts to just ninety-two in a year.—I wonder how posterity will be able to remember them all, or whether the people, in future times, will have any other business to mind, but that of getting the catalogue by heart.

Does the mayor of a corporation make a speech? he is instantly set down for a great man. Does a pedant digest his common place book into a folio? he quickly becomes great. Does a poet string up trite sentiments in rhyme? he also becomes the great man of the hour. How diminutive soever the object of

admiration, each is followed by a crowd of still more diminutive
admirers. The shout begins in his train, onward he marches
towards immortality, looks back at the pursuing crowd with
self-satisfaction; catching all the oddities, the whimsies, the
absurdities, and the littlenesses of conscious greatness, by the
way.

I was yesterday invited by a gentleman to dinner, who
promised that our entertainment should consist of an haunch of
venison, a turtle, and a great man. I came, according to appointment. The venison was fine, the turtle good, but the great man
insupportable. The moment I ventured to speak, I was at once
contradicted with a snap. I attempted, by a second and a third
assault, to retrieve my lost reputation, but was still beat back
with confusion. I was resolved to attack him once more from
entrenchment, and turned the conversation upon the government of China: but even here he asserted, snapped, and contradicted as before. Heavens, thought I, this man pretends to
know China even better than myself! I looked round to see who
was on my side, but every eye was fixed in admiration on the
great man; I therefore, at last thought proper to sit silent, and act
the pretty gentleman during the ensuing conversation.

When a man has once secured a circle of admirers, he may
be as ridiculous here as he thinks proper; and it all passes for
elevation of sentiment, or learned absence. If he transgresses
the common forms of breeding, mistakes even a teapot for a
tobacco-box, it is said, that his thoughts are fixed on more
important objects: to speak and act like the rest of mankind is to
be no greater than they. There is something of oddity in the very
idea of greatness; for we are seldom astonished at a thing very
much resembling ourselves.

When the Tartars make a Lama, their first care is to place him
in a dark corner of the temple; here he is to sit half concealed
from view, to regulate the motion of his hands, lips, and eyes;
but, above all, he is enjoined gravity and silence. This, however,
is but the prelude to his apotheosis: a set of emissaries are
despatched among the people to cry up his piety, gravity, and
love of raw flesh; the people take them at their word, approach
the Lama, now become an idol, with the most humble pros-
tration; he receives their addresses without motion, commences a
god, and is ever after fed by his priests with the spoon of immor-

tality. The same receipt in this country serves to make a great man. The idol only keeps close, sends out his little emissaries to be hearty in his praise; and straight, whether statesman or author, he is set down in the list of fame, continuing to be praised while it is fashionable to praise, or while he prudently keeps his minuteness concealed from the public.

I have visited many countries, and have been in cities without number, yet never did I enter a town which could not produce ten or twelve of those little great men; all fancying themselves known to the rest of the world, and complimenting each other upon their extensive reputation. It is amusing enough when two of those domestic prodigies of learning mount the stage of ceremony, and give and take praise from each other. I have been present when a German doctor, for having pronounced a panegyric upon a certain monk, was thought the most ingenious man in the world; till the monk soon after divided this reputation by returning the compliment; by which means they both marched off with universal applause.

The same degree of undeserved adulation that attends our great man while living, often also follows him to the tomb. It frequently happens that one of his little admirers sits down big with the important subject, and is delivered of the history of his life and writings. This may properly be called the revolutions of a life between the fireside and the easy-chair. In this we learn, the year in which he was born, at what an early age he gave symptoms of uncommon genius and application, together with some of his smart sayings, collected by his aunt and mother, while yet but a boy. The next book introduces him to the University, where we are informed of his amazing progress in learning, his excellent skill in darning stockings, and his new invention for papering books to save the covers. He next makes his appearance in the republic of letters, and publishes his folio. Now the colossus is reared, his works are eagerly bought up by all the purchasers of scarce books. The learned societies invite him to become a member; he disputes against some foreigner with a long Latin name, conquers in the controversy, is complimented by several authors of gravity and importance, is excessively fond of egg-sauce with his pig, becomes president of a literary club, and dies in the meridian of his glory. Happy they, who thus have some little faithful attendant, who never forsakes them, but pre-

pares to wrangle and to praise against every opposer; at once
ready to increase their pride while living, and their character
when dead. For you and I, my friend, who have no humble
admirer thus to attend us, we, who neither are, nor ever will be
great men, and who do not much care whether we are great men
or no, at least let us strive to be honest men, and to have common sense.

The Citizen of the World, 1762

On National Prejudices

As I am one of that sauntering tribe of mortals, who spend the
greatest part of their time in taverns, coffee-houses, and other
places of public resort, I have thereby an opportunity of observing an infinite variety of characters, which, to a person of a
contemplative turn, is a much higher entertainment than a view
of all the curiosities of art or nature. In one of these my late
rambles, I accidentally fell into the company of half a dozen
gentlemen, who were engaged in a warm dispute about some
political affair; the decision of which, as they were equally divided in their sentiments, they thought proper to refer to me,
which naturally drew me in for a share of the conversation.

Amongst a multiplicity of other topics, we took occasion to
talk of the different characters of the several nations of Europe;
when one of the gentlemen, cocking his hat, and assuming such
an air of importance as if he had possessed all the merit of the
English nation in his own person, declared that the Dutch were
a parcel of avaricious wretches; the French a set of flattering
sycophants; that the Germans were drunken sots, and beastly
gluttons; and the Spaniards proud, haughty and surly tyrants:
but that, in bravery, generosity, clemency, and in every other
virtue, the English excelled all the world.

This very learned and judicious remark was received with a
general smile of approbation by all the company—all, I mean,
but your humble servant; who, endeavouring to keep my gravity as well as I could, and reclining my head upon my arm,
continued for some time in a posture of affected thoughtfulness,

as if I had been musing on something else, and did not seem to attend to the subject of conversation; hoping, by this means, to avoid the disagreeable necessity of explaining myself, and thereby depriving the gentleman of his imaginary happiness.

But my pseudo-patriot had no mind to let me escape so easily: not satisfied that his opinion should pass without contradiction, he was determined to have it ratified by the suffrage of every one in the company; for which purpose, addressing himself to me with an air of inexpressible confidence, he asked me if I was not of the same way of thinking. As I am never forward in giving my opinion, especially when I have reason to believe that it will not be agreable; so, when I am obliged to give it, I always hold it for a maxim to speak my real sentiments. I therefore told him, that, for my own part, I should not have ventured to talk in such a peremptory strain, unless I had made the tour of Europe, and examined the manners of the several nations with great care and accuracy; that, perhaps, a more impartial judge would not scruple to affirm, that the Dutch were more frugal and industri-ous, the French more temperate and polite, the Germans more hardy and patient of labour and fatigue, and the Spaniards more staid and sedate, than the English; who, though undoubtedly brave and generous, were at the same time rash, headstrong, and impetuous, too apt to be elated with prosperity, and to despond in adversity.

I could easily perceive, that all the company began to regard me with a jealous eye before I had finished my answer; which I had no sooner done than the patriotic gentleman observed, with a contemptuous sneer, that he was greatly surprised how some people could have the conscience to live in a country which they did not love, and to enjoy the protection of a government, to which in their hearts they were inveterate enemies. Finding that, by this modest declaration of my sentiments, I had forfeited the good opinion of my companions, and given them occasion to call my political principles in question, and well knowing that it was in vain to argue with men who were so very full of them-selves, I threw down my reckoning, and retired to my own lodgings, reflecting on the absurd and ridiculous nature of national prejudice and prepossession.

Among all the famous sayings of antiquity, there is none that does greater honour to the author, or affords greater pleasure to

the reader, (at least if he be a person of a generous and benevol-
ent heart) than that of the philosopher, who, being asked what
countryman he was, replied that he was a citizen of the world.
How few are there to be found in modern times who can say the
same, or whose conduct is consistent with such a profession! We
are now become so much Englishmen, Frenchmen, Dutchmen,
Spaniards, or Germans, that we are no longer citizens of the
world; so much the natives of one particular spot, or members of
one petty society, that we no longer consider ourselves as the
general inhabitants of the globe, or members of that grand
society which comprehends the whole human kind.

Did these prejudices prevail only among the meanest and lowest of the people, perhaps they might be excused, as they have
few, if any, opportunities of correcting them by reading, travelling, or conversing with foreigners; but the misfortune is, that
they infect the minds, and influence the conduct even of our
gentlemen; of those, I mean, who have every title to this appellation but an exemption from prejudice, which, however, in my
opinion, ought to be regarded as the characteristical mark of a
gentleman: for let a man's birth be ever so high, his station ever
so exalted, or his fortune ever so large, yet, if he is not free from
the national and all other prejudices, I should make bold to tell
him, that he had a low and vulgar mind, and had no just claim to
the character of a gentleman. And, in fact, you will always find,
that those are most apt to boast of national merit, who have little
or no merit of their own to depend on, than which, to be sure,
nothing is more natural: the slender vine twists around the
sturdy oak for no other reason in the world, but because it has
not strength sufficient to support itself.

Should it be alleged in defence of national prejudice, that it is
the natural and necessary growth of love to our country, and
that therefore the former cannot be destroyed without hurting
the latter; I answer, that this is a gross fallacy and delusion. That
it is the growth of love to our country, I will allow; but that it is
the natural and necessary growth of it, I absolutely deny. Superstition and enthusiasm too are the growth of religion; but who
ever took it in his head to affirm, that they are the necessary
growth of this noble principle? They are, if you will, the bastard
sprouts of this heavenly plant; but not its natural and genuine
branches, and may safely enough be lopt off, without doing any

harm to the parent stock: nay, perhaps, till once they are lopt off, this goodly tree can never flourish in perfect health and vigour.

Is it not very possible that I may love my own country, without hating the natives of other countries? That I may exert the most heroic bravery, the most undaunted resolution, in defending its laws and liberty, without despising all the rest of the world as cowards and poltroons? Most certainly it is: and if it were not—but what need I suppose what is absolutely impossible?—but if it were not, I must own I should prefer the title of the ancient philosopher, namely, a citizen of the world, to that of an Englishman, a Frenchman, an European, or to any other appellation whatever.

1763

On War

WHILE viewing, as travellers usually do, the remarkable objects of curiosity at Venice, I was conducted through the different departments of the Arsenal; and as I contemplated that great storehouse of mortal engines, in which there is not only a large deposit of arms, but men are continually employed in making more, my thoughts *rebounded*, if I may use the expression, from what I beheld; and the effect was, that I was first as it were stunned into a state of amazement, and when I recovered from that, my mind expanded itself in reflections upon the horrid irrationality of war.

What those reflections were I do not precisely recollect. But the general impression dwells upon my memory; and however strange it may seem, my opinion of the irrationality of war is still associated with the Arsenal of Venice.

One particular however I well remember. When I saw workmen engaged with grave assiduity in fashioning weapons of death, I was struck with wonder at the shortsightedness, the *caecae mentes* of human beings, who were thus soberly preparing the instruments of destruction of their own species. I have since found upon a closer study of man, that my wonder might have been spared; because there are very few men whose minds are sufficiently enlarged to comprehend universal or even extensive good. The views of most individuals are limited to their own happiness; and the workmen whom I beheld so busy in the Arsenal of Venice saw nothing but what was good in the labour for which they received such wages as procured them the comforts of life. That their immediate satisfaction was not hindered by a view of the remote consequential and contingent evils for which alone their labours could be at all useful, would not surprise one who has had a tolerable share of experience in life. We must have the telescope of philosophy to make us perceive distant ills; nay, we know that there are individuals of our species to whom the immediate misery of others is nothing in comparison with their own advantage—for we know that in every age there have been found men very willing to perform the office of executioner even for a moderate hire.

To prepare instruments for the destruction of our species at large, is what I now see may very well be done by ordinary men, without starting, when they themselves are to run no risk. But I shall never forget, nor cease to wonder at a most extraordinary instance of thoughtless intrepidity which I had related to me by a cousin of mine, now a lieutenant-colonel in the British army, who was upon guard when it happened. A soldier of one of the regiments in garrison at Minorca, having been found guilty of a capital crime, was brought out to be hanged. They had neglected to have a rope in readiness, and the shocking business was at a stand. The fellow, with a spirit and alertness which in a general would, upon a difficult and trying emergency, have been very great presence of mind and conduct, stript the lace off his hat, said this will do, and actually made it serve as the fatal cord.

The irrationality of war is, I suppose, admitted by almost all men: I say almost all; because I have myself met with men who attempted seriously to maintain that it is an agreeable occupation and one of the chief means of human happiness. I must own that although I use the plural number here, I should have used the dual, had I been writing in Greek; for I never met with but two men who supported such a paradox; and one of them was a tragick poet, and one a Scotch Highlander. The first had his imagination so much in a blaze with heroic sentiments, with the 'pride, pomp and circumstance of glorious war,' that he did not advert to its miseries, as one dazzled with the pageantry of a magnificent funeral thinks not of the pangs of dissolution and the dismal corpse. The second had his attention so eagerly fixed on the advantage which accrued to his *clan* from 'the trade of war,' that he could think of it only as a good.

We are told by some writers, who assume the character of philosophers, that war is necessary to take off the superfluity of the human species, or at least to rid the world of numbers of idle and profligate men who are a burthen upon every community, and would grow an insupportable burthen, were they to live as long as men do in the usual course of nature. But there is unquestionably no reason to fear a superfluity of mankind, when we know that although perhaps the time 'when every rood of land maintain'd its man' is a poetical exaggeration, yet vigorous and well directed industry can raise sustenance for such a proportion of people in a certain space of territory, as is astonishing to us who are accustomed to see only moderate effects of labour; and when we also know what immense regions of the terrestrial globe in very good climates are uninhabited. In these there is room for millions to enjoy existence. In cultivating these, the idle and profligate, expelled from their original societies, might be employed and gradually reformed, which would be better surely, than continuing the practice of periodical destruction, which is also indiscriminate, and involves the best equally with the worst of men.

I have often thought that if war should cease over all the face of the earth, for a thousand years, its reality would not be believed at such a distance of time, notwithstanding the faith of authentick records in every nation. Were mankind totally free from every tincture of prejudice in favour of those gallant exertions which could not exist were there not the evil of violence to combat; had they never seen in their own days, or been told by their fathers or grandfathers, of battles, and were there no traces remaining of the *art of war*, I have no doubt that they would treat as fabulous or allegorical, the accounts in history, of prodigious armies being formed, of men who engaged themselves for an unlimited time, under the penalty of immediate death, to obey implicitly the orders of commanders to whom they were not attached either by affection or by interest; that those armies were sometimes led with toilsome expedition over vast tracts of land, sometimes crouded into ships, and obliged to endure tedious, unhealthy, and perilous voyages; and that the purpose of all this toil and danger was not to obtain any comfort or pleasure, but to be in a situation to encounter other armies; and that those opposite multitudes the individuals of which had no cause of quarrel, no ill-will to each other, continued for hours engaged with patient and obstinate perseverance, while thousands were slain, and thousands crushed and mangled by diversity of wounds.

We who have from our earliest years had our minds filled with scenes of war of which we have read in the books that we most

reverence and most admire, who have remarked it in every revolving century, and in every country that has been discovered
by navigators, even in the gentle and benign regions of the
southern oceans; we who have seen all the intelligence, power
and ingenuity of our own nation employed in war, who have
been accustomed to peruse Gazettes, and have had our friends
and relations killed or sent home to us wretchedly maimed; we
cannot without a steady effort of reflection be sensible of the
improbability that rational beings should act so irrationally as
to unite in deliberate plans, which must certainly produce the
direful effects which war is known to do. But I have no doubt
that if the project for a perpetual peace which the Abbé de
St Pierre sketched, and Rousseau improved, were to take place,
the incredibility of war would after the lapse of some ages be
universal.

Were there any good produced by war which could in any
degree compensate its direful effects; were better men to spring
up from the ruins of those who fall in battle, as more beautiful
material forms sometimes arise from the ashes of others; or were
those who escape from its destruction to have an increase of happiness; in short, were there any great beneficial effect to follow it,
the notion of its irrationality would be only the notion of narrow
comprehension. But we find that war is followed by no general
good whatever. The power, the glory, or the wealth of a very
few may be enlarged. But the people in general, upon both sides,
after all the sufferings are passed, pursue their ordinary occupations, with no difference from their former state. The evils
therefore of war, upon a general view of humanity are as the
French say, à pure perte, a mere loss without any advantage,
unless indeed furnishing subjects for history, poetry, and painting. And although it should be allowed that mankind have
gained enjoyment in these respects, I suppose it will not be
seriously said, that the misery is overbalanced. At any rate, there
is already such a store of subjects, that an addition to them
would be dearly purchased by more wars.

I am none of those who would set up their notions against the
opinion of the world; on the contrary, I have such a respect for
that authority, as to doubt of my own judgment when it opposes
that of numbers probably as wise as I am. But when I maintain
the irrationality of war, I am not contradicting the opinion, but

the practice of the world. For, as I have already observed, its
irrationality is generally admitted. Horace calls Hannibal, *demens*,
a madman; and Pope gives the same appellation to Alexander the
Great and Charles XII.

> From Macedonia's madman to the Swede.

How long war will continue to be practised, we have no
means of conjecturing. Civilization, which it might have been
expected would have abolished it, has only refined its savage
rudeness. The irrationality remains, though we have learnt
insanire certa ratione modoque, to have a method in our madness.

That amiable religion which 'proclaims peace on earth,' hath
not as yet made war to cease. The furious passions of men,
modified as they are by moral instruction, still operate with
much force; and by a perpetual fallacy, even the conscientious in
each contending nation think they may join in war, because they
each believe they are repelling an aggressor. Were the mild and
humane doctrine of those Christians, who are called Quakers,
which Mr Jenyns has lately embellished with his elegant pen, to
prevail, human felicity would gain more than we can well conceive. But perhaps it is necessary that mankind in this state of
existence, the purpose of which is so mysterious, should ever
suffer the woes of war.

To relieve my readers from reflections which they may think
too abstract, I shall conclude this paper with a few observations
upon actual war. In ancient times when a battle was fought man
to man, or as somebody has very well expressed it, was a group
of duels, there was an opportunity for individuals to distinguish
themselves by vigour and bravery. One who was '*robustus acri
militia*, hardy from keen warfare,' could gratify his ambition for
fame, by the exercise of his own personal qualities. It was therefore more reasonable then, for individuals to enlist, than it is in
modern times; for, a battle now is truly nothing else than a huge
conflict of opposite engines worked by men, who are themselves
as machines directed by a few; and the event is not so frequently
decided by what is intentionally done, as by accidents happening
in the dreadful confusion. It is as if two towns in opposite
territories should be set on fire at the same time, and victory
should be declared to the inhabitants of that in which the flames
were least destructive. We hear much of the conduct of generals;

and Addison himself has represented the Duke of Marlborough directing an army in battle, as an 'angel riding in a whirlwind and directing the storm.' Nevertheless I much doubt if upon many occasions the immediate schemes of a commander have had certain effect; and I believe Sir Callaghan O'Bralachan in Mr Macklin's *Love A la-mode* gives a very just account of a modern battle: 'There is so much doing every where that we cannot tell what is doing any where'.

1777

Dream Children; a Reverie

CHILDREN love to listen to stories about their elders, when *they* were children; to stretch their imagination to the conception of a traditionary great-uncle, or grandame, whom they never saw. It was in this spirit that my little ones crept about me the other evening to hear about their great-grandmother Field, who lived in a great house in Norfolk (a hundred times bigger than that in which they and papa lived) which had been the scene—so at least it was generally believed in that part of the country—of the tragic incidents which they had lately become familiar with from the ballad of the Children in the Wood. Certain it is that the whole story of the children and their cruel uncle was to be seen fairly carved out in wood upon the chimney-piece of the great hall, the whole story down to the Robin Redbreasts; till a foolish rich person pulled it down to set up a marble one of modern invention in its stead, with no story upon it. Here Alice put out one of her dear mother's looks, too tender to be called up-braiding. Then I went on to say, how religious and how good their great-grandmother Field was, how beloved and respected by everybody, though she was not indeed the mistress of this great house, but had only the charge of it (and yet in some respects she might be said to be the mistress of it too) committed to her by the owner, who preferred living in a newer and more fashionable mansion which he had purchased somewhere in the adjoining county; but still she lived in it in a manner as if it had been her own, and kept up the dignity of the great house in a sort while she lived, which afterwards came to decay, and was nearly pulled down, and all its old ornaments stripped and car-ried away to the owner's other house, where they were set up, and looked as awkward as if some one were to carry away the old tombs they had seen lately at the Abbey, and stick them up in Lady C.'s tawdry gilt drawing-room. Here John smiled, as much as to say, 'that would be foolish indeed.' And then I told how, when she came to die, her funeral was attended by a concourse of all the poor, and some of the gentry too, of the neighbourhood for many miles round, to show their respect for her memory, because she had been such a good and religious woman; so good indeed that she knew all the Psaltery by heart, ay, and a great part of the Testament besides. Here little Alice spread her hands. Then I told what a tall, upright, graceful person their great-grandmother Field once was; and how in her youth she was esteemed the best dancer—here Alice's little right foot played an involuntary movement, till, upon my looking grave, it desisted— the best dancer, I was saying, in the county, till a cruel disease, called a cancer, came, and bowed her down with pain; but it could never bend her good spirits, or make them stoop, but they were still upright, because she was so good and religious. Then I told how she was used to sleep by herself in a lone chamber of the great lone house; and how she believed that an apparition of two infants was to be seen at midnight gliding up and down the great staircase near where she slept, but she said 'those innocents would do her no harm'; and how frightened I used to be, though in those days I had my maid to sleep with me, because I was never half so good or religious as she—and yet I never saw the infants. Here John expanded all his eyebrows and tried to look courageous. Then I told how good she was to all her grand-children, having us to the great house in the holydays, where I in particular used to spend many hours by myself, in gazing upon the old busts of the twelve Caesars, that had been Emperors of Rome, till the old marble heads would seem to live again, or I to be turned into marble with them; how I never could be tired with roaming about that huge mansion, with its vast empty rooms, with their worn-out hangings, fluttering tapestry, and carved oaken panels, with the gilding almost rubbed out—some-times in the spacious old-fashioned gardens, which I had almost to myself, unless when now and then a solitary gardening man would cross me—and how the nectarines and peaches hung upon the walls, without my ever offering to pluck them, because they were forbidden fruit, unless now and then,—and because I had more pleasure in strolling about among the old melancholy-looking yew-trees, or the firs, and picking up the red berries, and the fir-apples, which were good for nothing but to look at—or in lying about upon the fresh grass with all the fine garden smells around me—or basking in the orangery, till I could almost fancy myself ripening too along with the oranges and the limes in that grateful warmth—or in watching the dace that darted to and fro in the fish-pond, at the bottom of the garden, with here and there a great sulky pike hanging midway down the water in silent state, as if it mocked at their impertinent friskings,—I had more pleasure in these busy-idle diversions than in all the sweet flavours of peaches, nectarines, oranges, and such-like common baits of children. Here John slyly deposited back upon the plate a bunch of grapes, which, not unobserved by Alice, he had meditated dividing with her, and both seemed willing to relinquish them for the present as irrelevant. Then, in somewhat a more heightened tone, I told how, though their great-grand-mother Field loved all her grandchildren, yet in an especial manner she might be said to love their uncle, John L——, because he was so handsome and spirited a youth, and a king to the rest of us; and, instead of moping about in solitary corners, like some of us, he would mount the most mettlesome horse he could get, when but an imp no bigger than themselves, and make it carry him half over the county in a morning, and join the hunters when there were any out—and yet he loved the old great house and gardens too, but had too much spirit to be always pent up within their boundaries—and how their uncle grew up to man's estate as brave as he was handsome, to the admiration of every-body, but of their great-grandmother Field most especially; and how he used to carry me upon his back when I was a lame-footed boy—for he was a good bit older than me—many a mile when I could not walk for pain;—and how in after life he became lame-footed too, and I did not always (I fear) make allowances enough for him when he was impatient and in pain, nor remember sufficiently how considerate he had been to me when I was lame-footed; and how when he died, though he had not been dead an hour, it seemed as if he had died a great while ago, such a distance there is betwixt life and death; and how I bore his death as I thought pretty well at first, but afterwards it haunted and haunted me; and though I did not cry or take it to heart as some do, and as I think he would have done if I had died, yet I missed him all day long, and knew not till then how much I had loved him. I missed his kindness, and I missed his crossness, and wished him to be alive again, to be quarrelling with him (for we quarrelled sometimes), rather than not have him again, and was as uneasy without him, as he, their poor uncle, must have been when the doctor took off his limb.—Here the children fell a-crying, and asked if their little mourning which they had on was not for uncle John, and they looked up, and prayed me not to go on about their uncle, but to tell them some stories about their pretty dead mother. Then I told how for seven long years, in hope sometimes, sometimes in despair, yet persisting ever, I courted the fair Alice W——n; and as much as children could understand, I explained to them what coyness, and difficulty, and denial, meant in maidens—when suddenly turning to Alice, the soul of the first Alice looked out at her eyes with such a reality of re-presentment, that I became in doubt which of them stood there before me, or whose that bright hair was; and while I stood gazing, both the children gradually grew fainter to my view, receding, and still receding, till nothing at last but two mournful features were seen in the uttermost distance, which, without speech, strangely impressed upon me the effects of speech: 'We are not of Alice, nor of thee, nor are we children at all. The children of Alice call Bartrum father. We are nothing; less than nothing, and dreams. We are only what might have been, and must wait upon the tedious shores of Lethe millions of ages before we have existence, and a name'—and immediately awaking, I found myself quietly seated in my bachelor armchair, where I had fallen asleep, with the faithful Bridget unchanged by my side—but John L. (or James Elia) was gone for ever.

1822

From *On Some of the Old Actors*

Of all the actors who flourished in my time—a melancholy phrase if taken aright, reader—Bensley had most of the swell of soul, was greatest in the delivery of heroic conceptions, the emotions consequent upon the presentment of a great idea to the fancy. He had the true poetical enthusiasm—the rarest faculty

among players. None that I remember possessed even a portion
of that fine madness which he threw out in Hotspur's famous
rant about glory, or the transports of the Venetian incendiary at
the vision of the fired city. His voice had the dissonance, and
at times the inspiriting effect, of the trumpet. His gait was un-
couth and stiff, but no way embarrassed by affectation; and the
thorough-bred gentleman was uppermost in every movement.
He seized the moment of passion with greatest truth; like a faithful clock, never striking before the time; never anticipating or
leading you to anticipate. He was totally destitute of trick and
artifice. He seemed come upon the stage to do the poet's message
simply, and he did it with as genuine fidelity as the nuncios in
Homer deliver the errands of the gods. He let the passion or
the sentiment do its own work without prop or bolstering. He
would have scorned to mountebank it; and betrayed none of that
cleverness which is the bane of serious acting. For this reason, his
Iago was the only endurable one which I remember to have seen.
No spectator, from his action, could divine more of his artifice
than Othello was supposed to do. His confessions in soliloquy
alone put you in possession of the mystery. There were no by-
intimations to make the audience fancy their own discernment so
much greater than that of the Moor—who commonly stands like
a great helpless mark, set up for mine Ancient, and a quantity of
barren spectators, to shoot their bolts at. The Iago of Bensley
did not go to work so grossly. There was a triumphant tone
about the character, natural to a general consciousness of power;
but none of that petty vanity which chuckles and cannot contain
itself upon any little successful stroke of its knavery—as is common with your small villains, and green probationers in mischief.
It did not clap or crow before its time. It was not a man setting
his wits at a child, and winking all the while at other children,
who are mightily pleased at being let into the secret; but a consummate villain entrapping a noble nature into toils against
which no discernment was available, where the manner was as
fathomless as the purpose seemed dark, and without motive. The
part of Malvolio, in the Twelfth Night, was performed by
Bensley with a richness and a dignity, of which (to judge from
some recent castings of that character) the very tradition must be
worn out from the stage. No manager in those days would have
dreamed of giving it to Mr Baddely, or Mr Parsons; when

Bensley was occasionally absent from the theatre, John Kemble thought it no derogation to succeed to the part. Malvolio is not essentially ludicrous. He becomes comic but by accident. He is cold, austere, repelling; but dignified, consistent, and, for what appears, rather of an over-stretched morality. Maria describes him as a sort of Puritan; and he might have worn his gold chain with honour in one of our old roundhead families, in the service of a Lambert, or a Lady Fairfax. But his morality and his manners are misplaced in Illyria. He is opposed to the proper *levities* of the piece, and falls in the unequal contest. Still his pride, or his gravity (call it which you will), is inherent, and native to the man, not mock or affected, which latter only are the fit objects to excite laughter. His quality is at the best unlovely, but neither buffoon nor contemptible. His bearing is lofty, a little above his station, but probably not much above his deserts. We see no reason why he should not have been brave, honourable, accomplished. His careless committal of the ring to the ground (which he was commissioned to restore to Cesario), bespeaks a generosity of birth and feeling. His dialect on all occasions is that of a gentleman and a man of education. We must not confound him with the eternal old, low steward of comedy. He is master of the household to a great princess; a dignity probably conferred upon him for other respects than age or length of service. Olivia, at the first indication of his supposed madness, declares that she 'would not have him miscarry for half of her dowry.' Does this look as if the character was meant to appear little or insignificant? Once, indeed, she accuses him to his face—of what?—of being 'sick of self-love,'—but with a gentleness and considerateness, which could not have been, if she had not thought that this particular infirmity shaded some virtues. His rebuke to the knight and his sottish revellers, is sensible and spirited; and when we take into consideration the unprotected condition of his mistress, and the strict regard with which her state of real or dissembled mourning would draw the eyes of the world upon her house-affairs, Malvolio might feel the honour of the family in some sort in his keeping; as it appears not that Olivia had any more brothers, or kinsmen, to look to it—for Sir Toby had dropped all such nice respects at the buttery-hatch. That Malvolio was meant to be represented as possessing estimable qualities, the expression of the Duke, in his anxiety to have him reconciled, almost infers: 'Pursue him, and entreat him to a peace.' Even in his abused state of chains and darkness, a sort of greatness seems never to desert him. He argues highly and well with the supposed Sir Topas, and philosophizes gallantly upon his straw. There must have been some shadow of worth about the man; he must have been something more than a mere vapour—a thing of straw, or Jack in office—before Fabian and Maria could have ventured sending him upon a courting-errand to Olivia. There was some consonancy (as he would say) in the undertaking, or the jest would have been too bold even for that house of misrule.

Bensley, accordingly, threw over the part an air of Spanish loftiness. He looked, spake, and moved like an old Castilian. He was starch, spruce, opinionated, but his superstructure of pride seemed bottomed upon a sense of worth. There was something in it beyond the coxcomb. It was big and swelling, but you could not be sure that it was hollow. You might wish to see it taken down, but you felt that it was upon an elevation. He was magnificent from the outset; but when the decent sobrieties of the character began to give way, and the poison of self-love, in his conceit of the Countess's affection, gradually to work, you would have thought that the hero of La Mancha in person stood before you. How he went smiling to himself! with what ineffable carelessness would he twirl his gold chain! what a dream it was! you were infected with the illusion, and did not wish that it should be removed! you had no room for laughter! if an unseasonable reflection of morality obtruded itself, it was a deep sense of the pitiable infirmity of man's nature, that can lay him open to such frenzies—but, in truth, you rather admired than pitied the lunacy while it lasted—you felt that an hour of such mistake was worth an age with the eyes open. Who would not wish to live but for a day in the conceit of such a lady's love as Olivia? Why, the Duke would have given his principality but for a quarter of a minute, sleeping or waking, to have been so deluded. The man seemed to tread upon air, to taste manna, to walk with his head in the clouds, to mate Hyperion. O! shake not the castles of his pride—endure yet for a season, bright moments of confidence—'stand still, ye watches of the element,' that Malvolio may be still in fancy fair Olivia's lord!—but fate and retribution say no—I hear the mischievous titter of Maria—the witty taunts of Sir Toby—the still more insupportable triumph of the foolish knight—the counterfeit Sir Topas is unmasked—and 'thus the whirligig of time,' as the true clown hath it, 'brings in his revenges.' I confess that I never saw the catastrophe of this character, while Bensley played it, without a kind of tragic interest.

1822

On the Pleasure of Hating

THERE is a spider crawling along the matted floor of the room where I sit (not the one which has been so well allegorised in the admirable *Lines to a Spider*, but another of the same edifying breed); he runs with heedless, hurried haste, he hobbles awkwardly towards me, he stops: he sees the giant shadow before him, and, at a loss whether to retreat or proceed, meditates his huge foe. But as I do not start up and seize upon the straggling caitiff, as he would upon a hapless fly within his toils, he takes heart, and ventures on with mingled cunning, impudence and fear. As he passes me, I lift up the matting to assist his escape, am glad to get rid of the unwelcome intruder, and shudder at the recollection after he is gone. A child, a woman, a clown, or a moralist a century ago, would have crushed the little reptile to death: my philosophy has got beyond that. I bear the creature no illwill, but still I hate the very sight of it. The spirit of malevolence survives the practical exertion of it. We learn to curb our will and keep our overt actions within the bounds of humanity, long before we can subdue our sentiments and imaginations to the same mild tone. We give up the external demonstration, the *brute* violence, but cannot part with the essence or principle of hostility. We do not tread upon the poor little animal in question (that seems barbarous and pitiful!) but we regard it with a sort of mystic horror and superstitious loathing. It will ask another hundred years of fine writing and hard thinking to cure us of the prejudice, and make us feel towards this ill-omened tribe with something of 'the milk of human kindness,' instead of their own shyness and venom.

Nature seems (the more we look into it) made up of antipathies: without something to hate, we should lose the very spring of thought and action. Life would turn to a stagnant pool, were it not ruffled by the jarring interests, the unruly passions, of men. The white streak in our own fortunes is brightened (or just rendered visible) by making all around it as dark as possible; so the rainbow paints its form upon the cloud. Is it pride? Is it envy? Is it the force of contrast? Is it weakness or malice? But so it is, that there is a secret affinity, a *hankering* after, evil in the human mind, and that it takes a perverse, but a fortunate delight in mischief, since it is a never-failing source of satisfaction. Pure good soon grows insipid, wants variety and spirit. Pain is a bitter-sweet, which never surfeits. Love turns, with a little indulgence, to indifference or disgust: hatred alone is immortal. Do we not see this principle at work everywhere? Animals torment and worry one another without mercy: children kill flies for sport: every one reads the accidents and offences in a newspaper as the cream of the jest: a whole town runs to be present at a fire, and the spectator by no means exults to see it extinguished. It is better to have it so, but it diminishes the interest; and our feelings take part with our passions rather than with our understandings. Men assemble in crowds, with eager enthusiasm, to witness a tragedy: but if there were an execution going forward in the next street, as Mr Burke observes, the theatre would be left empty. A strange cur in a village, an idiot, a crazy woman, are set upon and baited by the whole community. Public nuisances are in the nature of public benefits. How long did the Pope, the Bourbons, and the Inquisition keep the people of England in breath, and supply them with nicknames to vent their spleen upon! Had they done us any harm of late? No: but we have always a quantity of superfluous bile upon the stomach, and we wanted an object to let it out upon. How loth were we to give up our pious belief in ghosts and witches, because we liked to persecute the one, and frighten ourselves to death with the other! It is not the quality so much as the quantity of excitement that we are anxious about: we cannot bear a state of indifference and *ennui*: the mind seems to abhor a *vacuum* as much as ever nature was supposed to do. Even when the spirit of the age (that is, the progress of intellectual refinement, warring with our natural infirmities) no longer allows us to carry our vindictive and headstrong humours into effect, we try to revive them in description, and keep up the old bugbears, the phantoms of our terror and our hate, in imagination. We burn Guy Fawkes in effigy, and the hooting and buffeting and maltreating that poor tattered figure of rags and straw makes a festival in every village in England once a year. Protestants and Papists do not now burn one another at the stake: but we subscribe to new editions of Fox's *Book of Martyrs*; and the secret of the success of the *Scotch Novels* is much the same: they carry us back to the feuds, the heart-burnings, the havoc, the dismay, the wrongs and the revenge of a barbarous age and people—to the rooted prejudices and deadly animosities of sects and parties in politics and religion, and of contending chiefs and clans in war and intrigue. We feel the full force of the spirit of hatred with all of them in turn. As we read, we throw aside the trammels of civilization, the flimsy veil of humanity. 'Off, you lendings!' The wild beast resumes its sway within us, we feel like hunting-animals, and as the hound starts in his sleep and rushes on the chase in fancy, the heart rouses itself in its native lair, and utters a wild cry of joy, at being restored once more to freedom and lawless unrestrained impulses. Every one has his full swing, or goes to the Devil his own way. Here are no Jeremy Bentham Panopticons, none of Mr Owen's impassable Parallelograms (Rob Roy would have spurned and poured a thousand curses on them), no long calculations of self-interest: the will takes its instant way to its object, as the mountain-torrent flings itself over the precipice: the greatest possible good of each individual consists in doing all the mischief he can to his neighbour: that is charming, and finds a sure and sympathetic chord in every breast! So Mr Irving, the celebrated preacher, has rekindled the old, original, almost exploded, hell-fire in the aisles of the Caledonian Chapel, as they introduce the real water of the New River at Sadler's Wells, to the delight and astonishment of his fair audience. *'Tis pretty, though a plague*, to sit and peep into the pit of Tophet, to play at *snapdragon* with flames and brimstone (it gives a smart electrical shock, a lively fillip to delicate constitutions), and to see Mr Irving, like a huge Titan, looking as grim and swarthy as if he had to forge tortures for all the damned! What a strange being man is! Not content with doing all he can to vex and hurt his fellows here, 'upon this bank and shoal of time,' where one would think there were heart-aches, pain, disappointment, anguish, tears, sighs, and groans enough, the bigoted maniac takes him to the top of the high peak of school-divinity to hurl him down the yawning gulf of penal fire; his speculative malice asks eternity to wreak its infinite spite in, and calls on the Almighty to execute its relentless doom! The cannibals burn their enemies and eat them in good-fellowship with one another: meek Christian divines cast those who differ from them but a hair's-breadth, body and soul into hell-fire for the glory of God and the good of His creatures! It is well that the power of such persons is not co-ordinate with their wills: indeed, it is from the sense of their weakness and inability to control the opinions of others, that they thus 'outdo termagant,' and endeavour to frighten them into conformity by big words and monstrous denunciations.

The pleasure of hating, like a poisonous mineral, eats into the heart of religion, and turns it to rankling spleen and bigotry; it makes patriotism an excuse for carrying fire, pestilence and famine into other lands: it leaves to virtue nothing but the spirit of censoriousness, and a narrow, jealous, inquisitorial watchfulness over the actions and motives of others. What have the different sects, creeds, doctrines, in religion been but so many pretexts set up for men to wrangle, to quarrel, to tear one another in pieces about, like a target as a mark to shoot at? Does any one suppose that the love of country in an Englishman implies any friendly feeling or disposition to serve another bearing the same name? No, it means only hatred to the French or the inhabitants of any other country that we happen to be at war with for the time. Does the love of virtue denote any wish to discover or amend our own faults? No, but it atones for an obstinate adherence to our own vices by the most virulent intolerance to human frailties. This principle is of a most universal application. It extends to good as well as evil: if it makes us hate folly, it makes us no less dissatisfied with distinguished merit. If it inclines us to resent the wrongs of others, it impels us to be as impatient of their prosperity. We revenge injuries: we repay benefits with ingratitude. Even our strongest partialities and likings soon take this turn. 'That which was luscious as locusts, anon becomes bitter as coloquintida;' and love and friendship melt in their own fires. We hate old friends: we hate old books: we hate old opinions; and at last we come to hate ourselves.

I have observed that few of those whom I have formerly known most intimate, continue on the same friendly footing, or combine the steadiness with the warmth of attachment. I have been acquainted with two or three knots of inseparable companions, who saw each other 'six days in the week,' that have broken up and dispersed. I have quarrelled with almost all my old friends (they might say this is owing to my bad temper), but they have also quarrelled with one another. What is become of 'that set of whist-players,' celebrated by Elia in his notable *Epistle to Robert Southey, Esq.* (and now I think of it—that I myself have celebrated in this very volume) 'that for so many years called Admiral Burney friend'? They are scattered, like last year's snow. Some of them are dead, or gone to live at a distance, or pass one another in the street like strangers, or if they stop to speak, do it as coolly and try to *cut* one another as soon as possible. Some of us have grown rich, others poor. Some have got places under Government, others a *niche* in the *Quarterly Review*. Some of us have dearly earned a name in the world; whilst others remain in their original privacy. We despise the one, and envy and are glad to mortify the other. Times are changed; we cannot revive our old feelings; and we avoid the sight, and are uneasy in the presence of, those who remind us of our infirmity, and put us upon an effort at seeming cordiality which embarrasses ourselves, and does not impose upon our *quondam* associates. Old friendships are, like meats served up repeatedly, cold, comfortless and distasteful. The stomach turns against them. Either constant intercourse and familiarity breed weariness and contempt; or, if we meet again after an interval of absence, we appear no longer the same. One is too wise, another too foolish, for us; and we wonder we did not find this out before. We are disconcerted and kept in a state of continual alarm by the wit of one, or tired to death of the dullness of another. The *good things* of the first (besides leaving stings behind them) by repetition grow stale, and lose their startling effect; and the insipidity of the last becomes intolerable. The most amusing or instructive companion is at best like a favourite volume, that we wish after a time to *lay upon the shelf*; but as our friends are not willing to be laid there, this produces a misunderstanding and ill-blood between us. Or if the zeal and integrity of friendship is not abated, or its career interrupted by any obstacle arising out of its own nature, we look out for other subjects of complaint and sources of dissatisfaction. We begin to criticise each other's dress, looks, and general character. 'Such a one is a pleasant fellow, but it is a pity he sits so late!' Another fails to keep

his appointments, and that is a sore that never heals. We get
acquainted with some fashionable young men or with a mistress,
and wish to introduce our friend; but he is awkward and a
sloven, the interview does not answer, and this throws cold
water on our intercourse. Or he makes himself obnoxious to
opinion; and we shrink from our own convictions on the subject
as an excuse for not defending him. All or any of these causes
mount up in time to a ground of coolness or irritation; and at
last they break out into open violence as the only amends we can
make ourselves for suppressing them so long, or the readiest
means of banishing recollections of former kindness so little
compatible with our present feelings. We may try to tamper with
the wounds or patch up the carcase of departed friendship; but
the one will hardly bear the handling, and the other is not worth
the trouble of embalming! The only way to be reconciled to old
friends is to part with them for good: at a distance we may
chance to be thrown back (in a waking dream) upon old times
and old feelings: or at any rate we should not think of renewing
our intimacy, till we have fairly *spit our spite*, or said, thought and
felt all the ill we can of each other. Or if we can pick a quarrel
with some one else, and make him the scapegoat, this is an excellent contrivance to heal a broken bone. I think I must be friends
with Lamb again, since he has written that magnanimous Letter
to Southey, and told him a piece of his mind! I don't know what
it is that attaches me to H—— so much, except that he and I,
whenever we meet, sit in judgment on another set of old friends,
and 'carve them as a dish fit for the gods.' There was Leigh
Hunt, John Scott, Mrs Montagu, whose dark raven locks make a
picturesque background to our discourse, B——, who is grown
fat, and is, they say, married, Rickman; these had all separated
long ago, and their foibles are the common link that holds us
together. We do not affect to condole or whine over their follies;
we enjoy, we laugh at them, till we are ready to burst our sides,
'*sans* intermission, for hours by the dial.' We serve up a course of
anecdotes, *traits*, master-strokes of character, and cut and hack at
them till we are weary. Perhaps some of them are even with us.
For my own part, as I once said, I like a friend the better for
having faults that one can talk about. 'Then,' said Mrs Montagu,
'you will never cease to be a philanthropist!' Those in question
were some of the choice-spirits of the age, not 'fellows of no

mark or likelihood'; and we so far did them justice: but it is well they did not hear what we sometimes said of them. I care little what any one says of me, particularly behind my back, and in the way of critical and analytical discussion: it is looks of dislike and scorn that I answer with the worst venom of my pen. The expression of the face wounds me more than the expressions of the tongue. If I have in one instance mistaken this expression, or resorted to this remedy where I ought not, I am sorry for it. But the face was too fine over which it mantled, and I am too old to have misunderstood it! . . . I sometimes go up to Hume's; and as often as I do, resolve never to go again. I do not find the old homely welcome. The ghost of friendship meets me at the door, and sits with me all dinner-time. They have got a set of fine notions and new acquaintance. Allusions to past occurrences are thought trivial, nor is it always safe to touch upon more general subjects. H. does not begin as he formerly did every five minutes, 'Fawcett used to say,' etc. That topic is something worn. The girls are grown up, and have a thousand accomplishments. I perceive there is a jealousy on both sides. They think I give myself airs, and I fancy the same of them. Every time I am asked, 'If I do not think Mr Washington Irving a very fine writer?' I shall not go again till I receive an invitation for Christmas Day in company with Mr Liston. The only intimacy I never found to flinch or fade was a purely intellectual one. There was none of the cant of candour in it, none of the whine of mawkish sensibility. Our mutual acquaintance were considered merely as subjects of conversation and knowledge, not at all of affection. We regarded them no more in our experiments than 'mice in an air-pump': or like malefactors, they were regularly cut down and given over to the dissecting-knife. We spared neither friend nor foe. We sacrificed human infirmities at the shrine of tuth. The skeletons of character might be seen, after the juice was extracted, dangling in the air like flies in cobwebs: or they were kept for future inspection in some refined acid. The demonstration was as beautiful as it was new. There is no surfeiting on gall: nothing keeps so well as a decoction of spleen. We grow tired of every thing but turning others into ridicule, and congratulating ourselves on their defects.

We take a dislike to our favourite books, after a time, for the same reason. We cannot read the same works for ever. Our honeymoon, even though we wed the Muse, must come to an end; and is followed by indifference, if not by disgust. There are some works, those indeed that produce the most striking effect at first by novelty and boldness of outline, that will not bear reading twice: others of a less extravagant character, and that excite and repay attention by a greater nicety of details, have hardly interest enough to keep alive our continued enthusiasm. The popularity of the most successful writers operates to wean us from them, by the cant and fuss that is made about them, by hearing their names everlastingly repeated, and by the number of ignorant and indiscriminate admirers they draw after them:—we as little like to have to drag others from their unmerited obscurity, lest we should be exposed to the charge of affectation and singularity of taste. There is nothing to be said respecting an author that all the world have made up their minds about: it is a thankless as well as hopeless task to recommend one that nobody has ever heard of. To cry up Shakespeare as the god of our idolatry, seems like a vulgar national prejudice: to take down a volume of Chaucer, or Spenser, or Beaumont and Fletcher, or Ford, or Marlowe, has very much the look of pedantry and egotism. I confess it makes me hate the very name of Fame and Genius, when works like these are 'gone into the wastes of time,' while each successive generation of fools is busily employed in reading the trash of the day, and women of fashion gravely join with their waiting-maids in discussing the preference between the *Paradise Lost* and Mr Moore's *Loves of the Angels*. I was pleased the other day on going into a shop to ask, 'If they had any of the *Scotch Novels?*' to be told—'That they had just sent out the last, *Sir Andrew Wylie!*' Mr Galt will also be pleased with this answer! The reputation of some books is raw and *unaired*: that of others is worm-eaten and mouldy. Why fix our affections on that which we cannot bring ourselves to have faith in, or which others have long ceased to trouble themselves about? I am half afraid to look into *Tom Jones*, lest it should not answer my expectations at this time of day; and if it did not, I should certainly be disposed to fling it into the fire, and never look into another novel while I lived. But surely, it may be said, there are some works that, like nature, can never grow old; and that must always touch the imagination and passions alike! Or there are passages that seem as if we might brood over them all our lives, and not exhaust the

sentiments of love and admiration they excite: they become
favourites, and we are fond of them to a sort of dotage. Here is
one:

> Sitting in my window
> Printing my thoughts in lawn, I saw a god,
> I thought (but it was you), enter our gates;
> My blood flew out and back again, as fast
> As I had puffed it forth and sucked it in
> Like breath; then was I called away in haste
> To entertain you: never was a man
> Thrust from a sheepcote to a sceptre, raised
> So high in thoughts as I; you left a kiss
> Upon these lips then, which I mean to keep
> From you for ever. I did hear you talk
> Far above singing!

A passage like this, indeed, leaves a taste on the palate like
nectar, and we seem in reading it to sit with the gods at their
golden tables: but if we repeat it ofen in ordinary moods, it loses
its flavour, becomes vapid, 'the wine of *poetry* is drunk, and but
the lees remain.' Or, on the other hand, if we call in the aid of
extraordinary circumstances to set it off to advantage, as the
reciting it to a friend, or after having our feelings excited by a
long walk in some romantic situation, or while we

> play with Amaryllis in the shade,
> Or with the tangles of Neaera's hair

we afterwards miss the accompanying circumstances, and instead
of transferring the recollection of them to the favourable side,
regret what we have lost, and strive in vain to bring back 'the
irrevocable hour'—wondering in some instances how we survive
it, and at the melancholy blank that is left behind! The pleasure
rises to its height in some moment of calm solitude or intoxicating sympathy, declines ever after, and from the comparison and a
conscious falling-off, leaves rather a sense of satiety and irksome-
ness behind it.... 'Is it the same in pictures?' I confess it is,
with all but those from Titian's hand. I don't know why, but an
air breathes from his landscapes, pure, refreshing, as if it came
from other years; there is a look in his faces that never passes
away. I saw one the other day. Amidst the heartless desolation
and glittering finery of Fonthill, there is a portfolio of the

Dresden Gallery. It opens, and a young female head looks from it; a child, yet woman grown; with an air of rustic innocence and the graces of a princess, her eyes like those of doves, the lips about to open, a smile of pleasure dimpling the whole face, the jewels sparkling in her crisped hair, her youthful shape compressed in a rich antique dress, as the bursting leaves contain the April buds! Why do I not call up this image of gentle sweetness, and place it as a perpetual barrier between mischance and me? —It is because pleasure asks a greater effort of the mind to support it than pain; and we turn after a little idle dalliance from what we love to what we hate!

As to my old opinions, I am heartily sick of them. I have reason, for they have deceived me sadly. I was taught to think, and I was willing to believe, that genius was not a bawd, that virtue was not a mask, that liberty was not a name, that love had its seat in the human heart. Now I would care little if these words were struck out of the dictionary, or if I had never heard them. They are become to my ears a mockery and a dream. Instead of patriots and friends of freedom, I see nothing but the tyrant and the slave, the people linked with kings to rivet on the chains of despotism and superstition. I see folly join with knavery, and together make up public spirit and public opinions. I see the insolent Tory, the blind Reformer, the coward Whig! If mankind had wished for what is right, they might have had it long ago. The theory is plain enough; but they are prone to mischief, 'to every good work reprobate.' I have seen all that had been done by the mighty yearnings of the spirit and intellect of men, 'of whom the world was not worthy,' and that promised a proud opening to truth and good through the vista of future years, undone by one man, with just glimmering of understanding enough to feel that he was a king, but not to comprehend how he could be king of a free people! I have seen this triumph celebrated by poets, the friends of my youth and the friends of man, but who were carried away by the infuriate tide that, setting in from a throne, bore down every distinction of right reason before it; and I have seen all those who did not join in applauding this insult and outrage on humanity proscribed, hunted down (they and their friends made a byword of), so that it has become an understood thing that no one can live by his talents or knowledge who is not ready to prostitute those talents and that knowledge to betray his species, and prey upon his fellow-man. 'This was sometime a mystery: but the time gives evidence of it.' The echoes of liberty had awakened once more in Spain, and the morning of hope dawned again: but that dawn has been overcast by the foul breath of bigotry, and those reviving sounds stifled by fresh cries from the time-rent towers of the Inquisition: man yielding (as it is fit he should) first to brute force, but more to the innate perversity and dastard spirit of his own nature which leaves no room for farther hope or disappointment. And England, that arch-reformer, that heroic deliverer, that mouther about liberty and tool of power, stands gaping by, not feeling the blight and mildew coming over it, nor its very bones crack and turn to a paste under the grasp and circling folds of this new monster—Legitimacy! In private life do we not see hypocrisy, servility, selfishness, folly, and impudence succeed, while modesty shrinks from the encounter, and merit is trodden under foot? How often is 'the rose plucked from the forehead of a virtuous love to plant a blister there'! What chance is there of the success of real passion? What certainty of its continuance? Seeing all this as I do, and unravelling the web of human life into its various threads of meanness, spite, cowardice, want of feeling, and want of understanding, of indifference towards others and ignorance of ourselves—seeing custom prevail over all excellence, itself giving way to infamy—mistaken as I have been in my public and private hopes, calculating others from myself, and calculating wrong; always disappointed where I placed most reliance; the dupe of friendship, and the fool of love;—have I not reason to hate and to despise myself? Indeed I do; and chiefly for not having hated and despised the world enough.

1826

Brummelliana

W E look upon Beau Brummell as the greatest of small wits. Indeed, he may in this respect be considered, as Cowley says of Pindar, as 'a species alone,' and as forming a class by himself. He has arrived at the very *minimum* of wit, and reduced it, 'by happi-

ness or pains,' to an almost invisible point. All his *bons-mots* turn
upon a single circumstance, the exaggerating of the merest trifles
into matters of importance, or treating everything else with the
utmost *nonchalance* and indifference, as if whatever pretended to
pass beyond those limits was a *bore*, and disturbed the serene air
of high life. We have heard of

> A sound so fine,
> That nothing lived 'twixt it and silence.

So we may say of Mr Brummell's jests, that they are of a meaning so attenuated that 'nothing lives 'twixt them and nonsense':—they hover on the very brink of vacancy, and are in
their shadowy composition next of kin to nonentities. It is
impossible for anyone to go beyond him without falling flat into
insignificance and insipidity: he has touched the *ne plus ultra* that
divides the dandy from the dunce. But what a fine eye to discriminate: what a sure hand to hit this last and thinnest of all
intellectual partitions! *Exempli gratia*—for in so new a species, the
theory is unintelligible without furnishing the proofs:—

Thus, in the question addressed to a noble person (which we
quoted the other day), 'Do you call that *thing* a coat?' a distinction is taken as nice as it is startling. It seems all at once a vulgar
prejudice to suppose that a coat is a coat, the commonest of all
common things,—it is here lifted into an ineffable essence, so
that a coat is no longer a *thing*; or that it would take infinite
gradations of fashion, taste, and refinement, for a *thing* to aspire
to the undefined privileges, and mysterious attributes of a coat.
Finer 'fooling' than this cannot be imagined. What a cut upon
the Duke! The beau becomes an emperor among such insects!

The first anecdote in which Mr Brummell's wit dawned upon
us—and it really rises with almost every new instance—was the
following: A friend one day called upon him, and found him
confined to his room from a lameness in one foot, upon which he
expressed his concern at the accident. 'I am sorry for it too,'
answered Brummell very gravely, 'particularly as it's *my favourite
leg*!' Is not this as if a man of fashion had nothing else to do than
to sit and think of which of his legs he liked best; and in the
plenitude of his satisfactions, and the absence of all real wants, to
pamper this fanciful distinction into a serious sort of *pet* preference? Upon the whole, among so many beauties—*ubi tot nitent*, I

am inclined to give my suffrage in favour of this, as the most classical of all our contemporary's *jeux d'esprit*—there is an Horatian ease and elegance about it—a slippered negligence, a cushioned effeminacy—it would take years of careless study and languid enjoyment to strike out so quaint and ingenious a conceit—

> A subtler web Arachne cannot spin;
> Nor the fine nets which oft we woven see
> Of scorched dew, do not in the air more lightly flee!

It is truly the art of making something out of nothing.

We shall not go deeply into the common story of Mr Brummell's asking his servant, as he was going out for the evening, 'Where do I dine to-day, John?' This is little more than the common cant of a multiplicity of engagements, so as to make it impossible to bear them all in mind, and of an utter disinclination to all attention to one's own affairs; but the following is brilliant and original. Sitting one day at table between two other persons, Mr Brummell said to his servant, who stood behind his chair—'John!' 'Yes, sir.' 'Who is this at my right hand?' 'If you please, sir, it's the Marquis of Headfort.' 'And who is this at my left hand?' 'It's my Lord Yarmouth.' 'Oh, very well!' and the Beau then proceeded to address himself to the persons who were thus announced to him. Now, this is surely superb, and 'high fantastical.' No, the smallest fold of that nicely adjusted cravat was not to be deranged, the least deviation from that select posture was not to be supposed possible. Had his head been fastened in a vice, it could not have been more immovably fixed than by the 'great idea in his mind,' of how a coxcomb should sit: the air of fashion and affectation 'bound him with Styx nine times round him'; and the Beau preserved the perfection of an attitude—like a piece of incomprehensible *still-life*,—the whole of dinner-time. The *ideal* is everything, even in frivolity and folly.

It is not one of the least characteristic of our hero's answers to a lady, who asked him if he never tasted vegetables—'Madam, I once ate a pea!' This was reducing the quantity of offensive grossness to the smallest assignable fraction: anything beyond *that* his imagination was oppressed with; and even this he seemed to confess to, with a kind of remorse, and to hasten from the subject with a certain monosyllabic brevity of style.

I do not like the mere impudence (Mr Theodore Hook, with his extempore dullness, might do the same thing) of forcing himself into a lady's rout, who had not invited him to her parties, and the gabble about Hopkinses and Tomkinses; but there is something piquant enough in his answer to a city-fashionable, who asked him if he would dine with him on a certain day— 'Yes, if you won't mention it to anyone'; and in an altercation with the same person afterwards, about obligations, the assumption of superiority implied in the appeal—'Do you count my having *borrowed* a thousand pounds of you for nothing?' soars immediately above commonplace.

On one occasion, Mr Brummell falling ill, accounted for it by saying, 'They put me to bed to a damp ——!'* From what slight causes direst issues spring! So sensitive and apprehensive a constitution makes one sympathise with its delicate possessor, as much as if he had been shut up in the steam of a laundry, or 'his lodging had been on the cold ground.' Mr Brummell having been interrogated as to the choice of his present place of residence (Calais) as somewhat dull replied, 'He thought it hard if a gentleman could not pass his time agreeably between London and Paris.'

Some of Brummell's *bons-mots* have been attributed to Sir Lumley Skeffington, who is even said to have been the first in this minute and tender walk of wit. It is, for instance, reported of him that, being at table and talking of daisies, he should turn round to his valet, and say with sentimental *naïveté* and trivial fondness—'On what day of the month did I first see a daisy, Matthew?' 'On the 1st of February, sir.' There is here a kindred vein; but whoever was the inventor, Brummell has borne away the prize, as Pope eclipsed his master Dryden, and Titian surpassed Giorgione's fame. In fine, it was said, with equal truth and spirit by one of the parties concerned, that 'the year 1815 was fatal to three great men—Byron, Buonaparte, and Brummell!'

1828

* *Editor's note*: the word Hazlitt failed to supply is 'whore.'

LEIGH HUNT

Getting Up on Cold Mornings

A<small>N</small> Italian author—Giulio Cordara, a Jesuit—has written a poem upon insects, which he begins by insisting, that those troublesome and abominable little animals were created for our annoyance, and that they were certainly not inhabitants of Paradise. We of the north may dispute this piece of theology; but on the other hand, it is as clear as the snow on the housetops, that Adam was not under the necessity of shaving; and that when Eve walked out of her delicious bower, she did not step upon ice three inches thick.

Some people say it is a very easy thing to get up of a cold morning. You have only, they tell you, to take the resolution; and the thing is done. This may be very true; just as a boy at school has only to take a flogging, and the thing is over. But we have not at all made up our minds upon it; and we find it a very pleasant exercise to discuss the matter, candidly, before we get up. This at least is not idling, though it may be lying. It affords an excellent answer to those, who ask how lying in bed can be indulged in by a reasoning being,—a rational creature. How? Why with the argument calmly at work in one's head, and the clothes over one's shoulder. Oh—it is a fine way of spending a sensible, impartial half-hour.

If these people would be more charitable, they would get on with their argument better. But they are apt to reason so ill, and to assert so dogmatically, that one could wish to have them stand round one's bed of a bitter morning, and lie before their faces. They ought to hear both sides of the bed, the inside and out. If they cannot entertain themselves with their own thoughts for half an hour or so, it is not the fault of those who can. If their will is never pulled aside by the enticing arms of imagination, so much the luckier for the stage-coachman.

Candid inquiries into one's decumbency, besides the greater or less privileges to be allowed a man in proportion to his ability of keeping early hours, the work given his faculties, etc, will at least concede their due merits to such representations as the following. In the first place, says the injured but calm appealer, I have been warm all night, and find my system in a state perfectly suitable to a warm-blooded animal. To get out of this state into the cold, besides the inharmonious and uncritical abruptness of the transition, is so unnatural to such a creature, that the poets, refining upon the tortures of the damned, make one of their greatest agonies consist in being suddenly transported from heat to cold,—from fire to ice. They are 'haled' out of their 'beds,' says Milton, by 'harpyfooted furies,'—fellows who come to call them. On my first movement towards the anticipation of getting up, I find that such parts of the sheets and bolster, as are exposed to the air of the room, are stone-cold. On opening my eyes, the first thing that meets them is my own breath rolling forth, as if in the open air, like smoke out of a cottage chimney. Think of this symptom. Then I turn my eyes sideways and see the window all frozen over. Think of that. Then the servant comes in. 'It is very cold this morning, is it not?'—'Very cold, Sir.'—'Very cold indeed, isn't it?'—'Very cold indeed, Sir.'—'More than usually so, isn't it, even for this weather?' (Here the servant's wit and good-nature are put to a considerable test, and the inquirer lies on thorns for the answer.) 'Why, Sir ... I think it *is*.' (Good creature! There is not a better, or more truth-telling servant going.) 'I must rise, however—get me some warm water.'—Here comes a fine interval between the departure of the servant and the arrival of the hot water; during which, of course, it is of 'no use' to get up. The hot water comes. 'Is it quite hot?'—'Yes, Sir.' —'Perhaps too hot for shaving: I must wait a little?'—'No, Sir; it will just do.' (There is an over-nice propriety sometimes, an officious zeal of virtue, a little troublesome.) 'Oh—the shirt—you must air my clean shirt;—linen gets very damp this weather.' —'Yes, Sir.' Here another delicious five minutes. A knock at the door. 'Oh, the shirt—very well. My stockings—I think the stockings had better be aired too.'—'Very well, Sir.'—Here another interval. At length everything is ready, except myself. I now, continues our incumbent (a happy word, by the by, for a country vicar)—I now cannot help thinking a good deal—who can?—upon the unnecessary and villainous custom of shaving: it is a thing so unmanly (here I nestle closer)—so effeminate (here I recoil from an unlucky step into the colder part of the bed).—No wonder that the Queen of France took part with the rebels against the degenerate King, her husband, who first affronted her smooth visage with a face like her own. The Emperor Julian never showed the luxuriancy of his genius to better advantage than in reviving the flowing beard. Look at Cardinal Bembo's picture—at Michael Angelo's—at Titian's—at Shakespeare's—at Fletcher's—at Spenser's—at Chaucer's—at Alfred's—at Plato's —I could name a great man for every tick of my watch.—Look at the Turks, a grave and otiose people.—Think of Haroun Al Raschid and Bed-ridden Hassan.—Think of Wortley Montague, the worthy son of his mother, a man above the prejudice of his time.—Look at the Persian gentlemen, whom one is ashamed of meeting about the suburbs, their dress and appearance are so much finer than our own.—Lastly, think of the razor itself— how totally opposed to every sensation of bed—how cold, how edgy, how hard! how utterly different from anything like the warm and circling amplitude, which

> Sweetly recommends itself
> Unto our gentle senses.

Add to this, benumbed fingers, which may help you to cut yourself, a quivering body, a frozen towel, and a ewer full of ice; and he that says there is nothing to oppose in all this, only shows, at anyrate, that he has no merit in opposing it.

Thomson the poet, who exclaims in his *Seasons*:

> Falsely luxurious! Will not man awake?

used to lie in bed till noon, because he said he had no motive in getting up. He could imagine the good of rising; but then he could also imagine the good of lying still; and his exclamation, it must be allowed, was made upon summer-time, not winter. We must proportion the argument to the individual character. A money-getter may be drawn out of his bed by three or four pence; but this will not suffice for a student. A proud man may say, 'What shall I think of myself, if I don't get up?' but the more humble one will be content to waive this prodigious notion of himself, out of respect to his kindly bed. The mechanical man shall get up without any ado at all; and so shall the barometer.

An ingenious lier in bed will find hard matter of discussion even on the score of health and longevity. He will ask us for our proofs and precedents of the ill effects of lying later in cold weather; and sophisticate much on the advantages of an even temperature of body; of the natural propensity (pretty universal) to have one's way; and of the animals that roll themselves up, and sleep all the winter. As to longevity, he will ask whether the longest life is of necessity the best; and whether Holborn is the handsomest street in London.

We only know of one confounding, not to say confounded argument, fit to overturn the huge luxury, the 'enormous bliss'—of the vice in question. A lier in bed may be allowed to profess a disinterested indifference for his health or longevity; but while he is showing the reasonableness of consulting his own or one person's comfort, he must admit the proportionate claim of more than one; and the best way to deal with him is this, especially for a lady; for we earnestly recommend the use of that sex on such occasions, if not somewhat *over*-persuasive; since extremes have an awkward knack of meeting. First then, admit all the ingeniousness of what he says, telling him that the Bar has been deprived of an excellent lawyer. Then look at him in the most good-natured manner in the world, with a mixture of assent and appeal in your countenance, and tell him that you are waiting breakfast for him; that you never like to breakfast without him; that you really want it too; that the servants want theirs; that you shall not know how to get the house into order, unless he rises; and that you are sure he would do things twenty times worse, even than getting out of his warm bed, to put them all into good humour and a state of comfort. Then, after having said this, throw in the comparatively indifferent matter, to *him*, about his health; but tell him that it is no indifferent matter to you; that the sight of his illness makes more people suffer than one; but that if, nevertheless, he really does feel so very sleepy and so very much refreshed by—Yet stay; we hardly know whether the frailty of a—Yes, yes; say that too, especially if you say it with sincerity; for if the weakness of human nature on the one hand and the *vis inertiae* on the other, should lead him to take advantage of it once or twice, good-humour and sincerity form an irresistible junction at last: and are still better and warmer things than pillows and blankets.

Other little helps of appeal may be thrown in, as occasion requires. You may tell a lover, for instance, that lying in bed makes people corpulent; a father, that you wish him to complete the fine manly example he sets his children; a lady, that she will injure her bloom or her shape, which M. or W. admires so much; and a student or artist, that he is always so glad to have done a good day's work, in his best manner.

Reader. And pray, Mr Indicator, how do *you* behave yourself in this respect?

Indic. Oh, Madam, perfectly, of course; like all advisers.

Reader. Nay, I allow that your mode of argument does not look quite so suspicious as the old way of sermonising and severity, but I have my doubts, especially from that laugh of yours. If I should look in to-morrow morning—

Indic. Ah, Madam, the look in of a face like yours does anything with me. It shall fetch me up at nine, if you please—*six*, I meant to say.

1820

THOMAS DE QUINCEY

The Knocking at the Gate in Macbeth

From my boyish days I had always felt a great perplexity on one point in *Macbeth*. It was this: The knocking at the gate which succeeds to the murder of Duncan produced to my feelings an effect for which I never could account. The effect was that it reflected back upon the murderer a peculiar awfulness and a depth of solemnity: yet, however obstinately I endeavoured with my understanding to comprehend this, for many years I never could see *why* it should produce such an effect.

Here I pause for one moment, to exhort the reader never to pay any attention to his understanding when it stands in opposition to any other faculty of his mind. The mere understanding, however useful and indispensable, is the meanest faculty in the human mind, and the most to be distrusted; and yet the great majority of people trust to nothing else,—which may do for ordinary life, but not for philosophical purposes. Of this out of ten thousand instances that I might produce I will cite one. Ask of any person whatsoever who is not previously prepared for the demand by a knowledge of the perspective to draw in the rudest way the commonest appearance which depends upon the laws of that science,—as, for instance, to represent the effect of two walls standing at right angles to each other, or the appearance of the houses on each side of a street as seen by a person looking down the street from one extremity. Now, in all cases, unless the person has happened to observe in pictures how it is that artists produce these effects, he will be utterly unable to make the smallest approximation to it. Yet why? For he has actually seen the effect every day of his life. The reason is that he allows his understanding to overrule his eyes. His understanding, which includes no intuitive knowledge of the laws of vision, can furnish him with no reason why a line which is known and can be proved to be a horizontal line should not *appear* a horizontal line: a line that made any angle with the perpendicular less than a right angle would seem to him to indicate that his houses were all tumbling down together. Accordingly, he makes the line of his houses a horizontal line, and fails, of course, to produce the effect demanded. Here, then, is one instance out of many in which not only the understanding is allowed to overrule the eyes, but where the understanding is positively allowed to obliterate the eyes, as it were; for not only does the man believe the evidence of his understanding in opposition to that of his eyes, but (what is monstrous) the idiot is not aware that his eyes ever gave such evidence. He does not know that he has seen (and therefore *quoad* his consciousness has *not* seen) that which he *has* seen every day of his life.

But to return from this digression. My understanding could furnish no reason why the knocking at the gate in *Macbeth* should produce any effect, direct or reflected. In fact, my understanding said positively that it could *not* produce any effect. But I knew better; I felt that it did; and I waited and clung to the problem until further knowledge should enable me to solve it. At length, in 1812, Mr Williams made his *début* on the stage of Ratcliffe Highway, and executed those unparalleled murders which have procured for him such a brilliant and undying reputation. On which murders, by the way, I must observe that in one respect they have had an ill effect, by making the connoisseur in murder very fastidious in his taste, and dissatisfied by anything that has been since done in that line. All other murders look pale by the deep crimson of his; and, as an amateur once said to me in a querulous tone, 'There has been absolutely nothing *doing* since his time, or nothing that's worth speaking of.' But this is wrong; for it is unreasonable to expect all men to be great artists, and born with the genius of Mr Williams. Now, it will be remembered that in the first of these murders (that of the Marrs) the same incident (of a knocking at the door soon after the work of extermination was complete) did actually occur which the genius of Shakspere has invented; and all good judges, and the most eminent dilettanti, acknowledged the felicity of Shakspere's suggestion as soon as it was actually realized. Here, then was a fresh proof that I was right in relying on my own feeling, in opposition to my understanding; and I again set myself to study the problem. At length I solved it to my own satisfaction; and my solution is this:—Murder, in ordinary cases, where the sympathy is wholly directed to the case of the murdered person, is an incident of coarse and vulgar horror; and for this reason,—that it flings the interest exclusively upon the natural but ignoble instinct by which we cleave to life: an instinct which, as being indispensable to the primal law of self-preservation, is the same in kind (though different in degree) amongst all living creatures. This instinct, therefore, because it annihilates all distinctions, and degrades the greatest of men to the level of 'the poor beetle that we tread on,' exhibits human nature in its most abject and humiliating attitude. Such an attitude would little suit the purposes of the poet. What then must he do? He must throw the interest on the murderer. Our sympathy must be with *him* (of course I mean a sympathy of comprehension, a sympathy by which we enter into his feelings, and are made to understand them,—not a sympathy of pity or approbation).[1] In the murdered person, all strife of thought, all flux and reflux of passion and of purpose, are crushed by one overwhelming panic; the fear of instant death smites him 'with its petrific mace.' But in the murderer, such a murderer as a poet will condescend to, there must be raging some great storm of passion,—jealousy, ambition, vengeance, hatred,—which will create a hell within him; and into this hell we are to look.

In *Macbeth*, for the sake of gratifying his own enormous and teeming faculty of creation, Shakspere has introduced two murderers: and, as usual in his hands, they are remarkably discriminated: but,—though in Macbeth the strife of mind is greater than in his wife, the tiger spirit not so awake, and his feelings caught chiefly by contagion from her,—yet, as both were finally involved in the guilt of murder, the murderous mind of necessity is finally to be presumed in both. This was to be expressed: and, on its own account, as well as to make it a more proportionable antagonist to the unoffending nature of their victim, 'the gracious Duncan,' and adequately to expound 'the deep

---

[1] It seems almost ludicrous to guard and explain my use of a word in a situation where it would naturally explain itself. But it has become necessary to do so, in consequence of the unscholarlike use of the word sympathy, at present so general, by which, instead of taking it in its proper sense, as the act of reproducing in our minds the feelings of another, whether for hatred, indignation, love, pity, or approbation, it is made a mere synonym of the word *pity*; and hence, instead of saying 'sympathy *with* another,' many writers adopt the monstrous barbarism of 'sympathy *for* another.'

damnation of his taking off,' this was to be expressed with peculiar energy. We were to be made to feel that the human nature,—i.e. the divine nature of love and mercy, spread through the hearts of all creatures, and seldom utterly withdrawn from man,—was gone, vanished, extinct, and that the fiendish nature had taken its place. And, as this effect is marvellously accomplished in the *dialogues* and *soliloquies* themselves, so it is finally consummated by the expedient under consideration; and it is to this that I now solicit the reader's attention. If the reader has ever witnessed a wife, daughter, or sister in a fainting fit, he may chance to have observed that the most affecting moment in such a spectacle is *that* in which a sigh and a stirring announce the recommencement of suspended life. Or, if the reader has ever been present in a vast metropolis on the day when some great national idol was carried in funeral pomp to his grave, and, chancing to walk near the course through which it passed, has felt powerfully, in the silence and desertion of the streets, and in the stagnation of ordinary business, the deep interest which at that moment was possessing the heart of man,—if all at once he should hear the death-like stillness broken up by the sound of wheels rattling away from the scene, and making known that the transitory vision was dissolved, he will be aware that at no moment was his sense of the complete suspension and pause in ordinary human concerns so full and affecting as at that moment when the suspension ceases, and the goings on of human life are suddenly resumed. All action in any direction is best expounded, measured, and made apprehensible, by reaction. Now, applying this to the case in *Macbeth*: Here, as I have said, the retiring of the human heart and the entrance of the fiendish heart was to be expressed and made sensible. Another world has stept in; and the murderers are taken out of the region of human things, human purposes, human desires. They are transfigured: Lady Macbeth is 'unsexed'; Macbeth has forgot that he was born of woman; both are conformed to the image of devils; and the world of devils is suddenly revealed. But how shall this be conveyed and made palpable? In order that a new world may step in, this world must for a time disappear. The murderers and the murder must be insulated—cut off by an immeasurable gulf from the ordinary tide and succession of human affairs—locked up and sequestered in some deep recess; we must be made sensible that the world of ordinary life is suddenly arrested, laid asleep, tranced, racked into a dread armistice; time must be annihilated, relation to things without abolished; and all must pass self-withdrawn into a deep syncope and suspension of earthly passion. Hence it is that, when the deed is done, when the work of darkness is perfect, then the world of darkness passes away like a pageantry in the clouds: the knocking at the gate is heard, and it makes known audibly that the reaction has commenced; the human has made its reflux upon the fiendish; the pulses of life are beginning to beat again; and the re-establishment of the goings-on of the world in which we live first makes us profoundly sensible of the awful parenthesis that had suspended them.

O mighty poet! Thy works are not as those of other men, simply and merely great works of art, but are also like the phenomena of nature, like the sun and the sea, the stars and the flowers, like frost and snow, rain and dew, hail-storm and thunder, which are to be studied with entire submission of our own faculties, and in the perfect faith that in them there can be no too much or too little, nothing useless or inert, but that, the farther we press in our discoveries, the more we shall see proofs of design and self-supporting arrangement where the careless eye had seen nothing but accident!

1823

THOMAS CARLYLE

From *Signs of the Times*

W<small>ERE</small> we required to characterise this age of ours by any single epithet, we should be tempted to call it, not an Heroical, Devotional, Philosophical, or Moral Age, but, above all others, the Mechanical Age. It is the Age of Machinery, in every outward and inward sense of that word; the age which, with its whole undivided might, forwards, teaches and practises the great art of adapting means to ends. Nothing is now done directly, or by hand; all is by rule and calculated contrivance. For the simplest operation, some helps and accompaniments, some cunning abbreviating process is in readiness. Our old modes of exertion are all discredited, and thrown aside. On every hand, the living artisan is driven from his workshop, to make room for a speedier, inanimate one. The shuttle drops from the fingers of the weaver, and falls into iron fingers that ply it faster. The sailor furls his sail, and lays down his oar; and bids a strong, unwearied servant, on vaporous wings, bear him through the waters. Men have crossed oceans by steam; the Birmingham Fire-king has visited the fabulous East; and the genius of the Cape, were there any Camoens now to sing it, has again been alarmed, and with far stranger thunders than Gamas. There is no end to machinery. Even the horse is stripped of his harness, and finds a fleet fire-horse yoked in his stead. Nay, we have an artist that hatches chickens by steam; the very brood-hen is to be superseded! For all earthly, and for some unearthly purposes, we have machines and mechanic furtherances; for mincing our cabbages; for casting us into magnetic sleep. We remove mountains, and make seas our smooth highway; nothing can resist us. We war with rude Nature; and, by our resistless engines, come off always victorious, and loaded with spoils.

What wonderful accessions have thus been made, and are still making, to the physical power of mankind; how much better fed, clothed, lodged and, in all outward respects, accommodated men now are, or might be, by a given quantity of labour, is a grateful reflection which forces itself on every one. What changes, too, this addition of power is introducing into the Social System; how wealth has more and more increased, and at the same time gathered itself more and more into masses, strangely altering the old relations, and increasing the distance between the rich and the poor, will be a question for Political Economists, and a much more complex and important one than any they have yet engaged with.

But leaving these matters for the present, let us observe how the mechanical genius of our time has diffused itself into quite other provinces. Not the external and physical alone is now managed by machinery, but the internal and spiritual also. Here too nothing follows its spontaneous course, nothing is left to be accomplished by old natural methods. Everthing has its cunningly devised implements, its preestablished apparatus; it is not done by hand, but by machinery. Thus we have machines for Education: Lancastrian machines; Hamiltonian machines; monitors, maps and emblems. Instruction, that mysterious communing of Wisdom with Ignorance, is no longer an indefinable tentative process, requiring a study of individual aptitudes, and a perpetual variation of means and methods, to attain the same end; but a secure, universal, straightforward business, to be conducted in the gross, by proper mechanism, with such intellect as comes to hand. Then, we have Religious machines, of all imaginable varieties; the Bible-Society, professing a far higher and heavenly structure, is found, on inquiry, to be altogether an earthly contrivance: supported by collection of moneys, by fomenting of vanities, by puffing, intrigue and chicane; a machine for converting the Heathen. It is the same in all other departments. Has any man, or any society of men, a truth to speak, a piece of spiritual work to do ; they can nowise proceed at once and with the mere natural organs, but must first call a public meeting, appoint committees, issue prospectuses, eat a public dinner; in a word, construct or borrow machinery, wherewith to speak it and do it. Without machinery they were hopeless, helpless; a colony of Hindoo weavers squatting in the heart of Lancashire. Mark, too, how every machine must have its

moving power, in some of the great currents of society; every
little sect among us, Unitarians, Utilitarians, Anabaptists, Phreno-
logists, must have its Periodical, its monthly or quarterly Magazine;—hanging out, like its windmill, into the *popularis aura*, to
grind meal for the society.

With individuals, in like manner, natural strength avails little.
No individual now hopes to accomplish the poorest enterprise
single-handed and without mechanical aids; he must make
interest with some existing corporation, and till his field with
their oxen. In these days, more emphatically than ever, 'to live,
signifies to unite with a party, or to make one.' Philosophy, Science, Art, Literature, all depend on machinery. No Newton, by
silent meditation, now discovers the system of the world from
the falling of an apple; but some quite other than Newton stands
in his Museum, his Scientific Institution, and behind whole batteries of retorts, digesters, and galvanic piles imperatively 'inter-
rogates Nature,'—who, however, shows no haste to answer. In
defect of Raphaels, and Angelos, and Mozarts, we have Royal
Academies of Painting, Sculpture, Music; whereby the languishing spirit of Art may be strengthened, as by the more
generous diet of a Public Kitchen. Literature, too, has its
Paternoster-row mechanism, its Trade-dinners, its Editorial
conclaves, and huge subterranean, puffing bellows; so that books
are not only printed, but, in a great measure, written and sold, by
machinery.

National culture, spiritual benefit of all sorts, is under the
same management. No Queen Christina, in these times, needs to
send for her Descartes; no King Frederick for his Voltaire, and
painfully nourish him with pensions and flattery: any sovereign
of taste, who wishes to enlighten his people, has only to impose a
new tax, and with the proceeds establish Philosophic Institutes.
Hence the Royal and Imperial Societies, the Bibliothèques,
Glypothèques, Technothèques, which front us in all capital cities; like so many well-finished hives, to which it is expected the
stray agencies of Wisdom will swarm of their own accord, and
hive and make honey. In like manner, among ourselves, when
it is thought that religion is declining, we have only to vote
half-a-million's worth of bricks and mortar, and build new
churches. In Ireland it seems they have gone still farther, having
actually established a 'Penny-a-week Purgatory-Society'! Thus

does the Genius of Mechanism stand by to help us in all difficulties and emergencies, and with his iron back bears all our
burdens.

These things, which we state lightly enough here, are yet of
deep import, and indicate a mighty change in our whole manner
of existence. For the same habit regulates not our modes of
action alone, but our modes of thought and feeling. Men are
grown mechanical in head and in heart, as well as in hand. They
have lost faith in individual endeavour, and in natural force, of
any kind. Not for internal perfection, but for external combinations and arrangements, for institutions, constitutions,—for
Mechanism of one sort or other, do they hope and struggle.
Their whole efforts, attachments, opinions, turn on mechanism,
and are of a mechanical character.

We may trace this tendency in all the great manifestations of
our time; in its intellectual aspect, the studies it most favours and
its manner of conducting them; in its practical aspects, its politics, arts, religion, morals; in the whole sources, and throughout
the whole currents, of its spiritual, no less than its material
activity.

Consider, for example, the state of Science generally, in
Europe, at this period. It is admitted, on all sides, that the Meta-
physical and Moral Sciences are falling into decay, while the
Physical are engrossing, every day, more respect and attention.
In most of the European nations there is now no such thing as a
Science of Mind; only more or less advancement in the general
science, or the special sciences, of matter. The French were the
first to desert Metaphysics; and though they have lately affected
to revive their school, it has yet no signs of vitality. The land of
Malebranche, Pascal, Descartes and Fénelon, has now only its
Cousins and Villemains; while, in the department of Physics, it
reckons far other names. Among ourselves, the Philosophy of
Mind, after a rickety infancy, which never reached the vigour of
manhood, fell suddenly into decay, languished and finally died
out, with its last amiable cultivator, Professor Stewart. In no
nation but Germany has any decisive effort been made in
psychological science; not to speak of any decisive result. The
science of the age, in short, is physical, chemical, physiological;
in all shapes mechanical. Our favourite Mathematics, the highly
prized exponent of all these other sciences, has also become more

and more mechanical. Excellence in what is called its higher departments depends less on natural genius than on acquired expertness in wielding its machinery. Without undervaluing the wonderful results which a Lagrange or Laplace educes by means of it, we may remark, that their calculus, differential and integral, is little else than a more cunningly-constructed arithmetical mill; where the factors being put in, are, as it were, ground into the true product, under cover, and without other effort on our part than steady turning of the handle. We have more Mathematics than ever; but less Mathesis. Archimedes and Plato could not have read the *Mécanique Céleste*; but neither would the whole French Institute see aught in that saying, 'God geometrises!' but a sentimental rodomontade.

Nay, our whole Metaphysics itself, from Locke's time downwards, has been physical; not a spiritual philosophy, but a material one. The singular estimation in which his Essay was so long held as a scientific work (an estimation grounded, indeed, on the estimable character of the man) will one day be thought a curious indication of the spirit of these times. His whole doctrine is mechanical, in its aim and origin, in its method and its results. It is not a philosophy of the mind: it is a mere discussion concerning the origin of our consciousness, or ideas, or whatever else they are called; a genetic history of what we see *in* the mind. The grand secrets of Necessity and Freewill, of the Mind's vital or non-vital dependence on Matter, of our mysterious relations to Time and Space, to God, to the Universe, are not, in the faintest degree touched on in these inquiries; and seem not to have the smallest connexion with them.

The last class of our Scotch Metaphysicians had a dim notion that much of this was wrong; but they knew not how to right it. The school of Reid had also from the first taken a mechanical course, not seeing any other. The singular conclusions at which Hume, setting out from their admitted premises, was arriving, brought this school into being; they let loose Instinct, as an undiscriminating ban-dog, to guard them against these conclusions;—they tugged lustily at the logical chain by which Hume was so coldly towing them and the world into bottomless abysses of Atheism and Fatalism. But the chain somehow snapped between them; and the issue has been that nobody now cares about either,—any more than about Hartley's, Darwin's, or

Priestley's contemporaneous doings in England. Hartley's vibra-
tions and vibratiuncles, one would think, were material and
mechanical enough; but our Continental neighbours have gone
still farther. One of their philosophers has lately discovered, that
'as the liver secretes bile, so does the brain secrete thought';
which astonishing discovery Dr Cabanis, more lately still, in his
Rapports du Physique et du Moral de l'Homme, has pushed into its
minutest developments.

The metaphysical philosophy of this last inquirer is certainly
no shadowy or unsubstantial one. He fairly lays open our moral
structure with his dissecting-knives and real metal probes; and
exhibits it to the inspection of mankind, by Leuwenhoek micro-
scopes, and inflation with the anatomical blowpipe. Thought, he
is inclined to hold, is still secreted by the brain; but then Poetry
and Religion (and it is really worth knowing) are 'a product of
the smaller intestines'! We have the greatest admiration for this
learned doctor: with what scientific stoicism he walks through
the land of wonders, unwondering; like a wise man through
some huge, gaudy, imposing Vauxhall, whose fire-works, cas-
cades and symphonies, the vulgar may enjoy and believe in,
—but where he finds nothing real but the saltpetre, pasteboard
and catgut. His book may be regarded as the ultimatum of mechanical metaphysics in our time; a remarkable realisation of what
in Martinus Scriblerus was still only an idea, that 'as the jack had
a meat-roasting quality, so had the body a thinking quality,'
—upon the strength of which the Nurembergers were to build a
wood-and-leather man, 'who should reason as well as most
country parsons.' Vaucanson did indeed make a wooden duck,
that seemed to eat and digest; but that bold scheme of the
Nurembergers remained for a more modern virtuoso.

This condition of the two great departments of knowledge,—the outward, cultivated exclusively on mechanical principles; the inward, finally abandoned, because, cultivated on such
principles, it is found to yield no result,—sufficiently indicates
the intellectual bias of our time, its all-pervading disposition
towards that line of inquiry. In fact, an inward persuasion has
long been diffusing itself, and now and then even comes to utterance, that, except the external, there are no true sciences; that to
the inward world (if there be any) our only conceivable road is
through the outward; that, in short, what cannot be investigated

and understood mechanically, cannot be investigated and understood at all. We advert the more particularly to these intellectual propensities, as to prominent symptoms of our age, because Opinion is at all times doubly related to Action, first as cause, then as effect; and the speculative tendency of any age will therefore give us, on the whole, the best indications of its practical tendency.

Nowhere, for example, is the deep, almost exclusive faith we have in Mechanism more visible than in the Politics of this time. Civil government does by its nature include much that is mechanical, and must be treated accordingly. We term it indeed, in ordinary language, the Machine of Society, and talk of it as the grand working wheel from which all private machines must derive, or to which they must adapt, their movements. Considered merely as a metaphor, all this is well enough; but here, as in so many other cases, the 'foam hardens itself into a shell,' and the shadow we have wantonly evoked stands terrible before us and will not depart at our bidding. Government includes much also that is not mechanical, and cannot be treated mechanically; of which latter truth, as appears to us, the political speculations and exertions of our time are taking less and less cognisance.

Nay, in the very outset, we might note the mighty interest taken in *mere political arrangements*, as itself the sign of a mechanical age. The whole discontent of Europe takes this direction. The deep, strong cry of all civilised nations,—a cry which, every one now sees, must and will be answered, is: Give us a reform of Government! A good structure of legislation, a proper check upon the executive, a wise arrangement of the judiciary, is *all* that is wanting for human happiness. The Philosopher of this age is not a Socrates, a Plato, a Hooker, or Taylor, who inculcates on men the necessity and infinite worth of moral goodness, the great truth that our happiness depends on the mind which is within us, and not on the circumstances which are without us; but a Smith, a De Lolme, a Bentham, who chiefly inculcates the reverse of this,—that our happiness depends entirely on external circumstances; nay, that the strength and dignity of the mind within us is itself the creature and consequence of these. Were the laws, the government, in good order, all were well with us; the rest would care for itself! Dissentients from this opinion, expressed or implied, are now rarely to be met with; widely and angrily as men differ in its application, the principle is admitted by all.

Equally mechanical, and of equal simplicity, are the methods proposed by both parties for completing or securing this all-sufficient perfection of arrangement. It is no longer the moral, religious, spiritual condition of the people that is our concern, but their physical, practical, economical condition, as regulated by public laws. Thus is the Body-politic more than ever worshipped and tendered; but the Soul-politic less than ever. Love of country, in any high or generous sense, in any other than an almost animal sense, or mere habit, has little importance attached to it in such reforms, or in the opposition shown them. Men are to be guided only by their self-interests. Good government is a good balancing of these; and, except a keen eye and appetite for self-interest, requires no virtue in any quarter. To both parties it is emphatically a machine: to the discontented, a 'taxing-machine'; to the contented, a 'machine for securing property.' Its duties and its faults are not those of a father, but of an active parish-constable.

Thus it is by the mere condition of the machine, by preserving it untouched, or else by reconstructing it, and oiling it anew, that man's salvation as a social being is to be ensured and indefinitely promoted. Contrive the fabric of law aright, and without farther effort on your part, that divine spirit of Freedom, which all hearts venerate and long for, will of herself come to inhabit it; and under her healing wings every noxious influence will wither, every good and salutary one more and more expand. Nay, so devoted are we to this principle, and at the same time so curiously mechanical, that a new trade, specially grounded on it, has arisen among us, under the name of 'Codification,' or codemaking in the abstract; whereby any people, for a reasonable consideration, may be accommodated with a patent code;—more easily than curious individuals with patent breeches, for the people does *not* need to be measured first.

To us who live in the midst of all this, and see continually the faith, hope and practice of every one founded on Mechanism of one kind or other, it is apt to seem quite natural, and as if it could never have been otherwise. Nevertheless, if we recollect or reflect a little, we shall find both that it has been, and might again be otherwise. The domain of Mechanism,—meaning thereby political, ecclesiastical or other outward establishments,—was once considered as embracing, and we are persuaded can at any time embrace, but a limited portion of man's interests, and by no means the highest portion.

To speak a little pedantically, there is a science of *Dynamics* in man's fortunes and nature, as well as of *Mechanics*. There is a science which treats of, and practically addresses, the primary, unmodified forces and energies of man, the mysterious springs of Love, and Fear, and Wonder, of Enthusiasm, Poetry, Religion, all which have a truly vital and *infinite* character; as well as a science which practically addresses the finite, modified developments of these, when they take the shape of immediate 'motives,' as hope of reward, or as fear of punishment.

Now it is certain, that in former times the wise men, the enlightened lovers of their kind, who appeared generally as Moralists, Poets or Priests, did, without neglecting the Mechanical province, deal chiefly with the Dynamical; applying themselves chiefly to regulate, increase and purify the inward primary powers of man; and fancying that herein lay the main difficulty, and the best service they could undertake. But a wide difference is manifest in our age. For the wise men, who now appear as Political Philosophers, deal exclusively with the Mechanical province; and occupying themselves in counting-up and estimating men's motives, strive by curious checking and balancing, and other adjustments of Profit and Loss, to guide them to their true advantage: while, unfortunately, those same 'motives' are so innumerable, and so variable in every individual, that no really useful conclusion can ever be drawn from their enumeration. But though Mechanism, wisely contrived, has done much for man in a social and moral point of view, we cannot be persuaded that it has ever been the chief source of his worth or happiness. Consider the great elements of human enjoyment, the attainments and possessions that exalt man's life to its present height, and see what part of these he owes to institutions, to Mechanism of any kind; and what to the instinctive, unbounded force, which Nature herself lent him, and still continues to him. Shall we say, for example, that Science and Art are indebted principally to the founders of Schools and Universities? Did not Science originate rather, and gain advancement, in the obscure closets of the Roger

Bacons, Keplers, Newtons; in the workshops of the Fausts and the Watts; wherever, and in what guise soever Nature, from the first times downwards, had sent a gifted spirit upon the earth? Again, were Homer and Shakspeare members of any beneficed guild, or made Poets by means of it? Were Painting and Sculpture created by forethought, brought into the world by institutions for that end? No; Science and Art have, from first to last, been the free gift of Nature; an unsolicited, unexpected gift; often even a fatal one. These things rose up, as it were, by spontaneous growth, in the free soil and sunshine of Nature. They were not planted or grafted, nor even greatly multiplied or improved by the culture or manuring of institutions. Generally speaking, they have derived only partial help from these; often enough have suffered damage. They made constitutions for themselves. They originated in the Dynamical nature of man, not in his Mechanical nature.

Or, to take an infinitely higher instance, that of the Christian Religion, which, under every theory of it, in the believing or unbelieving mind, must ever be regarded as the crowning glory, or rather the life and soul, of our whole modern culture: How did Christianity arise and spread abroad among men? Was it by institutions, and establishments and well-arranged systems of mechanism? Not so; on the contrary, in all past and existing institutions for those ends, its divine spirit has invariably been found to languish and decay. It arose in the mystic deeps of man's soul; and was spread abroad by the 'preaching of the word,' by simple, altogether natural and individual efforts; and flew, like hallowed fire, from heart to heart, till all were purified and illuminated by it; and its heavenly light shone, as it still shines, and (as sun or star) will ever shine, through the whole dark destinies of man. Here again was no Mechanism; man's highest attainment was accomplished Dynamically, not Mechanically.

Nay, we will venture to say, that no high attainment, not even any far-extending movement among men, was ever accomplished otherwise. Strange as it may seem, if we read History with any degree of thoughtfulness, we shall find that the checks and balances of Profit and Loss have never been the grand agents with men; that they have never been roused into deep, thorough, all-pervading efforts by any computable prospect of Profit and

Loss, for any visible, finite object; but always for some invisible and infinite one. The Crusades took their rise in Religion; their visible object was, commercially speaking, worth nothing. It was the boundless Invisible world that was laid bare in the imaginations of those men; and in its burning light, the visible shrunk as a scroll. Not mechanical, nor produced by mechanical means, was this vast movement. No dining at Freemasons' Tavern, with the other long train of modern machinery; no cunning reconciliation of 'vested interests,' was required here: only the passionate voice of one man, the rapt soul looking through the eyes of one man; and rugged, steel-clad Europe trembled beneath his words, and followed him whither he listed. In later ages it was still the same. The Reformation had an invisible, mystic and ideal aim; the result was indeed to be embodied in external things; but its spirit, its worth, was internal, invisible, infinite. Our English Revolution too originated in Religion. Men did battle, in those old days, not for Purse-sake, but for Conscience-sake. Nay, in our own days, it is no way different. The French Revolution itself had something higher in it than cheap bread and a Habeas-corpus act. Here too was an Idea; a Dynamic, not a Mechanic force. It was a struggle, though a blind and at last an insane one, for the infinite, divine nature of Right, of Freedom, of Country.

Thus does man, in every age, vindicate, consciously or unconsciously, his celestial birthright. Thus does Nature hold on her wondrous, unquestionable course; and all our systems and theories are but so many froth-eddies or sand-banks, which from time to time she casts up, and washes away. When we can drain the Ocean into mill-ponds, and bottle-up the Force of Gravity, to be sold by retail, in gas jars; then may we hope to comprehend the infinitudes of man's soul under formulas of Profit and Loss; and rule over this too, as over a patent engine, by checks, and valves, and balances.

Nay, even with regard to Government itself, can it be necessary to remind any one that Freedom, without which indeed all spiritual life is impossible, depends on infinitely more complex influences than either the extension or the curtailment of the 'democratic interest'? Who is there that, 'taking the high *priori* road,' shall point out what these influences are; what deep, subtle, inextricably entangled influences they have been and may be? For man is not the creature and product of Mechanism; but, in a far truer sense, its creator and producer: it is the noble People that makes the noble Government; rather than conversely. On the whole, Institutions are much; but they are not all. The freest and highest spirits of the world have often been found under strange outward circumstances: Saint Paul and his brother Apostles were politically slaves; Epictetus was personally one. Again, forget the influences of Chivalry and Religion, and ask: What countries produced Columbus and Las Casas? Or, descending from virtue and heroism to mere energy and spiritual talent: Cortes, Pizarro, Alba, Ximenes? The Spaniards of the sixteenth century were indisputably the noblest nation of Europe: yet they had the Inquisition and Philip II. They have the same government at this day; and are the lowest nation. The Dutch too have retained their old constitution; but no Siege of Leyden, no William the Silent, not even an Egmont or De Witt any longer appears among them. With ourselves also, where much has changed, effect has nowise followed cause as it should have done: two centuries ago, the Commons Speaker addressed Queen Elizabeth on bended knees, happy that the virago's foot did not even smite him; yet the people were then governed, not by a Castlereagh, but by a Burghley; they had their Shakspeare and Philip Sidney, where we have our Sheridan Knowles and Beau Brummel.

These and the like facts are so familiar, the truths which they preach so obvious, and have in all past times been so universally believed and acted on, that we should almost feel ashamed for repeating them; were it not that, on every hand, the memory of them seems to have passed away, or at best died into a faint tradition, of no value as a practical principle. To judge by the loud clamour of our Constitution-builders, Statists, Economists, directors, creators, reformers of Public Societies; in a word, all manner of Mechanists, from the Cartwright up to the Code-maker; and by the nearly total silence of all Preachers and Teachers who should give a voice to Poetry, Religion and Morality, we might fancy either that man's Dynamical nature was, to all spiritual intents, extinct, or else so perfected that nothing more was to be made of it by the old means; and henceforth only in his Mechanical contrivances did any hope exist for him.

To define the limits of these two departments of man's activity, which work into one another, and by means of one another, so intricately and inseparably, were by its nature an impossible attempt. Their relative importance, even to the wisest mind, will vary in different times, according to the special wants and dispositions of those times. Meanwhile, it seems clear enough that only in the right coordination of the two, and the vigorous forwarding of *both*, does our true line of action lie. Undue cultivation of the inward or Dynamical province leads to idle, visionary, impracticable course, and, especially in rude eras, to Superstition and Fanaticism, with their long train of baleful and well-known evils. Undue cultivation of the outward, again, though less immediately prejudicial, and even for the time productive of many palpable benefits, must, in the long-run, by destroying Moral Force, which is the parent of all other Force, prove not less certainly, and perhaps still more hopelessly, pernicious. This, we take it, is the grand characteristic of our age. By our skill in Mechanism, it has come to pass, that in the management of external things we excel all other ages; while in whatever respects the pure moral nature, in true dignity of soul and character, we are perhaps inferior to most civilised ages.

1829

From *Lord Clive*

IT would have been easy for Clive, during his second administration in Bengal, to accumulate riches such as no subject in Europe possessed. He might indeed, without subjecting the rich inhabitants of the province to any pressure beyond that to which their mildest rulers had accustomed them, have received presents to the amount of three hundred thousand pounds a year. The neighbouring princes would gladly have paid any price for his favour. But he appears to have strictly adhered to the rules which he had laid down for the guidance of others. The Rajah of Benares offered him diamonds of great value. The Nabob of Oude pressed him to accept a large sum of money and a casket of costly jewels. Clive courteously, but peremptorily refused: and it should be observed that he made no merit of his refusal, and that the facts did not come to light till after his death. He kept an exact account of his salary, of his share of the profits accruing from the trade in salt, and of those presents which, according to the fashion of the East, it would be churlish to refuse. Out of the sum arising from these resources, he defrayed the expenses of his situation. The surplus he divided among a few attached friends who had accompanied him to India. He always boasted, and, as far as we can judge, he boasted with truth, that his last administration diminished instead of increasing his fortune.

One large sum indeed he accepted. Meer Jaffier had left him by will above sixty thousand pounds sterling in specie and jewels: and the rules which had been recently laid down extended only to presents from the living, and did not affect legacies from the dead. Clive took the money, but not for himself. He made the whole over to the Company, in trust for officers and soldiers invalided in their service. The fund which still bears his name owes its origin to this princely donation.

After a stay of eighteen months, the state of his health made it necessary for him to return to Europe. At the close of January, 1767, he quitted for the last time the country on whose destinies he had exercised so mighty an influence.

His second return from Bengal was not, like his first, greeted by the acclamations of his countrymen. Numerous causes were already at work which embittered the remaining years of his life, and hurried him to an untimely grave. His old enemies at the India House were still powerful and active; and they had been reinforced by a large band of allies whose violence far exceeded their own. The whole crew of pilferers and oppressors from whom he had rescued Bengal persecuted him with the implacable rancour which belongs to such abject natures. Many of them even invested their property in India stock, merely that they might be better able to annoy the man whose firmness had set bounds to their rapacity. Lying newspapers were set up for no purpose but to abuse him; and the temper of the public mind was then such, that these arts, which under ordinary circumstances would have been ineffectual against truth and merit, produced an extraordinary impression.

The great events which had taken place in India had called into existence a new class of Englishmen, to whom their countrymen gave the name of Nabobs. These persons had generally sprung from families neither ancient nor opulent; they had generally been sent at an early age to the East; and they had there acquired large fortunes, which they had brought back to their native land. It was natural that, not having had much opportunity of mixing with the best society, they should exhibit some of the awkwardness and some of the pomposity of upstarts. It was natural that, during their sojourn in Asia, they should have acquired some tastes and habits surprising, if not disgusting, to persons who never had quitted Europe. It was natural that, having enjoyed great consideration in the East, they should not be disposed to sink into obscurity at home; and as they had money, and had not birth or high connexion, it was natural that they should display a little obtrusively the single advantage which they possessed. Wherever they settled there was a kind of feud between them and the old nobility and gentry, similar to that which raged in France between the farmer-general and the marquess. This enmity to the aristocracy long continued to distinguish the servants of the Company. More than twenty years after the time of which we are now speaking, Burke pronounced that among the Jacobins might be reckoned 'the East Indians almost to a man, who cannot bear to find that their present importance does not bear a proportion to their wealth.'

The Nabobs soon became a most unpopular class of men. Some of them had in the East displayed eminent talents, and rendered great services to the state; but at home their talents were not shown to advantage, and their services were little known. That they had sprung from obscurity, that they had acquired great wealth, that they exhibited it insolently, that they spent it extravagantly, that they raised the price of every thing in their neighbourhood, from fresh eggs to rotten boroughs, that their liveries outshone those of dukes, that their coaches were finer than that of the Lord Mayor, that the examples of their large and ill governed households corrupted half the servants in the country, that some of them, with all their magnificence, could not catch the tone of good society, but, in spite of the stud and the crowd of menials, of the plate and the Dresden china, of the venison and the Burgundy, were still low men; these were things which excited, both in the class from which they had sprung and in the class into which they attempted to force themselves, the bitter aversion which is the effect of mingled envy and contempt. But when it was also rumoured that the fortune which had enabled its possessor to eclipse the Lord Lieutenant on the race-ground, or to carry the county against the head of a house as old as Domesday Book, had been accumulated by violating public faith, by deposing legitimate princes, by reducing whole provinces to beggary, all the higher and better as well as all the low and evil parts of human nature were stirred against the wretch who had obtained by guilt and dishonour the riches which he now lavished with arrogant and inelegant profusion. The unfortunate Nabob seemed to be made up of those foibles against which comedy has pointed the most merciless ridicule, and of those crimes which have thrown the deepest gloom over tragedy, of Turcaret and Nero, of Monsieur Jourdain and Richard the Third. A tempest of execration and derision, such as can be compared only to that outbreak of public feeling against the Puritans which took place at the time of the Restoration, burst on the servants of the Company. The humane man was horror-struck at the way in which they had got their money, the

thrifty man at the way in which they spent it. The Dilettante
sneered at their want of taste. The Maccaroni black-balled them
as vulgar fellows. Writers the most unlike in sentiment and style,
Methodists and libertines, philosophers and buffoons, were for
once on the same side. It is hardly too much to say that, during
a space of about thirty years, the whole lighter literature of
England was coloured by the feelings which we have described.
Foote brought on the stage an Anglo-Indian chief, dissolute, ungenerous, and tyrannical, ashamed of the humble friends of his
youth, hating the aristocracy, yet childishly eager to be numbered
among them, squandering his wealth on pandars and flatterers,
tricking out his chairmen with the most costly hothouse flowers,
and astounding the ignorant with jargon about rupees, lacs, and
jaghires. Mackenzie, with more delicate humour, depicted a plain
country family raised by the Indian acquisitions of one of its
members to sudden opulence, and exciting derision by an awkward mimicry of the manners of the great. Cowper, in that lofty
expostulation which glows with the very spirit of the Hebrew
poets, placed the oppression of India foremost in the list of those
national crimes for which God had punished England with years
of disastrous war, with discomfiture in her own seas, and with
the loss of her transatlantic empire. If any of our readers will take
the trouble to search in the dusty recesses of circulating libraries
for some novel published sixty years ago, the chance is that the
villain or sub-villain of the story will prove to be a savage old
Nabob, with an immense fortune, a tawny complexion, a bad
liver, and a worse heart.

Such, as far as we can now judge, was the feeling of the
country respecting Nabobs in general. And Clive was eminently
the Nabob, the ablest, the most celebrated, the highest in rank,
the highest in fortune, of all the fraternity. His wealth was exhibited in a manner which could not fail to excite odium. He
lived with great magnificence in Berkeley Square. He reared one
palace in Shropshire and another at Claremont. His parliamentary influence might vie with that of the greatest families. But in
all this splendour and power envy found something to sneer at.
On some of his relations wealth and dignity seem to have sat as
awkwardly as on Mackenzie's Margery Mushroom. Nor was he
himself, with all his great qualities, free from those weaknesses
which the satirists of that age represented as characteristic of his

whole class. In the field, indeed, his habits were remarkably simple. He was constantly on horseback, was never seen but in his uniform, never wore silk, never entered a palanquin, and was content with the plainest fare. But when he was no longer at the head of an army, he laid aside this Spartan temperance for the ostentatious luxury of a Sybarite. Though his person was ungraceful, and though his harsh features were redeemed from vulgar ugliness only by their stern, dauntless, and commanding expression, he was fond of rich and gay clothing, and replenished his wardrobe with absurd profusion. Sir John Malcolm gives us a letter worthy of Sir Matthew Mite, in which Clive orders 'two hundred shirts, the best and finest that can be got for love or money.' A few follies of this description, grossly exaggerated by report, produced an unfavourable impression on the public mind. But this was not the worst. Black stories, of which the greater part were pure inventions, were circulated touching his conduct in the East. He had to bear the whole odium, not only of those bad acts to which he had once or twice stooped, but of all the bad acts of all the English in India, of bad acts committed when he was absent, nay, of bad acts which he had manfully opposed and severely punished. The very abuses against which he had waged an honest, resolute, and successful war were laid to his account. He was, in fact, regarded as the personification of all the vices and weaknesses which the public, with or without reason, ascribed to the English adventurers in Asia. We have ourselves heard old men, who knew nothing of his history, but who still retained the prejudices conceived in their youth, talk of him as an incarnate fiend. Johnson always held this language. Brown, whom Clive employed to lay out his pleasure grounds, was amazed to see in the house of his noble employer a chest which had once been filled with gold from the treasury of Moorshedabad, and could not understand how the conscience of the criminal could suffer him to sleep with such an object so near to his bedchamber. The peasantry of Surrey looked with mysterious horror on the stately house which was rising at Claremont, and whispered that the great wicked lord had ordered the walls to be made so thick in order to keep out the devil, who would one day carry him away bodily. Among the gaping clowns who drank in this frightful story was a worthless ugly lad of the name of Hunt, since widely known as William Huntington, SS; and the superstition which was strangely mingled with the knavery of that remarkable impostor seems to have derived no small nutriment from the tales which he heard of the life and character of Clive.

In the meantime, the impulse which Clive had given to the administration of Bengal was constantly becoming fainter and fainter. His policy was to a great extent abandoned; the abuses which he had suppressed began to revive; and at length the evils which a bad government had engendered were aggravated by one of those fearful visitations which the best government cannot avert. In the summer of 1770, the rains failed; the earth was parched up; the tanks were empty; the rivers shrank within their beds; and a famine, such as is known only in countries where every household depends for support on its own little patch of cultivation, filled the whole valley of the Ganges with misery and death. Tender and delicate women, whose veils had never been lifted before the public gaze, came forth from the inner chambers in which Eastern jealousy had kept watch over their beauty, threw themselves on the earth before the passers-by, and, with loud wailings, implored a handful of rice for their children. The Hoogley every day rolled down thousands of corpses close to the porticoes and gardens of the English conquerors. The very streets of Calcutta were blocked up by the dying and the dead. The lean and feeble survivors had not energy enough to bear the bodies of their kindred to the funeral pile or to the holy river, or even to scare away the jackals and vultures, who fed on human remains in the face of day. The extent of the mortality was never ascertained; but it was popularly reckoned by millions. This melancholy intelligence added to the excitement which already prevailed in England on Indian subjects. The proprietors of East India stock were uneasy about their dividends. All men of common humanity were touched by the calamities of our unhappy subjects; and indignation soon began to mingle itself with pity. It was rumoured that the Company's servants had created the famine by engrossing all the rice of the country; that they had sold grain for eight, ten, twelve times the price at which they had bought it; that one English functionary who, the year before, was not worth a hundred guineas, had, during that season of misery, remitted sixty thousand pounds to London. These charges we believe to have been unfounded. That servants of the

Company had ventured, since Clive's departure, to deal in rice, is probable. That, if they dealt in rice, they must have gained by the scarcity, is certain. But there is no reason for thinking that they either produced or aggravated an evil which physical causes sufficiently explain. The outcry which was raised against them on this occasion was, we suspect, as absurd as the imputations which, in times of dearth at home, were once thrown by states-men and judges, and are still thrown by two or three old women, on the corn factors. It was, however, so loud and so general that it appears to have imposed even on an intellect raised so high above vulgar prejudices as that of Adam Smith. What was still more extraordinary, these unhappy events greatly increased the unpopularity of Lord Clive. He had been some years in England when the famine took place. None of his acts had the smallest tendency to produce such a calamity. If the servants of the Company had traded in rice, they had done so in direct contravention of the rule which he had laid down, and, while in power, had resolutely enforced. But, in the eyes of his countrymen, he was, as we have said, the Nabob, the Anglo-Indian character personified; and, while he was building and planting in Surrey, he was held responsible for all the effects of a dry season in Bengal.

Parliament had hitherto bestowed very little attention on our Eastern possessions. Since the death of George the Second, a rapid succession of weak administrations, each of which was in turn flattered and betrayed by the Court, had held the semblance of power. Intrigues in the palace, riots in the capital, and insurrectionary movements in the American colonies, had left the advisers of the Crown little leisure to study Indian politics. When they did interfere, their interference was feeble and irresol-ute. Lord Chatham, indeed, during the short period of his ascendency in the councils of George the Third, had meditated a bold attack on the Company. But his plans were rendered abort-ive by the strange malady which about that time began to over-cloud his splendid genius.

At length, in 1772, it was generally felt that Parliament could no longer neglect the affairs of India. The Government was stronger than any which had held power since the breach be-tween Mr Pitt and the great Whig connexion in 1761. No pressing question of domestic or European policy required the attention of public men. There was a short and delusive lull between two tempests. The excitement produced by the Middlesex election was over; the discontents of America did not yet threaten civil war; the financial difficulties of the Company brought on a crisis; the Ministers were forced to take up the subject; and the whole storm, which had long been gathering, now broke at once on the head of Clive.

His situation was indeed singularly unfortunate. He was hated throughout the country, hated at the India House, hated, above all, by those wealthy and powerful servants of the Company, whose rapacity and tyranny he had withstood. He had to bear the double odium of his bad and of his good actions, of every Indian abuse and of every Indian reform. The state of the political world was such that he could count on the support of no powerful connexion. The party to which he had belonged, that of George Grenville, had been hostile to the Government, and yet had never cordially united with the other sections of the Opposition, with the little band which still followed the fortunes of Lord Chatham, or with the large and respectable body of which Lord Rockingham was the acknowledged leader. George Grenville was now dead: his followers were scattered; and Clive, unconnected with any of the powerful factions which divided the Parliament, could reckon only on the votes of those members who were returned by himself. His enemies, particularly those who were the enemies of his virtues, were unscrupulous, ferocious, implacable. Their malevolence aimed at nothing less than the utter ruin of his fame and fortune. They wished to see him expelled from Parliament, to see his spurs chopped off, to see his estate confiscated; and it may be doubted whether even such a result as this would have quenched their thirst for revenge.

Clive's parliamentary tactics resembled his military tactics. Deserted, surrounded, outnumbered, and with every thing at stake, he did not even deign to stand on the defensive, but pushed boldly forward to the attack. At an early stage of the discussions on Indian affairs he rose, and in a long and elaborate speech vindicated himself from a large part of the accusations which had been brought against him. He is said to have produced a great impression on his audience. Lord Chatham who, now the ghost of his former self, loved to haunt the scene of his glory, was that night under the gallery of the House of Commons, and declared that he had never heard a finer speech. It was subsequently printed under Clive's direction, and, when the fullest allowance has been made for assistance which he may have obtained from literary friends, proves him to have possessed, not merely strong sense and a manly spirit, but talents both for disquisition and declamation which assiduous culture might have improved into the highest excellence. He confined his defence on this occasion to the measures of his last administration, and succeeded so far that his enemies thenceforth thought it expedient to direct their attacks chiefly against the earlier part of his life.

The earlier part of his life unfortunately presented some assailable points to their hostility. A committee was chosen by ballot to inquire into the affairs of India; and by this committee the whole history of that great revolution which threw down Surajah Dowlah and raised Meer Jaffier was sifted with malignant care. Clive was subjected to the most unsparing examination and cross-examination, and afterwards bitterly complained that he, the Baron of Plassey, had been treated like a sheepstealer. The boldness and ingenuousness of his replies would alone suffice to show how alien from his nature were the frauds to which, in the course of his Eastern negotiations, he had sometimes descended. He avowed the arts which he had employed to deceive Omichund, and resolutely said that he was not ashamed of them, and that, in the same circumstances, he would again act in the same manner. He admitted that he had received immense sums from Meer Jaffier; but he denied that, in doing so, he had violated any obligation of morality or honour. He laid claim, on the contrary, and not without some reason, to the praise of eminent disinterestedness. He described in vivid language the situation in which his victory had placed him; great princes dependent on his pleasure; an opulent city afraid of being given up to plunder; wealthy bankers bidding against each other for his smiles; vaults piled with gold and jewels thrown open to him alone. 'By God, Mr Chairman,' he exclaimed, 'at this moment I stand astonished at my own moderation.'

The inquiry was so extensive that the Houses rose before it had been completed. It was continued in the following session. When at length the committee had concluded its labours, enlightened and impartial men had little difficulty in making up their minds as to the result. It was clear that Clive had been guilty of some acts which it is impossible to vindicate without attacking the authority of all the most sacred laws which regulate the intercourse of individuals and of states. But it was equally clear that he had displayed great talents, and even great virtues; that he had rendered eminent services both to his country and to the people of India; and that it was in truth not for his dealings with Meer Jaffier nor for the fraud which he had practised on Omichund, but for his determined resistance to avarice and tyranny, that he was now called in question.

Ordinary criminal justice knows nothing of set-off. The greatest desert cannot be pleaded in answer to a charge of the slightest transgression. If a man has sold beer on Sunday morning, it is no defence that he has saved the life of a fellow-creature at the risk of his own. If he has harnessed a Newfoundland dog to his little child's carriage, it is no defence that he was wounded at Waterloo, But it is not in this way that we ought to deal with men who, raised far above ordinary restraints, and tried by far more than ordinary temptations, are entitled to a more than ordinary measure of indulgence. Such men should be judged by their contemporaries as they will be judged by posterity. Their bad actions ought not, indeed, to be called good; but their good and bad actions ought to be fairly weighed; and, if on the whole the good preponderate, the sentence ought to be one, not merely of acquittal, but of approbation. Not a single great ruler in history can be absolved by a judge who fixes his eye inexorably on one or two unjustifiable acts. Bruce the deliverer of Scotland, Maurice the deliverer of Germany, William the deliverer of Holland, his great descendant the deliverer of England, Murray the good regent, Cosmo the father of his country, Henry the Fourth of France, Peter the Great of Russia, how would the best of them pass such a scrutiny? History takes wider views: and the best tribunal for great political cases is the tribunal which anticipates the verdict of history.

Reasonable and moderate men of all parties felt this in Clive's case. They could not pronounce him blameless; but they were not disposed to abandon him to that low-minded and rancorous pack who had run him down and were eager to worry him to death. Lord North, though not very friendly to him, was not disposed to go to extremities against him. While the inquiry was

still in progress, Clive, who had some years before been created a
Knight of the Bath, was installed with great pomp in Henry the
Seventh's chapel. He was soon after appointed Lord Lieutenant
of Shropshire. When he kissed hands, George the Third, who
had always been partial to him, admitted him to a private audience, talked to him half an hour on Indian politics, and was
visibly affected when the persecuted general spoke of his services
and of the way in which they had been requited.

At length the charges came in a definite form before the
House of Commons. Burgoyne, chairman of the committee, and
a man of wit, fashion, and honour, an agreeable dramatic writer,
an officer whose courage was never questioned and whose skill
was at that time highly esteemed, appeared as the accuser. The
members of the administration took different sides; for in that
age all questions were open questions, except such as were
brought forward by the Government, or such as implied some
censure on the Government. Thurlow, the Attorney General,
was among the assailants. Wedderburne, the Solicitor General,
strongly attached to Clive, defended his friend with extraordinary force of argument and language. It is a curious circumstance
that, some years later, Thurlow was the most conspicuous champion of Warren Hastings, while Wedderburne was among the
most unrelenting persecutors of that great though not faultless
statesman. Clive spoke in his own defence at less length and with
less art than in the preceding year, but with much energy and
pathos. He recounted his great actions and his wrongs; and, after
bidding his hearers remember that they were about to decide not
only on his honour but on their own, he retired from the House.

The Commons resolved that acquisitions made by the arms of
the State belong to the State alone, and that it is illegal in the
servants of the State to appropriate such acquisitions to themselves. They resolved that this wholesome rule appeared to have
been systematically violated by the English functionaries in
Bengal. On a subsequent day they went a step farther, and
resolved that Clive had, by means of the power which he possessed as commander of the British forces in India, obtained
large sums from Meer Jaffier. Here the Commons stopped. They
had voted the major and minor of Burgoyne's syllogism; but they
shrank from drawing the logical conclusion. When it was moved
that Lord Clive had abused his powers, and set an evil example

to the servants of the public, the previous question was put and carried. At length, long after the sun had risen on an animated debate, Wedderburne moved that Lord Clive had at the same time rendered great and meritorious services to his country; and this motion passed without a division.

The result of this memorable inquiry appears to us, on the whole, honourable to the justice, moderation, and discernment of the Commons. They had indeed no great temptation to do wrong. They would have been very bad judges of an accusation brought against Jenkinson or against Wilkes. But the question respecting Clive was not a party question; and the House accordingly acted with the good sense and good feeling which may always be expected from an assembly of English gentlemen, not blinded by faction.

The equitable and temperate proceedings of the British Parliament were set off to the greatest advantage by a foil. The wretched government of Louis the Fifteenth had murdered, directly or indirectly, almost every Frenchman who had served his country with distinction in the east. Labourdonnais was flung into the Bastille, and, after years of suffering, left it only to die. Dupleix, stripped of his immense fortune, and broken-hearted by humiliating attendance in antechambers, sank into an obscure grave. Lally was dragged to the common place of execution with a gag between his lips. The Commons of England, on the other hand, treated their living captain with that discriminating justice which is seldom shown except to the dead. They laid down sound general principles; they delicately pointed out where he had deviated from those principles; and they tempered the gentle censure with liberal eulogy. The contrast struck Voltaire, always partial to England, and always eager to expose the abuses of the Parliaments of France. Indeed he seems, at this time, to have meditated a history of the conquest of Bengal. He mentioned his design to Dr Moore when that amusing writer visited him at Ferney. Wedderburne took great interest in the matter, and pressed Clive to furnish materials. Had the plan been carried into execution, we have no doubt that Voltaire would have produced a book containing much lively and picturesque narrative, many just and humane sentiments poignantly expressed, many grotesque blunders, many sneers at the Mosaic chronology, much scandal about the Catholic missionaries, and much sublime theo-philanthropy, stolen from the New Testament, and put into the mouths of virtuous and philosophical Brahmins.

Clive was now secure in the enjoyment of his fortune and his honours. He was surrounded by attached friends and relations; and he had not yet passed the season of vigorous bodily and mental exertion. But clouds had long been gathering over his mind, and now settled on it in thick darkness. From early youth he had been subject to fits of that strange melancholy 'which rejoiceth exceedingly and is glad when it can find the grave.' While still a writer at Madras, he had twice attempted to destroy himself. Business and prosperity had produced a salutary effect on his spirits. In India, while he was occupied by great affairs, in England, while wealth and rank had still the charm of novelty, he had borne up against his constitutional misery. But he had now nothing to do, and nothing to wish for. His active spirit in an inactive situation drooped and withered like a plant in an uncongenial air. The malignity with which his enemies had pursued him, the indignity with which he had been treated by the committee, the censure, lenient as it was, which the House of Commons had pronounced, the knowledge that he was regarded by a large portion of his countrymen as a cruel and perfidious tyrant, all concurred to irritate and depress him. In the meantime, his temper was tried by acute physical suffering. During his long residence in tropical climates, he had contracted several painful distempers. In order to obtain ease he called in the help of opium; and he was gradually enslaved by this treacherous ally. To the last, however, his genius occasionally flashed through the gloom. It was said that he would sometimes, after sitting silent and torpid for hours, rouse himself to the discussion of some great question, would display in full vigour all the talents of the soldier and the statesman, and would then sink back into his melancholy repose.

The disputes with America had now become so serious that an appeal to the sword seemed inevitable; and the Ministers were desirous to avail themselves of the services of Clive. Had he still been what he was when he raised the siege of Patna, and annihilated the Dutch army and navy at the mouth of the Ganges, it is not improbable that the resistance of the Colonists would have been put down, and that the inevitable separation would have been deferred for a few years. But it was too late.

His strong mind was fast sinking under many kinds of suffering. On the twenty-second of November, 1774, he died by his own hand. He had just completed his forty-ninth year.

In the awful close of so much prosperity and glory, the vulgar saw only a confirmation of all their prejudices; and some men of real piety and genius so far forgot the maxims both of religion and of philosophy as confidently to ascribe the mournful event to the just vengeance of God, and to the horrors of an evil conscience. It is with very different feelings that we contemplate the spectacle of a great mind ruined by the weariness of satiety, by the pangs of wounded honour, by fatal diseases, and more fatal remedies.

Clive committed great faults; and we have not attempted to disguise them. But his faults, when weighed against his merits, and viewed in connexion with his temptations, do not appear to us to deprive him of his right to an honourable place in the estimation of posterity.

From his first visit to India dates the renown of the English arms in the East. Till he appeared, his countrymen were despised as mere pedlars, while the French were revered as a people formed for victory and command. His courage and capacity dissolved the charm. With the defence of Arcot commences that long series of Oriental triumphs which closes with the fall of Ghizni. Nor must we forget that he was only twenty-five years old when he approved himself ripe for military command. This is a rare if not a singular distinction. It is true that Alexander, Condé, and Charles the Twelfth, won great battles at a still earlier age; but those princes were surrounded by veteran generals of distinguished skill, to whose suggestions must be attributed the victories of the Granicus, of Rocroi, and of Narva. Clive, an inexperienced youth, had yet more experience than any of those who served under him. He had to form himself, to form his officers, and to form his army. The only man, as far as we recollect, who at an equally early age ever gave equal proof of talents for war, was Napoleon Bonaparte.

From Clive's second visit to India dates the political ascendency of the English in that country. His dexterity and resolution realised, in the course of a few months, more than all the gorgeous visions which had floated before the imagination of Dupleix. Such an extent of cultivated territory, such an amount of revenue, such a multitude of subjects, was never added to the dominion of Rome by the most successful proconsul. Nor were such wealthy spoils ever borne under arches of triumph, down the Sacred Way, and through the crowded Forum, to the threshold of Tarpeian Jove. The fame of those who subdued Antiochus and Tigranes grows dim when compared with the splendour of the exploits which the young English adventurer achieved at the head of an army not equal in numbers to one half of a Roman legion. From Clive's third visit to India dates the purity of the administration of our Eastern empire. When he landed in Calcutta in 1765, Bengal was regarded as a place to which Englishmen were sent only to get rich, by any means, in the shortest possible time. He first made dauntless and unsparing war on that gigantic system of oppression, extortion, and corruption. In that war he manfully put to hazard his ease, his fame, and his splendid fortune. The same sense of justice which forbids us to conceal or extenuate the faults of his earlier days compels us to admit that those faults were nobly repaired. If the reproach of the Company and of its servants has been taken away, if in India the yoke of foreign masters, elsewhere the heaviest of all yokes, has been found lighter than that of any native dynasty, if to that gang of public robbers which formerly spread terror through the whole plain of Bengal has succeeded a body of functionaries not more highly distinguished by ability and diligence than by integrity, disinterestedness, and public spirit, if we now see such men as Munro, Elphinstone, and Metcalfe, after leading victorious armies, after making and deposing kings, return, proud of their honourable poverty, from a land which once held out to every greedy factor the hope of boundless wealth, the praise is in no small measure due to Clive. His name stands high on the roll of conquerors. But it is found in a better list, in the list of those who have done and suffered much for the happiness of mankind. To the warrior, history will assign a place in the same rank with Lucullus and Trajan. Nor will she deny to the reformer a share of that veneration with which France cherishes the memory of Turgot, and with which the latest generations of Hindoos will contemplate the statue of Lord William Bentinck.

1840

SIR   HENRY   TAYLOR

*On Secrecy**

W HOM a statesman trusts at all he should trust largely, not to say unboundedly; and he should avow his trust to the world. In nine cases out of ten of betrayed confidence in affairs of State, vanity is the traitor. When a man comes into possession of some chance secrets now and then—some one or two—he is tempted to parade them to this friend or that. But when he is known to be trusted with all manner of secrets, his vanity is interested, not to show them, but to show that he can keep them. And his fidelity of heart is also better secured.

A secret may be sometimes best kept by keeping the secret of its being a secret. It is not many years since a State secret of the greatest importance was printed without being divulged, merely by sending it to the press like any other matter, and trusting to the mechanical habits of the persons employed. They printed it piecemeal in ignorance of what it was about.

The only secrecy which is worthy of trust in matters of State—and indeed the same may be said of secrecy in private friendship—is that which not merely observes an *enjoined* silence, but which maintains a considerate and judicious reticence in matters in which silence is perceived to be expedient, though it have *not* been enjoined. Faithfulness to public interests and to official and to friendly confidence, demands a careful exercise of the judgment as to what shall be spoken and what not, on many occasions when there is no question of obedience to express injunctions of secrecy. And indeed, in dealing with a confidential officer or friend, a statesman would do well to avoid any fre-quency of injunction on this head on particular occasions, because it tends to impair, on the part of such officer or friend,

* A chapter from Taylor's disquisition on the art of government, *The Statesman*.

that general watchfulness which is produced in a man who feels that he is thrown upon his own judgment and caution.

Secrecy will hardly be perfectly preserved unless by one who makes it a rule to avoid the whole of a subject of which he has to retain a part. To flesh your friend's curiosity and then endeavour to leave him with a *hûc usque*, is exposing your faculty of reticence to an unnecessary trial.

The most difficult of all subjects to be kept secret are such as will furnish fair occasion for a jest; and a statesman should regulate his confidence accordingly; being especially sparing of it in regard to such matters, and where he must needs impart them, taking care not to imp their wings by any jest of his own imparted along with them.

Shy and unready men are great betrayers of secrets; for there are few wants more urgent for the moment than the want of something to say. Such men may stand in need of the assurance given in Ecclesiasticus,—'If thou hast heard a word, let it die with thee: and be bold, it will not burst thee.'

1836

Secular Knowledge not a Principle of Action*

P EOPLE say to me, that it is but a dream to suppose that Christianity should regain the organic power in human society which once it possessed. I cannot help that; I never said it could. I am not a politician; I am proposing no measures, but exposing a fallacy, and resisting a pretence. Let Benthamism reign, if men have no aspirations; but do not tell them to be romantic, and then solace them with glory; do not attempt by philosophy what once was done by religion. The ascendency of Faith may be impracticable, but the reign of Knowledge is incomprehensible. The problem for statesmen of this age is how to educate the masses, and literature and science cannot give the solution.

Not so deems Sir Robert Peel; his firm belief and hope is, 'that an increased sagacity will administer to an exalted faith; that it will make men not merely believe in the cold doctrines of Natural Religion, but that it will so prepare and temper the spirit and understanding, that they will be better qualified to comprehend the great scheme of human redemption.' He certainly thinks that scientific pursuits have some considerable power of impressing religion upon the mind of the multitude. I think not, and will now say why.

Science gives us the grounds or premises from which religious truths are to be inferred; but it does not set about inferring them, much less does it reach the inference;—that is not its province. It brings before us phenomena, and it leaves us, if we will, to call them works of design, wisdom, or benevolence; and further still, if we will, to proceed to confess an Intelligent Creator. We have to take its facts, and to give them a meaning, and

* One of a series of letters, originally published in *The Times*, in reply to a speech made by Sir Robert Peel at the opening of a reading room at Tamworth, in Staffordshire.

to draw our own conclusions from them. First comes Knowledge, then a view, then reasoning, and then belief. This is why Science has so little of a religious tendency; deductions have no power of persuasion. The heart is commonly reached, not through the reason, but through the imagination, by means of direct impressions, by the testimony of facts and events, by history, by description. Persons influence us, voices melt us, looks subdue us, deeds inflame us. Many a man will live and die upon a dogma: no man will be a martyr for a conclusion. A conclusion is but an opinion; it is not a thing which *is*, but which *we are 'certain about'*; and it has often been observed, that we never say we are certain without implying that we doubt. To say that a thing *must* be, is to admit that it *may not* be. No one, I say, will die for his own calculations; he dies for realities. This is why a literary religion is so little to be depended upon; it looks well in fair weather, but its doctrines are opinions, and, when called to suffer for them, it slips them between its folios, or burns them at its hearth. And this again is the secret of the distrust and raillery with which moralists have been so commonly visited. They say and do not. Why? Because they are contemplating the fitness of things, and they live by the square, when they should be realizing their high maxims in the concrete. Now Sir Robert thinks better of natural history, chemistry, and astronomy, than of such ethics; but they too, what are they more than divinity *in posse*? He protests against 'controversial divinity': is *inferential* much better?

I have no confidence, then, in philosophers who cannot help being religious, and are Christians by implication. They sit at home, and reach forward to distances which astonish us; but they hit without grasping, and are sometimes as confident about shadows as about realities. They have worked out by a calculation the lie of a country which they never saw, and mapped it by means of a gazetteer; and like blind men, though they can put a stranger on his way, they cannot walk straight themselves, and do not feel it quite their business to walk at all.

Logic makes but a sorry rhetoric with the multitude; first shoot round corners, and you may not despair of converting by a syllogism. Tell men to gain notions of a Creator from His works, and, if they were to set about it (which nobody does), they would be jaded and wearied by the labyrinth they were tracing.

Their minds would be gorged and surfeited by the logical operation. Logicians are more set upon concluding rightly, than on
right conclusions. They cannot see the end for the process. Few
men have that power of mind which may hold fast and firmly a
variety of thoughts. We ridicule 'men of one idea'; but a great
many of us are born to be such, and we should be happier if we
knew it. To most men argument makes the point in hand only
more doubtful, and considerably less impressive. After all, man
is *not* a reasoning animal; he is a seeing, feeling, contemplating,
acting animal. He is influenced by what is direct and precise. It is
very well to freshen our impressions and convictions from physics, but to create them we must go elsewhere. Sir Robert Peel
'never can think it possible that a mind can be so constituted,
that, after being familiarized with the wonderful discoveries
which have been made in every part of experimental science, it
can retire from such contemplations without more enlarged conceptions of God's providence, and a higher reverence for His
name.' If he speaks of religious minds, he perpetrates a truism; if
of irreligious, he insinuates a paradox.

Life is not long enough for a religion of inferences; we shall
never have done beginning, if we determine to begin with proof.
We shall ever be laying our foundations; we shall turn theology
into evidences, and divines into textuaries. We shall never get at
our first principles. Resolve to believe nothing, and you must
prove your proofs and analyze your elements, sinking further
and further, and finding 'in the lowest depth a lower deep,' till
you come to the broad bosom of scepticism. I would rather be
bound to defend the reasonableness of assuming that Christianity
is true, than to demonstrate a moral governance from the physical world. Life is for action. If we insist on proofs for everything, we shall never come to action: to act you must assume,
and that assumption is faith.

Let no one suppose that in saying this I am maintaining that
all proofs are equally difficult, and all propositions equally de-
batable. Some assumptions are greater than others, and some
doctrines involve postulates larger than others, and more numerous. I only say that impressions lead to action, and that reasonings lead from it. Knowledge of premises, and inferences upon
them,—this is not to *live*. It is very well as a matter of
liberal curiosity and of philosophy to analyze our modes of

thought; but let this come second, and when there is leisure for it, and then our examinations will in many ways even be subservient to action. But if we commence with scientific knowledge and argumentative proof, or lay any great stress upon it as the basis of personal Christianity, or attempt to make man moral and religious by Libraries and Museums, let us in consistency take chemists for our cooks, and mineralogists for our masons.

Now I wish to state all this as matter of fact, to be judged by the candid testimony of any persons whatever. Why we are so constituted that Faith, not Knowledge or Argument, is our principle of action, is a question with which I have nothing to do; but I think it is a fact, and if it be such, we must resign ourselves to it as best we may, unless we take refuge in the intolerable paradox, that the mass of men are created for nothing, and are meant to leave life as they entered it. So well has this practically been understood in all ages of the world, that no Religion has yet been a Religion of physics or of philosophy. It has ever been synonymous with Revelation. It never has been a deduction from what we know: it has ever been an assertion of what we are to believe. It has never lived in a conclusion; it has never been a message, or a history, or a vision. No legislator or priest ever dreamed of educating our moral nature by science or by argument. There is no difference here between true Religions and pretended. Moses was instructed, not to reason from the creation, but to work miracles. Christianity is a history supernatural, and almost scenic: it tells us what its Author is, by telling us what He has done. I have no wish at all to speak otherwise than respectfully of conscientious Dissenters, but I have heard it said by those who were not their enemies, and who had known much of their preaching, that they had often heard narrow-minded and bigoted clergymen, and often Dissenting ministers of a far more intellectual cast; but that Dissenting teaching came to nothing,— that it was dissipated in thoughts which had no point, and inquiries which converged to no centre, that it ended as it began, and sent away its hearers as it found them;—whereas the instruction in the Church, with all its defects and mistakes, comes to some end, for it started from some beginning. Such is the difference between the dogmatism of faith and the speculations of logic.

Lord Brougham himself, as we have already seen, has recognized the force of this principle. He has not left his philosophical religion to argument; he has committed it to the keeping of the imagination. Why should he depict a great republic of letters, and an intellectual Pantheon, but that he feels that instances and patterns, not logical reasonings, are the living conclusions which alone have a hold over the affections, or can form the character?

1841

RALPH WALDO EMERSON

The Conservative

THE two parties which divide the state, the party of Conservatism and that of Innovation, are very old, and have disputed the possession of the world ever since it was made. This quarrel is the subject of civil history. The conservative party established the reverend hierarchies and monarchies of the most ancient world. The battle of patrician and plebeian, of parent state and colony, of old usage and accommodation to new facts, of the rich and the poor, reappears in all countries and times. The war rages not only in battlefields, in national councils, and ecclesiastical synods, but agitates every man's bosom with opposing advantages every hour. On rolls the old world meantime, and now one, now the other gets the day, and still the fight renews itself as if for the first time, under new names and hot personalities.

Such an irreconcilable antagonism, of course, must have a correspondent depth of seat in the human constitution. It is the opposition of Past and Future, of Memory and Hope, of the Understanding and the Reason. It is the primal antagonism, the appearance in trifles of the two poles of nature.

There is a fragment of old fable which seems somehow to have been dropped from the current mythologies, which may deserve attention, as it appears to relate to this subject.

Saturn grew weary of sitting alone, or with none but the great Uranus or Heaven beholding him, and he created an oyster. Then he would act again, but he made nothing more, but went on creating the race of oysters. Then Uranus cried, 'a new work, O Saturn! the old is not good again.'

Saturn replied, 'I fear. There is not only the alternative of making and not making, but also of unmaking. Seest thou the great sea, how it ebbs and flows? so is it with me; my power ebbs; and if I put forth my hands, I shall not do, but undo.

Therefore I do what I have done; I hold what I have got; and so
I resist Night and Chaos.'

'O Saturn,' replied Uranus, 'thou canst not hold thine own,
but by making more. Thy oysters are barnacles and cockles, and
with the next flowing of the tide, they will be pebbles and sea-
foam.'

'I see,' rejoins Saturn, 'thou art in league with Night, thou
art become an evil eye; thou spakest from love; now thy words
smite me with hatred. I appeal to Fate, must there not be
rest?'—'I appeal to Fate also,' said Uranus, 'must there not be
motion?'—But Saturn was silent, and went on making oysters
for a thousand years.

After that, the word of Uranus came into his mind like a ray
of the sun, and he made Jupiter; and then he feared again; and
nature froze, the things that were made went backward, and, to
save the world, Jupiter slew his father Saturn.

This may stand for the earliest account of a conversation on
politics between a Conservative and a Radical, which has come
down to us. It is ever thus. It is the counteraction of the centri-
petal and the centrifugal forces. Innovation is the salient energy;
Conservatism the pause on the last movement. 'That which is
was made by God,' saith Conservatism. 'He is leaving that, he is
entering this other,' rejoins Innovation.

There is always a certain meanness in the argument of conservatism, joined with a certain superiority in its fact. It affirms
because it holds. Its fingers clutch the fact, and it will not open
its eyes to see a better fact. The castle, which conservatism is set
to defend, is the actual state of things, good and bad. The project
of innovation is the best possible state of things. Of course, conservatism always has the worst of the argument, is always apo-
logizing, pleading a necessity, pleading that to change would be
to deteriorate; it must saddle itself with the mountainous load
of the violence and vice of society, must deny the possibility of
good, deny ideas, and suspect and stone the prophet; whilst
innovation is always in the right, triumphant, attacking, and sure
of final success. Conservatism stands on man's confessed limitations; reform on his indisputable infinitude; conservatism on
circumstance; liberalism on power; one goes to make an adroit
member of the social frame; the other to postpone all things to
the man himself; conservatism is debonnair and social; reform is

individual and imperious. We are reformers in spring and summer; in autumn and winter, we stand by the old; reformers in the morning, conservers at night. Reform is affirmative, conservatism negative; conservatism goes for comfort, reform for truth. Conservatism is more candid to behold another's worth; reform more disposed to maintain and increase its own. Conservatism makes no poetry, breathes no prayer, has no invention; it is all memory. Reform has no gratitude, no prudence, no husbandry. It makes a great difference to your figure and to your thought, whether your foot is advancing or receding. Conservatism never puts the foot forward; in the hour when it does that, it is not establishment, but reform. Conservatism tends to universal seeming and treachery, believes in a negative fate; believes that men's temper governs them; that for me, it avails not to trust in principles; they will fail me; I must bend a little; it distrusts nature; it thinks there is a general law without a particular application,—law for all that does not include any one. Reform in its antagonism inclines to asinine resistance, to kick with hoofs; it runs to egotism and bloated self-conceit; it runs to a bodiless pretension, to unnatural refining and elevation, which ends in hypocrisy and sensual reaction.

And so whilst we do not go beyond general statements, it may be safely affirmed of these two metaphysical antagonists, that each is a good half, but an impossible whole. Each exposes the abuses of the other, but in a true society, in a true man, both must combine. Nature does not give the crown of its approbation, namely, beauty, to any action or emblem or actor, but to one which combines both these elements; not to the rock which resists the waves from age to age, nor to the wave which lashes incessantly the rock, but the superior beauty is with the oak which stands with its hundred arms against the storms of a century, and grows every year like a sapling; or the river which ever flowing, yet is found in the same bed from age to age; or, greatest of all, the man who has subsisted for years amid the changes of nature, yet has distanced himself, so that when you remember what he was, and see what he is, you say, what strides! what a disparity is here!

Throughout nature the past combines in every creature with the present. Each of the convolutions of the sea-shell, each node and spine marks one year of the fish's life, what was the mouth of the shell for one season, with the addition of new matter by the growth of the animal, becoming an ornamental node. The leaves and a shell of soft wood are all that the vegetation of this summer has made, but the solid columnar stem, which lifts that bank of foliage into the air to draw the eye and to cool us with its shade, is the gift and legacy of dead and buried years.

In nature, each of these elements being always present, each theory has a natural support. As we take our stand on Necessity, or on Ethics, shall we go for the conservative, or for the reformer. If we read the world historically, we shall say, Of all the ages, the present hour and circumstance is the cumulative result; this is the best throw of the dice of nature that has yet been, or that is yet possible. If we see it from the side of Will, or the Moral Sentiment, we shall accuse the Past and the Present, and require the impossible of the Future.

But although this bifold fact lies thus united in real nature, and so united that no man can continue to exist in whom both these elements do not work, yet men are not philosophers, but are rather very foolish children, who, by reason of their partiality, see everything in the most absurd manner, and are the victims at all times of the nearest object. There is even no philosopher who is a philosopher at all times. Our experience, our perception is conditioned by the need to acquire in parts and in succession, that is, with every truth a certain falsehood. As this is the invariable method of our training, we must give it allowance, and suffer men to learn as they have done for six millenniums, a word at a time, to pair off into insane parties, and learn the amount of truth each knows, by the denial of an equal amount of truth. For the present, then, to come at what sum is attainable to us, we must even hear the parties plead as parties.

That which is best about conservatism, that which, though it cannot be expressed in detail, inspires reverence in all, is the Inevitable. There is the question not only, what the conservative says for himself? but, why must he say it? What insurmountable fact binds him to that side? Here is the fact which men call Fate, and fate in dread degrees, fate behind fate, not to be disposed of by the consideration that the Conscience commands this or that, but necessitating the question, whether the faculties of man will play him true in resisting the facts of universal experience? For although the commands of the Conscience are *essentially* absolute, they are *historically* limitary. Wisdom does not seek a literal rectitude, but an useful, that is, a conditioned one, such a one as the faculties of man and the constitution of things will warrant. The reformer, the partisan loses himself in driving to the utmost some specialty of right conduct, until his own nature and all nature resist him; but Wisdom attempts nothing enormous and disproportioned to its powers, nothing which it cannot perform or nearly perform. We have all a certain intellection or presentiment of reform existing in the mind, which does not yet descend into the character, and those who throw themselves blindly on this lose themselves. Whatever they attempt in that direction, fails, and reacts suicidally on the actor himself. This is the penalty of having transcended nature. For the existing world is not a dream, and cannot with impunity be treated as a dream; neither is it a disease; but it is the ground on which you stand, it is the mother of whom you were born. Reform converses with possibilities, perchance with impossibilities; but here is sacred fact. This also was true, or it could not be: it had life in it, or it could not have existed; it has life in it, or it could not continue. Your schemes may be feasible, or may not be, but this has the endorsement of nature and a long friendship and cohabitation with the powers of nature. This will stand until a better cast of the dice is made. The contest between the Future and the Past is one between Divinity entering, and Divinity departing. You are welcome to try your experiments, and, if you can, to displace the actual order by that ideal republic you announce, for nothing but God will expel God. But plainly the burden of proof must lie with the projector. We hold to this, until you can demonstrate something better.

The system of property and law goes back for its origin to barbarous and sacred times; it is the fruit of the same mysterious cause as the mineral or animal world. There is a natural sentiment and prepossession in favor of age, of ancestors, of barbarous and aboriginal usages, which is a homage to the element of necessity and divinity which is in them. The respect for the old names of places, of mountains, and streams, is universal. The Indian and barbarous name can never be supplanted without loss. The ancients tell us that the gods loved the Ethiopians for their stable customs; and the Egyptians and Chaldeans, whose origin could not be explored, passed among the junior tribes of Greece and Italy for sacred nations.

Moreover, so deep is the foundation of the existing social system, that it leaves no one out of it. We may be partial, but Fate is not. All men have their root in it. You who quarrel with the arrangements of society, and are willing to embroil all, and risk the indisputable good that exists, for the chance of better, live, move, and have your being in this, and your deeds contradict your words every day. For as you cannot jump from the ground without using the resistance of the ground, nor put out the boat to sea, without shoving from the shore, nor attain liberty without rejecting obligation, so you are under the necessity of using the Actual order of things, in order to disuse it; to live by it, whilst you wish to take away its life. The past has baked your loaf, and in the strength of its bread you would break up the oven. But you are betrayed by your own nature. You also are conservatives. However men please to style themselves, I see no other than a conservative party. You are not only identical with us in your needs, but also in your methods and aims. You quarrel with my conservatism, but it is to build up one of your own; it will have a new beginning, but the same course and end, the same trials, the same passions; among the lovers of the new I observe that there is a jealousy of the newest, and that the seceder from the seceder is as damnable as the pope himself.

On these and the like grounds of general statement, conservatism plants itself without danger of being displaced. Especially before this *personal* appeal, the innovator must confess his weakness, must confess that no man is to be found good enough to be entitled to stand champion for the principle. But when this great tendency comes to practical encounters, and is challenged by young men, to whom it is no abstraction, but a fact of hunger, distress, and exclusion from opportunities, it must needs seem injurious. The youth, of course, is an innovator by the fact of his birth. There he stands, newly born on the planet, a universal beggar, with all the reason of things, one would say, on his side. In his first consideration how to feed, clothe, and warm himself, he is met by warnings on every hand, that this thing and that thing have owners, and he must go elsewhere. Then he says; If I am born into the earth, where is my part? have the goodness, gentlemen of this world, to show me my wood-lot, where I may fell my wood, my field where to plant my corn, my pleasant ground where to build my cabin.

'Touch any wood, or field, or house-lot, on your peril,' cry all the gentlemen of this world; 'but you may come and work in ours, for us, and we will give you a piece of bread.'

And what is that peril?

Knives and muskets, if we meet you in the act; imprisonment, if we find you afterward.

And by what authority, kind gentlemen?

By our law.

And your law,—is it just?

As just for you as it was for us. We wrought for others under this law, and got our lands so.

I repeat the question, Is your law just?

Not quite just, but necessary. Moreover, it is juster now than it was when we were born; we have made it milder and more equal.

I will none of your law, returns the youth; it encumbers me. I cannot understand, or so much as spare time to read that needless library of your laws. Nature has sufficiently provided me with rewards and sharp penalties, to bind me not to transgress. Like the Persian noble of old, I ask 'that I may neither command nor obey.' I do not wish to enter into your complex social system. I shall serve those whom I can, and they who can will serve me. I shall seek those whom I love, and shun those whom I love not, and what more can all your laws render me?

With equal earnestness and good faith, replies to this plaintiff an upholder of the establishment, a man of many virtues:

Your opposition is feather-brained and overfine. Young man, I have no skill to talk with you, but look at me; I have risen early and sat late, and toiled honestly, and painfully for very many years. I never dreamed about methods; I laid my bones to, and drudged for the good I possess; it was not got by fraud, nor by luck, but by work, and you must show me a warrant like these stubborn facts in your own fidelity and labor, before I suffer you, on the faith of a few fine words, to ride into my estate, and claim to scatter it as your own.

Now you touch the heart of the matter, replies the reformer. To that fidelity and labor, I pay homage. I am unworthy to arraign your manner of living, until I too have been tried. But I should be more unworthy, if I did not tell you why I cannot

walk in your steps. I find this vast network, which you call property, extended over the whole planet. I cannot occupy the
bleakest crag of the White Hills or the Alleghany Range, but
some man or corporation steps up to me to show me that it
is his. Now, though I am very peaceable, and on my private
account could well enough die, since it appears there was some
mistake in my creation, and that I have been *mis*sent to this earth,
where all the seats were already taken,—yet I feel called upon in
behalf of rational nature, which I represent, to declare to you my
opinion, that, if the Earth is yours, so also is it mine. All your
aggregate existences are less to me a fact than is my own; as I am
born to the Earth, so the Earth is given to me, what I want of it
to till and to plant; nor could I, without pusillanimity, omit to
claim so much. I must not only have a name to live, I must live.
My genius leads me to build a different manner of life from any
of yours. I cannot then spare you the whole world. I love you
better. I must tell you the truth practically; and take that which
you call yours. It is God's world and mine; yours as much as you
want, mine as much as I want. Besides, I know your ways; I
know the symptoms of the disease. To the end of your power,
you will serve this lie which cheats you. Your want is a gulf
which the possession of the broad earth would not fill. Yonder
sun in heaven you would pluck down from shining on the universe, and make him a property and privacy, if you could; and
the moon and the north star you would quickly have occasion
for in your closet and bedchamber. What you do not want for
use, you crave for ornament, and what your convenience could
spare, your pride cannot.

On the other hand, precisely the defence which was set up
for the British Constitution, namely, that with all its admitted
defects, rotten boroughs and monopolies, it worked well, and
substantial justice was somehow done; the wisdom and the
worth did get into parliament, and every interest did by right, or
might, or sleight, get represented;—the same defence is set up
for the existing institutions. They are not the best; they are not
just; and in respect to you, personally, O brave young man! they
cannot be justified. They have, it is most true, left you no acre
for your own, and no law but our law, to the ordaining of
which, you were no party. But they do answer the end, they are
really friendly to the good; unfriendly to the bad; they second

the industrious, and the kind; they foster genius. They really have so much flexibility as to afford your talent and character, on the whole, the same chance of demonstration and success which they might have, if there was no law and no property.

It is trivial and merely superstitious to say that nothing is given you, no outfit, no exhibition; for in this institution of *credit*, which is as universal as honesty and promise in the human countenance, always some neighbor stands ready to be bread and land and tools and stock to the young adventurer. And if in any one respect they have come short, see what ample retribution of good they have made. They have lost no time and spared no expense to collect libraries, museums, galleries, colleges, palaces, hospitals, observatories, cities. The ages have not been idle, nor kings slack, nor the rich niggardly. Have we not atoned for this small offence (which we could not help) of leaving you no right in the soil, by this splendid indemnity of ancestral and national wealth? Would you have been born like a gipsy in a hedge, and preferred your freedom on a heath, and the range of a planet which had no shed or boscage to cover you from sun and wind, —to this towered and citied world? to this world of Rome, and Memphis, and Constantinople, and Vienna, and Paris, and London, and New York? For thee Naples, Florence, and Venice, for thee the fair Mediterranean, the sunny Adriatic; for thee both Indies smile; for thee the hospitable North opens its heated palaces under the polar circle; for thee roads have been cut in every direction across the land, and fleets of floating palaces with every security for strength, and provision for luxury, swim by sail and by steam through all the waters of this world. Every island for thee has a town; every town a hotel. Though thou wast born landless, yet to thy industry and thrift and small condescension to the established usage,—scores of servants are swarming in every strange place with cap and knee to thy command, scores, nay hundreds and thousands, for thy wardrobe, thy table, thy chamber, thy library, thy leisure; and every whim is anticipated and served by the best ability of the whole population of each country. The king on the throne governs for thee, and the judge judges; the barrister pleads, the farmer tills, the joiner hammers, the postman rides. Is it not exaggerating a trifle to insist on a formal acknowledgment of your claims, when these substantial advantages have been secured to you? Now can your children be educated, your labor turned to their advantage, and its fruits secured to them after your death. It is frivolous to say, you have no acre, because you have not a mathematically measured piece of land. Providence takes care that you shall have a place, that you are waited for, and come accredited; and, as soon as you put your gift to use, you shall have acre or acre's worth according to your exhibition of desert,—acre, if you need land;—acre's worth, if you prefer to draw, or carve, or make shoes, or wheels, to the tilling of the soil.

Besides, it might temper your indignation at the supposed wrong which society has done you, to keep the question before you, how society got into this predicament? Who put things on this false basis? No single man, but all men. No man voluntarily and knowingly; but it is the result of that degree of culture there is in the planet. The order of things is as good as the character of the population permits. Consider it as the work of a great and beneficent and progressive necessity, which, from the first pulsation of the first animal life, up to the present high culture of the best nations, has advanced thus far. Thank the rude fostermother though she has taught you a better wisdom than her own, and has set hopes in your heart which shall be history in the next ages. You are yourself the result of this manner of living, this foul compromise, this vituperated Sodom. It nourished you with care and love on its breast, as it had nourished many a lover of the right, and many a poet, and prophet, and teacher of men. Is it so irremediably bad? Then again, if the mitigations are considered, do not all the mischiefs virtually vanish? The form is bad, but see you not how every personal character reacts on the form, and makes it new? A strong person makes the law and custom null before his own will. Then the principle of love and truth reappears in the strictest courts of fashion and property. Under the richest robes, in the darlings of the selectest circles of European or American aristocracy, the strong heart will beat with love of mankind, with impatience of accidental distinctions, with the desire to achieve its own fate, and make every ornament it wears authentic and real.

Moreover, as we have already shown that there is no pure reformer, so it is to be considered that there is no pure conservative, no man who from the beginning to the end of his life maintains the defective institutions; but he who sets his face like a flint against every novelty, when approached in the confidence of conversation, in the presence of friendly and generous persons, has also his gracious and relenting motions, and espouses for the time the cause of man; and even if this be a shortlived emotion, yet the remembrance of it in private hours mitigates his selfishness and compliance with custom.

The Friar Bernard lamented in his cell on Mount Cenis the crimes of mankind, and rising one morning before day from his bed of moss and dry leaves, he gnawed his roots and berries, drank of the spring, and set forth to go to Rome to reform the corruption of mankind. On his way he encountered many travellers who greeted him courteously; and the cabins of the peasants and the castles of the lords supplied his few wants. When he came at last to Rome, his piety and good will easily introduced him to many families of the rich, and on the first day he saw and talked with gentle mothers with their babes at their breasts, who told him how much love they bore their children, and how they were perplexed in their daily walk lest they should fail in their duty to them. 'What!' he said, 'and this on rich embroidered carpets, on marble floors, with cunning sculpture, and carved wood, and rich pictures, and piles of books about you?'—'Look at our pictures, and books,' they said, 'and we will tell you, good Father, how we spent the last evening. These are stories of godly children and holy families and romantic sacrifices made in old or in recent times by great and not mean persons; and last evening, our family was collected, and our husbands and brothers discoursed sadly on what we could save and give in the hard times.' Then came in the men, and they said, 'What cheer, brother? Does thy convent want gifts?' Then the Friar Bernard went home swiftly with other thoughts than he brought, saying, 'This way of life is wrong, yet these Romans, whom I prayed God to destroy, are lovers, they are lovers; what can I do?'

The reformer concedes that these mitigations exist, and that, if he proposed comfort, he should take sides with the establishment. Your words are excellent, but they do not tell the whole. Conservatism is affluent and openhanded, but there is a cunning juggle in riches. I observe that they take somewhat for everything they give. I look bigger, but am less; I have more clothes, but am not so warm; more armor, but less courage; more books, but less wit. What you say of your planted, builded and decorated world, is true enough, and I gladly avail myself of its convenience; yet I have remarked that what holds in particular, holds in general, that the plant Man does not require for his most glorious flowering this pomp of preparation and convenience, but the thoughts of some beggarly Homer who strolled, God knows when, in the infancy and barbarism of the old world; the gravity and sense of some slave Moses who leads away his fellow slaves from their masters; the contemplation of some Scythian Anacharsis; the erect, formidable valor of some Dorian towns-men in the town of Sparta; the vigor of Clovis the Frank, and Alfred the Saxon, and Alaric the Goth, and Mahomet, Ali, and Omar the Arabians, Saladin the Curd, and Othman the Turk, sufficed to build what you call society, on the spot and in the instant when the sound mind in a sound body appeared. Rich and fine is your dress. O conservatism! your horses are of the best blood; your roads are well cut and well paved; your pantry is full of meats and your cellar of wines, and a very good state and condition are you for gentlemen and ladies to live under; but every one of these goods steals away a drop of my blood. I want the necessity of supplying my own wants. All this costly culture of yours is not necessary. Greatness does not need it. Yonder peasant, who sits neglected there in a corner, carries a whole re-volution of man and nature in his head, which shall be a sacred history to some future ages. For man is the end of nature; nothing so easily organizes itself in every part of the universe as he; no moss, no lichen is so easily born; and he takes along with him and puts out from himself the whole apparatus of society and condition *extempore*, as an army encamps in a desert, and where all was just now blowing sand, creates a white city in an hour, a government, a market, a place for feasting, for conver-sation, and for love.

These considerations, urged by those whose characters and whose fortunes are yet to be formed, must needs command the sympathy of all reasonable persons. But beside that charity which should make all adult persons interested for the youth, and engage them to see that he has a free field and fair play on his entrance into life, we are bound to see that the society, of which we compose a part, does not permit the formation or continu-ance of views and practices injurious to the honor and welfare of mankind. The objection to conservatism, when embodied in a party, is, that in its love of acts, it hates principles; it lives in the senses, not in truth; it sacrifices to despair; it goes for availableness in its candidate, not for worth; and for expediency in its measures, and not for the right. Under pretence of allowing for friction, it makes so many additions and supplements to the machine of society, that it will play smoothly and softly, but will no longer grind any grist.

The conservative party in the universe concedes that the radical would talk sufficiently to the purpose, if we were still in the garden of Eden; he legislates for man as he ought to be; his theory is right, but he makes no allowance for friction; and this omission makes his whole doctrine false. The idealist retorts, that the conservative falls into a far more noxious error in the other extreme. The conservative assumes sickness as a necessity, and his social frame is a hospital, his total legislation is for the present distress, a universe in slippers and flannels, with bib and papspoon, swallowing pills and herb-tea. Sickness gets organized as well as health, the vice as well as the virtue. Now that a vicious system of trade has existed so long, it has stereotyped itself in the human generation, and misers are born. And now that sickness has got such a foothold, leprosy has grown cunning, has got into the ballot-box; the lepers outvote the clean; society has resolved itself into a Hospital Committee, and all its laws are quarantine. If any man resist, and set up a foolish hope he has entertained as good against the general despair, society frowns on him, shuts him out of her opportunities, her granaries, her refectories, her water and bread, and will serve him a sexton's turn. Conservatism takes as low a view of every part of human action and passion. Its religion is just as bad; a lozenge for the sick; a dolorous tune to beguile the distemper; mitigations of pain by pillows and anodynes; always mitigations, never remedies; pardons for sin, funeral honors,—never self-help, renovation, and virtue. Its social and political action has no better aim; to keep out wind and weather, to bring the day and year about, and make the world last our day; not to sit on the world and steer it; not to sink the memory of the past in the glory of a new and more excellent creation; a timid cobbler and patcher, it degrades whatever it touches. The cause of education is urged in this country with the utmost earnestness,—on what

ground? why on this, that the people have the power, and if they
are not instructed to sympathize with the intelligent, reading,
trading, and governing class, inspired with a taste for the same
competitions and prizes, they will upset the fair pageant of Judic-
ature, and perhaps lay a hand on the sacred muniments of
wealth itself, and new distribute the land. Religion is taught in
the same spirit. The contractors who were building a road out of
Baltimore, some years ago, found the Irish laborers quarrelsome
and refractory, to a degree that embarrassed the agents, and
seriously interrupted the progress of the work. The corporation
were advised to call off the police, and build a Catholic chapel;
which they did; the priest presently restored order, and the work
went on prosperously. Such hints, be sure, are too valuable to
be lost. If you do not value the Sabbath, or other religious
institutions, give yourself no concern about maintaining them.
They have already acquired a market value as conservators of
property; and if priest and church-member should fail, the
chambers of commerce and the presidents of the Banks, the very
innholders and landlords of the county would muster with fury
to their support.

Of course, religion in such hands loses its essence. Instead of
that reliance, which the soul suggests on the eternity of truth
and duty, men are misled into a reliance on institutions, which,
the moment they cease to be the instantaneous creations of
the devout sentiment, are worthless. Religion among the low
becomes low. As it loses its truth, it loses credit with the
sagacious. They detect the falsehood of the preaching, but when
they say so, all good citizens cry, Hush; do not weaken the state,
do not take off the strait jacket from dangerous persons. Every
honest fellow must keep up the hoax the best he can; must
patronize providence and piety, and wherever he sees anything
that will keep men amused, schools or churches or poetry, or
picture-galleries or music, or what not, he must cry 'Hist-a-boy,'
and urge the game on. What a compliment we pay to the good
Spirit with our superserviceable zeal!

But not to balance reasons for and against the establishment
any longer, and if it still be asked in this necessity of partial
organization, which party on the whole has the highest claims on
our sympathy? I bring it home to the private heart, where all
such questions must have their final arbitrement. How will every

strong and generous mind choose its ground,—with the defenders of the old? or with the seekers of the new? Which is that state which promises to edify a great, brave, and beneficent man; to throw him on his resources, and tax the strength of his character? On which part will each of us find himself in the hour of health and of aspiration?

I understand well the respect of mankind for war, because that breaks up the Chinese stagnation of society, and demonstrates the personal merits of all men. A state of war or anarchy, in which law has little force, is so far valuable, that it puts every man on trial. The man of principle is known as such, and even in the fury of faction is respected. In the civil wars of France, Montaigne alone, among all the French gentry, kept his castle gates unbarred, and made his personal integrity as good at least as a regiment. The man of courage and resources is shown, and the effeminate and base person. Those who rise above war, and those who fall below it, it easily discriminates, as well as those, who, accepting its rude conditions, keep their own head by their own sword.

But in peace and a commercial state we depend, not as we ought, on our knowledge and all men's knowledge that we are honest men, but we cowardly lean on the virtue of others. For it is always at last the virtue of some men in the society, which keeps the law in any reverence and power. Is there not something shameful that I should owe my peaceful occupancy of my house and field, not to the knowledge of my countrymen that I am useful, but to their respect for sundry other reputable persons, I know not whom, whose joint virtues still keep the law in good odor?

It will never make any difference to a hero what the laws are. His greatness will shine and accomplish itself unto the end, whether they second him or not. If he have earned his bread by drudgery, and in the narrow and crooked ways which were all an evil law had left him, he will make it at least honorable by his expenditure. Of the past he will take no heed; for its wrongs he will not hold himself responsible: he will say, all the meanness of my progenitors shall not bereave me of the power to make this hour and company fair and fortunate. Whatsoever streams of power and commodity flow to me, shall of me acquire healing virtue, and become fountains of safety. Cannot I too descend a

Redeemer into nature? Whosoever hereafter shall name my name, shall not record a malefactor, but a benefactor in the earth. If there be power in good intention, in fidelity, and in toil, the north wind shall be purer, the stars in heaven shall glow with a kindlier beam, that I have lived. I am primarily engaged to myself to be a public servant of all the gods, to demonstrate to all men that there is intelligence and good will at the heart of things, and ever higher and yet higher leadings. These are my engagements; how can your law further or hinder me in what I shall do to men? On the other hand, these dispositions establish their relations to me. Wherever there is worth, I shall be greeted. Wherever there are men, are the objects of my study and love. Sooner or later all men will be my friends, and will testify in all methods the energy of their regard. I cannot thank your law for my protection. I protect it. It is not in its power to protect me. It is my business to make myself revered. I depend on my honor, my labor, and my dispositions, for my place in the affections of mankind, and not on any conventions or parchments of yours.

But if I allow myself in derelictions, and become idle and dissolute, I quickly come to love the protection of a strong law, because I feel no title in myself to my advantages. To the intemperate and covetous person no love flows; to him mankind would pay no rent, no dividend, if force were once relaxed; nay, if they could give their verdict, they would say, that his self-indulgence and his oppression deserved punishment from society, and not that rich board and lodging he now enjoys. The law acts then as a screen of his unworthiness, and makes him worse the longer it protects him.

In conclusion, to return from this alternation of partial views, to the high platform of universal and necessary history, it is a happiness for mankind that innovation has got on so far, and has so free a field before it. The boldness of the hope men entertain transcends all former experience. It calms and cheers them with the picture of a simple and equal life of truth and piety. And this hope flowered on what tree? It was not imported from the stock of some celestial plant, but grew here on the wild crab of conservatism. It is much that this old and vituperated system of things has borne so fair a child. It predicts that amidst a planet peopled with conservatives, one Reformer may yet be born.

1841

The Haunted Mind

W HAT a singular moment is the first one, when you have hardly begun to recollect yourself, after starting from midnight slumber! By unclosing your eyes so suddenly, you seem to have surprised the personages of your dream in full convocation round your bed, and catch one broad glance at them before they can flit into obscurity. Or, to vary the metaphor, you find yourself, for a single instant, wide awake in that realm of illusions, whither sleep has been the passport, and behold its ghostly inhabitants and wondrous scenery, with a perception of their strangeness, such as you never attain while the dream is undisturbed. The distant sound of a church clock is borne faintly on the wind. You question with yourself, half seriously, whether it has stolen to your waking ear from some gray tower, that stood within the precincts of your dream. While yet in suspense, another clock flings its heavy clang over the slumbering town, with so full and distinct a sound, and such a long murmur in the neighboring air, that you are certain it must proceed from the steeple at the nearest corner. You count the strokes—one—two—and there they cease, with a booming sound, like the gathering of a third stroke within the bell.

If you could choose an hour of wakefulness out of the whole night, it would be this. Since your sober bedtime, at eleven, you have had rest enough to take off the pressure of yesterday's fatigue; while before you, till the sun comes from 'far Cathay' to brighten your window, there is almost the space of a summer night; one hour to be spent in thought, with the mind's eye half shut, and two in pleasant dreams, and two in that strangest of enjoyments, the forgetfulness alike of joy and woe. The moment of rising belongs to another period of time, and appears so distant, that the plunge out of a warm bed into the frosty air cannot yet be anticipated with dismay. Yesterday has already vanished

among the shadows of the past; to-morrow has not yet emerged
from the future. You have found an intermediate space, where
the business of life does not intrude; where the passing moment
lingers, and becomes truly the present; a spot where Father
Time, when he thinks nobody is watching him, sits down by the
way side to take breath. Oh, that he would fall asleep, and let
mortals live on without growing older!

Hitherto you have lain perfectly still, because the slightest
motion would dissipate the fragments of your slumber. Now,
being irrevocably awake, you peep through the half drawn window curtain, and observe that the glass is ornamented with fanciful devices in frost work, and that each pane presents something
like a frozen dream. There will be time enough to trace out the
analogy, while waiting the summons to breakfast. Seen through
the clear portion of the glass, where the silvery mountain peaks
of the frost scenery do not ascend, the most conspicuous object
is the steeple; the white spire of which directs you to the wintry
lustre of the firmament. You may almost distinguish the figures
on the clock that has just told the hour. Such a frosty sky, and
the snow covered roofs, and the long vista of the frozen street,
all white, and the distant water hardened into rock, might make
you shiver, even under four blankets and a woolen comforter.
Yet look at that one glorious star! Its beams are distinguishable
from all the rest, and actually cast the shadow of the casement on
the bed, with a radiance of deeper hue than moonlight, though
not so accurate an outline.

You sink down and muffle your head in the clothes, shivering
all the while, but less from bodily chill, than the bare idea of a
polar atmosphere. It is too cold even for the thoughts to venture
abroad. You speculate on the luxury of wearing out a whole
existence in bed, like an oyster in its shell, content with the sluggish ecstasy of inaction, and drowsily conscious of nothing but
delicious warmth, such as you now feel again. Ah! that idea has
brought a hideous one in its train. You think how the dead are
lying in their cold shrouds and narrow coffins, through the drear
winter of the grave, and cannot persuade your fancy that they
neither shrink nor shiver, when the snow is drifting over their
little hillocks, and the bitter blast howls against the door of the
tomb. That gloomy thought will collect a gloomy multitude, and
throw its complexion over your wakeful hour.

In the depths of every heart, there is a tomb and a dungeon, though the lights, the music, and revelry above may cause us to forget their existence, and the buried ones, or prisoners whom they hide. But sometimes, and oftenest at midnight, those dark receptacles are flung wide open. In an hour like this, when the mind has a passive sensibility, but no active strength; when the imagination is a mirror, imparting vividness to all ideas, without the power of selecting or controlling them; then pray that your griefs may slumber, and the brotherhood of remorse not break their chain. It is too late! A funeral train comes gliding by your bed, in which Passion and Feeling assume bodily shape, and things of the mind become dim spectres to the eye. There is your earliest Sorrow, a pale young mourner, wearing a sister's likeness to first love, sadly beautiful, with a hallowed sweetness in her melancholy features, and grace in the flow of her sable robe. Next appears a shade of ruined loveliness, with dust among her golden hair, and her bright garments all faded and defaced, stealing from your glance with drooping head, as fearful of reproach; she was your fondest Hope, but a delusive one; so call her Disappointment now. A sterner form succeeds, with a brow of wrinkles, a look and gesture of iron authority; there is no name for him unless it be Fatality, an emblem of the evil influence that rules your fortunes; a demon to whom you subjected yourself by some error at the outset of life, and were bound his slave forever, by once obeying him. See! those fiendish lineaments graven on the darkness, the writhed lip of scorn, the mockery of that living eye, the pointed finger, touching the sore place in your heart! Do you remember any act of enormous folly, at which you would blush, even in the remotest cavern of the earth? Then recognize your Shame.

Pass, wretched band! Well for the wakeful one, if, riotously miserable, a fiercer tribe do not surround him, the devils of a guilty heart, that holds its hell within itself. What if Remorse should assume the features of an injured friend? What if the fiend should come in woman's garments, with a pale beauty amid sin and desolation, and lie down by your side? What if he should stand at your bed's foot, in the likeness of a corpse, with a bloody stain upon the shroud? Sufficient without such guilt, is this nightmare of the soul; this heavy, heavy sinking of the spirits; this wintry gloom about the heart; this indistinct horror of the mind, blending itself with the darkness of the chamber.

By a desperate effort, you start upright, breaking from a sort of conscious sleep, and gazing wildly round the bed, as if the fiends were any where but in your haunted mind. At the same moment, the slumbering embers on the hearth send forth a gleam which palely illuminates the whole outer room, and flickers through the door of the bed-chamber, but cannot quite dispel its obscurity. Your eye searches for whatever may remind you of the living world. With eager minuteness, you take note of the table near the fire-place, the book with an ivory knife between its leaves, the unfolded letter, the hat and the fallen glove. Soon the flame vanishes, and with it the whole scene is gone, though its image remains an instant in your mind's eye, when darkness has swallowed the reality. Throughout the chamber, there is the same obscurity as before, but not the same gloom within your breast. As your head falls back upon the pillow, you think—in a whisper be it spoken—how pleasant in these night solitudes, would be the rise and fall of a softer breathing than your own, the slight pressure of a tenderer bosom, the quiet throb of a purer heart, imparting its peacefulness to your troubled one, as if the fond sleeper were involving you in her dream.

Her influence is over you, though she have no existence but in that momentary image. You sink down in a flowery spot, on the borders of sleep and wakefulness, while your thoughts rise before you in pictures, all disconnected, yet all assimilated by a pervading gladsomeness and beauty. The wheeling of gorgeous squadrons, that glitter in the sun, is succeeded by the merriment of children round the door of a school-house, beneath the glimmering shadow of old trees, at the corner of a rustic lane. You stand in the sunny rain of a summer shower, and wander among the sunny trees of an autumnal wood, and look upward at the brightest of all rainbows, over-arching the unbroken sheet of snow, on the American side of Niagara. Your mind struggles pleasantly between the dancing radiance round the hearth of a young man and his recent bride, and the twittering flight of birds in spring, about their new-made nest. You feel the merry bounding of a ship before the breeze; and watch the tuneful feet of rosy girls, as they twine their last and merriest dance, in a splendid ball room; and find yourself in the brilliant circle of a crowded theatre, as the curtain falls over a light and airy scene.

With an involuntary start, you seize hold on consciousness, and prove yourself but half awake, by running a doubtful parallel between human life and the hour which has now elapsed. In both you emerge from mystery, pass through a vicissitude that you can but imperfectly control, and are borne onward to another mystery. Now comes the peal of the distant clock, with fainter and fainter strokes as you plunge farther into the wilderness of sleep. It is the knell of a temporary death. Your spirit has departed, and strays like a free citizen, among the people of a shadowy world, beholding strange sights, yet without wonder or dismay. So calm, perhaps, will be the final change; so undisturbed, as if among familiar things, the entrance of the soul to its Eternal home!

1835

'Bentham and Coleridge'
(*from* Coleridge)

THE name of Coleridge is one of the few English names of our time which are likely to be oftener pronounced, and to become symbolical of more important things, in proportion as the inward workings of the age manifest themselves more and more in outward facts. Bentham excepted, no Englishman of recent date has left his impress so deeply in the opinions and mental tendencies of those among us who attempt to enlighten their practice by philosophical meditation. If it be true, as Lord Bacon affirms, that a knowledge of the speculative opinions of the men between twenty and thirty years of age is the great source of political prophecy, the existence of Coleridge will show itself by no slight or ambiguous traces in the coming history of our country; for no one has contributed more to shape the opinions of those among its younger men, who can be said to have opinions at all.

The influence of Coleridge, like that of Bentham, extends far beyond those who share in the peculiarities of his religious or philosophical creed. He has been the great awakener in this country of the spirit of philosophy, within the bounds of traditional opinions. He has been, almost as truly as Bentham, 'the great questioner of things established'; for a questioner needs not necessarily be an enemy. By Bentham, beyond all others, men have been led to ask themselves, in regard to any ancient or received opinion, Is it true? and by Coleridge, What is the meaning of it? The one took his stand outside the received opinion, and surveyed it as an entire stranger to it: the other looked at it from within, and endeavoured to see it with the eyes of a believer in it; to discover by what apparent facts it was at first suggested, and by what appearances it has ever since been rendered continually credible—has seemed, to a succession of persons, to be a faithful interpretation of their experience.

Bentham judged a proposition true or false as it accorded or not with the result of his own inquiries; and did not search very curiously into what might be meant by the proposition, when it obviously did not mean what he thought true. With Coleridge, on the contrary, the very fact that any doctrine had been believed by thoughtful men, and received by whole nations or generations of mankind, was part of the problem to be solved, was one of the phenomena to be accounted for. And as Bentham's short and easy method of referring all to the selfish interests of aristocracies, or priests, or lawyers, or some other species of impostors, could not satisfy a man who saw so much farther into the complexities of the human intellect and feelings—he considered the long or extensive prevalence of any opinion as a presumption that it was not altogether a fallacy; that, to its first authors at least, it was the result of a struggle to express in words something which had a reality to them, though perhaps not to many of those who have since received the doctrine by mere tradition. The long duration of a belief, he thought, is at least proof of an adaptation in it to some portion or other of the human mind; and if, on digging down to the root, we do not find, as is generally the case, some truth, we shall find some natural want or requirement of human nature which the doctrine in question is fitted to satisfy: among which wants the instincts of selfishness and of credulity have a place, but by no means an exclusive one. From this difference in the points of view of the two philosophers, and from the too rigid adherence of each to his own, it was to be expected that Bentham should continually miss the truth which is in the traditional opinions, and Coleridge that which is out of them, and at variance with them. But it was also likely that each would find, or show the way to finding, much of what the other missed.

It is hardly possible to speak of Coleridge, and his position among his contemporaries, without reverting to Bentham: they are connected by two of the closest bonds of association—resemblance and contrast. It would be difficult to find two persons of philosophic eminence more exactly the contrary of one another. Compare their modes of treatment of any subject, and you might fancy them inhabitants of different worlds. They seem to have scarcely a principle or a premise in common. Each of them sees scarcely anything but what the other does not see. Bentham would have regarded Coleridge with a peculiar measure of the good-humoured contempt with which he was accustomed to regard all modes of philosophizing different from his own. Coleridge would probably have made Bentham one of the exceptions to the enlarged and liberal appreciation which (to the credit of *his* mode of philosophizing) he extended to most thinkers of any eminence, from whom he differed. But contraries, as logicians say, are but *quae in eodem genere maxime distant*, the things which are farthest from one another in the same kind. These two agreed in being the men who, in their age and country, did most to enforce, by precept and example, the necessity of a philosophy. They agreed in making it their occupation to recall opinions to first principles; taking no proposition for granted without examining into the grounds of it, and ascertaining that it possessed the kind and degree of evidence suitable to its nature. They agreed in recognizing that sound theory is the only foundation for sound practice, and that whoever despises theory, let him give himself what airs of wisdom he may, is self-convicted of being a quack. If a book were to be compiled containing all the best things ever said on the rule-of-thumb school of political craftsmanship, and on the insufficiency for practical purposes of what the mere practical man calls experience, it is difficult to say whether the collection would be more indebted to the writings of Bentham or of Coleridge. They agreed, too, in perceiving that the groundwork of all other philosophy must be laid in the philosophy of the mind. To lay this foundation deeply and strongly, and to raise a superstructure in accordance with it, were the objects to which their lives were devoted. They employed, indeed, for the most part, different materials; but as the materials of both were real observations, the genuine product of experience—the results will in the end be found not hostile, but supplementary, to one another. Of their methods of philosophizing, the same thing may be said: they were different, yet both were legitimate logical processes. In every respect the two men are each other's 'completing counterpart': the strong points of each correspond to the weak points of the other. Whoever could master the premises and combine the methods of both, would possess the entire English philosophy of their age. Coleridge used to say that every one is born either a Platonist or an Aristotelian: it may be similarly affirmed, that every Englishman of the present day is by implication either a Benthamite or a Coleridgian; holds views of human affairs which can only be proved true on the principles either of Bentham or of Coleridge.

1840

City of London Churches

IF the confession that I have often travelled from this Covent Garden lodging of mine on Sundays, should give offence to those who never travel on Sundays, they will be satisfied (I hope) by my adding that the journeys in question were made to churches.

Not that I have any curiosity to hear powerful preachers. Time was, when I was dragged by the hair of my head, as one may say, to hear too many. On summer evenings, when every flower, and tree, and bird, might have better addressed my soft young heart, I have in my day been caught in the palm of a female hand by the crown, have been violently scrubbed from the neck to the roots of the hair as a purification for the Temple, and have then been carried off highly charged with saponaceous electricity, to be steamed like a potato in the unventilated breath of the powerful Boanerges Boiler and his congregation, until what small mind I had, was quite steamed out of me. In which pitiable plight I have been haled out of the place of meeting, at the conclusion of the exercises, and catechised respecting Boanerges Boiler, his fifthly, his sixthly, and his seventhly, until I have regarded that reverend person in the light of a most dismal and oppressive Charade. Time was, when I was carried off to platform assemblages at which no human child, whether of wrath or grace, could possibly keep its eyes open, and when I felt the fatal sleep stealing, stealing over me, and when I gradually heard the orator in possession, spinning and humming like a great top, until he rolled, collapsed, and tumbled over, and I discovered to my burning shame and fear, that as to that last stage it was not he, but I. I have sat under Boanerges when he has specifically addressed himself to us—us, the infants—and at this present writing I hear his lumbering jocularity (which never amused us, though we basely pretended that it did), and I behold his big round face, and I look up the inside of his outstretched coatsleeve as if it were a telescope with the stopper on, and I hate him with an unwholesome hatred for two hours. Through such means did it come to pass that I knew the powerful preacher from beginning to end, all over and all through, while I was very young, and that I left him behind at an early period of life. Peace be with him! More peace than he brought to me!

Now, I have heard many preachers since that time—not powerful; merely Christian, unaffected, and reverential—and I have had many such preachers on my roll of friends. But, it was not to hear these, any more than the powerful class, that I made my Sunday journeys. They were journeys of curiosity to the numerous churches in the City of London. It came into my head one day, here had I been cultivating a familiarity with all the churches of Rome, and I knew nothing of the insides of the old churches of London! This befel on a Sunday morning. I began my expeditions that very same day, and they lasted me a year.

I never wanted to know the names of the churches to which I went, and to this hour I am profoundly ignorant in that particular of at least nine-tenths of them. Indeed, saving that I know the church of old Gower's tomb (he lies in effigy with his head upon his books) to be the church of Saint Saviour's, Southwark; and the church of Milton's tomb to be the church of Cripplegate; and the church on Cornhill with the great golden keys to be the church of Saint Peter; I doubt if I could pass a competitive examination in any of the names. No question did I ever ask of living creature concerning these churches, and no answer to any antiquarian question on the subject that I ever put to books, shall harass the reader's soul. A full half of my pleasure in them arose out of their mystery; mysterious I found them; mysterious they shall remain for me.

Where shall I begin my round of hidden and forgotten old churches in the City of London?

It is twenty minutes short of eleven on a Sunday morning, when I stroll down one of the many narrow hilly streets in the City that tend due south to the Thames. It is my first experiment, and I have come to the region of Whittington in an omnibus, and we have put down a fierce-eyed spare old woman, whose slate-coloured gown smells of herbs, and who walked up

Aldersgate-street to some chapel where she comforts herself with
brimstone doctrine, I warrant. We have also put down a stouter
and sweeter old lady, with a pretty large prayer-book in an
unfolded pocket-handkerchief, who got out at a corner of a
court near Stationers' Hall, and who I think must go to church
there, because she is the widow of some deceased Old Company's Beadle. The rest of our freight were mere chance
pleasure-seekers and rural walkers, and went on to the Blackwall
railway. So many bells are ringing, when I stand undecided at a
street corner, that every sheep in the ecclesiastical fold might be
a bell-wether. The discordance is fearful. My state of indecision
is referable to, and about equally divisible among, four great
churches, which are all within sight and sound, all within the
space of a few square yards. As I stand at the street corner, I
don't see as many as four people at once going to church, though
I see as many as four churches with their steeples clamouring for
people. I choose my church, and go up the flight of steps to the
great entrance in the tower. A mouldy tower within, and like a
neglected washhouse. A rope comes through the beamed roof,
and a man in the corner pulls it and clashes the bell—a whity-
brown man, whose clothes were once black—a man with flue on
him, and cobweb. He stares at me, wondering how I come there,
and I stare at him, wondering how he comes there. Through a
screen of wood and glass, I peep into the dim church. About
twenty people are discernible, waiting to begin. Christening
would seem to have faded out of this church long ago, for the
font has the dust of desuetude thick upon it, and its wooden
cover (shaped like an old-fashioned tureen-cover) looks as if it
wouldn't come off, upon requirement. I perceive the altar to be
rickety, and the Commandments damp. Entering after this survey, I jostle the clergyman in his canonicals, who is entering too
from a dark lane behind a pew of state with curtains, where
nobody sits. The pew is ornamented with four blue wands, once
carried by four somebodys, I suppose, before somebody else, but
which there is nobody now to hold or receive honour from. I
open the door of a family pew, and shut myself in; if I could
occupy twenty family pews at once, I might have them. The
clerk, a brisk young man (how does *he* come here?), glances at
me knowingly, as who should say, 'You have done it now; you
must stop.' Organ plays. Organ-loft is in a small gallery across

the church; gallery congregation, two girls. I wonder within myself what will happen when we are required to sing.

There is a pale heap of books in the corner of my pew, and while the organ, which is hoarse and sleepy, plays in such fashion that I can hear more of the rusty working of the stops than of any music, I look at the books, which are mostly bound in faded baize and stuff. They belonged in 1754, to the Dowgate family; and who were they? Jane Comport must have married Young Dowgate, and come into the family that way; Young Dowgate was courting Jane Comport when he gave her her prayer-book, and recorded the presentation in the flyleaf; if Jane were fond of Young Dowgate, why did she die and leave the book here? Perhaps at the rickety altar, and before the damp Commandments, she, Comport, had taken him, Dowgate, in a flush of youthful hope and joy, and perhaps it had not turned out in the long run as great a success as was expected?

The opening of the service recals my wandering thoughts. I then find, to my astonishment, that I have been, and still am, taking a strong kind of invisible snuff, up my nose, into my eyes, and down my throat. I wink, sneeze, and cough. The clerk sneezes; the clergyman winks; the unseen organist sneezes and coughs (and probably winks); all our little party wink, sneeze, and cough. The snuff seems to be made of the decay of matting, wood, cloth, stone, iron, earth, and something else. Is the something else, the decay of dead citizens in the vaults below? As sure as Death it is! Not only in the cold damp February day, do we cough and sneeze dead citizens, all through the service, but dead citizens have got into the very bellows of the organ, and half choked the same. We stamp our feet to warm them, and dead citizens arise in heavy clouds. Dead citizens stick upon the walls, and lie pulverised on the sounding-board over the clergy-man's head, and, when a gust of air comes, tumble down upon him.

In this first experience I was so nauseated by too much snuff, made of the Dowgate family, the Comport branch, and other families and branches, that I gave but little heed to our dull man-ner of ambling through the service; to the brisk clerk's manner of encouraging us to try a note or two at psalm time; to the gallery-congregation's manner of enjoying a shrill duet, without a notion of time or tune; to the whity-brown man's manner of shutting the minister into the pulpit, and being very particular with the lock of the door, as if he were a dangerous animal. But, I tried again next Sunday, and soon accustomed myself to the dead citizens when I found that I could not possibly get on without them among the City churches.

Another Sunday.

After being again rung for by conflicting bells, like a leg of mutton or a laced hat a hundred years ago, I make selection of a church oddly put away in a corner among a number of lanes—a smaller church than the last, and an ugly: of about the date of Queen Anne. As a congregation, we are fourteen strong: not counting an exhausted charity school in a gallery, which has dwindled away to four boys, and two girls. In the porch, is a benefaction of loaves of bread, which there would seem to be nobody left in the exhausted congregation to claim, and which I saw an exhausted beadle, long faded out of uniform, eating with his eyes for self and family when I passed in. There is also an exhausted clerk in a brown wig, and two or three exhausted doors and windows have been bricked up, and the service books are musty, and the pulpit cushions are threadbare, and the whole of the church furniture is in a very advanced stage of exhaustion. We are three old women (habitual), two young lovers (accidental), two tradesmen, one with a wife and one alone, an aunt and nephew, again two girls (these two girls dressed out for church with everything about them limp that should be stiff, and *vice versâ*, are an invariable experience), and three sniggering boys. The clergyman is, perhaps, the chaplain of a civic company; he has the moist and vinous look, and eke the bulbous boots, of one acquainted with 'Twenty port, and comet vintages.

We are so quiet in our dulness that the three sniggering boys, who have got away into a corner by the altar-railing, give us a start, like crackers, whenever they laugh. And this reminds me of my own village church where, during sermon-time on bright Sundays when the birds are very musical indeed, farmers' boys patter out over the stone pavement, and the clerk steps out from his desk after them, and is distinctly heard in the summer repose to pursue and punch them in the churchyard, and is seen to return with a meditative countenance, making believe that nothing of the sort has happened. The aunt and nephew in this City church are much disturbed by the sniggering boys. The nephew is himself a boy, and the sniggerers tempt him to secular thoughts of marbles and string, by secretly offering such commodities to his distant contemplation. This young Saint Anthony for a while resists, but presently becomes a backslider, and in dumb show defies the sniggerers to 'heave' a marble or two in his direction. Herein he is detected by the aunt (a rigorous reduced gentlewoman who has the charge of offices), and I perceive that worthy relative to poke him in the side, with the corrugated hooked handle of an ancient umbrella. The nephew revenges himself for this, by holding his breath and terrifying his kinswoman with the dread belief that he has made up his mind to burst. Regardless of whispers and shakes, he swells and becomes discoloured, and yet again swells and becomes discoloured, until the aunt can bear it no longer, but leads him out, with no visible neck, and with his eyes going before him like a prawn's. This causes the sniggerers to regard flight as an eligible move, and I know which of them will go out first, because of the over-devout attention that he suddenly concentrates on the clergyman. In a little while, this hypocrite, with an elaborate demonstration of hushing his footsteps, and with a face generally expressive of having until now forgotten a religious appointment elsewhere, is gone. Number two gets out in the same way, but rather quicker. Number three getting safely to the door, there turns reckless, and banging it open, flies forth with a Whoop! that vibrates to the top of the tower above us.

The clergyman, who is of a prandial presence and a muffled voice, may be scant of hearing as well as of breath, but he only glances up, as having an idea that somebody has said Amen in a wrong place, and continues his steady jog-trot, like a farmer's wife going to market. He does all he has to do, in the same easy way, and gives us a concise sermon, still like the jog-trot of the farmer's wife on a level road. Its drowsy cadence soon lulls the three old women asleep, and the unmarried tradesman sits looking out at window, and the married tradesman sits looking at his wife's bonnet, and the lovers sit looking at one another, so superlatively happy, that I mind when I, turned of eighteen, went with my Angelica to a City church on account of a shower (by this special coincidence that it was in Huggin-lane), and when I said to my Angelica, 'Let the blessed event, Angelica, occur at no altar but this!' and when my Angelica consented that it should occur at no other—which it certainly never did, for it never occurred anywhere. And O, Angelica, what has become of you, this present Sunday morning when I can't attend to the sermon; and, more difficult question than that, what has become of Me as I was when I sat by your side!

But, we receive the signal to make that unanimous dive which surely is a little conventional—like the strange rustlings and settlings and clearings of throats and noses, which are never dispensed with, at certain points of the Church service, and are never held to be necessary under any other circumstances. In a minute more it is all over, and the organ expresses itself to be as glad of it as it can be of anything in its rheumatic state, and in another minute we are all of us out of the church, and Whity-brown has locked it up. Another minute or little more, and, in the neighbouring churchyard—not the yard of that church, but of another—a churchyard like a great shabby old mignonette-box, with two trees in it and one tomb—I meet Whity-brown, in his private capacity, fetching a pint of beer for his dinner from the public-house in the corner, where the keys of the rotting fire-ladders are kept and were never asked for, and where there is a ragged, white-seamed, out-at-elbowed bagatelle-board on the first floor.

In one of these City churches, and only in one, I found an individual who might have been claimed as expressly a City personage. I remember the church, by the feature that the clergyman couldn't get to his own desk without going through the clerk's, or couldn't get to the pulpit without going through the reading-desk—I forget which, and it is no matter—and by the presence of this personage among the exceedingly sparse congregation. I doubt if we were a dozen, and we had no exhausted charity school to help us out. The personage was dressed in black of square cut, and was stricken in years, and wore a black velvet cap, and cloth shoes. He was of a staid, wealthy, and dissatisfied aspect. In his hand, he conducted to church a mysterious child: a child of the feminine gender. The child had a beaver hat, with a stiff drab plume that surely never belonged to any bird of the air. The child was further attired in a nankeen frock and spencer, brown boxing-gloves, and a veil. It had a blemish, in the nature of currant jelly, on its chin; and was a thirsty child. Insomuch that the personage carried in his pocket a green bottle, from which, when the first psalm was given out, the child was openly refreshed. At all other times throughout the service it was motionless, and stood on the seat of the large pew, closely fitted into the corner, like a rain-water pipe.

The personage never opened his book, and never looked at the clergyman. *He* never sat down either, but stood with his arms leaning on the top of the pew, and his forehead sometimes shaded with his right hand, always looking at the church door. It was a long church for a church of its size, and he was at the upper end, but he always looked at the door. That he was an old book-keeper, or an old trader who had kept his own books, and that he might be seen at the Bank of England about Dividend times, no doubt. That he had lived in the City all his life and was disdainful of other localities, no doubt. Why he looked at the door, I never absolutely proved, but it is my belief that he lived in expectation of the time when the citizens would come back to live in the City, and its ancient glories would be renewed. He appeared to expect that this would occur on a Sunday, and that the wanderers would first appear, in the deserted churches, penitent and humbled. Hence, he looked at the door which they never darkened. Whose child the child was, whether the child of a disinherited daughter, or some parish orphan whom the personage had adopted, there was nothing to lead up to. It never played, or skipped, or smiled. Once, the idea occurred to me that it was an automaton, and that the personage had made it; but following the strange couple out one Sunday, I heard the personage say to it, 'Thirteen thousand pounds'; to which it added in a weak human voice, 'Seventeen and fourpence.' Four Sundays, I followed them out, and this is all I ever heard or saw them say. One Sunday, I followed them home. They lived behind a pump, and the personage opened their abode with an exceeding large key. The one solitary inscription on their house related to a fire-plug. The house was partly undermined by a deserted and closed gateway; its windows were blind with dirt; and it stood with its face disconsolately turned to a wall. Five great churches and two small ones rang their Sunday bells between this house and the church the couple frequented, so they must have had some special reason for going a quarter of a mile to it. The last time I saw them, was on this wise. I had been to explore another church at a distance, and happened to pass the church they frequented, at about two of the afternoon when that edifice was closed. But, a little side-door, which I had never observed before, stood open, and disclosed certain cellarous steps. Methought, 'They are airing the vaults to-day,' when the personage and the child silently arrived at the steps, and silently descended. Of course, I came to the conclusion that the personage had at last despaired of the looked-for return of the penitent citizens, and that he and the child went down to get themselves buried.

In the course of my pilgrimages I came upon one obscure church which had broken out in the melodramatic style, and was got up with various tawdry decorations, much after the manner of the extinct London maypoles. These attractions had induced several young priests or deacons in black bibs for waistcoats, and several young ladies interested in that holy order (the proportion being, as I estimated, seventeen young ladies to a deacon), to come into the City as a new and odd excitement. It was wonderful to see how these young people played out their little play in the heart of the City, all among themselves, without the deserted City's knowing anything about it. It was as if you should take an empty counting-house on a Sunday, and act one of the old Mysteries there. They had impressed a small school (from what neighbourhood I don't know) to assist in the performances, and it was pleasant to notice frantic garlands of inscription on the walls, especially addressing those poor innocents in characters impossible for them to decipher. There was a remarkably agreeable smell of pomatum in this congregation.

But, in other cases, rot and mildew and dead citizens formed the uppermost scent, while, infused into it in a dreamy way not at all displeasing, was the staple character of the neighbourhood. In the churches about Mark-lane, for example, there was a dry whiff of wheat; and I accidentally struck an airy sample of barley out of an aged hassock in one of them. From Rood-lane to Tower-street, and thereabouts, there was often a subtle flavour of wine: sometimes, of tea. One church near Mincing-lane smelt like a druggist's drawer. Behind the Monument, the service had a flavour of damaged oranges, which, a little further down towards the river, tempered into herrings, and gradually toned into a cosmopolitan blast of fish. In one church, the exact counterpart of the church in the Rake's Progress where the hero is being married to the horrible old lady, there was no speciality of atmosphere, until the organ shook a perfume of hides all over us from some adjacent warehouse.

Be the scent what it would, however, there was no speciality in the people. There were never enough of them to represent any calling or neighbourhood. They had all gone elsewhere over-night, and the few stragglers in the many churches languished there inexpressively.

Among the uncommercial travels in which I have engaged, this year of Sunday travel occupies its own place, apart from all the rest. Whether I think of the church where the sails of the oyster-boats in the river almost flapped against the windows, or of the church where the railroad made the bells hum as the train rushed by above the roof, I recall a curious experience. On sum-mer Sundays, in the gentle rain or the bright sunshine—either, deepening the idleness of the idle City—I have sat, in that singu-lar silence which belongs to resting-places usually astir, in scores of buildings at the heart of the world's metropolis, unknown to far greater numbers of people speaking the English tongue, than the ancient edifices of the Eternal City, or the Pyramids of Egypt. The dark vestries and registries into which I have peeped, and the little hemmed-in churchyards that have echoed to my feet, have left impressions on my memory as distinct and quaint as any it has in that way received. In all those dusty registers that the worms are eating, there is not a line but made some hearts leap, or some tears flow, in their day. Still and dry now, still and dry! and the old tree at the window with no room for its branches, has seen them all out. So with the tomb of the old Master of the old Company, on which it drips. His son restored it and died, his daughter restored it and died, and then he had been remembered long enough, and the tree took pos-session of him, and his name cracked out.

There are few more striking indications of the changes of manners and customs that two or three hundred years have brought about, than these deserted Churches. Many of them are handsome and costly structures, several of them were designed by Wren, many of them arose from the ashes of the great fire, others of them outlived the plague and the fire too, to die a slow death in these later days. No one can be sure of the coming time; but it is not too much to say of it that it has no sign in its outsetting tides, of the reflux to these churches of their congre-

gations and uses. They remain like the tombs of the old citizens
who lie beneath them and around them, Monuments of another
age. They are worth a Sunday-exploration now and then, for
they yet echo, not unharmoniously, to the time when the city of
London really was London; when the 'Prentices and Trained
Bands were of mark in the state; when even the Lord Mayor
himself was a Reality——not a Fiction conventionally be-puffed on
one day in the year by illustrious friends, who no less conventionally laugh at him on the remaining three hundred and sixty-
four days.

1860

Autour de mon Chapeau

NEVER have I seen a more noble tragic face. In the centre of the forehead there was a great furrow of care, towards which the brows rose piteously. What a deep solemn grief in the eyes! They looked blankly at the object before them, but through it, as it were, and into the grief beyond. In moments of pain, have you not looked at some indifferent object so? It mingles dumbly with your grief, and remains afterwards connected with it in your mind. It may be some indifferent thing—a book which you were reading at the time when you received her farewell letter (how well you remember the paragraph afterwards—the shape of the words, and their position on the page); the words you were writing when your mother came in, and said it was all over—she was MARRIED—Emily married—to that insignificant little rival at whom you have laughed a hundred times in her company. Well, well; my friend and reader, whoe'er you be—old man or young, wife or maiden—you have had your grief-pang. Boy, you have lain awake the first night at school, and thought of home. Worse still, man, you have parted from the dear ones with bursting heart; and, lonely boy, recall the bolstering an unfeeling comrade gave you; and, lonely man, just torn from your children—their little tokens of affection yet in your pocket—pacing the deck at evening in the midst of the roaring ocean, you can remember how you were told that supper was ready, and how you went down to the cabin and had brandy-and-water and biscuit. You remember the taste of them. Yes; for ever. You took them whilst you and your Grief were sitting together, and your Grief clutched you round the soul. Serpent, how you have writhed round me, and bitten me! Remorse, Remembrance, etc, come in the night season, and I feel you gnawing, gnawing! . . . I tell you that man's face was like Laocoon's (which, by the way, I always think over-rated. The real head is at Brussels, at the Duke Daremberg's, not at Rome).

That man! What man? That man of whom I said that this magnificent countenance exhibited the noblest tragic woe. He was not of European blood. He was handsome, but not of European beauty. His face white—not a Northern whiteness; his eyes protruding somewhat, and rolling in their grief. Those eyes had seen the Orient sun, and his beak was the eagle's. His lips were full. The beard, curling round them, was unkempt and tawny. The locks were of a deep, deep coppery red. The hands, swart and powerful, accustomed to the rough grasp of the wares in which he dealt, seemed unused to the flimsy artifices of the bath. He came from the Wilderness, and its sands were on his robe, his cheek, his tattered sandal, and the hardy foot it covered.

And his grief—whence came his sorrow? I will tell you. He bore it in his hand. He had evidently just concluded the compact by which it became his. His business was that of a purchaser of domestic raiment. At early dawn—nay, at what hour when the city is alive—do we not all hear the nasal cry of 'Clo'? In Paris, *Habits, Galons, Marchand d'habits*, is the twanging signal with which the wandering merchant makes his presence known. It was in Paris I saw this man. Where else have I not seen him? In the Roman Ghetto—at the Gate of David, in his fathers' once imperial city. The man I mean was an itinerant vendor and purchaser of wardrobes—what you call an ... Enough! You know his name.

On his left shoulder hung his bag; and he held in that hand a white hat, which I am sure he had just purchased, and which was the cause of the grief which smote his noble features. Of course I cannot particularize the sum, but he had given too much for that hat. He felt he might have got the thing for less money. It was not the amount, I am sure; it was the principle involved. He had given fourpence (let us say) for that which threepence would have purchased. He had been done: and a manly shame was upon him, that he, whose energy, acuteness, experience, point of hon-our, should have made him the victor in any mercantile duel in which he should engage, had been overcome by a porter's wife, who very likely sold him the old hat, or by a student who was tired of it. I can understand his grief. Do I seem to be speaking of it in a disrespectful or flippant way? Then you mistake me. He had been outwitted. He had desired, coaxed, schemed, haggled, got what he wanted, and now found he had paid too much for his bargain. You don't suppose I would ask you to laugh at that man's grief? It is you, clumsy cynic, who are disposed to sneer, whilst it may be tears of genuine sympathy are trickling down this nose of mine. What do you mean by laughing? If you saw a wounded soldier on the field of battle, would you laugh? If you saw a ewe robbed of her lamb, would you laugh, you brute? It is you who are the cynic, and have no feeling: and you sneer because that grief is unintelligible to you which touches my finer sensibility. The OLD-CLOTHES'-MAN had been defeated in one of the daily battles of his most interesting, chequered, adventurous life.

Have you ever figured to yourself what such a life must be? The pursuit and conquest of twopence must be the most eager and fascinating of occupations. We might all engage in that business if we would. Do not whist-players, for example, toil, and think, and lose their temper over sixpenny points? They bring study, natural genius, long forethought, memory, and careful historical experience to bear upon their favourite labour. Don't tell me that it is the sixpenny points, and five shillings the rub, which keeps them for hours over their painted pasteboard. It is the desire to conquer. Hours pass by. Night glooms. Dawn, it may be, rises unheeded; and they sit calling for fresh cards at the 'Portland,' or the 'Union,' while waning candles splutter in the sockets, and languid waiters snooze in the ante-room. Sol rises. Jones has lost four pounds; Brown has won two; Robinson lurks away to his family house and (mayhap indignant) Mrs R. Hours of evening, night, morning, have passed away whilst they have been waging this sixpenny battle. What is the loss of four pounds to Jones, the gain of two to Brown? B. is, perhaps, so rich that two pounds more or less are as naught to him; J. is so hopelessly involved that to win four pounds cannot benefit his creditors, or alter his condition; but they play for that stake: they put forward their best energies: they ruff, finesse (what are the technical words, and how do I know?) It is but a sixpenny game if you like; but they want to win it. So as regards my friend yonder with the hat. He stakes his money: he wishes to win the game, not the hat merely. I am not prepared to say that he is not

inspired by a noble ambition. Casesar wished to be first in a village. If first of a hundred yokels, why not first of two? And my
friend the old-clothes'-man wishes to win his game, as well as to
turn his little sixpence.

Suppose in the game of life—and it is but a twopenny game
after all—you are equally eager of winning. Shall you be
ashamed of your ambition, or glory in it? There are games, too,
which are becoming to particular periods of life. I remember in
the days of our youth, when my friend Arthur Bowler was an
eminent cricketer. Slim, swift, strong, well-built, he presented a
goodly appearance on the ground in his flannel uniform. *Militâsti
non sine gloria*, Bowler my boy! Hush! We tell no tales. Mum is the
word. Yonder comes Charley his son. Now Charley his son has
taken the field, and is famous among the eleven of his school.
Bowler senior, with his capacious waistcoat, etc, waddling after a
ball, would present an absurd object, whereas it does the eyes
good to see Bowler junior scouring the plain—a young exemplar
of joyful health, vigour, activity. The old boy wisely contents
himself with amusements more becoming his age and waist;
takes his sober ride; visits his farm soberly—busies himself about
his pigs, his ploughing, his peaches, or what not? Very small
routinier amusements interest him; and (thank goodness!) nature
provides very kindly for kindly-disposed fogies. We relish those
things which we scorned in our lusty youth. I see the young
folks of an evening kindling and glowing over their delicious
novels. I look up and watch the eager eye flashing down the
page, being, for my part, perfectly contented with my twaddling
old volume of *Howel's Letters* or the *Gentleman's Magazine*. I am
actually arrived at such a calm frame of mind that I like batter-
pudding. I never should have believed it possible; but it is so.
Yet a little while, and I may relish water-gruel. It will be the age
of *mon lait de poule et mon bonnet de nuit*. And then—the cotton
extinguisher is pulled over the old noddle, and the little flame of
life is popped out.

Don't you know elderly people who make learned notes in
Army Lists, Peerages, and the like? This is the batter-pudding,
water-gruel of old age. The worn-out old digestion does not care
for stronger food. Formerly it could swallow twelve-hours'
tough reading, and digest an encyclopaedia.

If I had children to educate, I would, at ten or twelve years of age, have a professor, or professoress, of whist for them, and cause them to be well grounded in that great and useful game. You cannot learn it well when you are old, any more than you can learn dancing or billiards. In our house at home we young-sters did not play whist because we were dear obedient children, and the elders said playing at cards was 'a waste of time.' A waste of time, my good people! *Allons!* What do elderly home-keeping people do of a night after dinner? Darby gets his newspaper; my dear Joan her *Missionary Magazine* or her volume of Cumming's Sermons—and don't you know what ensues? Over the arm of Darby's arm-chair the paper flutters to the ground unheeded, and he performs the trumpet obbligato *que vous savez* on his old nose. My dear old Joan's head nods over her sermon (awakening though the doctrine may be). Ding, ding, ding: can that be ten o'clock? It is time to send the servants to bed, my dear—and to bed master and mistress go too. But they have not wasted their time playing at cards. Oh, no! I belong to a Club where there is whist of a night; and not a little amusing is it to hear Brown speak of Thompson's play, and *vice versâ*. But there is one man —Greatorex let us call him—who is the acknowledged captain and primus of all the whist players. We all secretly admire him. I, for my part, watch him in private life, hearken to what he says, note what he orders for dinner, and have that feeling of awe for him that I used to have as a boy for the cock of the school. Not play at whist? *'Quelle triste vieillesse vous vous préparez!'* were the words of the great and good Bishop of Autun. I can't. It is too late now. Too late! too late! Ah! humiliating confession! That joy might have been clutched, but the life-stream has swept us by it—the swift life-stream rushing to the nearing sea. Too late! too late! Twentystone my boy! When you read in the papers 'Valse à deux temps,' and all the fashionable dances taught to adults by 'Miss Lightfoots,' don't you feel that you would like to go in and learn? Ah, it is too late! You have passed the *choreas*, Master Twentystone, and the young people are dancing without you.

I don't believe much of what my Lord Byron the poet says; but when he wrote, 'So, for a good old gentlemanly vice, I think I shall put up with avarice,' I think his lordshop meant what he wrote, and if he practised what he preached, shall not quarrel with him. As an occupation in declining years, I declare I think saving is useful, amusing, and not unbecoming. It must be a perpetual amusement. It is a game that can be played by day, by night, at home and abroad, and at which you must win in the long run. I am tired and want a cab. The fare to my house, say, is two shillings. The cabman will naturally want half-a-crown. I pull out my book. I show him the distance is exactly three miles and fifteen hundred and ninety yards. I offer him my card—my winning card. As he retires with the two shillings, blaspheming inwardly, every curse is a compliment to my skill. I have played him and beat him; and a sixpence is my spoil and just reward. This is a game, by the way, which women play far more cleverly than we do. But what an interest it imparts to life! During the whole drive home I know I shall have my game at the journey's end; am sure of my hand, and shall beat my adversary. Or I can play in another way. I won't have a cab at all, I will wait for the omnibus: I will be one of the damp fourteen in that steaming vehicle. I will wait about in the rain for an hour, and 'bus after 'bus shall pass, but I will not be beat. I *will* have a place, and get it at length, with my boots wet through, and an umbrella dripping between my legs. I have a rheumatism, a cold, a sore throat, a sulky evening—a doctor's bill to-morrow perhaps? Yes, but I have won my game, and am a gainer of a shilling on this rubber.

If you play this game all through life it is wonderful what daily interest it has, and amusing occupation. For instance, my wife goes to sleep after dinner over her volume of sermons. As soon as the dear soul is sound asleep, I advance softly and puff out her candle. Her pure dreams will be all the happier without that light; and, say she sleeps an hour, there is a penny gained.

As for clothes, *parbleu!* there is not much money to be saved in clothes, for the fact is, as a man advances in life—as he becomes an *Ancient Briton* (mark the pleasantry)—he goes without clothes. When my tailor proposes something in the way of a change of raiment, I laugh in his face. My blue coat and brass buttons will last these ten years. It is seedy? What then? I don't want to charm any body in particular. You say that my clothes are shabby? What do I care? When I wished to look well in somebody's eyes, the matter may have been different. But now, when I receive my bill of £10 (let us say) at the year's end, and contrast it with old tailors' reckonings, I feel that I have played the game with master tailor, and beat him; and my old clothes are a token of the victory.

I do not like to give servants board wages, though they are cheaper than household bills; but I know they save out of board wages; and so beat me. This shows that it is not the money but the game which interests me. So about wine. I have it good and dear. I will trouble you to tell me where to get it good and cheap. You may as well give me the address of a shop where I can buy meat for fourpence a pound, or sovereigns for fifteen shillings apiece. At the game of auctions, docks, shy wine-merchants, depend on it there is *no* winning; and I would as soon think of buying jewellery at an auction in Fleet Street as of pur-chasing wine from one of your dreadful needy wine-agents such as infest every man's door. Grudge myself good wine? As soon grudge my horse corn. *Merci!* that would be a very losing game indeed, and your humble servant has no relish for such.

But in the very pursuit of saving there must be a hundred harmless delights and pleasures which we who are careless necessarily forego. What do you know about the natural history of your household? Upon your honour and conscience, do you know the price of a pound of butter? Can you say what sugar costs, and how much your family consumes and ought to con-sume? How much lard do you use in your house? As I think on these subjects I own I hang down the head of shame. I suppose for a moment that you, who are reading this, are a middle-aged gentleman, and paterfamilias. Can you answer the above ques-tions? You know, sir, you cannot. Now turn round, lay down the book, and suddenly ask Mrs Jones and your daughters if *they* can answer? They cannot. They look at one another. They pre-tend they can answer. They can tell you the plot and principal characters of the last novel. Some of them know something about history, geology, and so forth. But of the natural history of home—*Nichts*, and for shame on you all! *Honnis soyez!* For shame on you? for shame on us!

In the early morning I hear a sort for call or *jodel* under my window: and know 'tis the matutinal milkman leaving his can at my gate. O household gods! have I lived all these years and don't know the price or the quantity of the milk which is delivered in that can? Why don't I know? As I live, if I live till to-morrow morning, as soon as I hear the call of Lactantius, I will dash out upon him. How many cows? How much milk, on an average, all the year round? What rent? What cost of food and dairy servants? What loss of animals, and average cost of purchase? If I interested myself properly about my pint (or hogshead, whatever it be) of milk, all this knowledge would ensue; all this additional interest in life. What is this talk of my friend, Mr Lewes, about objects at the seaside, and so forth? Objects at the seaside? Objects at the area-bell: objects before my nose: objects which the butcher brings me in his tray: which the cook dresses and puts down before me, and over which I say grace! My daily life is surrounded with objects which ought to interest me. The pudding I eat (or refuse, that is neither here nor there: and, between ourselves, what I have said about batter-pudding may be taken *cum grano*—we are not come to *that* yet, except for the sake of argument or illustration)—the pudding, I say, on my plate, the eggs that made it, the fire that cooked it, the tablecloth on which it is laid, and so forth—are each and all of these objects a knowledge of which I may acquire—a knowledge of the cost and production of which I might advantageously learn? To the man who *does* know these things, I say the interest of life is prodigiously increased. The milkman becomes a study to him; the baker a being he curiously and tenderly examines. Go, Lewes, and clap a hideous sea-anemone into a glass: I will put a cabman under mine, and make a vivisection of a butcher. O Lares, Penates, and gentle household gods, teach me to sympathize with all that comes within my doors! Give me an interest in the butcher's book. Let me look forward to the ensuing number of the grocer's account with eagerness. It seems ungrateful to my kitchen-chimney not to know the cost of sweeping it; and I trust that many a man who reads this, and muses on it, will feel, like the writer, ashamed of himself, and hang down his head humbly.

Now, if to this household game you could add a little money interest, the amusement would be increased far beyond the mere money value, as a game at cards for sixpence is better than a rubber for nothing. If you can interest yourself about sixpence, all life is invested with a new excitement. From sunrise to sleeping you can always be playing that game—with butcher, baker, coal-merchant, cabman, omnibus-man—nay, diamond-merchant and stockbroker. You can bargain for a guinea over the price of a diamond necklace, or for a sixteenth per cent in a transaction at the Stock Exchange. We all know men who have this faculty who are not ungenerous with their money. They give it on great occasions. They are more able to help than you and I who spend ours, and say to poor Prodigal who comes to us out at elbow, 'My dear fellow, I should have been delighted: but I have already anticipated my quarter, and am going to ask Screwby if he can do anything for me.'

In this delightful, wholesome, ever-novel twopenny game, there is a danger of excess, as there is in every other pastime or occupation, of life. If you grow too eager for your twopence, the acquisition or the loss of it may affect your peace of mind, and peace of mind is better than any amount of twopences. My friend, the old-clothes'-man, whose agonies over the hat have led to this rambling disquisition, has, I very much fear, by a too eager pursuit of small profits, disturbed the equanimity of a mind that ought to be easy and happy. 'Had I stood out,' he thinks, 'I might have had the hat for threepence,' and he doubts whether, having given fourpence for it, he will ever get back his money. My good Shadrach, if you go through life passionately deploring the irrevocable, and allow yesterday's transactions to embitter the cheerfulness of to-day and to-morrow—as lief walk down to the Seine, souse in, hats, body, clothes-bag and all, and put an end to your sorrow and sordid cares. Before and since Mr Franklin wrote his pretty apologue of the Whistle, have we not all made bargains of which we repented, and coveted and acquired objects for which we have paid too dearly? Who has not purchased his hat in some market or other? There is General M'Clellan's cocked-hat for example: I daresay he was eager enough to wear it, and he has learned that it is by no means cheerful wear. There were the military beavers of Messeigneurs of Orleans:[1] they wore them gallantly in the face of battle; but I suspect they were glad enough to pitch them into the James River and come home in mufti. Ah, *mes amis! à chacun son schakot!* I was looking at a bishop the other day, and thinking, 'My right reverend lord, that broad brim and rosette must bind your great broad forehead very tightly, and give you many a headache. A good easy wideawake were better for you, and I would like to see that honest face with a cutty pipe in the middle of it.' There is my Lord Mayor. My once dear lord, my kind friend, when

---

[1] Two cadets of the House of Orleans who served as Volunteers under General M'Clellan in his campaign against Richmond.

your two years' reign was over, did not you jump for joy and fling your chapeau-bras out of window: and hasn't *that* hat cost you a pretty bit of money? There, in a splendid travelling chariot, in the sweetest bonnet, all trimmed with orange-blossoms and Chantilly lace, sits my Lady Rosa, with old Lord Snowden by her side. Ah, Rosa! what a price have you paid for that hat which you wear; and is your ladyship's coronet not purchased too dear? Enough of hats. Sir, or Madam, I take off mine, and salute you with profound respect.

1863

ANTHONY TROLLOPE

The Plumber

THE plumber, painter and glazier of our youth has disappeared, and in lieu of him has come up the man who mends our kitchen furniture and destroys our roofs. Such, at least, is the reputation which our friend the plumber enjoys. We do not say that of our own knowledge it is deserved. We do not profess to declare that he plans the perforation of our leads. We cannot so far condemn the man who continually haunts our premises, and whose half-yearly bill is of all our torments the most regular, bearing a proportion to our rent which we should have regarded as formidable had we anticipated the necessity of these periodical visits. The plumber should be put down with the tax-gatherer as being as certain as fate and as inexorable,—almost as serious. You shall put your house into excellent order and think to have seen that the last of him for years; but he will be there again till the sight of him is a perpetual eyesore to you. You will come to have an unnatural hatred for the man and his myrmidons. He leaves nothing behind for you to eat as does the butcher, nothing to wear as does the tailor, nothing to delight you;—nothing, finally, in which you may exult among your acquaintance. Whoever spoke among his friends of his plumber, or boasted of his intimacy with that dark, silent, seemingly sullen man, who comes so frequently and on his coming has nothing to say for himself? The plumber is doubtless aware that he is odious. He feels himself, like Dickens's turnpike-man, to be the enemy of mankind. He has probably a wife at home and pretty children whom he fondles; but you, as you look at him, believe him to be alone in the world, and fancy him to be a man unblessed. How can one so saturnine press a wife to his bosom, or participate in the infantine gambols of children? Meet him in the street, and he does not, as your baker does, meet your eye with a half-ready and half-humble smile of acknowledgment. He walks by in silence, apparently engaged as to his thoughts in plotting some infernal hole among the roofs, or arranging for a future catastrophe with the water-pipes. You pass on, taking no heed of the obdurate sinner, but you turn him and his deeds over in your mind, and thank the Lord that in arranging for you your lot in life he did not make you a plumber.

We are far from saying that such is the true character of the man. We remember in a romance the story of one who was presumed to have made himself abominable to all his fellow-creatures. He was an executioner, and as such lived a miserable, a solitary, and a despised life. But he was in truth a general benefactor of the human race, and spent his whole time in doing magnificent deeds as to which he was content that the whole world should be ignorant of their existence. The nature of the mystery need not be explained here, but such was the fate of this hero. We have sometimes thought how possible it may be that our plumber is like that executioner, only that his mystery may be more easily solved. Can it be that he is really engaged in mending, according to the just rules of his trade, those leakages among the leads, in conquering those fugitive smells, in stopping those pernicious runnings of water, in reducing to order those rebellious bells, in ridding our rooms of the smoke that will not fly upwards, and that he does all this to the best of his ability with true workman-like assiduity and conscience? And he must be aware at the time of what the world is saying of him. That he is aware we are very certain. His looks betray him. We are perfectly sure that he knows the doom that has been pronounced against him. There is a something in his gait, in his manner, in his sullen indifference to all remonstrance, which assures us that it is so. He cannot have been about our house so often and have known so little of us, have been so little intimate with us or with our servants, have smiled so seldom,—seldom! nay, never,—had he not been the malevolent influence, as which the world regards him; or else that hero of romance, that unknown, mistaken, long-suffering, patriotic man, whom the world has conspired to condemn, but who knows himself to be pure.

On this subject we ourselves offer no opinion. We take the man simply as we find him, and leave the doubt to be decided according to the various idiosyncrasies of our readers. The man, though he be an angel of light, is undoubtedly a pest about our house; and, as far as our observation goes, the nearer to your street is that in which he lives the greater and more frequent becomes the annoyance. If perchance you live in the country, far removed from the resort of plumbers, where molten lead is a thing almost unknown, you shall hardly hear of him; and yet you live. You are neither killed by the smells, nor drowned by the water, nor destroyed by the weather. If once in three years a man pays you a visit from the neighbouring town, six or seven miles distant, he does what he has to do at one coming, and then departs. There is no time for him to leave an impression on your mind that he is your especial enemy. But in London he is to you as the skeleton in the cupboard, as the invisible guest who is present at all your feasts.

Needless to deny that the normal London plumber is a dishonest man. We do not even allow ourselves to think so. That question, as to the dishonesty of mankind generally, is one that disturbs us greatly;—whether a man in all grades of life will by degrees train his honesty to suit his own book, so that the course of life which he shall bring himself to regard as soundly honest shall, if known to his neighbours, subject him to their reproof. We own to a doubt whether the honesty of a bishop would shine bright as the morning star to the submissive ladies who now worship him, if the theory of life upon which he lives were understood by them in all its bearings. The Prime Minister with his Cabinet of compromising men would hardly do so to the eager politician who has welcomed him at the hustings, but who has himself never been called upon to compromise. And what of the honesty of the barrister, with his high feelings and noble sentiments in private life, whom we do not here condemn for taking a fee which he cannot earn, but who at any rate holds forth a brilliant example to men below him as to earning money, of which the men below him will not be slow to avail themselves?

Are we to expect much in the way of true honesty from the poor plumber, who has to live, and who thinks it to be the first duty of mankind to do so,—his first duty to take care that his wife and children shall have the means of living? When it comes in his way to make a complete job among your water-pipes,—a job so complete that neither he nor any of his confraternity shall be wanted there for the next five years,—is he to be expected to

make that job perfect, or to look forward for his wife and children, and perceive that for their sake, and for the sake of trade in
general,—by which he means the welfare of the wife and children of other plumbers,—he is bound to see that due preparation
may be made for another coming? As the barrister and the Cabinet Minister and the bishop think that they are, *per se*, blessings
of which a people and a country can hardly have too much, and
can hardly know too little as to their methods of work, why
should it not also be so with our friend the plumber? Is it
altogether impossible that he, too, should have taught himself to
think that a household can never be happy unless he be tapping
with his hammer among the leads? 'He makes work for himself,'
says Paterfamilias angrily. 'He absolutely prepares the hole which
shall be perforated through by the next storm. He purposely
arranges that within three months not a bell again shall ring in
the house. He is determined that at periodical intervals you shall
be a Sindbad to him as he sits upon your shoulders.' Then, muttering loud and deep, he pays the unexpected bill. On the next
morning, being by profession a respectable solicitor, he is hard at
work at Lincoln's Inn, paving the way for fresh litigation.

It is very hard to fight the plumber's battles for him, so
notorious are his defalcations and so destructive of our comfort.
It becomes a question to us whether it be possible that a plumber
should go to heaven. But on thinking over all the conditions of
his circumstances, reflecting that the question will have to be
settled by an unerring judgment, we do not see why he should be
debarred if barristers and Cabinet Ministers be allowed to enter.
Sound undeviating honesty, *totus teres atque rotundus*, is, as far as
we can see, to be found exclusively among writers for the press.
But men will surely be admitted on the score of faith if on no
other, even if they die among their lapses. But for the plumber
we earnestly recommend our readers to keep as far as possible
beyond his reach, and to submit rather to all the ills of the Arctic
and torrid zones than to have their patience troubled by the
coming of that much-hated individual.

1880

Night and Moonlight

CHANCING to take a memorable walk by moonlight some years ago, I resolved to take more such walks, and make acquaintance with another side of nature: I have done so.

According to Pliny, there is a stone in Arabia called Selenites, 'wherein is a white, which increases and decreases with the moon.' My journal for the last year or two has been *selenitic* in this sense.

Is not the midnight like Central Africa to most of us? Are we not tempted to explore it,—to penetrate to the shores of its lake Tchad, and discover the source of its Nile, perchance the Mountains of the Moon? Who knows what fertility and beauty, moral and natural, are there to be found? In the Mountains of the Moon, in the Central Africa of the night, there is where all Niles have their hidden heads. The expeditions up the Nile as yet extend but to the Cataracts, or perchance to the mouth of the White Nile; but it is the Black Nile that concerns us.

I shall be a benefactor if I conquer some realms from the night, if I report to the gazettes anything transpiring about us at that season worthy of their attention,—if I can show men that there is some beauty awake while they are asleep,—if I add to the domains of poetry.

Night is certainly more novel and less profane than day. I soon discovered that I was acquainted only with its complexion, and as for the moon, I had seen her only as it were through a crevice in a shutter, occasionally. Why not walk a little way in her light?

Suppose you attend to the suggestions which the moon makes for one month, commonly in vain, will it not be very different from anything in literature or religion? But why not study this Sanskrit? What if one moon has come and gone with its world of poetry, its weird teachings, its oracular suggestions,—so divine a creature freighted with hints for me, and I have not used her? One moon gone by unnoticed?

I think it was Dr Chalmers who said, criticising Coleridge, that for his part he wanted ideas which he could see all round, and not such as he must look at away up in the heavens. Such a man, one would say, would never look at the moon, because she never turns her other side to us. The light which comes from ideas which have their orbit as distant from the earth, and which is no less cheering and enlightening to the benighted traveler than that of the moon and stars, is naturally reproached or nicknamed as moonshine by such. They are moonshine, are they? Well, then do your night-traveling when there is no moon to light you; but I will be thankful for the light that reaches me from the star of least magnitude. Stars are lesser or greater only as they appear to us so. I will be thankful that I see so much as one side of a celestial idea,—one side of the rainbow,—and the sunset sky.

Men talk glibly enough about moonshine, as if they knew its qualities very well, and despised them; as owls might talk of sunshine,—None of your sunshine!—but this word commonly means merely something which they do not understand,—which they are abed and asleep to, however much it may be worth their while to be up and awake to it.

It must be allowed that the light of the moon, sufficient though it is for the pensive walker, and not disproportionate to the inner light we have, is very inferior in quality and intensity to that of the sun. But the moon is not to be judged alone by the quantity of light she sends to us, but also by her influence on the earth and its inhabitants. 'The moon gravitates toward the earth, and the earth reciprocally toward the moon.' The poet who walks by moonlight is conscious of a tide in his thought which is to be referred to lunar influence. I will endeavor to separate the tide in my thoughts from the current distractions of the day. I would warn my hearers that they must not try my thoughts by a daylight standard, but endeavor to realize that I speak out of the night. All depends on your point of view. In Drake's 'Collection of Voyages,' Wafer says of some Albinoes among the Indians of Darien, 'They are quite white, but their whiteness is like that of a horse, quite different from the fair or pale European, as they have not the least tincture of a blush or sanguine complexion.... Their eyebrows are milk-white, as is likewise the hair of their heads, which is very fine. . . . They seldom go abroad in the daytime, the sun being disagreeable to them, and causing their eyes, which are weak and poring, to water, especially if it shines towards them, yet they see very well by moonlight, from which we call them moon-eyed.'

Neither in our thoughts in these moonlight walks, methinks, is there 'the least tincture of a blush or sanguine complexion,' but we are intellectually and morally Albinoes,—children of Endymion,—such is the effect of conversing much with the moon.

I complain of Arctic voyagers that they do not enough remind us of the constant peculiar dreariness of the scenery, and the perpetual twilight of the Arctic night. So he whose theme is moonlight, though he may find it difficult, must, as it were, illustrate it with the light of the moon alone.

Many men walk by day; few walk by night. It is a very different season. Take a July night, for instance. About ten o'clock,—when man is asleep, and day fairly forgotten,—the beauty of moonlight is seen over lonely pastures where cattle are silently feeding. On all sides novelties present themselves. Instead of the sun there are the moon and stars, instead of the wood-thrush there is the whip-poor-will,—instead of butterflies in the meadows, fire-flies, winged sparks of fire! who would have believed it? What kind of cool deliberate life dwells in those dewy abodes associated with a spark of fire? So man has fire in his eyes, or blood, or brain. Instead of singing birds, the half-throttled note of a cuckoo flying over, the croaking of frogs, and the intenser dream of crickets. But above all, the wonderful trump of the bullfrog, ringing from Maine to Georgia. The potato-vines stand upright, the corn grows apace, the bushes loom, the grain-fields are boundless. On our open river terraces once cultivated by the Indian, they appear to occupy the ground like an army,—their heads nodding in the breeze. Small trees and shrubs are seen in the midst overwhelmed as by an inundation. The shadows of rocks and trees, and shrubs and hills, are more conspicuous than the objects themselves. The slightest irregularities in the ground are revealed by the shadows, and what the feet find comparatively smooth appears rough and diversified in consequence. For the same reason the whole landscape is more variegated and picturesque than by day. The smallest recesses in the rocks are dim and cavernous; the ferns in the wood appear of tropical size. The sweet fern and indigo in overgrown wood-paths wet you with dew up to your middle. The leaves of the shrub-oak are shining as if a liquid were flowing over them. The pools seen through the trees are as full of light as the sky. 'The light of the day takes refuge in their bosoms,' as the Purana says of the ocean. All white objects are more remarkable than by day. A distant cliff looks like a phosphorescent space on a hillside. The woods are heavy and dark. Nature slumbers. You see the moonlight reflected from particular stumps in the recesses of the forest, as if she selected what to shine on. These small fractions of her light remind one of the plant called moon-seed,—as if the moon were sowing it in such places.

In the night the eyes are partly closed or retire into the head. Other senses take the lead. The walker is guided as well by the sense of smell. Every plant and field and forest emits its odor now, swamp-pink in the meadow and tansy in the road; and there is the peculiar dry scent of corn which has begun to show its tassels. The senses both of hearing and smelling are more alert. We hear the tinkling of rills which we never detected before. From time to time, high up on the sides of hills, you pass through a stratum of warm air. A blast which has come up from the sultry plains of noon. It tells of the day, of sunny noon-tide hours and banks, of the laborer wiping his brow and the bee humming amid flowers. It is an air in which work has been done, —which men have breathed. It circulates about from woodside to hillside like a dog that has lost its master, now that the sun is gone. The rocks retain all night the warmth of the sun which they have absorbed. And so does the sand. If you dig a few inches into it you find a warm bed. You lie on your back on a rock in a pasture on the top of some bare hill at midnight, and speculate on the height of the starry canopy. The stars are the jewels of the night, and perchance surpass anything which day has to show. A companion with whom I was sailing one very windy but bright moonlight night, when the stars were few and faint, thought that a man could get along with *them*,—though he was considerably reduced in his circumstances,—that they were a kind of bread and cheese that never failed.

No wonder that there have been astrologers, that some have conceived that they were personally related to particular stars.

Dubartas, as translated by Sylvester, says he'll

> not believe that the great architect
> With all these fires the heavenly arches decked
> Only for show, and with these glistering shields,
> T' awake poor shepherds, watching in the fields.
> [He'll] not believe that the least flower which pranks
> Our garden borders, or our common banks,
> And the least stone, that in her warming lap
> Our mother· earth doth covetously wrap,
> Hath some peculiar virtue of its own,
> And that the glorious stars of heav'n have none.

And Sir Walter Raleigh well says, 'The stars are instruments of far greater use than to give an obscure light, and for men to gaze on after sunset'; and he quotes Plotinus as affirming that they 'are significant, but not efficient'; and also Augustine as saying, '*Deus regit inferiora corpora per superiora*': God rules the bodies below by those above. But best of all is this which another writer has expressed: '*Sapiens adjuvabit opus astrorum quemadmodum agricola terræ naturam*': a wise man assisteth the work of the stars as the husbandman helpeth the nature of the soil.

It does not concern men who are asleep in their beds, but it is very important to the traveler, whether the moon shines brightly or is obscured. It is not easy to realize the serene joy of all the earth, when she commences to shine unobstructedly, unless you have often been abroad alone in moonlight nights. She seems to be waging continual war with the clouds in your behalf. Yet we fancy the clouds to be *her* foes also. She comes on magnifying her dangers by her light, revealing, displaying them in all their hugeness and blackness, then suddenly casts them behind into the light concealed, and goes her way triumphant through a small space of clear sky.

In short, the moon traversing, or appearing to traverse, the small clouds which lie in her way, now obscured by them, now easily dissipating and shining through them, makes the drama of the moonlight night to all watchers and night-travelers. Sailors speak of it as the moon eating up the clouds. The traveler all alone, the moon all alone, except for his sympathy, overcoming with incessant victory whole squadrons of clouds above the forests and lakes and hills. When she is obscured he so sympathizes with her that he could whip a dog for her relief, as Indians do. When she enters on a clear field of great extent in the heavens, and shines unobstructedly, he is glad. And when she has fought her way through all the squadron of her foes, and rides majestic in a clear sky unscathed, and there are no more any obstructions in her path, he cheerfully and confidently pursues his way, and rejoices in his heart, and the cricket also seems to express joy in its song.

How insupportable would be the days, if the night with its dews and darkness did not come to restore the drooping world. As the shades begin to gather around us, our primeval instincts are aroused, and we steal forth from our lairs, like the inhabitants of the jungle, in search of those silent and brooding thoughts which are the natural prey of the intellect.

Richter says that 'the earth is every day overspread with the veil of night for the same reason as the cages of birds are darkened, viz.: that we may the more readily apprehend the higher harmonies of thought in the hush and quiet of darkness. Thoughts which day turns into smoke and mist stand about us in the night as light and flames; even as the column which fluctuates above the crater of Vesuvius, in the daytime appears a pillar of cloud, but by night a pillar of fire.'

There are nights in this climate of such serene and majestic beauty, so medicinal and fertilizing to the spirit, that methinks a sensitive nature would not devote them to oblivion, and perhaps there is no man but would be better and wiser for spending them out-of-doors, though he should sleep all the next day to pay for it; should sleep an Endymion sleep, as the ancients expressed it, —nights which warrant the Grecian epithet ambrosial, when, as in the land of Beulah, the atmosphere is charged with dewy fragrance, and with music, and we take our repose and have our dreams awake,—when the moon, not secondary to the sun,—

> gives us his blaze again,
> Void of its flame, and sheds a softer day.
> Now through the passing cloud she seems to stoop,
> Now up the pure cerulean rides sublime.

Diana still hunts in the New England sky.

> In Heaven queen she is among the spheres.
>   She, mistress-like, makes all things to be pure.
> Eternity in her oft change she bears;
>   She Beauty is; by her the fair endure.
>
> Time wears her not; she doth his chariot guide;
>   Mortality below her orb is placed;
> By her the virtues of the stars down slide;
>   By her is Virtue's perfect image cast.

The Hindoos compare the moon to a saintly being who has reached the last stage of bodily existence.

Great restorer of antiquity, great enchanter. In a mild night when the harvest or hunter's moon shines unobstructedly, the houses in our village, whatever architect they may have had by day, acknowledge only a master. The village street is then as wild as the forest. New and old things are confounded. I know not whether I am sitting on the ruins of a wall, or on the material which is to compose a new one. Nature is an instructed and impartial teacher, spreading no crude opinions, and flattering none; she will be neither radical nor conservative. Consider the moonlight, so civil, yet so savage!

The light is more proportionate to our knowledge than that of day. It is no more dusky in ordinary nights than our mind's habitual atmosphere, and the moonlight is as bright as our most illuminated moments are.

> In such a night let me abroad remain
> Till morning breaks, and all's confused again.

Of what significance the light of day, if it is not the reflection of an inward dawn?—to what purpose is the veil of night withdrawn, if the morning reveals nothing to the soul? It is merely garish and glaring.

When Ossian in his address to the sun exclaims,—

> Where has darkness its dwelling?
> Where is the cavernous home of the stars,
> When thou quickly followest their steps,
> Pursuing them like a hunter in the sky,—
> Thou climbing the lofty hills,
> They descending on barren mountains?

who does not in his thought accompany the stars to their 'cavernous home,' 'descending' with them 'on barren mountains'?

Nevertheless, even by night the sky is blue and not black, for we see through the shadow of the earth into the distant atmosphere of day, where the sunbeams are reveling.

1862

JAMES ANTHONY FROUDE

The Philosophy of Christianity

We should do our utmost to encourage the Beautiful, for
the Useful encourages itself.

GOETHE

A MOSS rose-bud hiding her face among the leaves one hot
summer morning, for fear the sun should injure her complexion, happened to let fall a glance towards her roots, and to see
the bed in which she was growing. What a filthy place! she cried.
What a home they have chosen for me! I, the most beautiful
of flowers, fastened down into so detestable a neighbourhood!
She threw her face into the air; thrust herself into the hands of
the first passer-by who stopped to look at her, and escaped in
triumph, as she thought, into the centre of a nosegay. But her
triumph was short-lived: in a few hours she withered and died.

I was reminded of this story when hearing a living thinker of
some eminence once say that he considered Christianity to have
been a misfortune. Intellectually it was absurd, and practically an
offence, over which he stumbled; and it would have been far
better for mankind, he thought, if they could have kept clear
of superstition, and followed on upon the track of the Grecian
philosophy, so little do men care to understand the conditions
which have made them what they are, and which have created
for them that very wisdom in which they themselves are so
contented. But it is strange, indeed, that a person who could
deliberately adopt such a conclusion should trouble himself any
more to look for truth. If a mere absurdity could make its way
out of a little fishing village in Galilee, and spread through the
whole civilized world; if men are so pitiably silly, that in an age
of great mental activity their strongest thinkers should have sunk
under an absorption of fear and folly, should have allowed it to
absorb into itself whatever of heroism, of devotion, self-sacrifice,

and moral nobleness there was among them; surely there were
nothing better for a wise man than to make the best of his time,
and to crowd what enjoyment he can find into it, sheltering himself in a very disdainful Pyrrhonism from all care for mankind or
for their opinions. For what better test of truth have we than the
ablest men's acceptance of it; and if the ablest men eighteen
centuries ago deliberately accepted what is now too absurd to
reason upon, what right have we to hope that with the same
natures, the same passions, the same understandings, no better
proof against deception, we, like they, are not entangled in what,
at the close of another era, shall seem again ridiculous? The scoff
of Cicero at the divinity of Liber and Ceres (bread and wine) may
be translated literally by the modern Protestant; and the sarcasms
which Clement and Tertullian flung at the Pagan creed, the modern sceptic returns upon their own. Of what use is it to destroy
an idol when another, or the same in another form takes immediate possession of the vacant pedestal?

But it is not so. Ptolemy was not perfect, but Newton had
been a fool if he had scoffed at Ptolemy. Newton could not have
been without Ptolemy, nor Ptolemy without the Chaldees; and as
it is with the minor sciences, so far more is it with the science of
sciences—the science of life, which has grown through all the
ages from the beginning of time. We speak of the errors of the
past. We, with this glorious present which is opening on us, we
shall never enter on it, we shall never understand it, till we have
learnt to see in that past, not error but instalment of truth, hard
fought-for truth, wrung out with painful and heroic effort. The
promised land is smiling before us, but we may not pass over
into possession of it while the bones of our fathers who laboured
through the wilderness lie bleaching on the sands, or a prey to
the unclean birds; we must gather them and bury them, and sum
up their labours, and inscribe the record of their actions on their
tombs as an honourable epitaph. If Christianity really is passing
away, if it has done its work, and if what is left of it is now holding us back from better things, it is not for our bitterness but for
our affectionate acknowledgment, not for our heaping contempt
on what it is, but for our reverent and patient examination of
what it has been, that it will be content to bid us farewell, and
give us God speed on our further journey.

In the Natural History of Religions certain broad phenomena

perpetually repeat themselves; they rise in the highest thought extant at the time of their origin; the conclusions of philosophy settle into a creed; art ornaments it, devotion consecrates it, time elaborates it. It grows through a long series of generations into the heart and habits of the people; and so long as no disturbing cause interferes, or so long as the idea at the centre of it survives; a healthy, vigorous, natural life shoots beautifully up out of it. But at last the idea becomes obsolete; the numbing influence of habit petrifies the spirit in the outside ceremonial, while quite new questions rise among the thinkers, and ideas enter into new and unexplained relations. The old formula will not serve; but new formulae are tardy in appearing; and habit and superstition cling to the past, and policy vindicates it, and statecraft upholds it forcibly as serviceable to order, till, from the combined action of folly, and worldliness, and ignorance, the once beautiful symbolism becomes at last no better than 'a whited sepulchre full of dead men's bones and all uncleanness.' So it is now. So it was in the era of the Caesars, out of which Christianity arose; and Christianity, in the form which it assumed at the close of the Arian controversy, was the deliberate solution which the most powerful intellects of that day could offer of the questions which had grown out with the growth of mankind, and on which Paganism had suffered shipwreck.

Paganism, as a creed, was entirely physical. When Paganism rose men had not begun to reflect upon themselves, or the infirmities of their own nature. The bad man was a bad man—the coward a coward—the liar a liar—individually hateful and despicable. But in hating and despising such unfortunates, the old Greeks were satisfied to have felt all that was necessary about them; and how such a phenomenon as a bad man came to exist in this world, they scarcely cared to inquire. There is no evil spirit in the mythology as an antagonist of the gods. There is the Erinnys as the avenger of monstrous villanies; a Tartarus where the darkest criminals suffer eternal tortures. But Tantalus and Ixion are suffering for enormous crimes, to which the small wickedness of common men offers no analogy. Moreover, these and other such stories are but curiously ornamented myths, representing physical phenomena. But with Socrates a change came over philosophy; a sign—perhaps a cause—of the decline of the existing religion. The study of man superseded the study of nature: a purer Theism came in with the higher ideal of perfection, and sin and depravity at once assumed an importance the intensity of which made every other question insignificant. How man could know the good and yet choose the evil; how God could be all pure and almighty, and yet evil have broken into his creation, these were the questions which thenceforth were the perplexity of every thinker. Whatever difficulty there might be in discovering how evil came to be, the leaders of all the sects agreed at last upon the seat of it—whether *matter* was eternal, as Aristotle thought, or created, as Plato thought, both Plato and Aristotle were equally satisfied that the secret of all the shortcomings in this world lay in the imperfection, reluctancy, or inherent grossness of this impracticable substance. God would have everything perfect, but the nature of the element in which He worked in some way defeated His purpose. Death, disease, decay, clung necessarily to everything which was created out of it; and pain, and want, and hunger, and suffering. Worse than all, the spirit in its material body was opposed and borne down, its aspirations crushed, its purity tainted by the passions and appetites of its companion, the fleshly lusts which waged perpetual war against it.

Matter was the cause of evil, and thenceforth the question was how to conquer it, or at least how to set free the spirit from its control.

The Greek language and the Greek literature spread behind the march of Alexander: but as his generals could only make their conquests permanent by largely accepting the Eastern manner, so philosophy could only make good its ground by becoming itself Orientalized.

The one pure and holy God whom Plato had painfully reasoned out for himself had existed from immemorial time in the traditions of the Jews, while the Persians who had before taught the Jews at Babylon the existence of an independent evil being now had him to offer to the Greeks as their account of the difficulties which had perplexed Socrates. Seven centuries of struggle, and many hundred thousand folios were the results of the remarkable fusion which followed. Out of these elements, united in various proportions, rose successively the Alexandrian philosophy, the Hellenists, the Therapeutæ, those strange Essene communists, with the innumerable sects of Gnostic or Christian heretics. Finally, the battle was limited to the two great rivals, under one or other of which the best of the remainder had ranged themselves—Manicheism and Catholic Christianity: Manicheism in which the Persian, Catholicism in which the Jewish element most preponderated. It did not end till the close of the fifth century, and it ended then rather by arbitration than by a decided victory which either side could claim. The Church has yet to acknowledge how large a portion of its enemy's doctrines it incorporated through the mediation of Augustine before the field was surrendered to it. Let us trace something of the real bearings of this section of the world's oriental history, which to so many moderns seems no better than an idle fighting over words and straws.

Facts witnessing so clearly that the especial strength of evil lay, as the philosophers had seen, in matter, so far it was a conclusion which both Jew and Persian were ready to accept. The naked Aristotelic view of it being most acceptable to the Persian, the Platonic to the Hellenistic Jew. But the purer theology of the Jew forced him to look for a solution of the question which Plato had left doubtful, and to explain how evil crept into matter. He could not allow that what God had created could be of its own nature imperfect. God made it very good; some other cause had broken in to spoil it. Accordingly, as before he had reduced the independent Arimanes, whose existence he had learnt at Babylon, into a subordinate spirit; so now, not questioning the facts of disease, of death, of pain, of the infirmity of the flesh which the natural strength of the spirit was unable to resist, he accounted for them under the supposition that the first man had deliberately sinned, and by his sin had brought a curse upon the whole material earth, and upon all which was fashioned out of it. The earth was created pure and lovely—a garden of delight of its own free accord, loading itself with fruit and flower, and everything most exquisite and beautiful. No bird or beast of prey broke the eternal peace which reigned over its hospitable surface. In calm and quiet intercourse, the leopard lay down by the kid, the lion browsed beside the ox, and the corporeal frame of man, knowing neither decay, nor death, nor unruly appetite, nor any change or infirmity, was pure as the pure immortal substance of the unfallen angels. But with the fatal apple all this fair scene passed away, and creation as it

seemed was hopelessly and irretrievably ruined. Adam sinned—
no matter how—he sinned; the sin was the one terrible fact:
moral evil was brought into the world by the only creature who
was capable of committing it. Sin entered in, and death by sin;
death and disease, storm and pestilence, earthquake and famine.
The imprisoned passions of the wild animals were let loose, and
earth and air became full of carnage; worst of all, man's animal
nature came out in gigantic strength, the carnal lusts, unruly
appetites, jealousies, hatred, rapine, and murder; and then the
law, and with it, of course, breaches of the law, and sin on sin.
The seed of Adam was infected in the animal change which had
passed over his person, and every child, therefore, thenceforth
naturally engendered in his posterity, was infected with the curse
which he had incurred. Every material organization thence-
forward contained in itself the elements of its own destruction,
and the philosophic conclusions of Aristotle were accepted and
explained by theology. Already, in the popular histories, those
who were infected by disease were said to be bound by Satan;
madness was a 'possession' by his spirit, and the whole creation
from Adam till Christ groaned and travailed under Satan's
power. The nobler nature in man still made itself felt; but it was
a slave when it ought to command. It might will to obey the
higher law, but the law in the members was over strong for it
and bore it down. This was the body of death which philosophy
detected but could not explain, and from which Christianity now
came forward with its magnificent promise of deliverance.

The carnal doctrine of the sacraments which they are compelled to acknowledge to have been taught as fully in the early
Church as it is now taught by the Roman Catholics, has long
been the stumbling-block to Protestants. It was the very essence
of Christianity itself. Unless the body could be purified, the soul
could not be saved; or, rather, as from the beginning, soul and
flesh were one man and inseparable, without his flesh, man was
lost, or would cease to be. But the natural organization of the
flesh was infected, and unless organization could begin again
from a new original, no pure material substance could exist at all.
He, therefore, by whom God had first made the world, entered
into the womb of the Virgin in the form (so to speak) of a new
organic cell, and around it, through the virtue of His creative
energy, a material body grew again of the substance of his

mother, pure of taint and clean as the first body of the first man when it passed out under His hand in the beginning of all things. In Him thus wonderfully born was the virtue which was to restore the lost power of mankind. He came to redeem man; and, therefore, he took a human body, and he kept it pure through a human life, till the time came when it could be applied to its marvellous purpose. He died, and then appeared what was the nature of a material human body when freed from the limitations of sin. The grave could not hold it, neither was it possible that it should see corruption. It was real, for the disciples were allowed to feel and handle it. He ate and drank with them to assure their senses. But space had no power over it, nor any of the material obstacles which limit an ordinary power. He willed and his body obeyed. He was here, He was there. He was visible, He was invisible. He was in the midst of his disciples and they saw Him, and then He was gone, whither who could tell? At last He passed away to heaven; but while in heaven, He was still on earth. His body became the body of His Church on earth, not in metaphor, but in fact. His very material body, in which and by which the faithful would be saved. His flesh and blood were thenceforth to be their food. They were to eat it as they would eat ordinary meat. They were to take it into their system, a pure material substance, to leaven the old natural substance and assimilate it to itself. As they fed upon it it would grow into them, and it would become their own real body. Flesh grown in the old way was the body of death, but the flesh of Christ was the life of the world, over which death had no power. Circumcision availed nothing, nor uncircumcision—but a *new creature*—this new creature, which the child first put on in baptism, being born again into Christ of water and the spirit. In the Eucharist he was fed and sustained and going on from strength to strength, and ever as the nature of his body changed, being able to render a more complete obedience, he would at last pass away to God through the gate of the grave, and stand holy and perfect in the presence of Christ. Christ had indeed been ever present with him; but because while life lasted some particles of the old Adam would necessarily cling to him, the Christian's mortal eye on earth cannot see Him. Hedged in by 'his muddy vesture of decay,' his eyes, like the eyes of the disciples of Emmaus, are holden, and only in faith he feels Him. But death, which till Christ had died had been the last victory of evil, in virtue of His submission to it, became its own destroyer, for it had power only over the tainted particles of the old substance, and there was nothing needed but that these should be washed away and the elect would stand out at once pure and holy, clothed in immortal bodies, like refined gold, the redeemed of God.

The being who accomplished a work so vast, a work compared to which the first creation appears but a trifling difficulty, what could He be but God? God Himself! Who but God could have wrested His prize from a power which half the thinking world believed to be His coequal and coeternal adversary. He was God. He was man also, for He was the second Adam—the second starting point of human growth. He was virgin born, that no original impurity might infect the substance which He assumed; and being Himself sinless, He showed in the nature of His person, after His resurrection, what the material body would have been in all of us except for sin, and what it will be when, after feeding on it in its purity, the bodies of each of us are transfigured after its likeness. Here was the secret of the spirit which set St Simeon on his pillar and sent St Anthony to the tombs—of the night watches, the weary fasts, the penitential scourgings, and life-long austerities which have been alternately the glory and the reproach of the mediaeval saints. They would overcome their animal bodies, and anticipate in life the work of death in uniting themselves more completely to Christ by the destruction of the flesh which lay as a veil between themselves and Him.

And such, I believe, to have been the central idea of the beautiful creed which, for 1800 years, has tuned the heart and formed the mind of the noblest of mankind. From this centre it radiated out and spread, as time went on, into the full circle of human activity, flinging its own philosophy and its own peculiar grace over the common detail of the common life of all of us. Like the seven lamps before the Throne of God, the seven mighty angels, and the seven stars, the seven sacraments shed over us a never ceasing stream of blessed influence. First there are the priests, a holy order set apart and endowed with mysterious power, representing Christ and administering his gifts. Christ, in his twelfth year, was presented in the temple, and first entered on His father's business; and the baptized child, when it has grown to an age to become conscious of its vow and of its privilege, again renews it in full knowledge of what it undertakes, and receives again sacramentally a fresh gift of grace to assist it forward on its way. In maturity it seeks a companion to share its pains and pleasures; and, again, Christ is present to consecrate the union. Marriage, which outside the church only serves to perpetuate the curse and bring fresh inheritors of misery into the world, He made holy by His presence at Cana, and chose it as the symbol to represent His own mystic union with His church.

Even saints cannot live without at times some spot adhering to them. The atmosphere in which we breathe and move is soiled, and Christ has anticipated our wants. Christ did penance forty days in the wilderness, not to subdue His own flesh, for that which was already perfect did not need subduing, but to give to penance a cleansing virtue to serve for our daily or our hourly ablution.

Christ consecrates our birth; Christ throws over us our baptismal robe of pure unsullied innocence. He strengthens us as we go forward. He raises us when we fall. He feeds us with the substance of His own most precious body. In the person of His minister he does all this for us, in virtue of that which in His own person he actually performed when a man living on this earth. Last of all, when all is drawing to its close with us, when life is past, when the work is done, and the dark gate is near, beyond which the garden of an eternal home is waiting to receive us, His tender care has not forsaken us. He has taken away the sting of death, but its appearance is still terrible; and He will not leave us without special help at our last need. He tried the agony of the moment; and He sweetens the cup for us before we drink it. We are dismissed to the grave with our bodies anointed with oil, which He made holy in His last anointing before His passion, and then all is over. We lie down and seem to decay—to decay—but not all. Our natural body decays, the last remains of which we have inherited from Adam, but the spiritual body, that glorified substance which has made our life, and is our real body as we are in Christ, that can never decay, but passes off into the kingdom which is prepared for it; that other world where there is no sin, and God is all in all! Such is the Philosophy of Christianity. It was worn and old when Luther found it. Our posterity will care less to respect Luther for rending it in pieces, when it has learnt to despise the miserable fabric which he stitched together out of its tatters.

1851

Thomas Carlyle

IT has been well said that the highest aim in education is analogous to the highest aim in mathematics, namely, to obtain not
results but *powers*, not particular solutions, but the means by
which endless solutions may be wrought. He is the most effective
educator who aims less at perfecting specific acquirements than
at producing that mental condition which renders acquirements
easy, and leads to their useful application; who does not seek to
make his pupils moral by enjoining particular courses of action,
but by bringing into activity the feelings and sympathies that
must issue in noble action. On the same ground it may be said
that the most effective writer is not he who announces a particular discovery, who convinces men of a particular conclusion,
who demonstrates that this measure is right and that measure
wrong; but he who rouses in others the activities that must issue
in discovery, who awakes men from their indifference to the
right and the wrong, who nerves their energies to seek for the
truth and live up to it at whatever cost. The influence of such a
writer is dynamic. He does not teach men how to use sword and
musket, but he inspires their souls with courage and sends a
strong will into their muscles. He does not, perhaps, enrich your
stock of data, but he clears away the film from your eyes that you
may search for data to some purpose. He does not, perhaps, convince you, but he strikes you, undeceives you, animates you.
You are not directly fed by his books, but you are braced as by a
walk up to an alpine summit, and yet subdued to calm and reverence as by the sublime things to be seen from that summit.

Such a writer is Thomas Carlyle. It is an idle question to ask
whether his books will be read a century hence: if they were all
burnt as the grandest of Suttees on his funeral pile, it would be
only like cutting down an oak after its acorns have sown a forest.
For there is hardly a superior or active mind of this generation

that has not been modified by Carlyle's writings; there has hardly
been an English book written for the last ten or twelve years that
would not have been different if Carlyle had not lived. The
character of his influence is best seen in the fact that many of the
men who have the least agreement with his opinions are those to
whom the reading of *Sartor Resartus* was an epoch in the history
of their minds. The extent of his influence may be best seen in
the fact that ideas which were startling novelties when he first
wrote them are now become common-places. And we think few
men will be found to say that this influence on the whole has not
been for good. There are plenty who question the justice of
Carlyle's estimates of past men and past times, plenty who quarrel with the exaggerations of the *Latter-Day Pamphlets*, and who
are as far as possible from looking for an amendment of things
from a Carlylian theocracy with the 'greatest man', as a Joshua
who is to smite the wicked (and the stupid) till the going down
of the sun. But for any large nature, those points of difference are
quite incidental. It is not as a theorist, but as a great and beautiful human nature, that Carlyle influences us. You may meet a
man whose wisdom seems unimpeachable, since you find him
entirely in agreement with yourself; but this oracular man of
unexceptionable opinions has a green eye, a wiry hand, and
altogether a *Wesen*, or demeanour, that makes the world look
blank to you, and whose unexceptionable opinions become a
bore; while another man who deals in what you cannot but think
'dangerous paradoxes', warms your heart by the pressure of his
hand, and looks out on the world with so clear and loving an
eye, that nature seems to reflect the light of his glance upon your
own feeling. So it is with Carlyle. When he is saying the very
opposite of what we think, he says it so finely, with so hearty
conviction—he makes the object about which we differ stand out
in such grand relief under the clear light of his strong and honest
intellect—he appeals so constantly to our sense of the manly and
the truthful—that we are obliged to say 'Hear! hear!' to the
writer before we can give the decorous 'Oh! oh!' to his opinions.

   Much twaddling criticism has been spent on Carlyle's style.
Unquestionably there are some genuine minds, not at all given to
twaddle, to whom his style is antipathetic, who find it as unen-
durable as an English lady finds peppermint. Against antipathies
there is no arguing; they are misfortunes. But instinctive repul-

sion apart, surely there is no one who can read and relish Carlyle without feeling that they could no more wish him to have written in another style than they could wish Gothic architecture not to be Gothic, or Raffaelle not to be Raffaellesque. It is the fashion to speak of Carlyle almost exclusively as a philosopher; but, to our thinking, he is yet more of an artist than a philosopher. He glances deep down into human nature, and shows the causes of human actions; he seizes grand generalisations, and traces them in the particular with wonderful acumen; and in all this he is a philosopher. But, perhaps, his greatest power lies in concrete presentation. No novelist has made his creations live for us more thoroughly than Carlyle has made Mirabeau and the men of the French Revolution, Cromwell and the Puritans. What humour in his pictures! Yet what depth of appreciation, what reverence for the great and godlike under every sort of earthly mummery!

1855

MATTHEW ARNOLD

Heine and the Philistines

(*from* Heinrich Heine)

HEINE is noteworthy, because he is the most important German successor and continuator of Goethe in Goethe's most important line of activity. And which of Goethe's lines of activity is this?—His line of activity as 'a soldier in the war of liberation of humanity.' . . .

Modern times find themselves with an immense system of institutions, established facts, accredited dogmas, customs, rules, which have come to them from times not modern. In this system their life has to be carried forward; yet they have a sense that this system is not of their own creation, that it by no means corresponds exactly with the wants of their actual life, that, for them, it is customary, not rational. The awakening of this sense is the awakening of the modern spirit. The modern spirit is now awake almost everywhere; the sense of want of correspondence between the forms of modern Europe and its spirit, between the new wine of the eighteenth and nineteenth centuries, and the old bottles of the eleventh and twelfth centuries, or even of the sixteenth and seventeenth, almost every one now perceives; it is no longer dangerous to affirm that this want of correspondence exists; people are even beginning to be shy of denying it. To remove this want of correspondence is beginning to be the settled endeavour of most persons of good sense. Dissolvents of the old European system of dominant ideas and facts we must all be, all of us who have any power of working; what we have to study is that we may not be acrid dissolvents of it.

And how did Goethe, that grand dissolvent in an age when there were fewer of them than at present, proceed in his task of dissolution, of liberation of the modern European from the old routine? He shall tell us himself. 'Through me the German poets have become aware that, as man must live from within outwards, so the artist must work from within outwards, seeing that, make what contortions he will, he can only bring to light his own individuality. I can clearly mark where this influence of mine has made itself felt; there arises out of it a kind of poetry of nature, and only in this way is it possible to be original.'

My voice shall never be joined to those which decry Goethe, and if it is said that the foregoing is a lame and impotent conclusion to Goethe's declaration that he had been the liberator of the Germans in general, and of the young German poets in particular, I say it is not. Goethe's profound, imperturbable naturalism is absolutely fatal to all routine thinking; he puts the standard, once for all, inside every man instead of outside him; when he is told, such a thing must be so, there is immense authority and custom in favour of its being so, it has been held to be so for a thousand years, he answers with Olympian politeness, 'But *is* it so? is it so to *me?*' Nothing could be more really subversive of the foundations on which the old European order rested; and it may be remarked that no persons are so radically detached from this order, no persons so thoroughly modern, as those who have felt Goethe's influence most deeply. If it is said that Goethe professes to have in this way deeply influenced but a few persons, and those persons poets, one may answer that he could have taken no better way to secure, in the end, the ear of the world; for poetry is simply the most beautiful, impressive, and widely effective mode of saying things, and hence its importance. Nevertheless the process of liberation, as Goethe worked it, though sure, is undoubtedly slow; he came, as Heine says, to be eighty years old in thus working it, and at the end of that time the old Middle-Age machine was still creaking on, the thirty German courts and their chamberlains subsisted in all their glory; Goethe himself was a minister, and the visible triumph of the modern spirit over prescription and routine seemed as far off as ever. It was the year 1830; the German sovereigns had passed the preceding fifteen years in breaking the promises of freedom they had made to their subjects when they wanted their help in the final struggle with Napoleon. Great events were happening in France; the revolution, defeated in 1815, had arisen from its defeat, and was wresting from its adversaries the power. Heinrich Heine, a young man of genius, born at Hamburg, and with all the culture of Germany, but by race a Jew; with warm

sympathies for France, whose revolution had given to his race
the rights of citizenship, and whose rule had been, as is well
known, popular in the Rhine provinces, where he passed his
youth; with a passionate admiration for the great French
Emperor, with a passionate contempt for the sovereigns who
had overthrown him, for their agents, and for their policy,
—Heinrich Heine was in 1830 in no humour for any such grad-
ual process of liberation from the old order of things as that
which Goethe had followed. His counsel was for open war.
Taking that terrible modern weapon, the pen, in his hand, he
passed the remainder of his life in one fierce battle. What was
that battle? the reader will ask. It was a life and death battle with
Philistinism.

Philistinism!—we have not the expression in English. Perhaps
we have not the word because we have so much of the thing.
At Soli, I imagine, they did not talk of solecisms; and here, at
the very headquarters of Goliath, nobody talks of Philistinism.
The French have adopted the term *épicier* (grocer), to designate
the sort of being whom the Germans designate by the term
Philistine; but the French term,—besides that it casts a slur upon
a respectable class, composed of living and susceptible members,
while the original Philistines are dead and buried long ago,—is
really, I think, in itself much less apt and expressive than the
German term. Efforts have been made to obtain in English some
term equivalent to *Philister* or *épicier*; Mr Carlyle has made
several such efforts: 'respectability with its thousand gigs,' he
says;—well, the occupant of every one of these gigs is, Mr
Carlyle means, a Philistine. However, the word *respectable* is far
too valuable a word to be thus perverted from its proper meaning; if the English are ever to have a word for the thing we are
speaking of,—and so prodigious are the changes which the modern spirit is introducing, that even we English shall perhaps one
day come to want such a word,—I think we had much better
take the term *Philistine* itself.

Philistine must have originally meant, in the mind of those
who invented the nickname, a strong, dogged, unenlightened
opponent of the chosen people, of the children of the light. The
party of change, the would-be remodellers of the old traditional
European order, the invokers of reason against custom, the representatives of the modern spirit in every sphere where it is

applicable, regarded themselves, with the robust self-confidence
natural to reformers as a chosen people, as children of the light.
They regarded their adversaries as humdrum people, slaves to
routine, enemies to light; stupid and oppressive, but at the same
time very strong. This explains the love which Heine, that Pala-
din of the modern spirit, has for France; it explains the preference which he gives to France over Germany: 'the French,' he
says, 'are the chosen people of the new religion, its first gospels
and dogmas have been drawn up in their language; Paris is the
new Jerusalem, and the Rhine is the Jordan which divides the
consecrated land of freedom from the land of the Philistines.' He
means that the French, as a people, have shown more accessib-
ility to ideas than any other people; that prescription and routine
have had less hold upon them than upon any other people; that
they have shown most readiness to move and to alter at
the bidding (real or supposed) of reason. This explains, too, the
detestation which Heine had for the English: 'I might settle in
England,' he says, in his exile, 'if it were not that I should find
there two things, coal-smoke and Englishmen; I cannot abide
either.' What he hated in the English was the 'ächtbrittische
Beschränktheit,' as he calls it,—the *genuine British narrowness*. In
truth, the English, profoundly as they have modified the old
Middle-Age order, great as is the liberty which they have secured
for themselves, have in all their changes proceeded, to use a
familiar expression, by the rule of thumb; what was intolerably
inconvenient to them they have suppressed, and as they have
suppressed it, not because it was irrational, but because it was
practically inconvenient, they have seldom in suppressing it
appealed to reason, but always, if possible, to some precedent, or
form, or letter, which served as a convenient instrument for their
purpose, and which saved them from the necessity of recurring
to general principles. They have thus become, in a certain sense,
of all people the most inaccessible to ideas and the most impatient of them; inaccessible to them, because of their want of
familiarity with them; and impatient of them because they have
got on so well without them, that they despise those who, not
having got on as well as themselves, still make a fuss for what
they themselves have done so well without. But there has certainly followed from hence, in this country, somewhat of a general depression of pure intelligence: Philistia has come to be

thought by us the true Land of Promise, and it is anything but that; the born lover of ideas, the born hater of commonplaces, must feel in this country, that the sky over his head is of brass and iron. The enthusiast for the idea, for reason, values reason, the idea, in and for themselves; he values them, irrespectively of the practical conveniences which their triumph may obtain for him; and the man who regards the possession of these practical conveniences as something sufficient in itself, something which compensates for the absence or surrender of the idea, of reason, is, in his eyes, a Philistine. This is why Heine so often and so mercilessly attacks the liberals; much as he hates conservatism he hates Philistinism even more, and whoever attacks conservatism itself ignobly, not as a child of light, not in the name of the idea, is a Philistine.

1863

From *Evolution and Ethics*

I SEE no reason to doubt that, at its origin, human society was as much a product of organic necessity as that of the bees. The human family, to begin with, rested upon exactly the same conditions as those which gave rise to similar associations among animals lower in the scale. Further, it is easy to see that every increase in the duration of the family ties, with the resulting co-operation of a larger and larger number of descendants for protection and defence, would give the families in which such modification took place a distinct advantage over the others. And, as in the hive, the progressive limitation of the struggle for existence between the members of the family would involve increasing efficiency as regards outside competition.

But there is this vast and fundamental difference between bee society and human society. In the former, the members of the society are each organically predestined to the performance of one particular class of functions only. If they were endowed with desires, each could desire to perform none but those offices for which its organization specially fits it; and which, in view of the good of the whole, it is proper it should do. So long as a new queen does not make her appearance, rivalries and competition are absent from the bee polity.

Among mankind, on the contrary, there is no such predestination to a sharply defined place in the social organism. However much men may differ in the quality of their intellects, the intensity of their passions, and the delicacy of their sensations, it cannot be said that one is fitted by his organization to be an agricultural labourer and nothing else, and another to be a landowner and nothing else. Moreover, with all their enormous differences in natural endowment, men agree in one thing, and that is their innate desire to enjoy the pleasures and to escape the pains of life; and, in short, to do nothing but that which it

pleases them to do, without the least reference to the welfare of
the society into which they are born. That is their inheritance
(the reality at the bottom of the doctrine of original sin) from the
long series of ancestors, human and semi-human and brutal, in
whom the strength of this innate tendency to self-assertion was
the condition of victory in the struggle for existence. That is
the reason of the *aviditas vitae*—the insatiable hunger for enjoyment—of all mankind, which is one of the essential conditions of
success in the war with the state of nature outside; and yet the
sure agent of the destruction of society if allowed free play
within.

The check upon this free play of self-assertion, or natural liberty, which is the necessary condition for the origin of human
society, is the product of organic necessities of a different kind
from those upon which the constitution of the hive depends.
One of these is the mutual affection of parent and offspring,
intensified by the long infancy of the human species. But the
most important is the tendency, so strongly developed in man, to
reproduce in himself actions and feelings similar to, or correlated
with, those of other men. Man is the most consummate of all
mimics in the animal world; none but himself can draw or
model; none comes near him in the scope, variety, and exactness
of vocal imitation; none is such a master of gesture; while he
seems to be impelled thus to imitate for the pure pleasure of it.
And there is no such another emotional chameleon. By a purely
reflex operation of the mind, we take the hue of passion of those
who are about us, or, it may be, the complementary colour. It is
not by any conscious 'putting one's self in the place' of a joyful
or a suffering person that the state of mind we call sympathy
usually arises;[1] indeed, it is often contrary to one's sense of right,
and in spite of one's will, that 'fellow-feeling makes us wondrous
kind,' or the reverse. However complete may be the indifference
to public opinion, in a cool, intellectual view, of the traditional
sage, it has not yet been my fortune to meet with any actual sage

---

[1] Adam Smith makes the pithy observation that the man who sympathises with a
woman in childbed, cannot be said to put himself in her place. ('The Theory of the Moral
Sentiments,' Part vii. sec. iii. chap. i.) Perhaps there is more humour than force in the
example; and, in spite of this and other observations of the same tenor, I think that the
one defect of the remarkable work in which it occurs is that it lays too much stress on
conscious substitution, too little on purely reflex sympathy.

who took its hostile manifestations with entire equanimity. Indeed, I doubt if the philosopher lives, or ever has lived, who could know himself to be heartily despised by a street boy without some irritation. And, though one cannot justify Haman for wishing to hang Mordecai on such a very high gibbet, yet, really, the consciousness of the Vizier of Ahasuerus, as he went in and out of the gate, that this obscure Jew had no respect for him, must have been very annoying.[1]

It is needful only to look around us, to see that the greatest restrainer of the anti-social tendencies of men is fear, not of the law, but of the opinion of their fellows. The conventions of honour bind men who break legal, moral, and religious bonds; and, while people endure the extremity of physical pain rather than part with life, shame drives the weakest to suicide.

Every forward step of social progress brings men into closer relations with their fellows, and increases the importance of the pleasures and pains derived from sympathy. We judge the acts of others by our own sympathies, and we judge our own acts by the sympathies of others, every day and all day long, from childhood upwards, until associations, as indissoluble as those of language, are formed between certain acts and the feelings of approbation or disapprobation. It becomes impossible to imagine some acts without disapprobation, or others without approbation of the actor, whether he be one's self, or any one else. We come to think in the acquired dialect of morals. An artificial personality, the 'man within,' as Adam Smith calls conscience, is built up beside the natural personality. He is the watchman of society, charged to restrain the anti-social tendencies of the natural man within the limits required by social welfare.

I have termed this evolution of the feelings out of which the primitive bonds of human society are so largely forged, into the organized and personified sympathy we call conscience, the ethical process. So far as it tends to make any human society more efficient in the struggle for existence with the state of nature, or

[1] Esther v. 9-13. '. . . but when Haman saw Mordecai in the king's gate, that he stood not up, nor moved for him, he was full of indignation against Mordecai. . . . And Haman told them of the glory of his riches . . . and all the things wherein the king had promoted him. . . . Yet all this availeth me nothing, so long as I see Mordecai the Jew sitting at the king's gate.' What a shrewd exposure of human weakness it is!

with other societies, it works in harmonious contrast with the cosmic process. But it is none the less true that, since law and morals are restraints upon the struggle for existence between men in society, the ethical process is in opposition to the principle of the cosmic process, and tends to the suppression of the qualities best fitted for success in that struggle.

It is further to be observed that, just as the self-assertion, necessary to the maintenance of society against the state of nature, will destroy that society if it is allowed free operation within; so the self-restraint, the essence of the ethical process, which is no less an essential condition of the existence of every polity, may, by excess, become ruinous to it.

Moralists of all ages and of all faiths, attending only to the relations of men towards one another in an ideal society, have agreed upon the 'golden rule,' 'Do as you would be done by.' In other words, let sympathy be your guide; put yourself in the place of the man towards whom your action is directed; and do to him what you would like to have done to yourself under the circumstances. However much one may admire the generosity of such a rule of conduct; however confident one may be that average men may be thoroughly depended upon not to carry it out to its full logical consequences; it is nevertheless desirable to recognize the fact that these consequences are incompatible with the existence of a civil state, under any circumstances of this world which have obtained, or, so far as one can see, are likely to come to pass.

For I imagine there can be no doubt that the great desire of every wrongdoer is to escape from the painful consequences of his actions. If I put myself in the place of the man who has robbed me, I find that I am possessed by an exceeding desire not to be fined or imprisoned; if in that of the man who has smitten me on one cheek, I contemplate with satisfaction the absence of any worse result than the turning of the other cheek for like treatment. Strictly observed, the 'golden rule' involves the negation of law by the refusal to put it in motion against lawbreakers; and, as regards the external relations of a polity, it is the refusal to continue the struggle for existence. It can be obeyed, even partially, only under the protection of a society which repudiates it. Without such shelter, the followers of the 'golden rule' may indulge in hopes of heaven, but they must reckon with the certainty that other people will be masters of the earth.

What would become of the garden if the gardener treated all the weeds and slugs and birds and trespassers as he would like to be treated, if he were in their place?

1894

Dull Government

PARLIAMENT is a great thing, but it is not a cheerful thing. Just reflect on the existence of 'Mr Speaker.' First, a small man speaks to him—then a shrill man speaks to him—then a man who cannot speak *will* speak to him. He leads a life of 'passing tolls,' joint-stock companies, and members out of order. Life is short, but the forms of the House are long. Mr Ewart complains that a multitude of members, including the Prime Minister himself, actually go to sleep. The very morning paper feels the weight of this leaden *régime*. Even in the dullest society you hear complaints of the dullness of Parliament—of the representative tedium of the nation.

That an Englishman should grumble is quite right, but that he should grumble at gravity is hardly right. He is rarely a lively being himself, and he should have a sympathy with those of his kind. And he should further be reminded that his criticism is out of place—that dullness in matters of government is a good sign, and not a bad one—that, in particular, dullness in parliamentary government is a test of its excellence, an indication of its success. The truth is, all the best business is a little dull. If you go into a merchant's counting-house, you see steel pens, vouchers, files, books of depressing magnitude, desks of awful elevation, staid spiders, and sober clerks moving among the implements of tedium. No doubt, to the parties engaged, much of this is very attractive. 'What,' it has been well said, 'are technicalities to those without, are realities to those within.' To every line in those volumes, to every paper on those damp files, there has gone doubt, decision, action—the work of a considerate brain, the touch of a patient hand. Yet even to those engaged, it is commonly the least interesting business which is the best. The more the doubt, the greater the liability to error—the longer the consideration, generally the worse the result—the more the pain of decision, the greater the likelihood of failure. In Westminster Hall, they have a legend of a litigant who stopped his case because the lawyers said it was 'interesting.' 'Ah,' he remarked afterwards, 'they were going up to the "Lords" with it, and I should never have seen my money.' To parties concerned in law, the best case is a plain case. To parties concerned in trade, the best transaction is a plain transaction—the sure result of familiar knowledge; in political matters, the best sign that things are going well is that there should be nothing difficult—nothing requiring deep contention of mind—no anxious doubt, no sharp resolution, no lofty and patriotic execution. The opportunity for these qualities is the danger of the commonwealth. You cannot have a Chatham in time of peace—you cannot storm a Redan in Somersetshire. There is no room for glorious daring in periods of placid happiness.

And if this be the usual rule, certainly there is nothing in the nature of parliamentary government to exempt it from its operation. If business is dull, business wrangling is no better. It is dull for an absolute minister to have to decide on passing tolls, but it is still duller to hear a debate on them—to have to listen to the two extremes and the *via media*. One honourable member considers that the existing ninepence ought to be maintained; another thinks it ought to be abolished; and a third—the independent thinker—has statistics of his own, and suggests that fourpence-halfpenny would 'attain the maximum of revenue with the minimum of inconvenience'—only he could wish there were a decimal coinage to 'facilitate the calculations of practical pilots.' Of course this is not the highest specimen of parliamentary speaking. Doubtless, on great questions, when the public mind is divided, when the national spirit is roused, when powerful interests are opposed, when large principles are working their way, when deep difficulties press for a decision, there is an opportunity for noble eloquence. But these very circumstances are the signs, perhaps of calamity, certainly of political difficulty and national doubt. The national spirit is not roused in happy times—powerful interests are not divided in years of peace—the path of great principles is marked through history by trouble, anxiety, and conflict. An orator requires a topic. 'Thoughts that breathe and words that burn' will not suit the 'liability of joint-stock companies'—you cannot shed tears over a 'toll'. Where can there be a better proof of national welfare than that Disraeli cannot be sarcastic, and that Lord Derby fails in a diatribe? Happy is the country which is at peace within its borders—yet stupid is the country when the opposition is without a cry.

Moreover, when parliamentary business is a bore, it is a bore which cannot be overlooked. There is much torpor secreted in the *bureaux* of an absolute government, but no one hears of it—no one knows of its existence. In England it is different. With pains and labour—by the efforts of attorneys—by the votes of freeholders—you collect more than six hundred gentlemen; and the question is, what are they to do? As they come together at a specific time, it would seem that they do so for a specific purpose—but what it is they do not know. It is the business of the Prime Minister to discover it for them. It is extremely hard on an effervescent First Lord to have to set people down to mere business—to bore them with slow reforms—to explain details they cannot care for—to abolish abuses they never heard of—to consume the hours of the night among the perplexing details of an official morning. But such is the Constitution. The Parliament is assembled—some work must be found for it—and this is all that there is. The details which an autocratic government most studiously conceals are exposed in open day—the national sums are done in public—finance is made the most of. If the war had not intervened, who knows that by this time Parliament would not be commonly considered 'The Debating Board of Trade'? Intelligent foreigners can hardly be brought to understand this. It puzzles them to imagine how any good or smooth result can be educed from so much jangling, talking, and arguing. M. de Montalembert has described amazement as among his predominant sensations in England. He felt, he says, as if he were in a manufactory—where wheels rolled, and hammers sounded, and engines crunched—where all was certainly noise, and where all seemed to be confusion—but from which, nevertheless, by a miracle of industrial art, some beautiful fabric issued, soft, complete, and perfect. Perhaps this simile is too flattering to the neatness of our legislation, but it happily expresses the depressing noise and tedious din by which its results are really arrived at.

As are the occupations, so are the men. Different kinds of government cause an endless variety in the qualities of statesmen. Not a little of the interest of political history consists in the singular degree in which it shows the mutability and flexibility of human nature. After various changes, we are now arrived at the business statesman—or rather, the business speaker. The details which have to be alluded to, the tedious reforms which have to be effected, the long figures which have to be explained, the slow arguments which require a reply—the heaviness of subjects, in a word—have caused a corresponding weight in our oratory. Our great speeches are speeches of exposition—our eloquence is an eloquence of detail. No one can read or hear the speeches of our ablest and most enlightened statemen without being struck with the contrast which they exhibit—we do not say to the orations of antiquity (which were delivered under circumstances too different to allow of a comparison), but to the great parliamentary displays of the last age—of Pitt, or Fox, or Canning. Differing from each other as the latter do in most of their characteristics, they all fall exactly within Sir James Mackintosh's definition of parliamentary oratory—'animated and continuous after-dinner conversation.' They all have a gentlemanly effervescence and lively agreeability. They are suitable to times when the questions discussed were few, simple, and large—when detail was not—when the first requisite was a pleasant statement of obvious considerations. We are troubled—at least our orators are troubled—with more complex and difficult topics. The patient exposition, the elaborate minuteness, the exhaustive disquisition, of modern parliamentary eloquence, would formerly have been out of place—they are now necessary on complicated subjects, which require the exercise of a laborious intellect, and a discriminating understanding. We have not gained in liveliness by the change, and those who remember the great speakers of the last age are the loudest in complaining of our tedium. The old style still lingers on the lips of Lord Palmerston; but it is daily yielding to a more earnest and practical, to a sober *before*-dinner style.

It is of no light importance that these considerations should be recognised, and their value carefully weighed. It has been the bane of many countries which have tried to obtain freedom, but failed in the attempt, that they have regarded popular government rather as a means of intellectual excitement than as an implement of political work. The preliminary discussion was more interesting than the consequent action. They found it pleasanter to refine arguments than to effect results—more glorious to expand the mind with general ratiocination than to contract it to actual business. They wished, in a word, to have a popular government, without, at the same time, having a dull government. The English people have never yet forgotten what some nations have scarcely ever remembered—that politics are a kind of business—that they bear the characteristics, and obey the laws, inevitably incident to that kind of human action. Steady labour and dull material—wrinkles on the forehead and figures on the tongue—these are the English admiration. We may prize more splendid qualities on uncommon occasions, but these are for daily wear. You cannot have an era *per annum*—if every year had something memorable for posterity, how would posterity ever remember it? Dullness is our line, as cleverness is that of the French. Woe to the English people if they ever forget that, all through their history, heavy topics and tedious talents have awakened the admiration and engrossed the time of their Parliament and their country.

1856

Talking about our Troubles

We may talk about our troubles to those persons who can give us direct help, but even in this case we ought as much as possible to come to a provisional conclusion before consultation; to be perfectly clear to ourselves within our own limits. Some people have a foolish trick of applying for aid before they have done anything whatever to aid themselves, and in fact try to talk themselves into perspicuity. The only way in which they can think is by talking, and their speech consequently is not the expression of opinion already and carefully formed, but the manufacture of it.

We may also tell our troubles to those who are suffering if we can lessen their own. It may be a very great relief to them to know that others have passed through trials equal to theirs and have survived. There are obscure, nervous diseases, hypochondriac fancies, almost uncontrollable impulses, which terrify by their apparent singularity. If we could believe that they are common, the worst of the fear would vanish.

But, as a rule, we should be very careful for our own sake not to speak much about what distresses us. Expression is apt to carry with it exaggeration, and this exaggerated form becomes henceforth that under which we represent our miseries to ourselves, so that they are thereby increased. By reserve, on the other hand, they are diminished, for we attach less importance to that which it was not worth while to mention. Secrecy, in fact, may be our salvation.

It is injurious to be always treated as if something were the matter with us. It is health-giving to be dealt with as if we were healthy, and the man who imagines his wits are failing becomes stronger and sounder by being entrusted with a difficult problem than by all the assurances of a doctor.

They are poor creatures who are always craving for pity. If we are sick, let us prefer conversation upon any subject rather than upon ourselves. Let it turn on matters that lie outside the dark chamber, upon the last new discovery, or the last new idea. So shall we seem still to be linked to the living world. By perpetually asking for sympathy an end is put to real friendship. The friend is afraid to intrude anything which has no direct reference to the patient's condition lest it should be thought irrelevant. No love even can long endure without complaint, silent it may be, against an invalid who is entirely self-centred; and what an agony it is to know that we are tended simply as a duty by those who are nearest to us, and that they will really be relieved when we have departed! From this torture we may be saved if we early apprentice ourselves to the art of self-suppression and sternly apply the gag to eloquence upon our own woes. Nobody who really cares for us will mind waiting on us even to the long-delayed last hour if we endure in fortitude.

There is no harm in confronting our disorders or misfortunes. On the contrary, the attempt is wholesome. Much of what we dread is really due to indistinctness of outline. If we have the courage to say to ourselves, What *is* this thing, then? let the worst come to the worst, and what then? we shall frequently find that after all it is not so terrible. What we have to do is to subdue tremulous, nervous, insane fright. Fright is often prior to an object; that is to say, the fright comes first and something is invented or discovered to account for it. There are certain states of body and mind which are productive of objectless fright, and the most ridiculous thing in the world is able to provoke it to activity. It is perhaps not too much to say that any calamity the moment it is apprehended by the reason alone loses nearly all its power to disturb and unfix us. The conclusions which are so alarming are not those of the reason, but, to use Spinoza's words, of the 'affects.'

1900

From *Autobiography*

NOBODY ever wrote a dull autobiography. If one may make such a bull, the very dullness would be interesting. The autobiographer has *ex officio* two qualifications of supreme importance in all literary work. He is writing about a topic in which he is keenly interested, and about a topic upon which he is the highest living authority. It may be reckoned, too, as a special felicity that an autobiography, alone of all books, may be more valuable in proportion to the amount of misrepresentation which it contains. We do not wonder when a man gives a false character to his neighbour, but it is always curious to see how a man contrives to present a false testimonial to himself. It is pleasant to be admitted behind the scenes and trace the growth of that singular phantom which, like the Spectre of the Brocken, is the man's own shadow cast upon the coloured and distorting mists of memory. Autobiography for these reasons is so generally interesting, that I have frequently thought with the admirable Benvenuto Cellini that it should be considered as a duty by all eminent men; and, indeed, by men not eminent. As every sensible man is exhorted to make his will, he should also be bound to leave to his descendants some account of his experience of life. The dullest of us would in spite of themselves say something profoundly interesting, if only by explaining how they came to be so dull—a circumstance which is sometimes in great need of explanation. On reflection, however, we must admit that autobiography done under compulsion would be in danger of losing the essential charm of spontaneity. The true autobiography is written by one who feels an irresistible longing for confidential expansion; who is forced by his innate constitution to unbosom himself to the public of the kind of matter generally reserved for our closest intimacy. Confessions dictated by a sense of duty, like many records of religious experience, have rarely the peculiar attractiveness of those which are prompted by the simple longing for human sympathy. Nothing, indeed, in all literature is more impressive than some of the writings in which great men have laid bare to us the working of their souls in the severest spiritual crises. But the solemnity and the loftiness of purpose generally remove such work to a rather different category. Augustine's 'Confessions' is an impassioned meditation upon great religious and philosophical questions which only condescends at intervals to autobiographical detail. Few books, to descend a little in the scale, are more interesting, whether to the fellow-believer or to the psychological observer, than Bunyan's 'Grace Abounding'. We follow this real pilgrim through a labyrinth of strange scruples invented by a quick brain placed for the time at the service of a self-torturing impulse, and peopled by the phantoms created by a poetical imagination under stress of profound excitement. Incidentally we learn to know and to love the writer, and certainly not the less because the spiritual fermentation reveals no morbid affectation. We give him credit for exposing the trial and the victory simply and solely for the reason which he alleges; that is to say, because he really thinks that his experience offers useful lessons to his fellow-creatures. He is no attitudiniser, proud at the bottom of his heart of the sensibility which he professes to lament, nor a sanctimonious sentimentalist simulating a false emotion for purposes of ostentation. He is as simple, honest, and soundhearted as he is tender and impassioned. But these very merits deprive the book of some autobiographical interest. It never enters his head that anybody will care about John Bunyan the tinker, or the details of his tinkering. He who painted the scenes in Vanity Fair could have drawn a vivid picture of Elstow and Bedford, of Puritanical preachers and Cromwellian soldiers, and the judges and gaolers under Charles II. Here and there, in scattered passages of his works, he gives us graphic anecdotes in passing which set the scene before us vividly as a bit of Pepys's diaries. The incidents connected with his commitment to prison are described with a dramatic force capable of exciting the envy of a practised reporter. But we see only enough to tantalise us with the possibilities. He tells us so little of his early life that his biographers cannot make up their minds as to whether he was, as Southey calls him, a 'blackguard,' or a few degrees above or below that zero-point of the scale of merit. Lord Macaulay takes it for granted that he was in the Parliamentary, and Mr Froude thinks it amost proved that he was in the Royalist army. He tells us nothing of the death of the first wife, whose love seems to have raised him from blackguardism; nor of his marriage to the second wife, who stood up for him so bravely before the judges, and was his faithful companion to the end of his pilgrimage. The book is therefore a profoundly interesting account of one phase in the development of the character of our great prose-poet; but hardly an autobiography. The narrative was worth writing, because his own heart, like his allegorical Mansoul, had been the scene of one incident in the everlasting struggle between the powers of light and darkness, not because the scene had any independent interest of its own.

In this one may be disposed to say Bunyan judged rightly. The wisest man, it is said, is he who realises most clearly the narrow limits of human knowledge; the greatest should be penetrated with the strongest conviction of his own insignificance. The higher we rise above the average mass of mankind, the more clearly we should see our own incapacity for acting the part of Providence. The village squire, who does not really believe in anything invisible from his own steeple, may fancy that he is of real importance to the world, for the world for him means his village. 'P. P. clerk of this parish' thought that all future generations would be interested in the fact that he had smoothed the dog's-ears in the great Bible. A genuine statesman who knows something of the forces by which the world is governed should have seen through the humbug of history. He should have learnt the fable of the fly and the chariot wheel, and be aware that what are called his achievements are really the events upon which, through some accident of position, he has been allowed to inscribe his name. One stage in a nation's life gets itself labelled Cromwell, and another William Pitt; but perhaps Pitt and Cromwell were really of little more importance than some contemporary P. P. This doctrine, however, is considered, I know not why, to be immoral, and to smack of fatalism, cynicism, jealousy of great men, and other objectionable tendencies. We are in a tacit conspiracy to flatter conspicuous men at the expense of their fellow-workers, and he is the most generous and appreciative who can heap the greatest number of superlatives upon growing reputations, and add a stone to the gigantic pile of eulogy under which the historical proportions of some great figures are pretty well buried. We must not complain, therefore, if we flatter the vanity which seems to be the most essential ingredient in the composition of a model *autobiographer*. A man who expects that future generations will be profoundly interested in the state of his interior seems to be drawing a heavy bill upon posterity. And yet it is generally honoured. We are flattered perhaps by this exhibition of confidence. We are touched by the demand for sympathy. There is something pathetic in this belief that we shall be moved by the record of past sufferings and aspirations as there is in a child's confidence that you will enter into its little fears and hopes. And perhaps vanity is so universal a weakness, and, in spite of good moralising, it so strongly resembles a virtue in some of its embodiments, that we cannot find it in our hearts to be angry with it. We can understand it too thoroughly. And then we make an ingenious compromise with our consciences. Our interest in Pepys's avowals of his own foibles, for example, is partly due to the fact that whilst we are secretly conscious of at least the germs of similar failings, the consciousness does not bring any sense of shame, because we set down the confession to the account of poor Pepys himself. The man who, like Goldsmith, is so running over with jealousy that he is forced to avow it openly, seems to be a sort of excuse to us for cherishing a less abundant stock of similar sentiment. This is one occult source of pleasure in reading autobiography. We have a delicate shade of conscious superiority in listening to the vicarious confession. 'I am sometimes troubled,' said Boswell, 'by a disposition to stinginess.' 'So am I,' replied Johnson, 'but I do not tell it.' That is our attitude in regard to the autobiographer. After all, we say to ourselves, this distinguished person is such a one as we are; and even more so, for he cannot keep it to himself. The conclusion is not quite fair, it may be, when applied to the case of a diarist like Pepys, who, poor man, meant only to confide his thoughts to his note-books. But it applies more or less to every genuine autobiographer—to every man, that is, who has deliberately written down a history of his own feelings and thoughts for the benefit of posterity.

1885

SAMUEL BUTLER

On Knowing what Gives us Pleasure

i

ONE can bring no greater reproach against a man than to say
that he does not set sufficient value upon pleasure, and there is
no greater sign of a fool than the thinking that he can tell at once
and easily what it is that pleases him. To know this is not easy,
and how to extend our knowledge of it is the highest and the
most neglected of all arts and branches of education. Indeed, if
we could solve the difficulty of knowing what gives us pleasure,
if we could find its springs, its inception and earliest *modus
operandi*, we should have discovered the secret of life and development, for the same difficulty has attended the development of
every sense from touch onwards, and no new sense was ever
developed without pains. A man had better stick to known and
proved pleasures, but, if he will venture in quest of new ones, he
should not do so with a light heart.

One reason why we find it so hard to know our own likings is
because we are so little accustomed to try; we have our likings
found for us in respect of by far the greater number of the
matters that concern us; thus we have grown all our limbs on the
strength of the likings of our ancestors and adopt these without
question.

Another reason is that, except in mere matters of eating and
drinking, people do not realise the importance of finding out
what it is that gives them pleasure if, that is to say, they would
make themselves as comfortable here as they reasonably can.
Very few, however, seem to care greatly whether they are comfortable or no. There are some men so ignorant and careless of
what gives them pleasure that they cannot be said ever to have
been really born as living beings at all. They present some of the
phenomena of having been born—they reproduce, in fact, so
many of the ideas which we associate with having been born that

it is hard not to think of them as living beings—but in spite of all appearances the central idea is wanting. At least one half of the misery which meets us daily might be removed or, at any rate, greatly alleviated, if those who suffer by it would think it worth their while to be at any pains to get rid of it. That they do not so think is proof that they neither know, nor care to know, more than in a very languid way, what it is that will relieve them most effectually or, in other words, that the shoe does not really pinch them so hard as we think it does. For when it really pinches, as when a man is being flogged, he will seek relief by any means in his power. So my great namesake said, 'Surely the pleasure is as great Of being cheated as to cheat'; and so, again, I remember to have seen a poem many years ago in *Punch* according to which a certain young lady, being discontented at home, went out into the world in quest to 'Some burden make or burden bear. But which she did not greatly care—Oh Miseree!' So long as there was discomfort somewhere it was all right.

To those, however, who are desirous of knowing what gives them pleasure but do not quite know how to set about it I have no better advice to give than that they must take the same pains about acquiring this difficult art as about any other, and must acquire it in the same way—that is by attending to one thing at a time and not being in too great a hurry. Proficiency is not to be attained here, any more than elsewhere, by short cuts or by getting other people to do work that no other than oneself can do. Above all things it is necessary here, as in all other branches of study, not to think we know a thing before we do know it—to make sure of our ground and be quite certain that we really do like a thing before we say we do. When you cannot decide whether you like a thing or not, nothing is easier than to say so and to hang it up among the uncertainties. Or when you know you do not know and are in such doubt as to see no chance of deciding, then you may take one side or the other provisionally and throw youself into it. This will sometimes make you uncomfortable, and you will feel you have taken the wrong side and thus learn that the other was the right one. Sometimes you will feel you have done right. Any way ere long you will know more about it. But there must have been a secret treaty with yourself to the effect that the decision was provisional only. For, after all, the most important first principle in this matter is the not lightly

thinking you know what you like till you have made sure of your
ground. I was nearly forty before I felt how stupid it was to pretend to know things that I did not know and I still often catch
myself doing so. Not one of my school-masters taught me this,
but altogether otherwise.

ii

I should like to like Schumann's music better than I do; I dare
say I could make myself like it better if I tried; but I do not like
having to try to make myself like things; I like things that make
me like them at once and no trying at all.

iii

To know whether you are enjoying a piece of music or not you
must see whether you find yourself looking at the advertisements
of Pears' soap at the end of the programme.

c.1880

Thoughts of God

How often we are moved to admit the intelligence exhibited in both the designing and the execution of some of His works. Take the fly, for instance. The planning of the fly was an application of pure intelligence, morals not being concerned. Not one of us could have planned the fly, not one of us could have constructed him; and no one would have considered it wise to try, except under an assumed name. It is believed by some that the fly was introduced to meet a long-felt want. In the course of ages, for some reason or other, there have been millions of these persons, but out of this vast multitude there has not been one who has been willing to explain what the want was. At least satisfactorily. A few have explained that there was need of a creature to remove disease-breeding garbage; but these being then asked to explain what long-felt want the disease-breeding garbage was introduced to supply, they have not been willing to undertake the contract.

There is much inconsistency concerning the fly. In all the ages he has not had a friend, there has never been a person in the earth who could have been persuaded to intervene between him and extermination; yet billions of persons have excused the Hand that made him—and this without a blush. Would they have excused a Man in the same circumstances, a man positively known to have invented the fly? On the contrary. For the credit of the race let us believe it would have been all day with that man. Would these persons consider it just to reprobate in a child, with its undeveloped morals, a scandal which they would overlook in the Pope?

When we reflect that the fly was not invented for pastime, but in the way of business; that he was not flung off in a heedless moment and with no object in view but to pass the time, but was the fruit of long and pains-taking labor and calculation, and with a definite and far-reaching purpose in view; that his character and conduct were planned out with cold deliberation; that his career was foreseen and foreordered, and that there was no want which he could supply, we are hopelessly puzzled, we cannot understand the moral lapse that was able to render possible the conceiving and the consummation of this squalid and malevolent creature.

Let us try to think the unthinkable; let us try to imagine a Man of a sort willing to invent the fly; that is to say, a man destitute of feeling; a man willing to wantonly torture and harass and persecute myriads of creatures who had never done him any harm and could not if they wanted to, and—the majority of them—poor dumb things not even aware of his existence. In a word, let us try to imagine a man with so singular and so lumbering a code of morals as this: that it is fair and right to send afflictions upon the *just*—upon the unoffending as well as upon the offending, without discrimination.

If we can imagine such a man, that is the man that could invent the fly, and send him out on his mission and furnish him his orders: 'Depart into the uttermost corners of the earth, and diligently do your appointed work. Persecute the sick child; settle upon its eyes, its face, its hands, and gnaw and pester and sting; worry and fret and madden the worn and tried mother who watches by the child, and who humbly prays for mercy and relief with the pathetic faith of the deceived and the unteachable. Settle upon the soldier's festering wounds in field and hospital and drive him frantic while he also prays, and betweentimes curses, with none to listen but you, Fly, who get all the petting and all the protection, without even praying for it. Harry and persecute the forlorn and forsaken wretch who is perishing of the plague, and in his terror and despair praying; bite, sting, feed upon his ulcers, dabble your feet in his rotten blood, gum them thick with plague-germs—feet cunningly designed and perfected for this function ages ago in the beginning—carry this freight to a hundred tables, among the just and the unjust, the high and the low, and walk over the food and gaum it with filth and death. Visit all; allow no man peace till he get it in the grave; visit and afflict the hard-worked and unoffending horse, mule, ox, ass, pester the patient cow, and all the kindly animals that

labor without fair reward here and perish without hope of it
hereafter; spare no creature, wild or tame; but wheresoever you
find one, make his life a misery, treat him as the innocent
deserve; and so please Me and increase My glory Who made the
fly.'

We hear much about His patience and forbearance and long-
suffering; we hear nothing about our own, which much exceeds
it. We hear much about His mercy and kindness and goodness—in words—the words of His Book and of His pulpit—and
the meek multitude is content with this evidence, such as it is,
seeking no further; but whoso searcheth after a concreted sample
of it will in time acquire fatigue. There being no instances of it.
For what are gilded as mercies are not in any recorded case more
than mere common justices, and *due*—due without thanks or
compliment. To rescue without personal risk a cripple from a
burning house is not a mercy, it is a mere commonplace duty;
anybody would do it that could. And not by proxy, either—
delegating the work but confiscating the credit for it. If men
neglected 'God's poor' and 'God's stricken and helpless ones' as
He does, what would become of them? The answer is to be
found in those dark lands where man follows His example and
turns his indifferent back upon them: they get no help at all; they
cry, and plead and pray in vain, they linger and suffer, and
miserably die. If you will look at the matter rationally and without prejudice, the proper place to hunt for the *facts* of His mercy,
is not where man does the mercies and He collects the praise, but
in those regions where He has the field to Himself.

It is plain that there is one moral law for heaven and another
for the earth. The pulpit assures us that wherever we see suffering and sorrow which we can relieve and do not do it, we sin,
heavily. *There was never yet a case of suffering or sorrow which God
could not relieve*. Does He sin, then? If He is the Source of Morals
He does—certainly nothing can be plainer than that, you will
admit. Surely the Source of law cannot violate law and stand
unsmirched; surely the judge upon the bench cannot forbid
crime and then revel in it himself unreproached. Nevertheless we
have this curious spectacle: daily the trained parrot in the pulpit
gravely delivers himself of these ironies, which he has acquired
at second-hand and adopted without examination, to a trained
congregation which accepts them without examination, and

neither the speaker nor the hearer laughs at himself. It does seem as if we ought to be humble when we are at a bench-show, and not put on airs of intellectual superiority there.

early 1890s

Sandro Botticelli

In Leonardo's treatise on painting only one contemporary is mentioned by name—Sandro Botticelli. This pre-eminence may be due to chance only, but to some will rather appear a result of deliberate judgment; for people have begun to find out the charm of Botticelli's work, and his name, little known in the last century, is quietly becoming important. In the middle of the fifteenth century he had already anticipated much of that meditative subtlety, which is sometimes supposed peculiar to the great imaginative workmen of its close. Leaving the simple religion which had occupied the followers of Giotto for a century, and the simple naturalism which had grown out of it, a thing of birds and flowers only, he sought inspiration in what to him were works of the modern world, the writings of Dante and Boccaccio, and in new readings of his own of classical stories: or, if he painted religious incidents, painted them with an undercurrent of original sentiment, which touches you as the real matter of the picture through the veil of its ostensible subject. What is the peculiar sensation, what is the peculiar quality of pleasure, which his work has the property of exciting in us, and which we cannot get elsewhere? For this, especially when he has to speak of a comparatively unknown artist, is always the chief question which a critic has to answer.

In an age when the lives of artists were full of adventure, his life is almost colourless. Criticism indeed has cleared away much of the gossip which Vasari accumulated, has touched the legend of Lippo and Lucrezia, and rehabilitated the character of Andrea del Castagno. But in Botticelli's case there is no legend to dissipate. He did not even go by his true name: Sandro is a nickname, and his true name is Filipepi, Botticelli being only the name of the goldsmith who first taught him art. Only two things

happened to him, two things which he shared with other artists:
—he was invited to Rome to paint in the Sistine Chapel, and he
fell in later life under the influence of Savonarola, passing apparently almost out of men's sight in a sort of religious melancholy,
which lasted till his death in 1515, according to the received date.
Vasari says that he plunged into the study of Dante, and even
wrote a comment on the *Divine Comedy*. But it seems strange that
he should have lived on inactive so long; and one almost wishes
that some document might come to light, which, fixing the date
of his death earlier, might relieve one, in thinking of him, of his
dejected old age.

He is before all things a poetical painter, blending the charm
of story and sentiment, the medium of the art of poetry, with the
charm of line and colour, the medium of abstract painting. So he
becomes the illustrator of Dante. In a few rare examples of the
edition of 1481, the blank spaces, left at the beginning of every
canto for the hand of the illuminator, have been filled, as far as
the nineteenth canto of the *Inferno*, with impressions of engraved
plates, seemingly by way of experiment, for in the copy in the
Bodleian Library, one of the three impressions it contains has
been printed upside down, and much awry, in the midst of the
luxurious printed page. Giotto, and the followers of Giotto, with
their almost childish religious aim, had not learned to put that
weight of meaning into outward things, light, colour, everyday
gesture, which the poetry of the *Divine Comedy* involves, and
before the fifteenth century Dante could hardly have found an
illustrator. Botticelli's illustrations are crowded with incident,
blending, with a naïve carelessness of pictorial propriety, three
phases of the same scene into one plate. The grotesques, so often
a stumbling-block to painters, who forget that the words of a
poet, which only feebly present an image to the mind, must be
lowered in key when translated into visible form, make one
regret that he has not rather chosen for illustration the more
subdued imagery of the *Purgatorio*. Yet in the scene of those who
'go down quick into hell,' there is an inventive force about the
fire taking hold on the upturned soles of the feet, which proves
that the design is no mere translation of Dante's words, but a
true painter's vision; while the scene of the Centaurs wins one at
once, for, forgetful of the actual circumstances of their appearance, Botticelli has gone off with delight on the thought of the

Centaurs themselves, bright, small creatures of the woodland, with arch baby faces and mignon forms, drawing tiny bows.

Botticelli lived in a generation of naturalists, and he might have been a mere naturalist among them. There are traces enough in his work of that alert sense of outward things, which, in the pictures of that period, fills the lawns with delicate living creatures, and the hillsides with pools of water, and the pools of water with flowering reeds. But this was not enough for him; he is a visionary painter, and in his visionariness he resembles Dante. Giotto, the tried companion of Dante, Masaccio, Ghirlandajo even, do but transcribe, with more or less refining, the outward image; they are dramatic, not visionary painters; they are almost impassive spectators of the action before them. But the genius of which Botticelli is the type usurps the data before it as the exponent of ideas, moods, visions of its own; in this interest it plays fast and loose with those data, rejecting some and isolating others, and always combining them anew. To him, as to Dante, the scene, the colour, the outward image or gesture, comes with all its incisive and importunate reality; but awakes in him, moreover, by some subtle law of his own structure, a mood which it awakes in no one else, for which it is the double or repetition, and which it clothes, that all may share it, with visible circumstance.

But he is far enough from accepting the conventional orthodoxy of Dante which, referring all human action to the simple formula of purgatory, heaven and hell, leaves an insoluble element of prose in the depths of Dante's poetry. One picture of his, with the portrait of the donor, Matteo Palmieri, below, had the credit or discredit of attracting some shadow of ecclesiastical censure. This Matteo Palmieri, (two dim figures move under that name in contemporary history,) was the reputed author of a poem, still unedited, *La Città Divina*, which represented the human race as an incarnation of those angels who, in the revolt of Lucifer, were neither for Jehovah nor for His enemies, a fantasy of that earlier Alexandrian philosophy about which the Florentine intellect in that century was so curious. Botticelli's picture may have been only one of those familiar compositions in which religious reverie has recorded its impressions of the various forms of beatified existence—*Glorias*, as they were called, like that in which Giotto painted the portrait of Dante; but

somehow it was suspected of embodying in a picture the wayward dream of Palmieri, and the chapel where it hung was
closed. Artists so entire as Botticelli are usually careless about
philosophical theories, even when the philosopher is a Florentine
of the fifteenth century, and his work a poem in *terza rima*. But
Botticelli, who wrote a commentary on Dante, and became the
disciple of Savonarola, may well have let such theories come and
go across him. True or false, the story interprets much of the
peculiar sentiment with which he infuses his profane and sacred
persons, comely, and in a certain sense like angels, but a sense of
displacement or loss about them—the wistfulness of exiles, conscious of a passion and energy greater than any known issue of
them explains, which runs through all his varied work with a
sentiment of ineffable melancholy.

So just what Dante scorns as unworthy alike of heaven and
hell, Botticelli accepts, that middle world in which men take no
side in great conflicts, and decide no great causes, and make
great refusals. He thus sets for himself the limits within which
art, undisturbed by any moral ambition, does its most sincere
and surest work. His interest is neither in the untempered goodness of Angelico's saints, nor the untempered evil of Orcagna's
Inferno; but with men and women, in their mixed and uncertain
condition, always attractive, clothed sometimes by passion with
a character of loveliness and energy, but saddened perpetually
by the shadow upon them of the great things from which they
shrink. His morality is all sympathy; and it is this sympathy,
conveying into his work somewhat more than is usual of the true
complexion of humanity, which makes him, visionary as he is, so
forcible a realist.

It is this which gives to his Madonnas their unique expression
and charm. He has worked out in them a distinct and peculiar
type, definite enough in his own mind, for he has painted it over
and over again, sometimes one might think almost mechanically,
as a pastime during that dark period when his thoughts were so
heavy upon him. Hardly any collection of note is without one of
these circular pictures, into which the attendant angels depress
their heads so naïvely. Perhaps you have sometimes wondered
why those peevish-looking Madonnas, conformed to no acknowledged or obvious type of beauty, attract you more and
more, and often come back to you when the Sistine Madonna

and the Virgins of Fra Angelico are forgotten. At first, contrasting them with those, you may have thought that there was something in them mean or abject even, for the abstract lines of the face have little nobleness, and the colour is wan. For with Botticelli she too, though she holds in her hands the 'Desire of all nations,' is one of those who are neither for Jehovah nor for His enemies; and her choice is on her face. The white light on it is cast up hard and cheerless from below, as when snow lies upon the ground, and the children look up with surprise at the strange whiteness of the ceiling. Her trouble is in the very caress of the mysterious child, whose gaze is always far from her, and who has already that sweet look of devotion which men have never been able altogether to love, and which still makes the born saint an object almost of suspicion to his earthly brethren. Once, indeed, he guides her hand to transcribe in a book the words of her exaltation, the *Ave*, and the *Magnificat*, and the *Gaude Maria*, and the young angels, glad to rouse her for a moment from her dejection, are eager to hold the inkhorn and to support the book. But the pen almost drops from her hand, and the high cold words have no meaning for her, and her true children are those others, among whom, in her rude home, the intolerable honour came to her, with that look of wistful inquiry on their irregular faces which you see in startled animals—gipsy children, such as those who, in Apennine villages, still hold out their long brown arms to beg of you, but on Sundays become *enfants du chœur*, with their thick black hair nicely combed, and fair white linen on their sunburnt throats.

What is strangest is that he carries this sentiment into classical subjects, its most complete expression being a picture in the *Uffizii*, of Venus rising from the sea, in which the grotesque emblems of the middle age, and a landscape full of its peculiar feeling, and even its strange draperies, powdered all over in the Gothic manner with a quaint conceit of daisies, frame a figure that reminds you of the faultless nude studies of Ingres. At first, perhaps, you are attracted only by a quaintness of design, which seems to recall all at once whatever you have read of Florence in the fifteenth century; afterwards you may think that this quaintness must be incongruous with the subject, and that the colour is cadaverous or at least cold. And yet, the more you come to understand what imaginative colouring really is, that all colour is no mere delightful quality of natural things, but a spirit upon them by which they become expressive to the spirit, the better you will like this peculiar quality of colour; and you will find that quaint design of Botticelli's a more direct inlet into the Greek temper than the works of the Greeks themselves even of the finest period. Of the Greeks as they really were, of their difference from ourselves, of the aspects of their outward life, we know far more than Botticelli, or his most learned contemporaries; but for us long familiarity has taken off the edge of the lesson, and we are hardly conscious of what we owe to the Hellenic spirit. But in pictures like this of Botticelli's you have a record of the first impression made by it on minds turned back towards it, in almost painful aspiration, from a world in which it had been ignored so long; and in the passion, the energy, the industry of realisation, with which Botticelli carries out his intention, is the exact measure of the legitimate influence over the human mind of the imaginative system of which this is perhaps the central myth. The light is indeed cold—mere sunless dawn; but a later painter would have cloyed you with sunshine; and you can see the better for that quietness in the morning air each long promontory, as it slopes down to the water's edge. Men go forth to their labours until the evening; but she is awake before them, and you might think that the sorrow in her face was at the thought of the whole long day of love yet to come. An emblematical figure of the wind blows hard across the grey water, moving forward the dainty-lipped shell on which she sails, the sea 'showing his teeth,' as it moves, in thin lines of foam, and sucking in, one by one, the falling roses, each severe in outline, plucked off short at the stalk, but embrowned a little, as Botticelli's flowers always are. Botticelli meant all this imagery to be altogether pleasurable; and it was partly an incompleteness of resources, inseparable from the art of that time, that subdued and chilled it. But this predilection for minor tones counts also; and what is unmistakable is the sadness with which he has conceived the goddess of pleasure, as the depositary of a great power over the lives of men.

I have said that the peculiar character of Botticelli is the result of a blending in him of a sympathy for humanity in its uncertain condition, its attractiveness, its investiture at rarer moments in a character of loveliness and energy, with his consciousness of the shadow upon it of the great things from which it shrinks, and that this conveys into his work somewhat more than painting usually attains of the true complexion of humanity. He paints the story of the goddess of pleasure in other episodes besides that of her birth from the sea, but never without some shadow of death in the grey flesh and wan flowers. He paints Madonnas, but they shrink from the pressure of the divine child, and plead in unmistakable undertones for a warmer, lower humanity. The same figure—tradition connects it with Simonetta, the Mistress of Giuliano de'Medici—appears again as Judith, returning home across the hill country, when the great deed is over, and the moment of revulsion come, when the olive branch in her hand is becoming a burthen; as *Justice*, sitting on a throne, but with a fixed look of self-hatred which makes the sword in her hand seem that of a suicide; and again as *Veritas*, in the allegorical picture of *Calumnia*, where one may note in passing the suggestiveness of an accident which identifies the image of Truth with the person of Venus. We might trace the same sentiment through his engravings; but his share in them is doubtful, and the object of this brief study has been attained, if I have defined aright the temper in which he worked.

But, after all, it may be asked, is a painter like Botticelli—a secondary painter, a proper subject for general criticism? There are a few great painters, like Michelangelo or Leonardo, whose work has become a force in general culture, partly for this very reason that they have absorbed into themselves all such workmen as Sandro Botticelli; and, over and above mere technical or antiquarian criticism, general criticism may be very well employed in that sort of interpretation which adjusts the position of these men to general culture, whereas smaller men can be the proper subjects only of technical or antiquarian treatment. But, besides those great men, there is a certain number of artists who have a distinct faculty of their own by which they convey to us a peculiar quality of pleasure which we cannot get elsewhere; and these too have their place in general culture, and must be interpreted to it by those who have felt their charm strongly, and are often the object of a special diligence and a consideration wholly affectionate, just because there is not about them the stress of a great name and authority. Of this select number Botticelli is one. He has the freshness, the uncertain and diffident promise, which belong to the earlier Renaissance itself, and make it perhaps the most interesting period in the history of the mind. In studying his work one begins to understand to how great a place in human culture the art of Italy had been called.

1870

Wasps

ONE rough day in early autumn I paused in my walk in a Surrey orchard to watch a curious scene in insect life—a pretty little insect comedy I might have called it had it not brought back to remembrance old days when my mind was clouded with doubts, and the ways of certain insects, especially of wasps, were much in my thoughts. For we live through and forget many a tempest that shakes us; but long afterwards a very little thing—the scent of a flower, the cry of a wild bird, even the sight of an insect—may serve to bring it vividly back and to revive a feeling that seemed dead and gone.

In the orchard there was an old pear-tree which produced very large late pears, and among the fruit the September wind had shaken down that morning there was one over-ripe in which the wasps had eaten a deep cup-shaped cavity. Inside the cavity six or seven wasps were revelling in the sweet juice, lying flat and motionless, crowded together. Outside the cavity, on the pear, thirty or forty blue-bottle flies had congregated, and were hungry for the juice, but apparently afraid to begin feeding on it; they were standing round in a compact crowd, the hindmost pressing on and crowding over the others: but still, despite the pressure, the foremost row of flies refused to advance beyond the rim of the eaten-out part. From time to time one of a more venturesome spirit would put out his proboscis and begin sucking at the edge; the slight tentative movement would instantly be detected by a wasp, and he would turn quickly round to face the presumptuous fly, lifting his wings in a threatening manner, and the fly would take his proboscis off the rim of the cup. Occasionally hunger would overcome their fear; a general movement of the flies would take place, and several would begin sucking at the same time; then the wasp, seeming to think that more than a mere menacing look or gesture was required in such a case, would start up with an angry buzz, and away the whole crowd of flies would go to whirl round and round in a little blue cloud with a loud, excited hum, only to settle again in a few moments on the big yellow pear and begin crowding round the pit as before.

Never once during the time I spent observing them did the guardian wasp relax his vigilance. When he put his head down to suck with the others his eyes still appeared able to reflect every movement in the surrounding crowd of flies into his little spiteful brain. They could crawl round and crawl round as much as they liked on the very rim, but let one begin to suck and he was up in arms in a moment.

The question that occurred to me was: How much of all this behaviour could be set down to instinct and how much to intelligence and temper? The wasp certainly has a waspish disposition, a quick resentment, and is most spiteful and tyrannical towards other inoffensive insects. He is a slayer and devourer of them, too, as well as a feeder with them on nectar and sweet juices; but when he kills, and when the solitary wasp paralyses spiders, caterpillars, and various insects and stores them in cells to provide a horrid food for the grubs which will eventually hatch from the still undeposited eggs, the wasp then acts automatically, or by instinct, and is driven, as it were, by an extraneous force. In a case like the one of the wasp's behaviour on the pear, and in innumerable other cases which one may read of or see for himself, there appears to be a good deal of the element of mind. Doubtless it exists in all insects, but differs in degree; and some Orders appear to be more intelligent than others. Thus, any person accustomed to watch insects closely and note their little acts would probably say that there is less mind in the beetles and more in the Hymenoptera than in other insects; also that in the last-named Order the wasps rank highest.

The scene in the orchard also served to remind me of a host of wasps, greatly varying in size, colour, and habits, although in their tyrannical temper very much alike, which I had been accustomed to observe in boyhood and youth in a distant region. They attracted me more, perhaps, than any other insects on account of their singular and brilliant coloration and their formidable character. They were beautiful but painful creatures; the pain they caused me was first bodily, when I interfered in their concerns or handled them carelessly, and was soon over; later it was mental and more enduring.

To the very young colour is undoubtedly the most attractive quality in nature, and these insects were enamelled in colours that made them the rivals of butterflies and shining metallic beetles. There were wasps with black and yellow rings and with black and scarlet rings; wasps of a uniform golden brown; others like our demoiselle dragon-fly that looked as if fresh from a bath of splendid metallic blue; others with steel-blue bodies and bright red wings; others with crimson bodies, yellow head and legs, and bright blue wings; others black and gold, with pink head and legs; and so on through scores and hundreds of species 'as Nature list to play with her little ones,' until one marvelled at so great a variety, so many singular and beautiful contrasts, produced by half-a-dozen brilliant colours.

It was when I began to find out the ways of wasps with other insects on which they nourish their young that my pleasure in them became mixed with pain. For they did not, like spiders, ants, dragon-flies, tiger-beetles, and other rapacious kinds, kill their prey at once, but paralysed it by stinging its nerve centres to make it incapable of resistance, and stored it in a closed cell, so that the grub to be hatched by and by should have fresh meat to feed on—not fresh-killed but live meat.

Thus the old vexed question—How reconcile these facts with the idea of a beneficent Being who designed it all—did not come to me from reading, nor from teachers, since I had none, but was thrust upon me by nature itself. In spite, however, of its having come in that sharp way, I, like many another, succeeded in putting the painful question from me and keeping to the old traditions. The noise of the battle of Evolution, which had been going on for years, hardly reached me; it was but a faintly heard murmur, as of storms immeasurably far away 'on alien shores.' This could not last.

One day an elder brother, on his return from travel in distant lands, put a copy of the famous *Origin of Species* in my hands and advised me to read it. When I had done so, he asked me what I thought of it. 'It's false!' I exclaimed in a passion, and he laughed, little knowing how important a matter this was to me, and told me I could have the book if I liked. I took it without thanks and read it again and thought a good deal about it, and was nevertheless able to resist its teachings for years, solely because I could not endure to part with a philosophy of life, if I may so describe it, which could not logically be held, if Darwin was right, and without which life would not be worth having. So I thought at the time; it is a most common delusion of the human mind, for we see that the good which is so much to us is taken forcibly away, and that we get over our loss and go on very much as before.

It is curious to see now that Darwin himself gave the first comfort to those who, convinced against their will, were anxious to discover some way of escape which would not involve the total abandonment of their cherished beliefs. At all events, he suggested the idea, which religious minds were quick to seize upon, that the new explanation of the origin of the innumerable forms of life which people the earth from one or a few primordial organisms afforded us a nobler conception of the creative mind than the traditional one. It does not bear examination, probably it originated in the author's kindly and compassionate feelings rather than in his reasoning faculties; but it gave temporary relief and served its purpose. Indeed, to some, to very many perhaps, it still serves as a refuge—this poor, hastily made straw shelter, which lets in the rain and wind, but seems better to them than no shelter at all.

But of the intentionally consoling passages in the book, the most impressive to me was that in which he refers to instincts and adaptation such as those of the wasp, which writers on natural history subjects are accustomed to describe, in a way that seems quite just and natural, as *diabolical*. That, for example, of the young cuckoo ejecting its foster-brothers from the nest; of slave-making ants, and of the larvae of the Ichneumonidae feeding on the live tissues of the caterpillars in whose bodies they have been hatched. He said that it was not perhaps a logical conclusion, but it seemed to him more satisfactory to regard such things 'not as specially endowed or created instincts, but as small consequences of one general law'—the law of variation and the survival of the fittest.

1905

Disintroductions

T HE devil is a citizen of every country, but only in our own are we in constant peril of an introduction to him. That is democracy. All men are equal; the devil is a man; therefore, the devil is equal. If that is not a good and sufficient syllogism I should be pleased to know what is the matter with it.

To write in riddles when one is not prophesying is too much trouble; what I am affirming is the horror of the characteristic American custom of promiscuous, unsought and unauthorized introductions.

You incautiously meet your friend Smith in the street; if you had been prudent you would have remained indoors. Your help-lessness makes you desperate and you plunge into conversation with him, knowing entirely well the disaster that is in cold storage for you.

The expected occurs: another man comes along and is promptly halted by Smith and you are introduced! Now, you have not given to the Smith the right to enlarge your circle of acquaintance and select the addition himself; why did he do this thing? The person whom he has condemned you to shake hands with may be an admirable person, though there is a strong numerical presumption against it; but for all that the Smith knows he may be your bitterest enemy. The Smith has never thought of that. Or you may have evidence (independent of the fact of the introduction) that he is some kind of thief—there are one thousand and fifty kinds of thieves. But the Smith has never thought of that. In short, the Smith has never thought. In a Smithocracy all men, as aforesaid, being equal, all are equally agreeable to one another.

That is a logical extension of the Declaration of American Independence. If it is erroneous the assumption that a man will be pleasing to me because he is pleasing to another is erroneous too, and to introduce me to one that I have not asked nor consented to know is an invasion of my rights—a denial and limitation of my liberty to a voice in my own affairs. It is like determining what kind of clothing I shall wear, what books I shall read, or what my dinner shall be.

In calling promiscuous introducing an American custom I am not unaware that it obtains in other countries than ours. The difference is that in those it is mostly confined to persons of no consequence and no pretensions to respectability; here it is so nearly universal that there is no escaping it. Democracies are naturally and necessarily gregarious. Even the French of to-day are becoming so, and the time is apparently not distant when they will lose that fine distinctive social sense that has made them the most punctilious, because the most considerate, of all nations excepting the Spanish and the Japanese. By those who have lived in Paris since I did I am told that the chance introduction is beginning to devastate the social situation, and men of sense who wish to know as few persons as possible can no longer depend on the discretion of their friends.

To say so is not the same thing as to say 'Down with the republic!' The republic has its advantages. Among these is the liberty to say, 'Down with the republic!'

It is to be wished that some great social force, say a billionaire, would set up a system of disintroductions. It should work somewhat like this:

MR WHITE—Mr Black, knowing the low esteem in which you hold each other, I have the honor to disintroduce you from Mr Green.

MR BLACK (*bowing*)—Sir, I have long desired the advantage of your unacquaintance.

MR GREEN (*bowing*)—Charmed to unmeet you, sir. Our acquaintance (the work of a most inconsiderate and unworthy person) has distressed me beyond expression. We are greatly indebted to our good friend here for his tact in repairing the mischance.

MR WHITE—Thank you. I'm sure you will become very good strangers.

This is only the ghost of a suggestion; of course the plan is capable of an infinite elaboration. Its capital defect is that the

persons who are now so liberal with their unwelcome introductions, will be equally lavish with their disintroductions, and
will estrange the best of friends with as little ceremony as they
now observe in their more fiendish work.

1902

The Ph.D. Octopus

SOME years ago we had at our Harvard Graduate School a very brilliant student of Philosophy, who, after leaving us and supporting himself by literary labor for three years, received an appointment to teach English Literature at a sister-institution of learning. The governors of this institution, however, had no sooner communicated the appointment than they made the awful discovery that they had enrolled upon their staff a person who was unprovided with the Ph.D. degree. The man in question had been satisfied to work at Philosophy for her own sweet (or bitter) sake, and had disdained to consider that an academic bauble should be his reward.

His appointment had thus been made under a misunderstanding. He was not the proper man; and there was nothing to do but to inform him of the fact. It was notified to him by his new President that his appointment must be revoked, or that a Harvard doctor's degree must forthwith be procured.

Although it was already the spring of the year, our Subject, being a man of spirit, took up the challenge, turned his back upon literature (which in view of his approaching duties might have seemed his more urgent concern) and spent the weeks that were left him, in writing a metaphysical thesis and grinding his psychology, logic and history of philosophy up again, so as to pass our formidable ordeals.

When the thesis came to be read by our committee, we could not pass it. Brilliancy and originality by themselves won't save a thesis for the doctorate; it must also exhibit a heavy technical apparatus of learning; and this our candidate had neglected to bring to bear. So, telling him that he was temporarily rejected, we advised him to pad out the thesis properly, and return with it next year, at the same time informing his new President that this signified nothing as to his merits, that he was of ultra Ph.D. quality, and one of the strongest men with whom we had ever had to deal.

To our surprise we were given to understand in reply that the quality *per se* of the man signified nothing in this connection, and that three magical letters were the thing seriously required. The College had always gloried in a list of faculty members who bore the doctor's title, and to make a gap in the galaxy, and admit a common fox without a tail, would be a degradation impossible to be thought of. We wrote again, pointing out that a Ph.D. in philosophy would prove little anyhow as to one's ability to teach literature; we sent separate letters in which we outdid each other in eulogy of our candidate's powers, for indeed they were great; and at last, *mirabile dictu*, our eloquence prevailed. He was allowed to retain his appointment provisionally, on condition that one year later at the farthest his miserably naked name should be prolonged by the sacred appendage the lack of which had given so much trouble to all concerned.

Accordingly he came up here the following spring with an adequate thesis (known since in print as a most brilliant contribution to metaphysics), passed a first-rate examination, wiped out the stain, and brought his college into proper relations with the world again. Whether his teaching, during that first year, of English Literature was made any the better by the impending examination in a different subject, is a question which I will not try to solve.

I have related this incident at such length because it is so characteristic of American academic conditions at the present day. Graduate schools still are something of a novelty, and higher diplomas something of a rarity. The latter, therefore, carry a vague sense of preciousness and honor, and have a particularly 'up-to-date' appearance, and it is no wonder if smaller institutions, unable to attract professors already eminent, and forced usually to recruit their faculties from the relatively young, should hope to compensate for the obscurity of the names of their officers of instruction by the abundance of decorative titles by which those names are followed on the pages of the catalogues where they appear. The dazzled reader of the list, the parent or student, says to himself, 'This must be a terribly distinguished crowd,— their titles shine like the stars in the firmament; Ph.D.'s, S.D.'s, and Litt.D.'s, bespangle the page as if they were sprinkled over it from a pepper caster.'

Human nature is once for all so childish that every reality becomes a sham somewhere, and in the minds of Presidents and Trustees the Ph.D. degree is in point of fact already looked upon as a mere advertising resource, a manner of throwing dust in the Public's eyes. 'No instructor who is not a Doctor' has become a maxim in the smaller institutions which represent demand; and in each of the larger ones which represent supply, the same belief in decorated scholarship expresses itself in two antagonistic passions, one for multiplying as much as possible the annual output of doctors, the other for raising the standard of difficulty in passing, so that the Ph.D. of the special institution shall carry a higher blaze of distinction than it does elsewhere. Thus we at Harvard are proud of the number of candidates whom we reject, and of the inability of men who are not *distingués* in intellect to pass our tests.

America is thus as a nation rapidly drifting towards a state of things in which no man of science or letters will be acccounted respectable unless some kind of badge or diploma is stamped upon him, and in which bare personality will be a mark of outcast estate. It seems to me high time to rouse ourselves to consciousness, and to cast a critical eye upon this decidedly grotesque tendency. Other nations suffer terribly from the Mandarin disease. Are we doomed to suffer like the rest?

Our higher degrees were instituted for the laudable purpose of stimulating scholarship, especially in the form of 'original research.' Experience has proved that great as the love of truth may be among men, it can be made still greater by adventitious rewards. The winning of a diploma certifying mastery and marking a barrier successfully passed, acts as a challenge to the ambitious; and if the diploma will help to gain bread-winning positions also, its power as a stimulus to work is tremendously increased. So far, we are on innocent ground; it is well for a country to have research in abundance, and our graduate schools do but apply a normal psychological spur. But the institutionizing on a large scale of any natural combination of need and motive always tends to run into technicality and to develop a tyrannical Machine with unforeseen powers of exclusion and corruption. Observation of the workings of our Harvard system for twenty years past has brought some of these drawbacks home to my consciousness, and I should like to call the attention of my readers to this disadvantageous aspect of the picture, and to make a couple of remedial suggestions, if I may.

In the first place, it would seem that to stimulate study, and to increase the *gelehrtes Publikum*, the class of highly educated men in our country, is the only positive good, and consequently the sole direct end at which our graduate schools, with their diploma-giving powers, should aim. If other results have developed they should be deemed secondary incidents, and if not desirable in themselves, they should be carefully guarded against.

To interfere with the free development of talent, to obstruct the natural play of supply and demand in the teaching profession, to foster academic snobbery by the prestige of certain privileged institutions, to transfer accredited value from essential manhood to an outward badge, to blight hopes and promote invidious sentiments, to divert the attention of aspiring youth from direct dealings with truth to the passing of examinations, —such consequences, if they exist, ought surely to be regarded as drawbacks to the system, and an enlightened public consciousness ought to be keenly alive to the importance of reducing their amount. Candidates themselves do seem to be keenly conscious of some of these evils, but outside of their ranks or in the general public no such consciousness, so far as I can see, exists; or if it does exist, it fails to express itself aloud. Schools, Colleges, and Universities, appear enthusiastic over the entire system, just as it stands, and unanimously applaud all its developments.

I beg the reader to consider some of the secondary evils which I have enumerated. First of all, is not our growing tendency to appoint no instructors who are not also doctors an instance of pure sham? Will any one pretend for a moment that the doctor's degree is a guarantee that its possessor will be successful as a teacher? Notoriously his moral, social and personal characteristics may utterly disqualify him for success in the class-room; and of these characteristics his doctor's examination is unable to take any account whatever. Certain bare human beings will always be better candidates for a given place than all the doctor-applicants on hand; and to exclude the former by a rigid rule, and in the end to have to sift the latter by private inquiry into their personal peculiarities among those who know them, just as if they were not doctors at all, is to stultify one's own procedure. You may say that at least you guard against ignorance of the subject by considering only the candidates who are doctors; but how then about making doctors in one subject teach a different subject? This happened in the instance by which I introduced this article, and it happens daily and hourly in all our colleges. The truth is that the Doctor-Monopoly in teaching, which is becoming so rooted an American custom, can show no serious grounds whatsoever for itself in reason. As it actually prevails and grows in vogue among us, it is due to childish motives exclusively. In reality it is but a sham, a bauble, a dodge, whereby to decorate the catalogues of schools and colleges.

Next, let us turn from the general promotion of a spirit of academic snobbery to the particular damage done to individuals by the system.

There are plenty of individuals so well endowed by nature that they pass with ease all the ordeals with which life confronts them. Such persons are born for professional success. Examinations have no terrors for them, and interfere in no way with their spiritual or worldly interests. There are others, not so gifted who nevertheless rise to the challenge, get a stimulus from the difficulty, and become doctors, not without some baleful nervous wear and tear and retardation of their purely inner life, but on the whole successfully, and with advantage. These two classes form the natural Ph.D.'s for whom the degree is legitimately instituted. To be sure, the degree is of no consequence one way or the other for the first sort of man, for in him the personal worth obviously outshines the title. To the second set of persons, however, the doctor ordeal may contribute a touch of energy and solidity of scholarship which otherwise they might have lacked, and were our candidates all drawn from these classes, no oppression would result from the institution.

But there is a third class of persons who are genuinely, and in the most pathetic sense, the institution's victims. For this type of character the academic life may become, after a certain point, a virulent poison. Men without marked originality or native force, but fond of truth and especially of books and study, ambitious of reward and recognition, poor often, and needing a degree to get a teaching position, weak in the eyes of their examiners,—among these we find the veritable *chair à canon* of the wars of learning, the unfit in the academic struggle for existence. There are individuals of this sort for whom to pass one degree after another seems the limit of earthly aspiration. Your private advice does not discourage them. They will fail, and go away to recuperate, and then present themselves for another ordeal, and sometimes prolong the process into middle life. Or else, if they are less heroic morally they will accept the failure as a sentence of doom that they are not fit, and are broken-spirited men thereafter.

We of the university faculties are responsible for deliberately creating this new class of American social failures, and heavy is the responsibility. We advertise our 'schools' and send out our degree-requirements, knowing well that aspirants of all sorts will be attracted, and at the same time we set a standard which intends to pass no man who has not native intellectual distinction. We know that there is no test, however absurd, by which, if a title or decoration, a public badge or mark, were to be won by it, some weakly suggestible or hauntable persons would not feel challenged, and remain unhappy if they went without it. We dangle our three magic letters before the eyes of these predestined victims, and they swarm to us like moths to an electric light. They come at a time when failure can no longer be repaired easily and when the wounds it leaves are permanent; and we say deliberately that mere work faithfully performed, as they perform it, will not by itself save them, they must in addition put in evidence the one thing they have not got, namely this quality of intellectual distinction. Occasionally, out of sheer human pity, we ignore our high and mighty standard and pass them. Usually, however, the standard, and not the candidate, commands our fidelity. The result is caprice, majorities of one on the jury, and on the whole a confession that our pretensions about the degree cannot be lived up to consistenly. Thus, partiality in the favored cases; in the unfavored, blood on our hands; and in both a bad conscience,—are the results of our administration.

The more widespread becomes the popular belief that our diplomas are indispensable hall-marks to show the sterling metal of their holders, the more widespread these corruptions will become. We ought to look to the future carefully, for it takes generations for a national custom, once rooted, to be grown away from. All the European countries are seeking to diminish the check upon individual spontaneity which state examinations with their tyrannous growth have brought in their train. We have had to institute state examinations too; and it will perhaps be fortunate if some day hereafter our descendants, comparing machine with machine, do not sigh with regret for old times and American freedom, and wish that the *régime* of the dear old bosses might be reinstalled, with plain human nature, the glad hand and the marble heart, liking and disliking, and man-to-man relations grown possible again. Meanwhile, whatever evolution our state-examinations are destined to undergo, our universities at least should never cease to regard themselves as the jealous custodians of personal and spiritual spontaneity. They are indeed its only organized and recognized custodians in America to-day. They ought to guard against contributing to the increase of officialism and snobbery and insincerity as against a pestilence; they ought to keep truth and disinterested labor always in the foreground, treat degrees as secondary incidents, and in season and out of season make it plain that what they live for is to help men's souls, and not to decorate their persons with diplomas.

There seem to be three obvious ways in which the increasing hold of the Ph.D. Octopus upon American life can be kept in check.

The first way lies with the universities. They can lower their fantastic standards (which here at Harvard we are so proud of) and give the doctorate as a matter of course, just as they give the bachelor's degree, for a due amount of time spent in patient labor in a special department of learning, whether the man be a brilliantly gifted individual or not. Surely native distinction needs no official stamp, and should disdain to ask for one. On the other hand, faithful labor, however commonplace, and years devoted to a subject, always deserve to be acknowledged and requited.

The second way lies with both the universities and colleges. Let them give up their unspeakably silly ambition to bespangle their lists of officers with these doctorial titles. Let them look more to substance and less to vanity and sham.

The third way lies with the individual student, and with his personal advisers in the faculties. Every man of native power, who might take a higher degree, and refuses to do so, because examinations interfere with the free following out of his more immediate intellectual aims, deserves well of his country, and in a rightly organized community, would not be made to suffer for his independence. With many men the passing of these extraneous tests is a very grievous interference indeed. Private letters of recommendation from their instructors, which in any event are ultimately needful, ought, in these cases, completely to offset the lack of the bread-winning degree; and instructors ought to be ready to advise students against it upon occasion, and to pledge themselves to back them later personally, in the market-struggle which they have to face.

It is indeed odd to see this love of titles—and such titles—growing up in a country of which the recognition of individuality and bare manhood have so long been supposed to be the very soul. The independence of the State, in which most of our colleges stand, relieves us of those more odious forms of academic politics which continental European countries present. Anything like the elaborate university machine of France, with its throttling influences upon individuals is unknown here. The spectacle of the 'Rath' distinction in its innumerable spheres and grades, with which all Germany is crawling to-day, is displeasing to American eyes; and displeasing also in some respects is the institution of knighthood in England, which, aping as it does an aristocratic title, enables one's wife as well as one's self so easily to dazzle the servants at the house of one's friends. But are we Americans ourselves destined after all to hunger after similar vanities on an infinitely more contemptible scale? And is individuality with us also going to count for nothing unless stamped and licensed and authenticated by some title-giving machine? Let us pray that our ancient national genius may long preserve vitality enough to guard us from a future so unmanly and so unbeautiful!

1903

From *London*

I

THERE is a certain evening that I count as virtually a first
impression—the end of a wet, black Sunday, twenty years ago,
about the first of March. There had been an earlier vision, but it
had turned gray, like faded ink, and the occasion I speak of was a
fresh beginning. No doubt I had a mystic prescience of how
fond of the murky modern Babylon I was one day to become;
certain it is that as I look back I find every small circumstance of
those hours of approach and arrival still as vivid as if the solemnity of an opening era had breathed upon it. The sense of
approach was already almost intolerably strong at Liverpool,
where, as I remember, the perception of the English character of
everything was as acute as a surprise, though it could only be a
surprise without a shock. It was expectation exquisitely gratified,
superabundantly confirmed. There was a kind of wonder indeed
that England should be as English as, for my entertainment, she
took the trouble to be; but the wonder would have been greater,
and all the pleasure absent, if the sensation had not been violent.
It seems to sit there again like a visiting presence, as it sat
opposite to me at breakfast at a small table in a window of the
old coffee-room of the Adelphi Hotel—the unextended (as it
then was), the unimproved, the unblushingly local Adelphi.
Liverpool is not a romantic city, but that smoky Saturday returns to me as a supreme success, measured by its association
with the kind of emotion in the hope of which, for the most part,
we betake ourselves to far countries.

It assumed this character at an early hour—or rather indeed
twenty-four hours before—with the sight, as one looked across
the wintry ocean, of the strange, dark, lonely freshness of the
coast of Ireland. Better still, before we could come up to the city,
were the black steamers knocking about in the yellow Mersey,

under a sky so low that they seemed to touch it with their funnels, and in the thickest, windiest light. Spring was already in the air, in the town; there was no rain, but there was still less sun—one wondered what had become, on this side of the world, of the big white splotch in the heavens; and the gray mildness, shading away into black at every pretext, appeared in itself a promise. This was how it hung about me, between the window and the fire, in the coffee-room of the hotel—late in the morning for breakfast, as we had been long disembarking. The other passengers had dispersed, knowingly catching trains for London (we had only been a handful); I had the place to myself, and I felt as if I had an exclusive property in the impression. I prolonged it, I sacrificed to it, and it is perfectly recoverable now, with the very taste of the national muffin, the creak of the waiter's shoes as he came and went (could anything be so English as his intensely professional back? it revealed a country of tradition), and the rustle of the newspaper I was too excited to read.

I continued to sacrifice for the rest of the day; it didn't seem to me a sentient thing, as yet, to inquire into the means of getting away. My curiosity must indeed have languished, for I found myself on the morrow in the slowest of Sunday trains, pottering up to London with an interruptedness which might have been tedious without the conversation of an old gentleman who shared the carriage with me and to whom my alien as well as comparatively youthful character had betrayed itself. He instructed me as to the sights of London, and impressed upon me that nothing was more worthy of my attention than the great cathedral of St Paul. 'Have you seen St Peter's in Rome? St Peter's is more highly embellished, you know; but you may depend upon it that St Paul's is the better building of the two.' The impression I began with speaking of was, strictly, that of the drive from Euston, after dark, to Morley's Hotel in Trafalgar Square. It was not lovely—it was in fact rather horrible; but as I move again through dusky, tortuous miles, in the greasy four-wheeler to which my luggage had compelled me to commit myself, I recognise the first step in an initiation of which the subsequent stages were to abound in pleasant things. It is a kind of humiliation in a great city not to know where you are going, and Morley's Hotel was then, to my imagination, only a vague ruddy spot in the general immensity. The immensity was the great fact, and that was a charm; the miles of housetops and viaducts, the complication of junctions and signals through which the train made its way to the station had already given me· the scale. The weather had turned to wet, and we went deeper and deeper into the Sunday night. The sheep in the fields, on the way from Liverpool, had shown in their demeanour a certain consciousness of the day; but this momentous cab-drive was an introduction to rigidities of custom. The low black houses were as inanimate as so many rows of coal-scuttles, save where at frequent corners, from a gin-shop, there was a flare of light more brutal still than the darkness. The custom of gin—that was equally rigid, and in this first impression the public-houses counted for much.

Morley's Hotel proved indeed to be a ruddy spot; brilliant, in my recollection, is the coffee-room fire, the hospitable mahogany, the sense that in the stupendous city this, at any rate for the hour, was a shelter and a point of view. My remembrance of the rest of the evening—I was probably very tired—is mainly a remembrance of a vast four-poster. My little bedroom-candle, set in its deep basin, caused this monument to project a huge shadow and to make me think, I scarce knew why, of *The Ingoldsby Legends*. If at a tolerably early hour the next day I found myself approaching St Paul's, it was not wholly in obedience to the old gentleman in the railway-carriage: I had an errand in the City, and the City was doubtless prodigious. But what I mainly recall is the romantic consciousness of passing under Temple Bar and the way two lines of *Henry Esmond* repeated themselves in my mind as I drew near the masterpiece of Sir Christopher Wren. 'The stout, red-faced woman' whom Esmond had seen tearing after the staghounds over the slopes at Windsor was not a bit like the effigy 'which turns its stony back upon St Paul's and faces the coaches struggling up Ludgate Hill.' As I looked at Queen Anne over the apron of my hansom—she struck me as very small and dirty, and the vehicle ascended the mild incline without an effort—it was a thrilling thought that the statue had been familiar to the hero of the incomparable novel. All history appeared to live again and the continuity of things to vibrate through my mind.

To this hour, as I pass along the Strand, I take again the walk I took there that afternoon. I love the place to-day, and that was the commencement of my passion. It appeared to me to present phenomena, and to contain objects, of every kind, of an inexhaustible interest: in particular it struck me as desirable and even indispensable that I should purchase most of the articles in most of the shops. My eyes rest with a certain tenderness on the places where I resisted and on those where I succumbed. The fragrance of Mr Rimmel's establishment is again in my nostrils; I see the slim young lady (I hear her pronunciation), who waited upon me there. Sacred to me to-day is the particular aroma of the hairwash that I bought of her. I pause before the granite portico of Exeter Hall (it was unexpectedly narrow and wedge-like), and it evokes a cloud of associations which are none the less impressive because they are vague; coming from I don't know where —from *Punch*, from Thackeray, from old volumes of *The Illustrated London News* turned over in childhood; seeming connected with Mrs Beecher Stowe and *Uncle Tom's Cabin*. Memorable is a rush I made into a glover's at Charing Cross—the one you pass going eastward, just before you turn into the station; that, however, now that I think of it, must have been in the morning, as soon as I issued from the hotel. Keen within me was a sense of the importance of deflowering, of despoiling the shop.

A day or two later, in the afternoon, I found myself staring at my fire, in a lodging of which I had taken possession on foreseeing that I should spend some weeks in London. I had just come in, and, having attended to the distribution of my luggage, sat down to consider my habitation. It was on the ground-floor, and the fading daylight reached it in a sadly damaged condition. It struck me as stuffy and unsocial, with its mouldy smell and its decoration of lithographs and wax-flowers—an impersonal black hole in the huge general blackness. The uproar of Piccadilly hummed away at the end of the street, and the rattle of a heartless hansom passed close to my ears. A sudden horror of the whole place came over me, like a tiger-pounce of homesickness which had been watching its moment. London was hideous, vicious, cruel, and above all overwhelming; whether or no she was 'careful of the type' she was as indifferent as nature herself to the single life. In the course of an hour I should have to go out to my dinner, which was not supplied on the premises, and that effort assumed the form of a desperate and dangerous quest. It appeared to me that I would rather remain dinnerless, would

rather even starve than sally forth into the infernal town, where
the natural fate of an obscure stranger would be to be trampled
to death in Piccadilly and his carcass thrown into the Thames.
I did not starve, however, and I eventually attached myself
by a hundred human links to the dreadful, delightful city. That
momentary vision of its smeared face and stony heart has
remained memorable to me, but I am happy to say that I can
easily summon up others.

II

It is no doubt not the taste of every one, but for the real
London-lover the mere immensity of the place is a large part of
its merit. A small London would be an abomination, as it fortunately is an impossibility, for the idea and the name are beyond
everything an expression of extent and number. Practically, of
course, one lives in a quarter, in a plot; but in imagination and
by a constant mental act of reference the sympathising resident
inhabits the whole—and it is only of him that I deem it worth
while to speak. He fancies himself, as they say, for being a particle in so unequalled an aggregation; and its immeasurable
circumference, even though unvisited and lost in smoke, gives
him the sense of a social, an intellectual margin. There is a luxury
in the knowledge that he may come and go without being
noticed, even when his comings and goings have no nefarious
end. I don't mean by this that the tongue of London is not a very
active member; the tongue of London would indeed be worthy
of a chapter by itself. But the eyes which at least in some measure
feed its activity are fortunately for the common advantage
solicited at any moment by a thousand different objects. If the
place is big, everything it contains is certainly not so; but this
may at least be said, that if small questions play a part there, they
play it without illusions about its importance. There are too
many questions, small or great; and each day, as it arrives, leads
its children, like a kind of mendicant mother, by the hand.
Therefore perhaps the most general characteristic is the absence
of insistence. Habits and inclinations flourish and fall, but intensity is never one of them. The spirit of the great city is not ana-
lytic, and, as they come up, subjects rarely receive at its hands a
treatment offensively earnest or indiscreetly thorough. There are
not many—of those of which London disposes with the assur-

ance begotten of its large experience—that wouldn't lend themselves to a tenderer manipulation elsewhere. It takes a very great affair, a turn of the Irish screw or a divorce-case lasting many days, to be fully threshed out. The mind of Mayfair, when it aspires to show what it really can do, lives in the hope of a new divorce-case, and an indulgent providence—London is positively in certain ways the spoiled child of the world—abundantly recognises this particular aptitude and humours the whim.

The compensation is that material does arise; that there is great variety, if not morbid subtlety; and that the whole of the procession of events and topics passes across your stage. For the moment I am speaking of the inspiration there may be in the sense of far frontiers; the London-lover loses himself in this swelling consciousness, delights in the idea that the town which incloses him is after all only a paved country, a state by itself. This is his condition of mind quite as much if he be an adoptive as if he be a matter-of-course son. I am by no means sure even that he need be of Anglo-Saxon race and have inherited the birthright of English speech; though on the other hand I make no doubt that these advantages minister greatly to closeness of allegiance. The great city spreads her dusky mantle over innumerable races and creeds, and I believe there is scarcely a known form of worship that has not some temple there—have I not attended at the Church of Humanity, in Lamb's Conduit, in company with an American lady, a vague old gentleman and several seamstresses?—or any communion of men that has not some club or guild. London is indeed an epitome of the round world, and just as it is a commonplace to say that there is nothing one can't 'get' there, so it is equally true that there is nothing one can't study at first hand.

One doesn't test these truths every day, but they form part of the air one breathes (and welcome, says the London-hater—for there *is* such a benighted animal—to the pestilent compound). They colour the thick, dim distances which in my opinion are the most romantic town-vistas in the world; they mingle with the troubled light to which the straight, ungarnished aperture in one's dull, undistinctive house-front affords a passage and which makes an interior of friendly corners, mysterious tones and unbetrayed ingenuities, as well as with the low, magnificent medium of the sky, where the smoke and the fog and the weather in general, the strangely undefined hour of the day and season of the year, the emanations of industries and the reflection of furnaces, the red gleams and blurs that may or may not be of sunset—as you never see any *source* of radiance you can't in the least tell—all hang together in a confusion, a complication, a shifting but irremovable canopy. They form the undertone of the deep perpetual voice of the place. One remembers them when one's loyalty is on the defensive; when it is a question of introducing as many striking features as possible into the list of fine reasons one has sometimes to draw up, that eloquent catalogue with which one confronts the hostile indictment—the array of *other* reasons which may easily be as long as one's arm. According to these other reasons it plausibly and conclusively stands that as a place to be happy in London will never do. I don't say it is necessary to meet so absurd an allegation except for one's personal complacency. If indifference, in so gorged an organism, is still livelier than curiosity, you may avail yourself of your own share in it simply to feel that since such and such a person doesn't care for real greatness so much the worse for such and such a person. But once in a while the best believer recognises the impulse to set his religion in order, to sweep the temple of his thoughts and trim the sacred lamp. It is at such hours as this that he reflects with elation that the British capital is the particular spot in the world which communicates the greatest sense of life.

III

The reader will perceive that I do not shrink even from the extreme concession of speaking of our capital as British, and this in a shameless connection with the question of loyalty on the part of an adoptive son. For I hasten to explain that if half the source of one's interest in it comes from feeling that it is the property and even the home of the human race—Hawthorne, that best of Americans, says so somewhere, and places it in this sense side by side with Rome—one's appreciation of it is really a large sympathy, a comprehensive love of humanity. For the sake of such a charity as this one may stretch one's allegiance; and the most alien of the cockneyfied, though he may bristle with every protest at the intimation that England has set its stamp upon him, is free to admit with conscious pride that he has submitted

to Londonisation. It is a real stroke of luck for a particulaı
country that the capital of the human race happens to be British.
Surely every other people would have it theirs if they could.
Whether the English deserve to hold it any longer might be an
interesting field of inquiry; but as they have not yet let it slip the
writer of these lines professes without scruple that the arrangement is to his personal taste. For after all if the sense of life is
greatest there, it is a sense of the life of people of our incomparable English speech. It is the headquarters of that strangely
elastic tongue; and I make this remark with a full sense of the
terrible way in which the idiom is misused by the populace in
general, than whom it has been given to few races to impart to
conversation less of the charm of tone. For a man of letters who
endeavours to cultivate, however modestly, the medium of
Shakespeare and Milton, of Hawthorne and Emerson, who
cherishes the notion of what it has achieved and what it may
even yet achieve, London must ever have a great illustrative and
suggestive value and indeed a kind of sanctity. It is the single
place in which most readers, most possible lovers are gathered
together; it is the most inclusive public and the largest social
incarnation of the language, of the tradition. Such a personage
may well let it go for this and leave the German and the Greek to
speak for themselves, to express the grounds of *their* predilection, presumably very different.

When a social product is so vast and various it may be
approached on a thousand different sides, and liked and disliked
for a thousand different reasons. The reasons of Piccadilly are
not those of Camden Town, nor are the curiosities and discouragements of Kilburn the same as those of Westminster and
Lambeth. The reasons of Piccadilly—I mean the friendly
ones—are those of which, as a general thing, the rooted visitor
remains most conscious; but it must be confessed that even
these, for the most part, do not lie upon the surface. The absence
of style, or rather of the intention of style, is certainly the most
general characteristic of the face of London. To cross to Paris
under this impression is to find one's self surrounded with far
other standards. There everything reminds you that the idea of
beautiful and stately arrangement has never been out of fashion,
that the art of composition has always been at work or at play.
Avenues and squares, gardens and quays have been distributed

for effect, and to-day the splendid city reaps the accumulation of
all this ingenuity. The result is not in every quarter interesting,
and there is a tiresome monotony of the 'fine' and the symmetrical, above all of the deathly passion for making things 'to match.'
On the other hand the whole air of the place is architectural. On
the banks of the Thames it is a tremendous chapter of accidents—the London-lover has to confess to the existence of miles
upon miles of the dreariest, stodgiest commonness. Thousands
of acres are covered by low black houses, of the cheapest construction, without ornament, without grace, without character or
even identity. In fact there are many, even in the best quarters, in
all the region of Mayfair and Belgravia, of so paltry and inconvenient and above all of so diminutive a type (those that are let
in lodgings—such poor lodgings as they make—may serve as an
example), that you wonder what peculiarly limited domestic need
they were constructed to meet. The great misfortune of London,
to the eye (it is true that this remark applies much less to the
City), is the want of elevation. There is no architectural impression without a certain degree of height, and the London
street-vista has none of that sort of pride.

All the same, if there be not the intention, there is at least the
accident, of style, which, if one looks at it in a friendly way,
appears to proceed from three sources. One of these is simply
the general greatness, and the manner in which that makes a
difference for the better in any particular spot, so that though
you may often perceive yourself to be in a shabby corner it never
occurs to you that this is the end of it. Another is the atmosphere, with its magnificent mystifications, which flatters and
superfuses, makes everything brown, rich, dim, vague, magnifies
distances and minimises details, confirms the inference of vastness by suggesting that, as the great city makes everything, it
makes its own system of weather and its own optical laws. The
last is the congregation of the parks, which constitute an ornament not elsewhere to be matched and give the place a superiority that none of its uglinesses overcome. They spread themselves
with such a luxury of space in the centre of the town that they
form a part of the impression of any walk, of almost any view,
and, with an audacity altogether their own, make a pastoral landscape under the smoky sky. There is no mood of the rich
London climate that is not becoming to them—I have seen them

look delightfully romantic, like parks in novels, in the wettest winter—and there is scarcely a mood of the appreciative resident to which they have not something to say. The high things of London, which here and there peep over them, only make the spaces vaster by reminding you that you are after all not in Kent or Yorkshire; and these things, whatever they be, rows of 'eligible' dwellings, towers of churches, domes of institutions, take such an effective gray-blue tint that a clever watercolourist would seem to have put them in for pictorial reasons.

The view from the bridge over the Serpentine has an extraordinary nobleness, and it has often seemed to me that the Londoner twitted with his low standard may point to it with every confidence. In all the town-scenery of Europe there can be few things so fine; the only reproach it is open to is that it begs the question by seeming—in spite of its being the pride of five millions of people—not to belong to a town at all. The towers of Notre Dame, as they rise, in Paris, from the island that divides the Seine, present themselves no more impressively than those of Westminster as you see them looking doubly far beyond the shining stretch of Hyde Park water. Equally admirable is the large, river-like manner in which the Serpentine opens away between its wooded shores. Just after you have crossed the bridge (whose very banisters, old and ornamental, of yellowish-brown stone, I am particularly fond of), you enjoy on your left, through the gate of Kensington Gardens as you go towards Bayswater, an altogether enchanting vista—a footpath over the grass, which loses itself beneath the scattered oaks and elms exactly as if the place were a 'chase.' There could be nothing less like London in general than this particular morsel, and yet it takes London, of all cities, to give you such an impression of the country.

1888

Under the Early Stars

PLAY is not for every hour of the day, or for any hour taken at random. There is a tide in the affairs of children. Civilization is cruel in sending them to bed at the most stimulating time of dusk. Summer dusk, especially, is the frolic moment for children, baffle them how you may. They may have been in a pottering mood all day, intent upon all kinds of close industries, breathing hard over choppings and poundings. But when late twilight comes, there comes also the punctual wildness. The children will run and pursue, and laugh for the mere movement—it does so jolt their spirits.

What remembrances does this imply of the hunt, what of the predatory dark? The kitten grows alert at the same hour, and hunts for moths and crickets in the grass. It comes like an imp, leaping on all fours. The children lie in ambush and fall upon one another in the mimicry of hunting.

The sudden outbreak of action is complained of as a defiance and a rebellion. Their entertainers are tired, and the children are to go home. But, with more or less of life and fire, the children strike some blow for liberty. It may be the impotent revolt of the ineffectual child, or the stroke of the conqueror; but something, something is done for freedom under the early stars.

This is not the only time when the energy of children is in conflict with the weariness of men. But it is less tolerable that the energy of men should be at odds with the weariness of children, which happens at some time of their jaunts together, especially, alas! in the jaunts of the poor.

Of games for the summer dusk when it rains, cards are most beloved by children. Three tiny girls were to be taught 'old maid' to beguile the time. One of them, a nut-brown child of five, was persuading another to play. 'Oh come,' she said, 'and play with me at new maid.'

The time of falling asleep is a child's immemorial and incalculable hour. It is full of traditions, and beset by antique habits. The habit of prehistoric races has been cited as the only explanation of the fixity of some customs in mankind. But if the inquirers who appeal to that beginning remembered better their own infancy, they would seek no further. See the habits in falling to sleep which have children in their thralldom. Try to overcome them in any child, and his own conviction of their high antiquity weakens your hand.

Childhood is antiquity, and with the sense of time and the sense of mystery is connected for ever the hearing of a lullaby. The French sleep-song is the most romantic. There is in it such a sound of history as must inspire any imaginative child, falling to sleep, with a sense of the incalculable; and the songs themselves are old. 'Le Bon Roi Dagobert' has been sung over French cradles since the legend was fresh. The nurse knows nothing more sleepy than the tune and the verse that she herself slept to when a child. The gaiety of the thirteenth century, in 'Le Pont d' Avignon,' is put mysteriously to sleep, away in the *tête à tête* of child and nurse, in a thousand little sequestered rooms at night. 'Malbrook' would be comparatively modern, were not all things that are sung to a drowsing child as distant as the day of Abraham.

If English children are not rocked to many such aged lullabies, some of them are put to sleep to strange cradle-songs. The affectionate races that are brought into subjection sing the primitive lullaby to the white child. Asiatic voices and African persuade him to sleep in the tropical night. His closing eyes are filled with alien images.

1897

The Acorn-Gatherer

Black rooks, yellow oak leaves, and a boy asleep at the foot of
the tree. His head was lying on a bulging root close to the stem:
his feet reached to a small sack or bag half full of acorns. In his
slumber his forehead frowned—they were fixed lines, like the
grooves in the oak bark. There was nothing else in his features
attractive or repellent: they were such as might have belonged
to a dozen hedge children. The set angry frown was the only
distinguishing mark—like the dents on a penny made by a hob-
nail boot, by which it can be known from twenty otherwise precisely similar. His clothes were little better than sacking, but
clean, tidy, and repaired. Any one would have said, 'Poor, but
carefully tended.' A kind heart might have put a threepenny-bit
in his clenched little fist, and sighed. But that iron set frown on
the young brow would not have unbent even for the silver. Caw!
Caw!

The happiest creatures in the world are the rooks at the
acorns. It is not only the eating of them, but the finding: the fluttering up there and hopping from branch to branch, the sidling
out to the extreme end of the bough, and the inward chuckling
when a friend lets his acorn drop tip-tap from bough to bough.
Amid such plenty they cannot quarrel or fight, having no cause
of battle, but they can boast of success, and do so to the loudest
of their voices. He who has selected a choice one flies with it as if
it were a nugget in his beak, out to some open spot of ground,
followed by a general Caw!

This was going on above while the boy slept below. A thrush
looked out from the hedge, and among the short grass there was
still the hum of bees, constant sun-worshippers as they are. The
sunshine gleamed on the rooks' black feathers overhead, and
on the sward sparkled from hawkweed, some lotus and yellow
weed, as from a faint ripple of water. The oak was near a corner

formed by two hedges, and in the angle was a narrow thorny
gap. Presently an old woman, very upright, came through this
gap carrying a faggot on her shoulder and a stout ash stick in her
hand. She was very clean, well dressed for a labouring woman,
hard of feature, but superior in some scarcely defined way to
most of her class. The upright carriage had something to do with
it, the firm mouth, the light blue eyes that looked every one
straight in the face. Possibly these, however, had less effect than
her conscious righteousness. Her religion lifted her above the
rest, and I do assure you that it was perfectly genuine. That hard
face and cotton gown would have gone to the stake.

When she had got through the gap she put the faggot down in
it, walked a short distance out into the field, and came back
towards the boy, keeping him between her and the corner. Caw!
said the rooks, Caw! Caw! Thwack, thwack, bang, went the ash
stick on the sleeping boy, heavily enough to have broken his
bones. Like a piece of machinery suddenly let loose, without
a second of dubious awakening and without a cry, he darted
straight for the gap in the corner. There the faggot stopped him,
and before he could tear it away the old woman had him again,
thwack, thwack, and one last stinging slash across his legs as he
doubled past her. Quick as the wind as he rushed he picked up
the bag of acorns and pitched it into the mound, where the
acorns rolled down into a pond and were lost—a good round
shilling's worth. Then across the field, without his cap, over
the rising ground, and out of sight. The old woman made no
attempt to hold him, knowing from previous experience that it
was useless, and would probably result in her own overthrow.
The faggot, brought a quarter of a mile for the purpose, enabled
her, you see, to get two good chances at him.

A wickeder boy never lived: nothing could be done with the
reprobate. He was her grandson—at least, the son of her daughter, for he was not legitimate. The man drank, the girl died, as
was believed, of sheer starvation: the granny kept the child, and
he was now between ten and eleven years old. She had done and
did her duty, as she understood it. A prayer-meeting was held in
her cottage twice a week, she prayed herself aloud among them,
she was a leading member of the sect. Neither example, precept,
nor the rod could change that boy's heart. In time perhaps she
got to beat him from habit rather than from any particular anger

of the moment, just as she fetched water and filled her kettle, as one of the ordinary events of the day. Why did not the father interfere? Because if so he would have had to keep his son: so many shillings a week the less for ale.

In the garden attached to the cottage there was a small shed with a padlock, used to store produce or wood in. One morning, after a severe beating, she drove the boy in there and locked him in the whole day without food. It was no use, he was as hardened as ever.

A footpath which crossed the field went by the cottage, and every Sunday those who were walking to church could see the boy in the window with granny's Bible open before him. There he had to sit, the door locked, under terror of stick, and study the page. What was the use of compelling him to do that? He could not read. 'No,' said the old woman, 'he won't read, but I makes him look at his book.'

The thwacking went on for some time, when one day the boy was sent on an errand two or three miles, and for a wonder started willingly enough. At night he did not return, nor the next day, nor the next, and it was as clear as possible that he had run away. No one thought of tracking his footsteps, or following up the path he had to take, which passed a railway, brooks, and a canal. He had run away, and he might stop away: it was beautiful summer weather, and it would do him no harm to stop out for a week. A dealer who had business in a field by the canal thought indeed that he saw something in the water, but he did not want any trouble, nor indeed did he know that some one was missing. Most likely a dead dog; so he turned his back and went to look again at the cow he thought of buying. A barge came by, and the steerswoman, with a pipe in her mouth, saw something roll over and come up under the rudder: the length of the barge having passed over it. She knew what it was, but she wanted to reach the wharf and go ashore and have a quart of ale. No use picking it up, only make a mess on deck, there was no reward —'Gee-up! Neddy.' The barge went on, turning up the mud in the shallow water, sending ripples washing up to the grassy meadow shores, while the moorhens hid in the flags till it was gone. In time a labourer walking on the towing-path saw 'it,' and fished it out, and with it a slender ash sapling, with twine and hook, a worm still on it. This was why the dead boy had gone so willingly, thinking to fish in the 'river,' as he called the canal. When his feet slipped and he fell in, his fishing-line somehow became twisted about his arms and legs, else most likely he would have scrambled out, as it was not very deep. This was the end; nor was he even remembered. Does any one sorrow for the rook, shot, and hung up as a scarecrow? The boy had been talked to, and held up as a scarecrow all his life: he was dead, and that is all. As for granny, she felt no twinge: she had done her duty.

1884

ROBERT LOUIS STEVENSON

*Aes Triplex**

THE changes wrought by death are in themselves so sharp and final, and so terrible and melancholy in their consequences, that the thing stands alone in man's experience, and has no parallel upon earth. It outdoes all other accidents because it is the last of them. Sometimes it leaps suddenly upon its victims, like a Thug; sometimes it lays a regular siege and creeps upon their citadel during a score of years. And when the business is done, there is sore havoc made in other people's lives, and a pin knocked out by which many subsidiary friendships hung together. There are empty chairs, solitary walks, and single beds at night. Again, in taking away our friends, death does not take them away utterly, but leaves behind a mocking, tragical, and soon intolerable residue, which must be hurriedly concealed. Hence a whole chapter of sights and customs striking to the mind, from the pyramids of Egypt to the gibbets and dule trees of mediaeval Europe. The poorest persons have a bit of pageant going towards the tomb; memorial stones are set up over the least memorable; and, in order to preserve some show of respect for what remains of our old loves and friendships, we must accompany it with much grimly ludicrous ceremonial, and the hired undertaker parades before the door. All this, and much more of the same sort, accompanied by the eloquence of poets, has gone a great way to put humanity in error; nay, in many philosophies the error has been embodied and laid down with every circumstance of logic; although in real life the bustle and swiftness, in leaving people little time to think, have not left them time enough to go dangerously wrong in practice.

As a matter of fact, although few things are spoken of with more fearful whisperings than this prospect of death, few have less influence on conduct under healthy circumstances. We have all heard of cities in South America built upon the side of fiery

---

* *Editor's note*: 'Aes triplex'—a phrase from Horace—means 'triple brass'.

mountains, and how, even in this tremendous neighbourhood, the inhabitants are not a jot more impressed by the solemnity of mortal conditions than if they were delving gardens in the greenest corner of England. There are serenades and suppers and much gallantry among the myrtles overhead; and meanwhile the foundation shudders underfoot, the bowels of the mountain growl, and at any moment living ruin may leap sky-high into the moonlight, and tumble man and his merry-making in the dust. In the eyes of very young people, and very dull old ones, there is something indescribably reckless and desperate in such a picture. It seems not credible that respectable married people, with umbrellas, should find appetite for a bit of supper within quite a long distance of a fiery mountain; ordinary life begins to smell of high-handed debauch when it is carried on so close to a catastrophe; and even cheese and salad, it seems, could hardly be relished in such circumstances without something like a defiance of the Creator. It should be a place for nobody but hermits dwelling in prayer and maceration, or mere born-devils drowning care in a perpetual carouse.

And yet, when one comes to think upon it calmly, the situation of these South American citizens forms only a very pale figure for the state of ordinary mankind. This world itself, travelling blindly and swiftly in overcrowded space, among a million other worlds travelling blindly and swiftly in contrary directions, may very well come by a knock that would set it into explosion like a penny squib. And what, pathologically looked at, is the human body with all its organs, but a mere bagful of petards? The least of these is as dangerous to the whole economy as the ship's powder-magazine to the ship; and with every breath we breathe, and every meal we eat, we are putting one or more of them in peril. If we clung as devotedly as some philosophers pretend we do to the abstract idea of life, or were half as frightened as they make out we are, for the subversive accident that ends it all, the trumpets might sound by the hour and no one would follow them into battle—the blue-peter might fly at the truck, but who would climb into a sea-going ship? Think (if these philosophers were right) with what a preparation of spirit we should affront the daily peril of the dinner-table: a deadlier spot than any battle-field in history, where the far greater proportion of our ancestors have miserably left their bones! What woman would ever be lured into marriage, so much more dangerous than the wildest sea? And what would it be to grow old? For, after a certain distance, every step we take in life we find the ice growing thinner below our feet, and all around us and behind us we see our contemporaries going through. By the time a man gets well into the seventies, his continued existence is a mere miracle; and when he lays his old bones in bed for the night, there is an overwhelming probability that he will never see the day. Do the old men mind it, as a matter of fact? Why, no. They were never merrier; they have their grog at night, and tell the raciest stories; they hear of the death of people about their own age, or even younger, not as if it was a grisly warning, but with a simple child-like pleasure at having outlived some one else; and when a draught might puff them out like a guttering candle, or a bit of a stumble shatter them like so much glass, their old hearts keep sound and unaffrighted, and they go on, bubbling with laughter, through years of man's age compared to which the valley at Balaklava was as safe and peaceful as a village cricket-green on Sunday. It may fairly be questioned (if we look to the peril only) whether it was a much more daring feat for Curtius to plunge into the gulf, than for any old gentleman of ninety to doff his clothes and clamber into bed.

Indeed, it is a memorable subject for consideration, with what unconcern and gaiety mankind pricks on along the Valley of the Shadow of Death. The whole way is one wilderness of snares, and the end of it, for those who fear the last pinch, is irrevocable ruin. And yet we go spinning through it all, like a party for the Derby. Perhaps the reader remembers one of the humorous devices of the deified Caligula: how he encouraged a vast concourse of holiday-makers on to his bridge over Baiae bay; and when they were in the height of their enjoyment, turned loose the Praetorian guards among the company, and had them tossed into the sea. This is no bad miniature of the dealings of nature with the transitory race of man. Only, what a chequered picnic we have of it, even while it lasts! and into what great waters, not to be crossed by any swimmer, God's pale Praetorian throws us over in the end!

We live the time that a match flickers; we pop the cork of a ginger-beer bottle, and the earthquake swallows us on the instant. Is it not odd, is it not incongruous, is it not, in the highest sense of human speech, incredible, that we should think so highly of the ginger-beer, and regard so little the devouring earthquake? The love of Life and the fear of Death are two famous phrases that grow harder to understand the more we think about them. It is a well-known fact that an immense proportion of boat accidents would never happen if people held the sheet in their hands instead of making it fast; and yet, unless it be some martinet of a professional mariner or some landsman with shattered nerves, every one of God's creatures makes it fast. A strange instance of man's unconcern and brazen boldness in the face of death!

We confound ourselves with metaphysical phrases, which we import into daily talk with noble inappropriateness. We have no idea of what death is, apart from its circumstances and some of its consequences to others; and although we have some experience of living, there is not a man on earth who has flown so high into abstraction as to have any practical guess at the meaning of the word *life*. All literature, from Job and Omar Khayam to Thomas Carlyle or Walt Whitman, is but an attempt to look upon the human state with such largeness of view as shall enable us to rise from the consideration of living to the Definition of Life. And our sages give us about the best satisfaction in their power when they say that it is a vapour, or a show, or made out of the same stuff with dreams. Philosophy, in its more rigid sense, has been at the same work for ages; and after a myriad bald heads have wagged over the problem, and piles of words have been heaped one upon another into dry and cloudy volumes without end, philosophy has the honour of laying before us, with modest pride, her contribution towards the subject: that life is a Permanent Possibility of Sensation. Truly a fine result! A man may very well love beef, or hunting, or a woman; but surely, surely, not a Permanent Possibility of Sensation! He may be afraid of a precipice, or a dentist, or a large enemy with a club, or even an undertaker's man; but not certainly of abstract death. We may trick with the word life in its dozen senses until we are weary of tricking; we may argue in terms of all the philosophies on earth, but one fact remains true throughout—that we do not love life, in the sense that we are greatly preoccupied about its conservation; that we do not, properly speaking, love life at all, but living. Into the views of the least careful there will

enter some degree of providence; no man's eyes are fixed entirely
on the passing hour; but although we have some anticipation of
good health, good weather, wine, active employment, love, and
self-approval, the sum of these anticipations does not amount to
anything like a general view of life's possibilities and issues; nor
are those who cherish them most vividly, at all the most scrupu-
lous of their personal safety. To be deeply interested in the
accidents of our existence, to enjoy keenly the mixed texture of
human experience, rather leads a man to disregard precautions,
and risk his neck against a straw. For surely the love of living is
stronger in an Alpine climber roping over a peril, or a hunter
riding merrily at a stiff fence, than in a creature who lives upon
a diet and walks a measured distance in the interest of his
constitution.

There is a great deal of very vile nonsense talked upon both
sides of the matter: tearing divines reducing life to the dimensions of a mere funeral procession, so short as to be hardly
decent; and melancholy unbelievers yearning for the tomb as if it
were a world too far away. Both sides must feel a little ashamed
of their performances now and again when they draw in their
chairs to dinner. Indeed, a good meal and a bottle of wine is an
answer to most standard works upon the question. When a
man's heart warms to his viands, he forgets a great deal of soph-
istry, and soars into a rosy zone of contemplation. Death may be
knocking at the door, like the Commander's statue; we have
something else in hand, thank God, and let him knock. Passing
bells are ringing all the world over. All the world over, and
every hour, some one is parting company with all his aches and
ecstasies. For us also the trap is laid. But we are so fond of life
that we have no leisure to entertain the terror of death. It is a
honeymoon with us all through, and none of the longest. Small
blame to us if we give our whole hearts to this glowing bride of
ours, to the appetites, to honour, to the hungry curiosity of the
mind, to the pleasure of the eyes in nature, and the p. de of our
own nimble bodies.

We all of us appreciate the sensations; but as for caring about
the Permanence of the Possibility, a man's head is generally very
bald, and his senses very dull, before he comes to that. Whether
we regard life as a lane leading to a dead wall—a mere bag's end,
as the French say—or whether we think of it as a vestibule or

gymnasium, where we wait our turn and prepare our faculties for some more noble destiny; whether we thunder in a pulpit, or pule in little atheistic poetry-books, about its vanity and brevity; whether we look justly for years of health and vigour, or are about to mount into a bath-chair, as a step towards the hearse; in each and all of these views and situations there is but one conclusion possible: that a man should stop his ears against paralysing terror, and run the race that is set before him with a single mind. No one surely could have recoiled with more heartache and terror from the thought of death than our respected lexicographer; and yet we know how little it affected his conduct, how wisely and boldly he walked, and in what a fresh and lively vein he spoke of life. Already an old man, he ventured on his Highland tour; and his heart, bound with triple brass, did not recoil before twenty-seven individual cups of tea. As courage and intelligence are the two qualities best worth a good man's cultivation, so it is the first part of intelligence to recognise our precarious estate in life, and the first part of courage to be not at all abashed before the fact. A frank and somewhat headlong carriage, not looking too anxiously before, not dallying in maudlin regret over the past, stamps the man who is well armoured for this world.

And not only well armoured for himself, but a good friend and a good citizen to boot. We do not go to cowards for tender dealing; there is nothing so cruel as panic; the man who has least fear for his own carcase, has most time to consider others. That eminent chemist who took his walks abroad in tin shoes, and subsisted wholly upon tepid milk, had all his work cut out for him in considerate dealings with his own digestion. So soon as prudence has begun to grow up in the brain, like a dismal fungus, it finds its first expression in a paralysis of generous acts. The victim begins to shrink spiritually; he develops a fancy for parlours with a regulated temperature, and takes his morality on the principle of tin shoes and tepid milk. The care of one important body or soul becomes so engrossing, that all the noises of the outer world begin to come thin and faint into the parlour with the regulated temperature; and the tin shoes go equably forward over blood and rain. To be overwise is to ossify; and the scruple-monger ends by standing stockstill. Now the man who has his heart on his sleeve, and a good whirling weathercock of a brain, who reckons his life as a thing to be dashingly used and cheerfully hazarded, makes a very different acquaintance of the world, keeps all his pulses going true and fast, and gathers impetus as he runs, until, if he be running towards anything better than wildfire, he may shoot up and become a constellation in the end. Lord look after his health, Lord have a care of his soul, says he; and he has at the key of the position, and swashes through incongruity and peril towards his aim. Death is on all sides of him with pointed batteries, as he is on all sides of all of us; unfortunate suprises gird him round; mim-mouthed friends and relations hold up their hands in quite a little elegiacal synod about his path: and what cares he for all this? Being a true lover of living, a fellow with something pushing and spontaneous in his inside, he must, like any other soldier, in any other stirring, deadly warfare, push on at his best pace until he touch the goal. 'A peerage or Westminster Abbey!' cried Nelson in his bright, boyish, heroic manner. These are great incentives; not for any of these, but for the plain satisfaction of living, of being about their business in some sort or other, do the brave, serviceable men of every nation tread down the nettle danger, and pass flyingly over all the stumbling-blocks of prudence. Think of the heroism of Johnson, think of that superb indifference to mortal limitation that set him upon his dictionary, and carried him through triumphantly until the end! Who, if he were wisely considerate of things at large, would ever embark upon any work much more considerable than a halfpenny postcard? Who would project a serial novel, after Thackeray and Dickens had each fallen in mid-course? Who would find heart enough to begin to live, if he dallied with the consideration of death?

And, after all, what sorry and pitiful quibbling all this is ! To forego all the issues of living in a parlour with a regulated temperature—as if that were not to die a hundred times over, and for ten years at a stretch! As if it were not to die in one's own lifetime, and without even the sad immunities of death! As if it were not to die, and yet be the patient spectators of our own pitiable change! The Permanent Possibility is preserved, but the sensations carefully held at arm's length, as if one kept a photographic plate in a dark chamber. It is better to lose health like a spendthrift than to waste it like a miser. It is better to live and be done with it, than to die daily in the sick-room. By all means begin your folio; even if the doctor does not give you a year, even if he hesitates about a month, make one brave push and see what can be accomplished in a week. It is not only in finished undertakings that we ought to honour useful labour. A spirit goes out of the man who means execution, which outlives the most untimely ending. All who have meant good work with their whole hearts, have done good work, although they may die before they have the time to sign it. Every heart that has beat strong and cheerfully has left a hopeful impulse behind it in the world, and bettered the tradition of mankind. And even if death catch people, like an open pitfall, and in mid-career, laying out vast projects, and planning monstrous foundations, flushed with hope, and their mouths full of boastful language, they should be at once tripped up and silenced: is there not something brave and spirited in such a termination? and does not life go down with a better grace, foaming in full body over a precipice, than miserably straggling to an end in sandy deltas? When the Greeks made their fine saying that those whom the gods love die young, I cannot help believing they had this sort of death also in their eye. For surely, at whatever age it overtake the man, this is to die young. Death has not been suffered to take so much as an illusion from his heart. In the hot-fit of life, a-tiptoe on the highest point of being, he passes at a bound on to the other side. The noise of the mallet and chisel is scarcely quenched, the trumpets are hardly done blowing, when, trailing with him clouds of glory, this happy-starred, full-blooded spirit shoots into the spiritual land.

1878

OSCAR WILDE

'The True Critic'

(*from* The Critic as Artist)

*E*RNEST: Well, now that you have settled that the critic has at his disposal all objective forms, I wish you would tell me what are the qualities that should characterise the true critic.

Gilbert: What would you say they were?

Ernest: Well, I should say that a critic should above all things be fair.

Gilbert: Ah! not fair. A critic cannot be fair in the ordinary sense of the word. It is only about things that do not interest one that one can give a really unbiassed opinion, which is no doubt the reason why an unbiassed opinion is always absolutely value-less. The man who sees both sides of a question, is a man who sees absolutely nothing at all. Art is a passion, and, in matters of art, Thought is inevitably coloured by emotion, and so is fluid rather than fixed, and, depending upon fine moods and exquisite moments, cannot be narrowed into the rigidity of a scientific for-mula or a theological dogma. It is to the soul that Art speaks, and the soul may be made the prisoner of the mind as well as of the body. One should, of course, have no prejudices; but, as a great Frenchman remarked a hundred years ago, it is one's busi-ness in such matters to have preferences, and when one has preferences one ceases to be fair. It is only an auctioneer who can equally and impartially admire all schools of Art. No; fairness is not one of the qualities of the true critic. It is not even a con-dition of criticism. Each form of Art with which we come in contact dominates us for the moment to the exclusion of every other form. We must surrender ourselves absolutely to the work in question, whatever it may be, if we wish to gain its secret. For the time, we must think of nothing else, can think of nothing else, indeed.

Ernest: The true critic will be rational, at any rate, will he not?

Gilbert: Rational? There are two ways of disliking Art, Ernest. One is to dislike it. The other, to like it rationally. For Art, as Plato saw, and not without regret, creates in listener and spectator a form of divine madness. It does not spring from inspiration, but it makes others inspired. Reason is not the faculty to which it appeals. If one loves Art at all, one must love it beyond all other things in the world, and against such love, the reason, if one listened to it, would cry out. There is nothing sane about the worship of beauty. It is too splendid to be sane. Those of whose lives it forms the dominant note will always seem to the world to be pure visionaries.

Ernest: Well, at least, the critic will be sincere.

Gilbert: A little sincerity is a dangerous thing, and a great deal of it is absolutely fatal. The true critic will, indeed, always be sincere in his devotion to the principle of beauty, but he will seek for beauty in every age and in each school, and will never suffer himself to be limited to any settled custom of thought, or stereotyped mode of looking at things. He will realise himself in many forms, and by a thousand different ways, and will ever be curious of new sensations and fresh points of view. Through constant change, and through constant change alone, he will find his true unity. He will not consent to be the salve of his own opinions. For what is mind but motion in the intellectual sphere? The essence of thought, as the essence of life, is growth. You must not be frightened by words, Ernest. What people call insincerity is simply a method by which we can multiply our personalities.

Ernest: I am afraid I have not been fortunate in my suggestions.

Gilbert: Of the three qualifications you mentioned, two, sincerity and fairness, were, if not actually moral, at least on the borderland of morals, and the first condition of criticism is that the critic should be able to recognise that the sphere of Art and the sphere of Ethics are absolutely distinct and separate. When they are confused, Chaos has come again. They are too often confused in England now, and though our modern Puritans cannot destroy a beautiful thing, yet, by means of their extraordinary prurience, they can almost taint beauty for a moment. It is chiefly, I regret to say, through journalism that such people find expression. I regret it because there is much to be said in favour of modern journalism. By giving us the opinions of the uneducated, it keeps us in touch with the ignorance of the community. By carefully chronicling the current events of contemporary life, it shows us of what very little importance such events really are. By invariably discussing the unnecessary, it makes us understand what things are requisite for culture, and what are not. But it should not allow poor Tartuffe to write articles upon modern art. When it does this it stultifies itself. And yet Tartuffe's articles and Chadband's notes do this good, at least. They serve to show how extremely limited is the area over which eithics, and ethical considerations, can claim to exercise influence. Science is out of the reach of morals, for her eyes are fixed upon eternal truths. Art is out of the reach of morals, for her eyes are fixed upon things beautiful and immortal and everchanging. To morals belong the lower and less intellectual spheres. However, let these mouthing Puritans pass; they have their comic side. Who can help laughing when an ordinary journalist seriously proposes to limit the subject-matter at the disposal of the artist? Some limitation might well, and will soon, I hope, be placed upon some of our newspapers and newspaper writers. For they give us the bald, sordid, disgusting facts of life. They chronicle, with degrading avidity, the sins of the second-rate, and with the conscientiousness of the illiterate give us accurate and prosaic details of the doings of people of absolutely no interest whatsoever. But the artist, who accepts the facts of life, and yet transforms them into shapes of beauty, and makes them vehicles of pity or of awe, and shows their colour-element, and their wonder, and their true ethical import also, and builds out of them a world more real than reality itself, and of loftier and more noble import—who shall set limits to him? Not the apostles of that new Journalism which is but the old vulgarity 'writ large.' Not the apostles of that new Puritanism, which is but the whine of the hypocrite, and is both writ and spoken badly. The mere suggestion is ridiculous.

1891

Sir George Grove

Beethoven and his Nine Symphonies
by George Grove, C.B.

O<small>N</small> cold Saturday afternoons in winter, as I sit in the theatrical desert, making my bread with great bitterness by chronicling insignificant plays and criticizing incompetent players, it sometimes comes upon me that I have forgotten something—omitted something—missed some all-important appointment. This is a legacy from my old occupation of musical critic. All my old occupations leave me such legacies. When I was in my teens I had certain official duties to perform, which involved every day the very strict and punctual discharge of certain annual payments, which were set down in a perpetual diary. I sometimes dream now that I am back at those duties again, but with an amazed consciousness of having allowed them to fall into ruinous arrear for a long time past. My Saturday afternoon misgivings are just like that. They mean that for several years I passed those afternoons in that section of the gallery of the Crystal Palace concert-room which is sacred to Sir George Grove and to the Press. There were two people there who never grew older—Beethoven and Sir George. August Manns's hair changed from raven black to swan white as the years passed; young critics grew middle-aged and middle-aged critics grew old; Rossini lost caste and was shouldered into the promenade; the fire-new overture to Tannhäuser began to wear as threadbare as William Tell; Arabella Goddard went and Sophie Menter came; Joachim, Hallé, Norman Neruda, and Santley no longer struck the rising generations with the old sense of belonging to tomorrow, like Isaÿe, Paderewski, and Bispham; the men whom I had shocked as an iconoclastic upstart Wagnerian, braying derisively when they observed that 'the second subject, appearing in the key of the dominant, contrasts effectively with its predecessor, not only in tonality, but by its suave, melodious character,' lived to see me shocked and wounded in my turn by the audacities of J. F. Runciman; new evening papers launched into musical criticism, and were read publicly by Mr Smith, the eminent drummer, whenever he had fifty bars rest; a hundred trifles marked the flight of time; but Sir George Grove fed on Beethoven's symphonies as the gods in Das Rheingold fed on the apples of Freia, and grew no older. Sometimes, when Mendelssohn's Scotch symphony, or Schubert's Ninth in C, were in the program, he got positively younger, clearing ten years backward in as many minutes when Manns and the band were at their best. I remonstrated with him more than once on this unnatural conduct; and he was always extremely apologetic, assuring me that he was getting on as fast as he could. He even succeeded in producing a wrinkle or two under stress of Berlioz and Raff, Liszt and Wagner; but presently some pianist would come along with the concerto in E flat; and then, if I sat next him, strangers would say to me 'Your son, sir, appears to be a very enthusiastic musician.' And I could not very well explain that the real bond between us was the fact that Beethoven never ceased to grow on us. In my personality, my views, and my style of criticism there was so much to forgive that many highly amiable persons never quite succeeded in doing it. To Sir George I must have been a positively obnoxious person, not in the least because I was on the extreme left in politics and other matters, but because I openly declared that the finale of Schubert's symphony in C could have been done at half the length and with twice the effect by Rossini. But I knew Beethoven's symphonies from the opening bar of the first to the final chord of the ninth, and yet made new discoveries about them at every fresh performance. And I am convinced that 'G' regarded this as evidence of a fundamental rectitude in me which would bear any quantity of superficial aberrations. Which is quite my own opinion too.

It may be asked why I have just permitted myself to write of so eminent a man as Sir George Grove by his initial. That question would not have been asked thirty years ago, when 'G,' the rhapsodist who wrote the Crystal Palace programs, was one of the best ridiculed men in London. At that time the average programmist would unblushingly write, 'Here the composer, by one of those licenses which are, perhaps, permissible under exceptional circumstances to men of genius, but which cannot be too carefully avoided by students desirous of forming a legitimate style, has abruptly introduced the dominant seventh of the key of C major into the key of A flat, in order to recover, by a forced modulation, the key relationship proper to the second subject of a movement in F: an awkward device which he might have spared himself by simply introducing his second subject in its true key of C.' 'G,' who was 'no musician,' cultivated this style in vain. His most conscientious attempts at it never brought him any nearer than 'The lovely melody then passes, by a transition of remarkable beauty, into the key of C major, in which it seems to go straight up to heaven.' Naturally the average Englishman was profoundly impressed by the inscrutable learning of the first style (which I could teach to a poodle in two hours), and thought 'G's' obvious sentimentality idiotic. It did not occur to the average Englishman that perhaps Beethoven's symphonies were an affair of sentiment and nothing else. This, of course, was the whole secret of them. Beethoven was the first man who used music with absolute integrity as the expression of his own emotional life. Others had shewn how it could be done—had done it themselves as a curiosity of their art in rare, self-indulgent, *unprofessional* moments—but Beethoven made this, and nothing else, his business. Stupendous as the resultant difference was between his music and any other ever heard in the world before his time, the distinction is not clearly apprehended to this day, because there was nothing new in the musical expression of emotion: every progression in Bach is sanctified by emotion; and Mozart's subtlety, delicacy, and exquisite tender touch and noble feeling were the despair of all the musical world. But Bach's theme was not himself, but his religion; and Mozart was always the dramatist and story-teller, making the men and women of his imagination speak, and dramatizing even the instruments in his orchestra, so that you know their very sex the moment their voices reach you. Haydn really came nearer to Beethoven, for he is neither the praiser of God nor the dramatist, but, always within the limits of good manners and of his primary function as a purveyor of formal decorative music, a man of moods. This is how he created the symphony and put it ready-made into Beethoven's hand. The revolutionary giant at once seized it, and throwing supernatural religion, conventional good manners, dramatic fiction, and all external standards and objects into the lumber room, took his own humanity as the material of his music, and expressed it all without compromise, from his roughest jocularity to his holiest aspiration after that purely human reign of intense life—of Freude—when

> Alle Menschen werden Brüder
> Wo dein sanfter Flügel weilt.

In thus fearlessly expressing himself, he has, by his common humanity, expressed us as well, and shewn us how beautifully, how strongly, how trustworthily we can build with our own real selves. This is what is proved by the immense superiority of the Beethoven symphony to any oratorio or opera.

In this light all Beethoven's work becomes clear and simple; and the old nonsense about his obscurity and eccentricity and stage sublimity and so on explains itself as pure misunderstanding. His criticisms, too, become quite consistent and inevitable: for instance, one is no longer tempted to resent his declaration that Mozart wrote nothing worth considering but parts of Die Zauberflöte (those parts, perhaps, in which the beat of dein sanfter Flügel is heard), and to retort upon him by silly comparisons of his tunes with Non piu andrai and Deh vieni alla finestra. The man who wrote the Eighth symphony has a right to rebuke the man who put his raptures of elation, tenderness, and nobility into the mouths of a drunken libertine, a silly peasant girl, and a conventional fine lady, instead of confessing them to himself, glorying in them, and uttering them without motley as the universal inheritance.

I must not make 'G' responsible for my own opinions; but I leave it to his old readers whether his huge success as a program writer was not due to the perfect simplicity with which he seized and followed up this clue to the intention of Beethoven's symphonies. He seeks always for the mood, and is not only delighted at every step by the result of his search, but escapes quite easily and unconsciously from the boggling and blundering of the men who are always wondering why Beethoven did not do what any professor would have done. He is always joyous,

always successful, always busy and interesting, never tedious
even when he is superfluous (not that the adepts ever found him
so), and always as pleased as Punch when he is not too deeply
touched. Sometimes, of course, I do not agree with him. Where
he detects anger in the Eighth symphony, I find nothing but
boundless, thundering elation. In his right insistence on the joc-
ular element in the symphonies, I think he is occasionally led by
his personal sense that octave skips on the bassoon and drum are
funny to conclude too hastily that Beethoven was always joking
when he used them. And I will fight with him to the death
on the trio of the Eighth symphony, maintaining passionately
against him and against all creation that those 'cello arpeggios
which steal on tiptoe round the theme so as not to disturb its
beauty are only 'fidgety' when they are played 'à la Mendelssohn,'
and that they are perfectly tender and inevitable when they are
played 'à la Wagner.' The passage on this point in Wagner's
essay on Conducting is really not half strong enough; and when
'G' puts it down to 'personal bias' and Wagner's 'poor opinion
of Mendelssohn,' it is almost as if someone had accounted in
the same way for Beethoven's opinion of Mozart. Wagner was
almost as fond of Mendelssohn's music as 'G' is; but he had
suffered unbearably, as we all have, from the tradition established by Mendelssohn's conducting of Beethoven's symphonies.
Mendelssohn's music is all *nervous music*: his allegros, expressing
only excitement and impetuosity without any ground, have fire
and motion without substance. Therefore the conductor must,
above all things, *keep them going*; if he breaks their lambent flight
to dwell on any moment of them, he is lost. With Beethoven the
longer you dwell on any moment the more you will find in it.
Provided only you do not sacrifice his splendid energetic rhythm
and masterly self-possessed emphasis to a maudlin preoccupation with his feeling, you cannot possibly play him too senti-
mentally; for Beethoven is no reserved gentleman, but a man
proclaiming the realities of life. Consequently, when for generations they played Beethoven's allegros exactly as it is necessary to
play the overture to Ruy Blas, or Stone him to death—a practice
which went on until Wagner's righteous ragings stopped it—our
performances of the symphonies simply spoiled the tempers of
those who really understood them. For the sake of redeeming
that lovely trio from 'fidgetiness,' 'G' must let us face this fact

even at the cost of admitting that Wagner was right where Mendelssohn was wrong.

But though it is possible thus to differ here and there from 'G,' he is never on the wrong lines. He is always the true musician: that is, the man the professors call 'no musician'—just what they called Beethoven himself. It is delightful to have all the old programs bound into a volume, with the quotations from the score all complete, and the information brought up to date, and largely supplemented. It is altogether the right sort of book about the symphonies, made for practical use in the concert room under the stimulus of a heartfelt need for bringing the public to Beethoven. I hope it will be followed by another volume or two dealing with the pianoforte concertos—or · say with the G, the E flat, the choral fantasia, and the three classical violin concertos: Beethoven, Mendelssohn, and Brahms. And then a Schubert-Mendelssohn-Schumann volume. Why, dear 'G,' should these things be hidden away in old concert programs which never circulate beyond Sydenham?

1896

JOSEPH CONRAD

The Censor of Plays

AN APPRECIATION

A COUPLE of years ago I was moved to write a one-act play
—and I lived long enough to accomplish the task. We live and
learn. When the play was finished I was informed that it had to
be licensed for performance. Thus I learned of the existence of
the Censor of Plays. I may say without vanity that I am intelligent enough to have been astonished by that piece of information: for facts must stand in some relation to time and space,
and I was aware of being in England—in the twentieth-century
England. The fact did not fit the date and the place. That was my
first thought. It was, in short, an improper fact. I beg you to
believe that I am writing in all seriousness and am weighing my
words scrupulously.

Therefore I don't say inappropriate. I say improper—that is:
something to be ashamed of. And at first this impression was
confirmed by the obscurity in which the figure embodying this
after all considerable fact had its being. The Censor of Plays! His
name was not in the mouths of all men. Far from it. He seemed
stealthy and remote. There was about that figure the scent of the
far East, like the peculiar atmosphere of a Mandarin's back yard,
and the mustiness of the Middle Ages, that epoch when mankind
tried to stand still in a monstrous illusion of final certitude
attained in morals, intellect and conscience.

It was a disagreeable impression. But I reflected that probably
the censorship of plays was an inactive monstrosity; not exactly a
survival, since it seemed obviously at variance with the genius of
the people, but an heirloom of past ages, a bizarre and imported
curiosity preserved because of that weakness one has for one's
old possessions apart from any intrinsic value; one more object
of exotic *virtù*, an Oriental *potiche*, a *magot chinois* conceived by a

childish and extravagant imagination, but allowed to stand in stolid impotence in the twilight of the upper shelf.

Thus I quieted my uneasy mind. Its uneasiness had nothing to do with the fate of my one-act play. The play was duly produced, and an exceptionally intelligent audience stared it coldly off the boards. It ceased to exist. It was a fair and open execution. But having survived the freezing atmosphere of that auditorium I continued to exist, labouring under no sense of wrong. I was not pleased, but I was content. I was content to accept the verdict of a free and independent public, judging after its conscience the work of its free, independent and conscientious servant—the artist.

Only thus can the dignity of artistic servitude be preserved—not to speak of the bare existence of the artist and the self-respect of the man. I shall say nothing of the self-respect of the public. To the self-respect of the public the present appeal against the censorship is being made and I join in it with all my heart.

For I have lived long enough to learn that the monstrous and outlandish figure, the *magot chinois* whom I believed to be but a memorial of our forefathers' mental aberration, that grotesque *potiche*, works! The absurd and hollow creature of clay seems to be alive with a sort of (surely) unconscious life worthy of its traditions. It heaves its stomach, it rolls its eyes, it brandishes a monstrous arm: and with the censorship, like a Bravo of old Venice with a more carnal weapon, stabs its victim from behind in the twilight of its upper shelf. Less picturesque than the Venetian in cloak and mask, less estimable, too, in this, that the assassin plied his moral trade at his own risk deriving no countenance from the powers of the Republic, it stands more malevolent, inasmuch that the Bravo striking in the dusk killed but the body, whereas the grotesque thing nodding its mandarin head may in its absurd unconsciousness strike down at any time the spirit of an honest, of an artistic, perhaps of a sublime creation.

This Chinese monstrosity, disguised in the trousers of the Western Barbarian and provided by the State with the immortal Mr Stiggins's plug hat and umbrella, is with us. It is an office. An office of trust. And from time to time there is found an official to fill it. He is a public man. The least prominent of public men, the most unobtrusive, the most obscure if not the most modest.

But however obscure, a public man may be told the truth if only once in his life. His office flourishes in the shade; not in the rustic shade beloved of the violet but in the muddled twilight of mind, where tyranny of every sort flourishes. Its holder need not have either brain or heart, no sight, no taste, no imagination, not even bowels of compassion. He needs not these things. He has power. He can kill thought, and incidentally truth, and incidentally beauty, providing they seek to live in a dramatic form. He can do it, without seeing, without understanding, without feeling anything; out of mere stupid suspicion, as an irresponsible Roman Caesar could kill a senator. He can do that and there is no one to say him nay. He may call his cook (Molière used to do that) from below and give her five acts to judge every morning as a matter of constant practice and still remain the unquestioned destroyer of men's honest work. He may have a glass too much. This accident has happened to persons of unimpeachable morality—to gentlemen. He may suffer from spells of imbecility like Clodius. He may . . . what might he not do! I tell you he is the Caesar of the dramatic world. There has been since the Roman Principate nothing in the way of irresponsible power to compare with the office of the Censor of Plays.

Looked at in this way it has some grandeur, something colossal in the odious and the absurd. This figure in whose power it is to suppress an intellectual conception—to kill thought (a dream for a mad brain, my masters!)—seems designed in a spirit of bitter comedy to bring out the greatness of a Philistine's conceit and his moral cowardice.

But this is England in the twentieth century, and one wonders that there can be found a man courageous enough to occupy the post. It is a matter for meditation. Having given it a few minutes I come to the conclusion in the serenity of my heart and the peace of my conscience that he must be either an extreme megalomaniac or an utterly unconscious being.

He must be unconscious. It is one of the qualifications for his magistracy. Other qualifications are equally easy. He must have done nothing, expressed nothing, imagined nothing. He must be obscure, insignificant and mediocre—in thought, act, speech and sympathy. He must know nothing of art, of life—and of himself. For if he did he would not dare to be what he is. Like that much questioned and mysterious bird, the phoenix, he sits amongst the cold ashes of his predecessor upon the altar of morality, alone of his kind in the sight of wondering generations.

And I will end with a quotation reproducing not perhaps the exact words but the true spirit of a lofty conscience.

'Often when sitting down to write the notice of a play, especially when I felt it antagonistic to my canons of art, to my tastes or my convictions, I hesitated in the fear lest my conscientious blame might check the development of a great talent, my sincere judgment condemn a worthy mind. With the pen poised in my hand I hesitated, whispering to myself "What if I were perchance doing my part in killing a masterpiece."'

Such were the lofty scruples of M. Jules Lemaître—dramatist and dramatic critic, a great citizen and a high magistrate in the Republic of Letters; a Censor of Plays exercising his august office openly in the light of day, with the authority of a European reputation. But then M. Jules Lemaître is a man possessed of wisdom, of great fame, of a fine conscience—not an obscure hollow Chinese monstrosity ornamented with Mr Stiggins's plug hat and cotton umbrella by its anxious grandmother—the State.

Frankly, is it not time to knock the improper object off its shelf? It has stood too long there. Hatched in Pekin (I should say) by some Board of Respectable Rites, the little caravan monster has come to us by way of Moscow—I suppose. It is outlandish. It is not venerable. It does not belong here. Is it not time to knock it off its dark shelf with some implement appropriate to its worth and status? With an old broom handle for instance.

1907

JAMES G. HUNEKER

A Visit to Walt Whitman

MY edition of Walt Whitman's Leaves of Grass is dated 1867, the third, if I am not mistaken, the first appearing in 1855. Inside is pasted a card upon which is written in large, clumsy letters: 'Walt Whitman, Camden, New Jersey, July, 1877.' I value this autograph, because Walt gave it to me; rather I paid him for it, the proceeds, two dollars (I think that was the amount), going to some asylum in Camden. In addition, the 'good grey poet' was kind enough to add a woodcut of himself as he appeared in the 1855 volume, 'hankering, gross, mystical, nude,' and another of his old mother, with her shrewd, kindly face. Walt is in his shirt-sleeves, a hand on his hip, the other in his pocket, his neck bare, the pose that of a nonchalant workman—though in actual practice he was always opposed to work of any sort; on his head is a slouch-hat, and you recall his line: 'I wear my hat as I please, indoors or out.' The picture is characteristic, even to the sensual mouth and Bowery-boy pose. You almost hear him say: 'I find no sweeter fat than sticks to my own bones.' Altogether a different man from the later bard, the heroic apparition of Broadway, Pennsylvania Avenue, and Chestnut Street. I had convalesced from a severe attack of Edgar Allan Poe only to fall desperately ill with Whitmania. Youth is ever in revolt, age alone brings resignation. My favourite reading was Shelley, my composer among composers, Wagner. Chopin came later. This was in 1876, when the Bayreuth apotheosis made Wagner's name familiar to us, especially in Philadelphia, where his empty, sonorous Centennial March was first played by Theodore Thomas at the Exposition. The reading of a magazine article by Moncure D. Conway caused me to buy a copy, at an extravagant price for my purse, of The Leaves of Grass, and so uncritical was I that I wrote a parallel between Wagner and Whitman; between the most consciously artistic of men and the wildest among impro-

visators. But then it seemed to me that both had thrown off the
'shackles of convention.' (What prison-like similes we are given
to in the heady, generous impulses of green adolescence.) I was a
boy, and seeing Walt on Market Street, as he came from the
Camden Ferry, I resolved to visit him. It was some time after the
Fourth of July, 1877, and I soon found his little house on Mickle
Street. A policeman at the ferry-house directed me. I confess I
was scared after I had given the bell one of those pulls that we
tremblingly essay at a dentist's door. To my amazement the old
man soon stood before me, and cordially bade me enter.

'Walt,' I said, for I had heard that he disliked a more cere-
monious prefix, 'I've come to tell you how much the Leaves
have meant to me.' 'Ah!' he simply replied, and asked me to take
a chair. To this hour I can see the humble room, but when I try
to recall our conversation I fail. That it was on general literary
subjects I know, but the main theme was myself. In five minutes
Walt had pumped me dry. He did it in his quiet, sympathetic
way, and, with the egoism of my age, I was not averse from
relating to him the adventures of my soul. That Walt was a fluent
talker one need but read his memoirs by Horace Traubel. Witness his tart allusion to Swinburne's criticism of himself: 'Isn't he
the damnedest simulacrum?' But he was a sphinx the first time I
met him. I do recall that he said Poe wrote too much in a dark
cellar, and that music was his chief recreation—of which art he
knew nothing; it served him as a sounding background for his
pencilled improvisations. I begged for an autograph. He told me
of his interest in a certain asylum or hospital, whose name has
gone clean out of my mind, and I paid my few dollars for the
treasured signature. It is now one of my literary treasures.

If I forget the tenor of our discourse I have not forgotten the
immense impression made upon me by the man. As vain as a
peacock, Walt looked like a Greek rhapsodist. Tall, imposing in
bulk, his regular features, mild, light-blue or grey eyes, clear
ruddy skin, plentiful white hair and beard, evoked an image of
the magnificently fierce old men he chants in his book. But he
wasn't fierce, his voice was a tenor of agreeable timbre, and he
was gentle, even to womanliness. Indeed, he was like a receptive,
lovable old woman, the kind he celebrates so often. He never
smoked, his only drink was water. I doubt if he ever drank
spirits. His old friends say 'No,' although he is a terrible rake in

print. Without suggesting effeminacy, he gave me the impression
of a feminine soul in a masculine envelope. When President Lincoln first saw him he said: 'Well, he *looks* like a man!' Perhaps
Lincoln knew, for his remark has other connotations than the
speech of Napoleon when he met Goethe: 'Voilà un homme!'
Hasn't Whitman asked in Calamus, the most revealing section
of Leaves: 'Do you suppose yourself advancing on real ground
toward a real heroic man?' He also wrote of Calamus: 'Here the
frailest leaves of me . . . . Here I shade down and hide my
thoughts. I do not express them. And yet they expose me more
than all my other poems.' Mr Harlan, Secretary of the Interior,
when he dismissed Walt from his department because of Leaves,
did not know about the Calamus section—I believe they were
not incorporated till later—but Washington was acquainted with
Walt and his idiosyncrasies, and, despite W. D. Connor's spirited
vindication, certain rumours would not be stifled. Walt was
thirty-six when Leaves appeared; forty-one when Calamus was
written.

I left the old man after a hearty hand-shake, a So long! just as
in his book, and returned to Philadelphia. Full of the day, I told
my policeman at the ferry that I had seen Walt. 'That old gas-
bag comes here every afternoon. He gets free rides across the
Delaware,' and I rejoiced to think that a soulless corporation had
some appreciation of a great poet, though the irreverence of
this 'powerful uneducated person' shocked me. When I reached
home I also told my mother of my visit. She was plainly disturbed. She said that the writings of the man were immoral, but
she was pleased at my report of Walt's sanity, sweetness, mellow
optimism, and his magnetism, like some natural force. I forgot,
in my enthusiasm, that it was Walt who listened, I who gabbled.
My father, who had never read Leaves, had sterner criticism to
offer: 'If I ever hear of you going to see that fellow you'll be
sorry!' This coming from the most amiable of parents, surprised
me. Later I discovered the root of his objection, for, to be quite
frank, Walt did not bear a good reputation in Philadelphia, and I
have heard him spoken of so contemptuously that it would bring
a blush to the shining brow of a Whitmaniac. Yet dogs followed
him and children loved him. I saw Walt accidentally at intervals,
though never again in Camden. I met him on the streets, and
several times took him from the Carl Gaertner String Quartet

Concerts in the foyer of the Broad Street Academy of Music to the Market Street cars. He lumbered majestically, his hairy breast exposed, but was a feeble old man, older than his years; paralysis had maimed him. He is said to have incurred it from his unselfish labours as nurse in the camp hospitals at Washington during the Civil War; however, it was in his family on the paternal side, and at thirty he was quite grey. The truth is, Walt was not the healthy hero he celebrates in his book. That he never dissipated we know; but his husky masculinity, his posing as the Great God Priapus in the garb of a Bowery boy is discounted by the facts. Parsiphallic, he was, but not of Pan's breed. In the Children of Adam, the part most unfavourably criticised of Leaves, he is the Great Bridegroom, and in no literature, ancient or modern, have been the 'mysteries' of the temple of love so brutally exposed. With all his genius in naming certain unmentionable matters, I don't believe in the virility of these pieces, scintillating with sexual images. They leave one cold despite their erotic vehemence; the abuse of the vocative is not persuasive, their raptures are largely rhetorical. This exaltation, this ecstasy, seen at its best in William Blake, is sexual ecstasy, but only when the mood is married to the mot lumière is there authentic conflagration. Then his 'barbaric yawp is heard across the roofs of the world'; but in the underhumming harmonics of Calamus, where Walt really loafs and invites his soul, we get the real man, not the inflated humbuggery of These States, Camerados, or My Message, which fills Leaves with their patriotic frounces. His philosophy is fudge. It was an artistic misfortune for Walt that he had a 'mission,' it is a worse one that his disciples endeavour to ape him. He was an unintellectual man who wrote conventionally when he was plain Walter Whitman, living in Brooklyn. But he imitated Ossian and Blake, and their singing robes ill-befitted his burly frame. If, in Poe, there is much 'rant and rococo,' Whitman is mostly yawping and yodling. He is destitute of humour, like the majority of 'prophets' and uplifters, else he might have realised that a Democracy based on the 'manly love of comrades' is an absurdity. Not alone in Calamus, but scattered throughout Leaves, there are passages that fully warrant unprejudiced psychiatrists in styling this book the bible of the third sex.

But there is rude red music in the versicles of Leaves. They stimulate, and, for some young hearts, they are as a call to battle. The book is a capital hunting-ground for quotations. Such massive head-lines—that soon sink into platitudinous prose; such robust swinging rhythms, Emerson told Walt that he must have had a 'long foreground.' It is true. Notwithstanding his catalogues of foreign countries, he was hardly a cosmopolitan. Whitman's so-called 'mysticism' is a muddled echo of New England Transcendentalism; itself a pale dilution of an outworn German idealism—what Coleridge called 'the holy jungle of Transcendental metaphysics.' His concrete imagination automatically rejected metaphysics. His chief asset is an extraordinary sensitiveness to the sense of touch; it is his distinguishing passion, and tactile images flood his work; this, and an eye that records appearances, the surface of things, and registers in phrases of splendour the picturesque, yet seldom fuses matter and manner into a poetical synthesis. The community of interest between his ideas and images is rather affiliated than cognate. He has a tremendous, though ill-assorted vocabulary. His prose is jolting, rambling, tumid, invertebrate. An 'arrant artist,' as Mr Brownell calls him, he lacks formal sense and the diffuseness and vagueness of his supreme effort—the Lincoln burial hymn—serves as a nebulous buffer between sheer overpraise and serious criticism. He contrives atmosphere with facility, and can achieve magical pictures of the sea and the 'mad naked summer night.' His early poem, Walt Whitman, is for me his most spontaneous offering. He has at times the primal gift of the poet—ecstasy; but to attain it he often wades through shallow, ill-smelling sewers, scales arid hills, traverses dull drab levels where the slag covers rich ore, or plunges into subterrene pools of nocturnal abominations—veritable regions of the 'mother of dead dogs.' Probably the sexlessness of Emerson's, Poe's, and Hawthorne's writings sent Whitman to an orgiastic extreme, and the morbid, nasty-nice puritanism that then tainted English and American letters received its first challenge to come out into the open and face natural facts. Despite his fearlessness, one must subscribe to Edmund Clarence Stedman's epigram: 'There are other lights in which a dear one may be regarded than as the future mother of men.' Walt let in a lot of fresh air on the stuffy sex question of his day, but, in demanding equal sexual rights for women, he meant it in the reverse sense as propounded by our old grannies' purity leagues. Continence is not the sole virtue or charm in womanhood; nor, by the same token, is unchastity a brevet of feminine originality. But women, as a rule, have not rallied to his doctrines, instinctively feeling that he is indifferent to them, notwithstanding the heated homage he pays to their physical attractions. Good old Walt sang of his camerados, capons, Americanos, deck-hands, stagecoach-drivers, machinists, brakemen, firemen, sailors, butchers, bakers, and candlestick makers, and he associated with them; but they never read him or understood him. They prefer Longfellow. It is the cultured class he so despises that discovered, lauded him, believing that he makes vocal the underground world; above all, believing that he truly represents America and the dwellers thereof—which he decidedly does not. We are, if you will, a commonplace people, but normal, and not enamoured of 'athletic love of comrades.' I remember a dinner given by the Whitman Society about twenty years ago, at the St Denis Hotel, which was both grotesque and pitiable. The guest of honour was 'Pete' Doyle, the former car-conductor and 'young rebel friend of Walt's,' then a middle-aged person. John Swinton, who presided, described Whitman as a troglodyte, but a cave-dweller he never was; rather the avatar of the hobo. As John Jay Chapman wittily wrote: 'He patiently lived on cold pie, and tramped the earth in triumph.' Instead of essaying the varied, expressive, harmonious music of blank verse, he chose the easier, more clamorous, and disorderly way; but if he had not so chosen we should have missed the salty tang of the true Walt Whitman. Toward the last there was too much Camden in his Cosmos. Quite appropriately his dying word was le mot de Cambronne. It was the last victory of an organ over an organism. And he was a gay old pagan who never called a sin a sin when it was a pleasure.

1915

William James

NONE of us will ever see a man like William James again: there is no doubt about that. And yet it is hard to state what it was in him that gave him either his charm or his power, what it was that penetrated and influenced us, what it is that we lack and feel the need of, now that he has so unexpectedly and incredibly died. I always thought that William James would continue forever; and I relied upon his sanctity as if it were sunlight.

I should not have been abashed at being discovered in some mean action by William James; because I should have felt that he would understand and make allowances. The abstract and sublime quality of his nature was always enough for two; and I confess to having always trespassed upon him and treated him with impertinence, without gloves, without reserve, without ordinary, decent concern for the sentiments and weaknesses of human character. Knowing nothing about philosophy, and having the dimmest notions as to what James's books might contain, I used occasionally to write and speak to him about his specialties in a tone of fierce contempt; and never failed to elicit from him in reply the most spontaneous and celestial gayety. Certainly he was a wonderful man.

He was so devoid of selfish aim or small personal feeling that your shafts might pierce, but could never wound him. You could not 'diminish one dowle that's in his plume.' Where he walked, nothing could touch him; and he enjoyed the Emersonian immunity of remaining triumphant even after he had been vanquished. The reason was, as it seems to me, that what the man really meant was always something indestructible and persistent; and that he knew this inwardly. He had not the gift of expression, but rather the gift of suggestion. He said things which meant one thing to him and something else to the reader or listener. His mind was never quite in focus, and there was always something left over after each discharge of the battery, something which now became the beginning of a new thought. When he found out his mistake or defect of expression, when he came to see that he had not said quite what he meant, he was the first to proclaim it, and to move on to a new position, a new mis-statement of the same truth,—a new, debonair apperception, clothed in non-conclusive and suggestive figures of speech.

How many men have put their shoulders out of joint in strik-ing at the phantasms which James projected upon the air! James was always in the right, because what he meant was true. The only article of his which I ever read with proper attention was 'The Will to Believe,' a thing that exasperated me greatly until I began to see, or to think I saw, what James meant, and at the same time to acknowledge to myself that he had said something quite different. I hazard this idea about James as one might haz-ard an idea about astronomy, fully aware that it may be very foolish.

In private life and conversation there was the same radiation of thought about him. The center and focus of his thought fell within his nature, but not within his intellect. You were thus played upon by a logic which was not the logic of intellect, but a far deeper thing, limpid and clear in itself, confused and refract-ory only when you tried to deal with it intellectually. You must take any fragment of such a man by itself, for his whole meaning is in the fragment. If you try to piece the bits together, you will endanger their meaning. In general talk on life, literature, and politics James was always throwing off sparks that were cognate only in this, that they came from the same central fire in him. It was easy to differ from him; it was easy to go home thinking that James had talked the most arrant rubbish, and that no educated man had a right to be so ignorant of the first principles of thought and of the foundations of human society. Yet it was impossible not to be morally elevated by the smallest contact with William James. A refining, purgatorial influence came out of him.

I believe that in his youth, James dedicated himself to the glory of God and the advancement of Truth, in the same spirit that a young knight goes to seek the Grail, or a young military hero dreams of laying down his life for his country. What his early leanings towards philosophy or his natural talent for it may have

been, I do not know; but I feel as if he had first taken up philosophy out of a sense of duty,—the old Puritanical impulse,—in
his case illumined, however, with a humor and genius not at all
of the Puritan type. He adopted philosophy as his lance and
buckler,—psychology, it was called in his day,—and it proved to
be as good as the next thing,—as pliable as poetry or fiction or
politics or law would have been,—or anything else that he might
have adopted as a vehicle through which his nature could work
upon society.

He, himself, was all perfected from the beginning, a selfless
angel. It is this quality of angelic unselfishness which gives the
power to his work. There may be some branches of human
study—mechanics perhaps—where the personal spirit of the
investigator does not affect the result; but philosophy is not one
of them. Philosophy is a personal vehicle; and every man makes
his own, and through it he says what he has to say. It is all personal: it is all human: it is all non-reducible to science, and
incapable of being either repeated or continued by another man.

Now James was an illuminating ray, a dissolvent force. He
looked freshly at life, and read books freshly. What he had to say
about them was not entirely articulated, but was always spontaneous. He seemed to me to have too high an opinion of
everything. The last book he had read was always 'a great book';
the last person he had talked with, a wonderful being. If I may
judge from my own standpoint, I should say that James saw too
much good in everything, and felt towards everything a too
indiscriminating approval. He was always classing things up into
places they didn't belong and couldn't remain in.

Of course, we know that Criticism is proverbially an odious
thing; it seems to deal only in shadows,—it acknowledges only
varying shades of badness in everything. And we know, too, that
Truth is light; Truth cannot be expressed in shadow, except by
some subtle art which proclaims the shadow-part to be the lie,
and the non-expressed part to be the truth. And it is easy to look
upon the whole realm of Criticism and see in it nothing but a science which concerns itself with the accurate statement of lies.
Such, in effect, it is in the hands of most of its adepts. Now
James's weakness as a critic was somehow connected with the
peculiar nature of his mind, which lived in a consciousness of
light. The fact is that James was non-critical, and therefore

divine. He was forever hovering, and never could alight; and this is a quality of truth and a quality of genius.

The great religious impulse at the back of all his work, and which pierces through at every point, never became expressed in conclusive literary form, or in dogmatic utterance. It never became formulated in his own mind into a statable belief. And yet it controlled his whole life and mind, and accomplished a great work in the world. The spirit of a priest was in him,—in his books and in his private conversation. He was a sage, and a holy man; and everybody put off his shoes before him. And yet in spite of this,—in conjunction with this, he was a sportive, wayward, Gothic sort of spirit, who was apt, on meeting a friend, to burst into foolery, and whose wit was always three parts poetry. Indeed his humor was as penetrating as his seriousness. Both of these two sides of James's nature—the side that made a direct religious appeal, and the side that made a veiled religious appeal—became rapidly intensified during his latter years; so that, had the process continued much longer, the mere sight of him must have moved beholders to amend their lives.

I happened to be at Oxford at one of his lectures in 1908; and it was remarkable to see the reverence which that very unrevering class of men—the University dons—evinced towards James, largely on account of his appearance and personality. The fame of him went abroad, and the Sanhedrim attended. A quite distinguished, and very fussy scholar, a member of the old guard of Nil-admirari Cultivation,—who would have sniffed nervously if he had met Moses—told me that he had gone to a lecture of James's 'though the place was so crowded, and stank so that he had to come away immediately.'—'But,' he added, 'he certainly has the face of a sage.'

There was, in spite of his playfulness, a deep sadness about James. You felt that he had just stepped out of this sadness in order to meet you, and was to go back into it the moment you left him. It may be that sadness inheres in some kinds of profoundly religious characters,—in dedicated persons who have renounced all, and are constantly hoping, thinking, acting, and (in the typical case) praying for humanity. Lincoln was sad, and Tolstoi was sad, and many sensitive people, who view the world as it is, and desire nothing for themselves except to become of use to others, and to become agents in the spread of truth and happiness,—such people are often sad. It has sometimes crossed my mind that James wanted to be a poet and an artist, and that there lay in him, beneath the ocean of metaphysics, a lost Atlantis of the fine arts; that he really hated philosophy and all its works, and pursued them only as Hercules might spin, or as a prince in a fairy tale might sort seeds for an evil dragon, or as anyone might patiently do some careful work for which he had no aptitude. It would seem most natural, if this were the case between James and the metaphysical sciences; for what is there in these studies that can drench and satisfy a tingling mercurial being who loves to live on the surface, as well as in the depths of life? Thus we reason, forgetting that the mysteries of temperament are deeper than the mysteries of occupation. If James had had the career of Molière, he would still have been sad. He was a victim of divine visitation: the Searching Spirit would have winnowed him in the same manner, no matter what avocation he might have followed.

The world watched James as he pursued through life his search for religious truth; the world watched him, and often gently laughed at him, asking, 'When will James arise and fly? When will "he take the wings of the morning, and dwell in the uttermost parts of the sea"?' And in the meantime, James was there already. Those were the very places that he was living in. Through all the difficulties of polyglot metaphysics and of modern psychology he waded for years, lecturing and writing and existing,—and creating for himself a public which came to see in him only the saint and the sage, which felt only the religious truth which James was in search of, yet could never quite grasp in his hand. This very truth constantly shone out through him,—shone, as it were, straight through his waistcoat,—and distributed itself to everyone in the drawing-room, or in the lecture-hall where he sat. Here was the familiar paradox, the old parable, the psychological puzzle of the world. 'But what went ye out for to see?' In the very moment that the world is deciding that a man was no prophet and had nothing to say, in that very moment perhaps is his work perfected, and he himself is gathered to his fathers, after having been a lamp to his own generation, and an inspiration to those who come after.

1915

Intellectual Ambition

WHEN we consider the situation of the human mind in nature, its limited plasticity and few channels of communication with the outer world, we need not wonder that we grope for light, or that we find incoherence and instablility in human systems of ideas. The wonder rather is that we have done so well, that in the chaos of sensations and passions that fills the mind we have found any leisure for self-concentration and reflection, and have succeeded in gathering even a light harvest of experience from our distracted labours. Our occasional madness is less wonderful than our occasional sanity. Relapses into dreams are to be expected in a being whose brief existence is so like a dream; but who could have been sure of this sturdy and indomitable perseverance in the work of reason in spite of all checks and discouragements?

The resources of the mind are not commensurate with its ambition. Of the five senses, three are of little use in the formation of permanent notions: a fourth, sight, is indeed vivid and luminous, but furnishes transcripts of things so highly coloured and deeply modified by the medium of sense, that a long labour of analysis and correction is needed before satisfactory conceptions can be extracted from it. For this labour, however, we are endowed with the requisite instrument. We have memory and we have certain powers of synthesis, abstraction, reproduction, invention,—in a word, we have understanding. But this faculty of understanding has hardly begun its work of deciphering the hieroglyphics of sense and framing an idea of reality, when it is crossed by another faculty—the imagination. Perceptions do not remain in the mind, as would be suggested by the trite simile of the seal and the wax, passive and changeless, until time wear off their sharp edges and make them fade. No, perceptions fall into the brain rather as seeds into a furrowed field or even as sparks into a keg of powder. Each image breeds a hundred more, sometimes slowly and subterraneously, sometimes (when a passionate train is started) with a sudden burst of fancy. The mind, exercised by its own fertility and flooded by its inner lights, has infinite trouble to keep a true reckoning of its outward perceptions. It turns from the frigid problems of observation to its own visions; it forgets to watch the courses of what should be its 'pilot stars.' Indeed, were it not for the power of convention in which, by a sort of mutual cancellation of errors, the more practical and normal conceptions are enshrined, the imagination would carry men wholly away,—the best men first and the vulgar after them. Even as it is, individuals and ages of fervid imagination usually waste themselves in dreams, and must disappear before the race, saddened and dazed, perhaps, by the memory of those visions, can return to its plodding thoughts.

Five senses, then, to gather a small part of the infinite influences that vibrate in nature, a moderate power of understanding to interpret those senses, and an irregular, passionate fancy to overlay that interpretation—such is the endowment of the human mind. And what is its ambition? Nothing less than to construct a picture of all reality, to comprehend its own origin and that of the universe, to discover the laws of both and prophesy their destiny. Is not the disproportion enormous? Are not confusions and profound contradictions to be looked for in an attempt to build so much out of so little?

1900

Intuitive Morality

To one brought up in a sophisticated society, or in particular under an ethical religion, morality seems at first an external command, a chilling and arbitrary set of requirements and prohibitions which the young heart, if it trusted itself, would not reckon at a penny's worth. Yet while this rebellion is brewing in the secret conclave of the passions, the passions themselves are prescribing a code. They are inventing gallantry and kindness and honour; they are discovering friendship and paternity. With maturity comes the recognition that the authorized precepts of morality were essentially not arbitrary; that they expressed the genuine aims and interests of a practised will; that their alleged alien and supernatural basis (which if real would have deprived them of all moral authority) was but a mythical cover for their forgotten natural springs. Virtue is then seen to be admirable essentially, and not merely by conventional imputation. If traditional morality has much in it that is out of proportion, much that is unintelligent and inert, nevertheless it represents on the whole the verdict of reason. It speaks for a typical human will chastened by a typical human experience.

Gnomic wisdom, however, is notoriously polychrome, and proverbs depend for their truth entirely on the occasion they are applied to. Almost every wise saying has an opposite one, no less wise, to balance it; so that a man rich in such lore, like Sancho Panza, can always find a venerable maxim to fortify the view he happens to be taking. In respect to foresight, for instance, we are told, Make hay while the sun shines, A stitch in time saves nine, Honesty is the best policy, Murder will out, Woe unto you, ye hypocrites, Watch and pray, Seek salvation with fear and trembling, and *Respice finem*. But on the same authorities exactly we have apposite maxims, inspired by a feeling that mortal prudence is fallible, that life is shorter than policy, and that only the present is real; for we hear, A bird in the hand is worth two in the bush, *Carpe diem*, *Ars longa*, *vita brevis*, Be not righteous overmuch, Enough for the day is the evil thereof, Behold the lilies of the field, Judge not, that ye be not judged, Mind your own business, and It takes all sorts of men to make a world. So when some particularly shocking thing happens one man says, *Cherchez la femme*, and another says, Great is Allah.

That these maxims should be so various and partial is quite intelligible when we consider how they spring up. Every man, in moral reflection, is animated by his own intent; he has something in view which he prizes, he knows not why, and which wears to him the essential and unquestionable character of a good. With this standard before his eyes, he observes easily—for love and hope are extraordinarily keen-sighted—what in action or in circumstances forwards his purpose and what thwarts it; and at once the maxim comes, very likely in the language of the particular instance before him. Now the interests that speak in a man are different at different times; and the outer facts or measures which in one case promote that interest may, where other less obvious conditions have changed, altogether defeat it. Hence all sorts of precepts looking to all sorts of results.

Prescriptions of this nature differ enormously in value; for they differ enormously in scope. By chance intuitive maxims may be so central, so expressive of ultimate aims, so representative, I mean, of all aims in fusion, that they merely anticipate what moral science would have come to if it had existed. This happens much as in physics ultimate truths may be divined by poets long before they are discovered by investigators; the *vivida vis animi* taking the place of much recorded experience, because much unrecorded experience has secretly fed it. Such, for instance, is the central maxim of Christianity, Love thy neighbour as thyself. On the other hand, what is usual in intuitive codes is a mixture of some elementary precepts, necessary to any society, with others representing local traditions or ancient rites: so Thou shalt not kill, and Thou shalt keep holy the Sabbath day, figure side by side in the Decalogue. When Antigone, in her sublimest exaltation, defies human enactments and appeals to laws which are not of to-day nor yesterday, no man knowing whence they have arisen, she mixes various types of obligation in a most instructive fashion; for a superstitious horror at leaving a body unburied—something decidedly of yesterday—gives poignancy in her mind to natural affection for a brother—something indeed universal, yet having a well-known origin. The passionate assertion of right is here, in consequence, more dramatic than spiritual; and even its dramatic force has suffered somewhat by the change in ruling ideals.

Intuitive ethics has nothing to offer in the presence of discord except an appeal to force and to ultimate physical sanctions. It can instigate, but cannot resolve, the battle of nations and the battle of religions. Precisely the same zeal, the same patriotism, the same readiness for martyrdom fires adherents to rival societies, and fires them especially in view of the fact that the adversary is no less uncompromising and fierce. It might seem idle, if not cruel and malicious, to wish to substitute one historical allegiance for another, when both are equally arbitrary, and the existing one is the more congenial to those born under it; but to feel this aggression to be criminal demands some degree of imagination and justice, and sectaries would not be sectaries if they possessed it.

Truly religious minds, while eager perhaps to extirpate every religion but their own, often rise above national jealousies; for spirituality is universal, whatever churches may be. Similarly politicians often understand very well the religious situation; and of late it has become again the general practice among prudent governments to do as the Romans did in their conquests, and to leave people free to exercise what religion they have, without pestering them with a foreign one. On the other hand the same politicians are the avowed agents of a quite patent iniquity; for what is their ideal? To substitute their own language, commerce, soldiers, and tax-gatherers for the tax-gatherers, soldiers, commerce, and language of their neighbours; and no means is thought illegitimate, be it fraud in policy or bloodshed in war, to secure this absolutely nugatory end. Is not one country as much a country as another? Is it not as dear to its inhabitants? What then is gained by oppressing its genius or by seeking to destroy it altogether?

Here are two flagrant instances where pre-rational morality defeats the ends of morality. Viewed from within, each religious or national fanaticism stands for a good; but in its outward operation it produces and becomes an evil. It is possible, no doubt, that its agents are really so far apart in nature and ideals that, like men and mosquitoes, they can stand in physical relations only, and if they meet can meet only to poison or to crush one another. More probably, however, humanity in them is no merely nominal essence; it is definable ideally by a partially identical function and intent. In that case, by studying their own nature, they could rise above their mutual opposition, and feel that in their fanaticism they were taking too contracted a view for their own souls and were hardly doing justice to themselves when they did such great injustice to others.

1905

Cordova

SEEN from the further end of the Moorish bridge by the Calahorra, where the road starts to Seville, Cordova is a long brown line between the red river and the purple hills, an irregular, ruinous line, following the windings of the river, and rising to the yellow battlements and great middle bulk of the Cathedral. It goes up sheer from the river-side, above a broken wall, and in a huddle of mean houses, with so lamentably picturesque an air that no one would expect to find, inside that rough exterior, such neat, clean, shining streets, kept, even in the poorest quarters, with so admirable a care, and so bright with flowers and foliage, in patios and on upper balconies. From the bridge one sees the Moorish mills, rising yellow out of the yellow water, and, all day long, there is a slow procession along it of mules and donkeys, with their red saddles, carrying their burdens, and sometimes men heavily draped in great blanket-cloaks. Cross the city, and come out on the Paseo de la Victoria, open to the Sierra Morena, and you are in an immense village-green with red and white houses on one side, and black wooded hills on every other side; the trees, when I saw it for the first time at the beginning of winter, already shivering, and the watchers sitting on their chairs with their cloaks across their faces.

All Cordova seems to exist for its one treasure, the mosque, and to exist for it in a kind of remembrance; it is white, sad, delicately romantic, set in the midst of a strange, luxuriant country, under the hills, and beside the broad Guadalquivir, which, seen at sunset from the Ribera, flows with so fantastic a violence down its shallow weirs, between the mills and beneath the arches of the Moors. The streets are narrow and roughly paved, and they turn on themselves like a maze, around blank walls, past houses with barred windows and open doors, through which one sees a flowery patio, and by little irregular squares, in which the grass is sometimes growing between the stones, and outside the doors of great shapeless churches, mounting and descending steeply, from the river-bank to the lanes and meadows beyond the city walls. Turn and turn long enough through the white solitude of these narrow streets, and you come on the dim arcades and tall houses of the market-place, and on alleys of shops and bazaars, bright with coloured things, crimson umbrellas, such as every one carries here, cloaks lined with crimson velvet, soft brown leather, shining silver-work. The market is like a fair; worthless, picturesque lumber is heaped all over the ground, and upon stalls, and in dark shops like caves: steel and iron and leather goods, vivid crockery-ware, roughly burnt into queer, startling patterns, old clothes, cheap bright handkerchiefs and scarves. Passing out through the market-place, one comes upon sleepier streets, dwindling into the suburbs; grass grows down the whole length of the street, and the men and women sit in the middle of the road in their chairs, the children, more solemnly, in their little chairs. Vehicles pass seldom, and only through certain streets, where a board tells them it is possible to pass; but mules and donkeys are always to be seen, in long tinkling lines, nodding their wise little heads, as they go on their own way by themselves. At night Cordova sleeps early; a few central streets are still busy with people, but the rest are all deserted, the houses look empty, there is an almost oppressive silence. Only, here and there, as one passes heedlessly along a quiet street, one comes suddenly upon a cloaked figure, with a broad-brimmed hat, leaning against the bars of a window, and one may catch, through the bars, a glimpse of a vivid face, dark hair, and a rose (an artifical rose) in the hair. Not in any part of Spain have I seen the traditional Spanish love-making, the cloak and hat at the barred window, so frankly and so delightfully on view. It brings a touch of genuine romance, which it is almost difficult for those who know comic opera better than the countries in which life is still, in its way, a serious travesty, to take quite seriously. Lovers' faces, on each side of the bars of a window, at night, in a narrow street of white houses: that, after all, and not even the miraculous mosque, may perhaps be the most vivid recollection that one brings away with one from Cordova.

1898

HILAIRE BELLOC

On the Departure of a Guest

C'est ma Jeunesse qui s'en va.
  Adieu! la très gente compagne—
Oncques ne suis moins gai pour ça
(C'est ma Jeunessse qui s'en va)
Et lon-lon-laire, et lon-lon-là
  Peut-être perds; peut-être gagne.
C'est ma Jeunesse qui s'en va.

(From the Author's MSS.
In the library of the Abbey of Theleme.)

Host: Well, Youth, I see you are about to leave me, and since it is in the terms of your service by no means to exceed a certain period in my house, I must make up my mind to bid you farewell.

Youth: Indeed, I would stay if I could; but the matter lies as you know in other hands, and I may not stay.

Host: I trust, dear Youth, that you have found all comfortable while you were my guest, that the air has suited you and the company?

Youth: I thank you, I have never enjoyed a visit more; you may say that I have been most unusually happy.

Host: Then let me ring for the servant who shall bring down your things.

Youth: I thank you civilly! I have brought them down already—see, they are here. I have but two, one very large bag and this other small one.

Host: Why, you have not locked the small one! See it gapes!

Youth (*somewhat embarrassed*): My dear Host ... to tell the truth ... I usually put it off till the end of my visits ... but the truth ... to tell the truth, my luggage is of two kinds.

Host: I do not see why that need so greatly confuse you.

Youth (*still more embarrassed*): But you see—the fact is—I stay with people so long that—well, that very often they forget which things are mine and which belong to the house.... And—well, the truth is that I have to take away with me a number of things which ... which, in a word, you may possibly have thought your own.

HOST (*coldly*): Oh!

YOUTH (*eagerly*): Pray do not think the worse of me—you know how strict are my orders.

HOST (*sadly*): Yes, I know; you will plead that Master of yours, and no doubt you are right.... But tell me, Youth, what are those things?

YOUTH: They fill this big bag. But I am not so ungracious as you think. See, in this little bag, which I have purposely left open, are a number of things properly mine, yet of which I am allowed to make gifts to those with whom I lingered—you shall choose among them, or if you will, you shall have them all.

HOST: Well, first tell me what you have packed in the big bag and mean to take away.

YOUTH: I will open it and let you see. (*He unlocks it and pulls the things out.*) I fear they are familiar to you.

HOST: Oh! Youth! Youth! Must you take away all of these? Why, you are taking away, as it were, my very self! Here is the love of women, as deep and changeable as an opal; and here is carelessness that looks like a shower of pearls. And here I see—Oh! Youth, for shame!—you are taking away that silken stuff which used to wrap up the whole and which you once told me had no name, but which lent to everything it held plenitude and satisfaction. Without it surely pleasures are not all themselves. Leave me that at least.

YOUTH: No, I must take it, for it is not yours, though from courtesy I forbore to tell you so till now. These also go: Facility, the ointment; Sleep, the drug; Full Laughter, that tolerated all follies. It was the only musical thing in the house. And I must take—yes, I fear I must take Verse.

HOST: Then there is nothing left!

YOUTH: Oh! yes! See this little open bag which you may choose from! Feel it!

HOST (*lifting it*): Certainly it is very heavy, but it rattles and is uncertain.

YOUTH: That is because it is made up of divers things having no similarity; and you may take all or leave all, or choose as you will. Here (*holding up a clout*) is Ambition: Will you have that? . . .

HOST (doubtfully): I cannot tell. . . . It has been mine and yet . . . without those other things . . .

YOUTH (*cheerfully*): Very well, I will leave it. You shall decide on it a few years hence. Then, here is the perfume Pride. Will you have that?

HOST: No; I will have none of it. It is false and corrupt, and only yesterday I was for throwing it out of window to sweeten the air in my room.

YOUTH: So far you have chosen well; now pray choose more.

HOST: I will have this—and this—and this. I will take Health (*takes it out of the bag,*) not that it is of much use to me without those other things, but I have grown used to it. Then I will take this (*takes out a plain steel purse and chain*), which is the tradition of my family, and which I desire to leave to my son. I must have it cleaned. Then I will take this (*pulls out a trinket*), which is the Sense of Form and Colour. I am told it is of less value later on, but it is a pleasant ornament. . . . And so, Youth, goodbye.

YOUTH (*with a mysterious smile*): Wait—I have something else for you (*he feels in his ticket pocket*); no less a thing (*he feels again in his watch pocket*) than (*he looks a trifle anxious and feels in his waistcoat pockets*) a promise from my Master, signed and sealed, to give you back all I take and more in Immortality! (*He feels in his handkerchief pocket.*)

HOST: Oh! Youth!

YOUTH (*still feeling*): Do not thank me! It is my Master you should thank. (*Frowns.*) Dear me! I hope I have not lost it! (*Feels in his trousers pockets.*)

HOST (*loudly*): Lost it?

YOUTH (*pettishly*): I did not say I had lost it! I said I hoped I had not . . . (*feels in his great-coat pocket, and pulls out an envelope*). Ah! Here it is! (*His face clouds over.*) No, that is the message to Mrs George, telling her the time has come to get a wig . . . (*Hopelessly*): Do you know I am afraid I have lost it! I am really very sorry—I cannot wait. (*He goes off.*)

1908

On Being Modern-Minded

Our age is the most parochial since Homer. I speak not of any geographical parish: the inhabitants of Mudcombe-in-the-Meer are more aware than at any former time of what is being done and thought at Praha, at Gorki, or at Peiping. It is in the chronological sense that we are parochial: as the new names conceal the historic cities of Prague, Nijni-Novgorod, and Pekin, so new catchwords hide from us the thoughts and feelings of our ancestors, even when they differed little from our own. We imagine ourselves at the apex of intelligence, and cannot believe that the quaint clothes and cumbrous phrases of former times can have invested people and thoughts that are still worthy of our attention. If *Hamlet* is to be interesting to a really modern reader, it must first be translated into the language of Marx or of Freud, or, better still, into a jargon inconsistently compounded of both. I read some years ago a contemptuous review of a book by Santayana, mentioning an essay on Hamlet 'dated, in every sense, 1908'—as if what has been discovered since then made any earlier appreciation of Shakespeare irrelevant and comparatively superficial. It did not occur to the reviewer that his review was 'dated, in every sense, 1936.' Or perhaps this thought did occur to him, and filled him with satisfaction. He was writing for the moment, not for all time; next year he will have adopted the new fashion in opinions, whatever it may be, and he no doubt hopes to remain up to date as long as he continues to write. Any other ideal for a writer would seem absurd and old-fashioned to the modern-minded man.

The desire to be contemporary is of course new only in degree; it has existed to some extent in all previous periods that believed themselves to be progressive. The Renaissance had a contempt for the Gothic centuries that had preceded it; the seventeenth and eighteenth centuries covered priceless mosaics with whitewash; the Romantic movement despised the age of the heroic couplet. Eighty years ago Lecky reproached my mother for being led by intellectual fashion to oppose fox-hunting: 'I am sure,' he wrote, 'you are not really at all sentimental about foxes or at all shocked at the prettiest of all the assertions of women's rights, riding across country. But you always look upon politics and intellect as a fierce race and are so dreadfully afraid of not being sufficiently advanced or intellectual.' But in none of these former times was the contempt for the past nearly as complete as it is now. From the Renaissance to the end of the eighteenth century men admired Roman antiquity; the Romantic movement revived the Middle Ages; my mother, for all her belief in nineteenth-century progress, constantly read Shakespeare and Milton. It is only since the 1914–18 war that it has been fashionable to ignore the past *en bloc*.

The belief that fashion alone should dominate opinion has great advantages. It makes thought unnecessary and puts the highest intelligence within the reach of everyone. It is not difficult to learn the correct use of such words as 'complex,' 'sadism,' 'Oedipus,' 'bourgeois,' 'deviation,' 'left'; and nothing more is needed to make a brilliant writer or talker. Some, at least, of such words represented much thought on the part of their inventors; like paper money they were originally convertible into gold. But they have become for most people inconvertible, and in depreciating have increased nominal wealth in ideas. And so we are enabled to despise the paltry intellectual fortunes of former times.

The modern-minded man, although he believes profoundly in the wisdom of his period, must be presumed to be very modest about his personal powers. His highest hope is to think first what is about to be thought, to say what is about to be said, and to feel what is about to be felt; he has no wish to think better thoughts than his neighbours, to say things showing more insight, or to have emotions which are not those of some fashionable group, but only to be slightly ahead of others in point of time. Quite deliberately he suppresses what is individual in himself for the sake of the admiration of the herd. A mentally solitary life, such as that of Copernicus, or Spinoza, or Milton after the Restoration, seems pointless according to modern standards. Copernicus should have delayed his advocacy of the

Copernican system until it could be made fashionable; Spinoza should have been either a good Jew or a good Christian; Milton should have moved with the times, like Cromwell's widow, who asked Charles II for a pension on the ground that she did not agree with her husband's politics. Why should an individual set himself up as an independent judge? Is it not clear that wisdom resides in the blood of the Nordic race or, alternatively, in the proletariat? And in any case what is the use of an eccentric opinion, which never can hope to conquer the great agencies of publicity?

The money rewards and widespread though ephemeral fame which those agencies have made possible place temptations in the way of able men which are difficult to resist. To be pointed out, admired, mentioned constantly in the press, and offered easy ways of earning much money is highly agreeable; and when all this is open to a man, he finds it difficult to go on doing the work that he himself thinks best and is inclined to subordinate his judgment to the general opinion.

Various other factors contribute to this result. One of these is the rapidity of progress which has made it difficult to do work which will not soon be superseded. Newton lasted till Einstein; Einstein is already regarded by many as antiquated. Hardly any man of science, nowadays, sits down to write a great work, because he knows that, while he is writing it, others will discover new things that will make it obsolete before it appears. The emotional tone of the world changes with equal rapidity, as wars, depressions, and revolutions chase each other across the stage. And public events impinge upon private lives more forcibly than in former days. Spinoza, in spite of his heretical opinions, could continue to sell spectacles and meditate, even when his country was invaded by foreign enemies; if he had lived now, he would in all likelihood have been conscripted or put in prison. For these reasons a greater energy of personal conviction is required to lead a man to stand out against the current of his time than would have been necessary in any previous period since the Renaissance.

The change has, however, a deeper cause. In former days men wished to serve God. When Milton wanted to exercise 'that one talent which is death to hide,' he felt that his soul was 'bent to serve therewith my Maker.' Every religiously minded artist was convinced that God's aesthetic judgements coincided with his own; he had therefore a reason, independent of popular applause, for doing what he considered his best, even if his style was out of fashion. The man of science in pursuing truth, even if he came into conflict with current superstition, was still setting forth the wonders of Creation and bringing men's imperfect beliefs more nearly into harmony with God's perfect knowledge. Every serious worker, whether artist, philosopher, or astronomer, believed that in following his own convictions he was serving God's purposes. When with the progress of enlightenment this belief began to grow dim, there still remained the True, the Good, and the Beautiful. Non-human standards were still laid up in heaven, even if heaven had no topographical existence.

Throughout the nineteenth century the True, the Good, and the Beautiful preserved their precarious existence in the minds of earnest atheists. But their very earnestness was their undoing, since it made it impossible for them to stop at a halfway house. Pragmatists explained that Truth is what it pays to believe. Historians of morals reduced the Good to a matter of tribal custom. Beauty was abolished by the artists in a revolt against the sugary insipidities of a philistine epoch and in a mood of fury in which satisfaction is to be derived only from what hurts. And so the world was swept clear not only of God as a person but of God's essence as an ideal to which man owed an ideal allegiance; while the individual, as a result of a crude and uncritical interpretation of sound doctrines, was left without any inner defence against social pressure.

All movements go too far, and this is certainly true of the movement toward subjectivity, which began with Luther and Descartes as an assertion of the individual and has culminated by an inherent logic in his complete subjection. The subjectivity of truth is a hasty doctrine not validly deducible from the premises which have been thought to imply it; and the habits of centuries have made many things seem dependent upon theological belief which in fact are not so. Men lived with one kind of illusion, and when they lost it they fell into another. But it is not by old error that new error can be combated. Detachment and objectivity, both in thought and in feeling, have been historically but not logically associated with certain traditional beliefs; to preserve them without these beliefs is both possible and important. A certain degree of isolation both in space and time is essential to generate the independence required for the most important work; there must be something which is felt to be of more importance than the admiration of the contemporary crowd. We are suffering not from the decay of theological beliefs but from the loss of solitude.

1950

'*A Clergyman*'

FRAGMENTARY, pale, momentary; almost nothing; glimpsed and gone; as it were, a faint human hand thrust up, never to reappear, from beneath the rolling waters of Time, he forever haunts my memory and solicits my weak imagination. Nothing is told of him but that once, abruptly, he asked a question, and received an answer.

This was on the afternoon of April 7th, 1778, at Streatham, in the well-appointed house of Mr Thrale. Johnson, on the morning of that day, had entertained Boswell at breakfast in Bolt Court, and invited him to dine at Thrale Hall. The two took coach and arrived early. It seems that Sir John Pringle had asked Boswell to ask Johnson 'what were the best English sermons for style.' In the interval before dinner, accordingly, Boswell reeled off the names of several divines whose prose might or might not win commendation. 'Atterbury?' he suggested. 'JOHNSON: Yes, Sir, one of the best. BOSWELL: Tillotson? JOHNSON: Why, not now. I should not advise any one to imitate Tillotson's style; though I don't know; I should be cautious of censuring anything that has been applauded by so many suffrages.—South is one of the best, if you except his peculiarities, and his violence, and sometimes coarseness of language.—Seed has a very fine style; but he is not very theological. Jortin's sermons are very elegant. Sherlock's style, too, is very elegant, though he has not made it his principal study.—And you may add Smalridge. BOSWELL: I like Ogden's Sermons on Prayer very much, both for neatness of style and subtility of reasoning. JOHNSON: I should like to read all that Ogden has written. BOSWELL: What I want to know is, what sermons afford the best specimen of English pulpit eloquence. JOHNSON: We have no sermons addressed to the passions, that are good for anything; if you mean that kind of eloquence. A CLERGYMAN, whose name I do not recollect: Were not

Dodd's sermons addressed to the passions? JOHNSON: They were
nothing, Sir, be they addressed to what they may.'

The suddenness of it! Bang!—and the rabbit that had popped
from its burrow was no more.

I know not which is the more startling—the début of the
unfortunate clergyman, or the instantaneousness of his end. Why
hadn't Boswell told us there was a clergyman present? Well, we
may be sure that so careful and acute an artist had some good
reason. And I suppose the clergyman was left to take us un-
awares because just so did he take the company. Had we been
told he was there, we might have expected that sooner or later he
would join in the conversation. He would have had a place in
our minds. We may assume that in the minds of the company
around Johnson he had no place. He sat forgotten, overlooked;
so that his self-assertion startled every one just as on Boswell's
page it startles us. In Johnson's massive and magnetic presence
only some very remarkable man, such as Mr Burke, was sharply
distinguishable from the rest. Others might, if they had something in them, stand out slightly. This unfortunate clergyman
may have had something in him, but I judge that he lacked the
gift of seeming as if he had. That deficiency, however, does not
account for the horrid fate that befell him. One of Johnson's
strongest and most inveterate feelings was his veneration for the
Cloth. To any one in Holy Orders he habitually listened with a
grave and charming deference. To-day moreover, he was in
excellent good humour. He was at the Thrales', where he so
loved to be; the day was fine; a fine dinner was in close prospect;
and he had had what he always declared to be the sum of human
felicity—a ride in a coach. Nor was there in the question put by
the clergyman anything likely to enrage him. Dodd was one
whom Johnson had befriended in adversity; and it had always
been agreed that Dodd in his pulpit was very emotional. What
drew the blasting flash must have been not the question itself,
but the manner in which it was asked. And I think we can guess
what that manner was.

Say the words aloud: 'Were not Dodd's sermons addressed to
the passions?' They are words which, if you have any dramatic
and histrionic sense, *cannot* be said except in a high, thin voice.

You may, from sheer perversity, utter them in a rich and sonorous baritone or bass. But if you do so, they sound utterly

unnatural. To make them carry the conviction of human utterance, you have no choice: you must pipe them.

Remember, now, Johnson was very deaf. Even the people
whom he knew well, the people to whose voices he was accustomed, had to address him very loudly. It is probable that this
unregarded, young, shy clergyman, when at length he suddenly
mustered courage to 'cut in,' let his high, thin voice soar *too*
high, insomuch that it was a kind of scream. On no other
hypothesis can we account for the ferocity with which Johnson
turned and rended him. Johnson didn't, we may be sure, mean
to be cruel. The old lion, startled, just struck out blindly. But the
force of paw and claws was not the less lethal. We have endless
testimony to the strength of Johnson's voice; and the very
cadence of those words, 'They were nothing, Sir, be they
addressed to what they may,' convinces me that the old lion's
jaws never gave forth a louder roar. Boswell does not record that
there was any further conversation before the announcement
of dinner. Perhaps the whole company had been temporarily
deafened. But I am not bothering about *them*. My heart goes out
to the poor dear clergyman exclusively.

I said a moment ago that he was young and shy; and I admit
that I slipped those epithets in without having justified them to
you by due process of induction. Your quick mind will have
already supplied what I omitted. A man with a high, thin voice,
and without power to impress any one with a sense of his importance, a man so null in effect that even the retentive mind of
Boswell did not retain his very name, would assuredly not be a
self-confident man. Even if he were not naturally shy, social
courage would soon have been sapped in him, and would in time
have been destroyed, by experience. That he had not yet given
himself up as a bad job, that he still had faint wild hopes, is
proved by the fact that he did snatch the opportunity for asking
that question. He must, accordingly, have been young. Was he
the curate of the neighbouring church? I think so. It would
account for his having been invited. I see him as he sits there
listening to the great Doctor's pronouncement on Atterbury and
those others. He sits on the edge of a chair in the background.
He has colourless eyes, fixed earnestly, and a face almost as pale
as the clerical bands beneath his somewhat receding chin. His
forehead is high and narrow, his hair mouse-coloured. His hands

are clasped tight before him, the knuckles standing out sharply. This constriction does not mean that he is steeling himself to speak. He has no positive intention of speaking. Very much, nevertheless, is he wishing in the back of his mind that he *could* say something—something whereat the great Doctor would turn on him and say, after a pause for thought, 'Why yes, Sir. That is most justly observed' or 'Sir, this has never occurred to me. I thank you'—thereby fixing the observer for ever high in the esteem of all. And now in a flash the chance presents itself. 'We have,' shouts Johnson, 'no sermons addressed to the passions, that are good for anything.' I see the curate's frame quiver with sudden impulse, and his mouth fly open, and—no, I can't bear it, I shut my eyes and ears. But audible, even so, is something shrill, followed by something thunderous.

Presently I re-open my eyes. The crimson has not yet faded from that young face yonder, and slowly down either cheek falls a glistening tear. Shades of Atterbury and Tillotson! Such weakness shames the Established Church. What would Jortin and Smalridge have said?—what Seed and South? And, by the way, who *were* they, these worthies? It is a solemn thought that so little is conveyed to us by names which to the palaeo-Georgians conveyed so much. We discern a dim, composite picture of a big man in a big wig and a billowing black gown, with a big congregation beneath him. But we are not anxious to hear what he is saying. We know it is all very elegant. We know it will be printed and be bound in finely-tooled full calf, and no palaeo-Georgian gentleman's library will be complete without it. Literate people in those days were comparatively few; but, bating that, one may say that sermons were as much in request as novels are to-day. I wonder, will mankind continue to be capricious? It is a very solemn thought indeed that no more than a hundred-and-fifty years hence the novelists of our time, with all their moral and political and sociological outlook and influence, will perhaps shine as indistinctly as do those old preachers, with all their elegance, now. 'Yes, Sir,' some great pundit may be telling a disciple at this moment, 'Wells is one of the best. Galsworthy is one of the best, if you except his concern for delicacy of style. Mrs Ward has a very firm grasp of problems, but is not very creational.—Caine's books are very edifying. I should like to read all that Caine has written. Miss Corelli, too, is very edifying.

—And you may add Upton Sinclair.' 'What I want to know,' says the disciple, 'is, what English novels may be selected as specially enthralling.' The pundit answers: 'We have no novels addressed to the passions that are good for anything, if you mean that kind of enthralment.' And here some poor wretch (whose name the disciple will not remember) inquires: 'Are not Mrs Glyn's novels addressed to the passions?' and is in due form annihilated. Can it be that a time will come when readers of this passage in our pundit's Life will take more interest in the poor nameless wretch than in all the bearers of those great names put together, being no more able or anxious to discriminate between (say) Mrs Ward and Mr Sinclair than we are to set Ogden above Sherlock, or Sherlock above Ogden? It seems impossible. But we must remember that things are not always what they seem.

Every man illustrious in his day, however much he may be gratified by his fame, looks with an eager eye to posterity for a continuance of past favours, and would even live the remainder of his life in obscurity if by so doing he could insure that future generations would preserve a correct attitude towards him forever. This is very natural and human, but, like so many very natural and human things, very silly. Tillotson and the rest need not, after all, be pitied for our neglect of them. They either know nothing about it, or are above such terrene trifles. Let us keep our pity for the seething mass of divines who were *not* elegantly verbose, and had no fun or glory while they lasted. And let us keep a specially large portion for one whose lot was so much worse than merely undistinguished. If that nameless curate had not been at the Thrales' that day, or, being there, had kept the silence that so well became him, his life would have been drab enough, in all conscience. But at any rate an unpromising career would not have been nipped in the bud. And that is what in fact happened, I'm sure of it. A robust man might have rallied under the blow. Not so our friend. Those who knew him in infancy had not expected that he would be reared. Better for him had they been right. It is well to grow up and be ordained, but not if you are delicate and very sensitive, and shall happen to annoy the greatest, the most stentorian and roughest of contemporary personages. 'A Clergyman' never held up his head or smiled again after the brief encounter recorded for us by Boswell. He

sank into a rapid decline. Before the next blossoming of Thrale
Hall's almond trees he was no more. I like to think that he died
forgiving Dr Johnson.

1918

The Dream

ONE foggy afternoon in November 1947 I was painting in my studio at the cottage down the hill at Chartwell. Someone had sent me a portrait of my father which had been painted for one of the Belfast Conservative Clubs about the time of his visit to Ulster in the Home Rule crisis of 1886. The canvas had been badly torn, and though I am very shy of painting human faces I thought I would try to make a copy of it.

My easel was under a strong daylight lamp, which is necessary for indoor painting in the British winter. On the right of it stood the portrait I was copying, and behind me was a large looking glass, so that one could frequently study the painting in reverse. I must have painted for an hour and a half, and was deeply concentrated on my subject. I was drawing my father's face, gazing at the portrait, and frequently turning round right-handed to check progress in the mirror. Thus I was intensely absorbed, and my mind was freed from all other thoughts except the impressions of that loved and honoured face now on the canvas, now on the picture, now in the mirror.

I was just trying to give the twirl to his moustache when I suddenly felt an odd sensation. I turned round with my palette in my hand, and there, sitting in my red leather upright armchair, was my father. He looked just as I had seen him in his prime, and as I had read about him in his brief year of triumph. He was small and slim, with the big moustache I was just painting, and all his bright, captivating, jaunty air. His eyes twinkled and shone. He was evidently in the best of tempers. He was engaged in filling his amber cigarette-holder with a little pad of cotton-wool before putting in the cigarette. This was in order to stop the nicotine, which used to be thought deleterious. He was so exactly like my memories of him in his most charming moods that I could hardly believe my eyes. I felt no alarm, but I thought I would stand where I was and go no nearer.

'Papa!' I said.

'What are you doing, Winston?'

'I am trying to copy your portrait, the one you had done when you went over to Ulster in 1886.'

'I should never have thought it,' he said.

'I only do it for amusement,' I replied.

'Yes, I am sure you could never earn your living that way.'

There was a pause.

'Tell me,' he asked, 'what year is it?'

'Nineteen forty-seven.'

'Of the Christian era, I presume?'

'Yes, that all goes on. At least, they still count that way.'

'I don't remember anything after ninety-four. I was very confused that year.... So more than fifty years have passed. A lot must have happened.'

'It has indeed, Papa.'

'Tell me about it.'

'I really don't know where to begin,' I said.

'Does the Monarchy go on?' he asked.

'Yes, stronger than in the days of Queen Victoria.'

'Who is King?'

'King George the Sixth.'

'What! Two more Georges?'

'But, Papa, you remember the death of the Duke of Clarence.'

'Quite true; that settled the name. They must have been clever to keep the Throne.'

'They took the advice of the Ministers who had majorities in the House of Commons.'

'That all goes on still? I suppose they still use the Closure and the Guillotine?'

'Yes, indeed.'

'Does the Carlton Club go on?'

'Yes, they are going to rebuild it.'

'I thought it would have lasted longer; the structure seemed quite solid. What about the Turf Club?'

'It's OK.'

'How do you mean, OK?'

'It's an American expression, Papa. Nowadays they use initials for all sorts of things, like they used to say RSPCA and HMG.'

'What does it mean?'

'It means all right.'

'What about racing? Does that go on?'

'You mean horse-racing?'

'Of course,' he said, 'What other should there be?'

'It all goes on.'

'What, the Oaks, the Derby, the Leger?'

'They have never missed a year.'

'And the Primrose League?'

'They have never had more members.'

He seemed to be pleased at this.

'I always believed in Dizzy, that old Jew. He saw into the future. He had to bring the British working man into the centre of the picture.' And here he glanced at my canvas.

'Perhaps I am trespassing on your art?' he said, with that curious, quizzical smile of his, which at once disarmed and disconcerted.

Palette in hand, I made a slight bow.

'And the Church of England?'

'You made a very fine speech about it in eighty-four.' I quoted, ' "And, standing out like a lighthouse over a stormy ocean, it marks the entrance to a port wherein the millions and masses of those who at times are wearied with the woes of the world and tired of the trials of existence may seek for, and may find, that peace which passeth all understanding".'

'What a memory you have got! But you always had one. I remember Dr Welldon telling me how you recited the twelve hundred lines of Macaulay without a single mistake.'

After a pause. 'You are still a Protestant?' he said.

'Episcopalian.'

'Do the bishops still sit in the House of Lords?'

'They do indeed, and make a lot of speeches.'

'Are they better than they used to be?'

'I never heard the ones they made in the old days.'

'What party is in power now? Liberals or Tories?'

'Neither, Papa. We have a Socialist Government, with a very large majority. They have been in office for two years, and will probably stay for two more. You know we have changed the Septennial Act to five years.'

'Socialist!' he exclaimed. 'But I thought you said we still have a Monarchy.'

'The Socialists are quite in favour of the Monarchy, and make generous provisions for it.'

'You mean in regard to Royal grants, the Civil List, and so forth? How can they get those through the Commons?'

'Of course they have a few rebels, but the old Republicanism of Dilke and Labby is dead as mutton. The Labour men and the trade unions look upon the Monarchy not only as a national but a nationalised institution. They even go to the parties at Buckingham Palace. Those who have very extreme principles wear sweaters.'

'How very sensible. I am glad all that dressing up has been done away with.'

'I am sorry, Papa,' I said, 'I like the glitter of the past.'

'What does the form matter if the facts remain? After all, Lord Salisbury was once so absent-minded as to go to a levée in uniform with carpet slippers. What happened to old Lord Salisbury?'

'Lord Salisbury leads the Conservative party in the House of Lords.'

'What!' he said. 'He must be a Methuselah!'

'No. It is his grandson.'

'Ah, and Arthur Balfour? Did he ever become Prime Minister?'

'Oh, yes. He was Prime Minister, and came an awful electoral cropper. Afterwards he was Foreign Secretary and held other high posts. He was well in the eighties when he died.'

'Did he make a great mark?'

'Well, Ramsay MacDonald, the Prime Minister of the first Socialist Government, which was in office at his death, said he "saw a great deal of life from afar".'

'How true! But who was Ramsay MacDonald?'

'He was the leader of the first and second Labour-Socialist Governments, in a minority.'

'The first Socialist Government? There has been more than one?'

'Yes, several. But this is the first that had a majority.'

'What have they done?'

'Not much. They have nationalised the mines and railways and a few other services, paying full compensation. You know, Papa, though stupid, they are quite respectable, and increasingly bourgeois. They are not nearly so fierce as the old Radicals, though of course they are wedded to economic fallacies.'

'What is the franchise?'

'Universal,' I replied. 'Even the women have votes.'

'Good gracious!' he exclaimed.

'They are a strong prop to the Tories.'

'Arthur was always in favour of female suffrage.'

'It did not turn out as badly as I thought,' I said.

'You don't allow them in the House of Commons?' he inquired.

'Oh, yes. Some of them have even been Ministers. There are not many of them. They have found their level.'

'So Female Suffrage has not made much difference?'

'Well, it has made politicians more mealy-mouthed than in your day. And public meetings are much less fun. You can't say the things you used to.'

'What happened to Ireland? Did they get Home Rule?'

'The South got it, but Ulster stayed with us.'

'Are the South a republic?'

'No one knows what they are. They are neither in nor out of the Empire. But they are much more friendly to us than they used to be. They have built up a cultured Roman Catholic system in the South. There has been no anarchy or confusion. They are getting more happy and prosperous. The bitter past is fading.'

'Ah,' he said, 'how vexed the Tories were with me when I observed that there was no English statesman who had not had his hour of Home Rule.' Then, after a pause, 'What about the Home Rule meaning "Rome Rule"?'

'It certainly does, but they like it. And the Catholic Church has now become a great champion of individual liberty.'

'You must be living in a very happy age. A Golden Age, it seems.'

His eye wandered round the studio, which is entirely panelled with scores of my pictures. I followed his travelling eye as it rested now on this one and on that. After a while: 'Do you live in this cottage?'

'No,' I said, 'I have a house up on the hill, but you cannot see it for the fog.'

'How do you get a living?' he asked. 'Not, surely, by these?' indicating the pictures.

'No, indeed, Papa. I write books and articles for the Press.'

'Ah, a reporter. There is nothing discreditable in that. I myself wrote articles for the *Daily Graphic* when I went to South Africa. And well I was paid for them. A hundred pounds an article!'

Before I could reply: 'What has happened to Blenheim? Blandford (his brother) always said it could only become a museum for Oxford.'

'The Duke and Duchess of Marlborough are still living there.'

He paused again for a while, and then: 'I always said "Trust the people". Tory democracy alone could link the past with the future.'

'They are only living in a wing of the Palace,' I said. 'The rest is occupied by MI5.'

'What does that mean?'

'A Government department formed in the war.'

'War?' he said, sitting up with a startled air. 'War, do you say? Has there been a war?'

'We have had nothing else but wars since democracy took charge.'

'You mean real wars, not just frontier expeditions? Wars where tens of thousands of men lose their lives?'

'Yes, indeed, Papa,' I said. 'That's what has happened all the time. Wars and rumours of war ever since you died.'

'Tell me about them.'

'Well, first there was the Boer War.'

'Ah, I would have stopped that. I never agreed with "Avenge Majuba". Never avenge anything, especially if you have the power to do so. I always mistrusted Joe.'

'You mean Mr Chamberlain?'

'Yes. There is only one Joe, or only one I ever heard of. A Radical turned Jingo is an ugly and dangerous thing. But what happened in the Boer War?'

'We conquered the Transvaal and the Orange Free State.'

'England should never have done that. To strike down two independent republics must have lowered our whole position in the world. It must have stirred up all sorts of things. I am sure the Boers made a good fight. When I was there I saw lots of them. Men of the wild, with rifles, on horseback. It must have taken a lot of soldiers. How many? Forty thousand?'

'No, over a quarter of a million.'

'Good God! What a shocking drain on the Exchequer!'

'It was,' I said. 'The Income Tax went up to one and three-pence.' He was visibly disturbed. So I said that they got it down to eightpence afterwards.

'Who was the General who beat the Boers?' he asked.

'Lord Roberts,' I answered.

'I always believed in him. I appointed him Commander-in-Chief in India when I was Secretary of State. That was the year I annexed Burma. The place was in utter anarchy. They were just butchering one another. We had to step in, and very soon there was an ordered, civilised Government under the vigilant control of the House of Commons.' There was a sort of glare in his eyes as he said 'House of Commons'.

'I have always been a strong supporter of the House of Commons, Papa. I am still very much in favour of it.'

'You had better be, Winston, because the will of the people must prevail. Give me a fair arrangement of the constituencies, a wide franchise, and free elections—say what you like, and one part of Britain will correct and balance the other.'

'Yes, you brought me up to that.'

'I never brought you up to anything. I was not going to talk politics with a boy like you ever. Bottom of the school! Never passed any examinations, except into the Cavalry! Wrote me stilted letters. I could not see how you would make your living on the little I could leave you and Jack, and that only after your mother. I once thought of the Bar for you but you were not clever enough. Then I thought you might go to South Africa. But of course you were very young, and I loved you dearly. Old people are always very impatient with young ones. Fathers always expect their sons to have their virtues without their faults. You were very fond of playing soldiers, so I settled for the Army. I hope you had a successful military career.'

'I was a Major in the Yeomanry.'

He did not seem impressed.

'However, here you are. You must be over 70. You have a roof over your head. You seem to have plenty of time on your hands to mess about with paints. You have evidently been able to keep yourself going. Married?'

'Forty years.'

'Children?'

'Four.'

'Grandchildren?'

'Four.'

'I am so glad. But tell me more about these other wars.'

'They were the wars of nations, caused by demagogues and tyrants.'

'Did we win?'

'Yes, we won all our wars. All our enemies were beaten down. We even made them surrender unconditionally.'

'No one should be made to do that. Great people forget sufferings, but not humiliations.'

'Well, that was the way it happened, Papa.'

'How did we stand after it all? Are we still at the summit of the world, as we were under Queen Victoria?'

'No, the world grew much bigger all around us.'

'Which is the leading world-power?'

'The United States.'

'I don't mind that. You are half American yourself. Your mother was the most beautiful woman ever born. The Jeromes were a deep-rooted American family.'

'I have always,' I said, 'worked for friendship with the United States, and indeed throughout the English-speaking world.'

'English-speaking world,' he repeated, weighing the phrase. 'You mean, with Canada, Australia and New Zealand, and all that?'

'Yes, all that.'

'Are they still loyal?'

'They are our brothers.'

'And India, is that all right? And Burma?'

'Alas! They have gone down the drain.'

He gave a groan. So far he had not attempted to light the cigarette he had fixed in the amber holder. He now took his matchbox from his watch-chain, which was the same as I was wearing. For the first time I felt a sense of awe. I rubbed my brush in the paint on the palette to make sure that everything was real. All the same I shivered. To relieve his consternation I said:

'But perhaps they will come back and join the English-speaking world. Also, we are trying to make a world organisation in which we and America will be quite important.'

But he remained sunk in gloom, and huddled back in the chair. Presently: 'About these wars, the ones after the Boer War, I mean. What happened to the great States of Europe? Is Russia still the danger?'

'We are all very worried about her.'

'We always were in my day, and in Dizzy's before me. Is there still a Tsar?'

'Yes, but he is not a Romanoff. It's another family. He is much more powerful, and much more despotic.'

'What of Germany? What of France?'

'They are both shattered. Their only hope is to rise together.'

'I remember,' he said, 'taking you through the Place de la Concorde when you were only nine years old, and you asked me about the Strasbourg monument. You wanted to know why this one was covered in flowers and crape. I told you about the lost provinces of France. What flag flies in Strasbourg now?'

'The Tricolor flies there.'

'Ah, so they won. They had their revanche. That must have been a great triumph for them.'

'It cost them their life blood,' I said.

'But wars like these must have cost a million lives. They must have been as bloody as the American Civil War.'

'Papa,' I said, 'in each of them about thirty million men were killed in battle. In the last one seven million were murdered in cold blood, mainly by the Germans. They made human slaughter-pens like the Chicago stockyards. Europe is a ruin. Many of her cities have been blown to pieces by bombs. Ten capitals in Eastern Europe are in Russian hands. They are Communists now, you know—Karl Marx and all that. It may well be that an even worse war is drawing near. A war of the East against the West. A war of liberal civilisation against the Mongol hordes. Far gone are the days of Queen Victoria and a settled world order. But, having gone through so much, we do not despair.'

He seemed stupefied, and fumbled with his matchbox for what seemed a minute or more. Then he said:

'Winston, you have told me a terrible tale. I would never have believed that such things could happen. I am glad I did not live to see them. As I listened to you unfolding these fearful facts you seemed to know a great deal about them. I never expected that you would develop so far and so fully. Of course you are too old now to think about such things, but when I hear you talk I really wonder you didn't go into politics. You might have done a lot to help. You might even have made a name for yourself.'

He gave me a benignant smile. He then took the match to light his cigarette and struck it. There was a tiny flash. He vanished. The chair was empty. The illusion had passed. I rubbed my brush again in my paint, and turned to finish the moustache. But so vivid had my fancy been that I felt too tired to go on. Also my cigar had gone out, and the ash had fallen among all the paints.

1947

A Defence of Penny Dreadfuls

ONE of the strangest examples of the degree to which ordinary life is undervalued is the example of popular literature, the vast mass of which we contentedly describe as vulgar. The boy's novelette may be ignorant in a literary sense, which is only like saying that a modern novel is ignorant in the chemical sense, or the economic sense, or the astronomical sense; but it is not vulgar intrinsically—it is the actual centre of a million flaming imaginations.

In former centuries the educated class ignored the ruck of vulgar literature. They ignored, and therefore did not, properly speaking, despise it. Simple ignorance and indifference does not inflate the character with pride. A man does not walk down the street giving a haughty twirl to his moustaches at the thought of his superiority to some variety of deep-sea fishes. The old scholars left the whole under-world of popular compositions in a similar darkness.

Today, however, we have reversed this principle. We do despise vulgar compositions, and we do not ignore them. We are in some danger of becoming petty in our study of pettiness; there is a terrible Circean law in the background that if the soul stoops too ostentatiously to examine anything it never gets up again. There is no class of vulgar publications about which there is, to my mind, more utterly ridiculous exaggeration and misconception than the current boys' literature of the lowest stratum. This class of composition has presumably always existed, and must exist. It has no more claim to be good literature than the daily conversation of its readers to be fine oratory, or the lodging-houses and tenements they inhabit to be sublime architecture. But people must have conversation, they must have houses, and they must have stories. The simple need for some kind of ideal world in which fictitious persons play an unhampered part is infinitely deeper and older than the rules of good art, and much more important. Every one of us in childhood has constructed such an invisible *dramatis personae*; but it never occurred to our nurses to correct the composition by careful comparison with Balzac. In the East the professional story-teller goes from village to village with a small carpet; and I wish sincerely that any one had the moral courage to spread that carpet and sit on it in Ludgate Circus. But it is not probable that all the tales of the carpet-bearer are little gems of original artistic workmanship. Literature and fiction are two entirely different things. Literature is a luxury; fiction is a necessity. A work of art can hardly be too short, for its climax is its merit. A story can never be too long, for its conclusion is merely to be deplored, like the last halfpenny or the last pipelight. And so, while the increase of the artistic conscience tends in more ambitious works to brevity and impressionism, voluminous industry still marks the producer of the true romantic trash. There was no end to the ballads of Robin Hood; there is no end to the volumes about Dick Deadshot and the Avenging Nine. These two heroes are deliberately conceived as immortal.

But instead of basing all discussion of the problem upon the common-sense recognition of this fact—that the youth of the lower orders always has had and always must have formless and endless romantic reading of some kind, and then going on to make provision for its wholesomeness—we begin, generally speaking, by fantastic abuse of this reading as a whole and indignant surprise that the errand-boys under discussion do not read *The Egoist* and *The Master Builder*. It is the custom, particularly among magistrates, to attribute half the crimes of the Metropolis to cheap novelettes. If some grimy urchin runs away with an apple, the magistrate shrewdly points out that the child's knowledge that apples appease hunger is traceable to some curious literary researches. The boys themselves, when penitent, frequently accuse the novelettes with great bitterness, which is only to be expected from young people possessed of no little native humour. If I had forged a will, and could obtain sympathy by tracing the incident to the influence of Mr George Moore's novels, I should find the greatest entertainment in the diversion. At any rate, it is firmly fixed in the minds of most people that gutter-boys, unlike everybody else in the community, find their principal motives for conduct in printed books.

Now it is quite clear that this objection, the objection brought by magistrates, has nothing to do with literary merit. Bad story writing is not a crime. Mr Hall Caine walks the streets openly, and cannot be put in prison for an anti-climax. The objection rests upon the theory that the tone of the mass of boys' novelettes is criminal and degraded, appealing to low cupidity and low cruelty. This is the magisterial theory, and this is rubbish.

So far as I have seen them, in connexion with the dirtiest bookstalls in the poorest districts, the facts are simply these: The whole bewildering mass of vulgar juvenile literature is concerned with adventures, rambling, disconnected, and endless. It does not express any passion of any sort, for there is no human character of any sort. It runs eternally in certain grooves of local and historical type: the medieval knight, the eighteenth-century duellist, and the modern cowboy recur with the same stiff simplicity as the conventional human figures in an Oriental pattern. I can quite as easily imagine a human being kindling wild appetites by the contemplation of his Turkey carpet as by such dehumanized and naked narrative as this.

Among these stories there are a certain number which deal sympathetically with the adventures of robbers, outlaws, and pirates, which present in a dignified and romantic light thieves and murderers like Dick Turpin and Claude Duval. That is to say, they do precisely the same thing as Scott's *Ivanhoe*, Scott's *Rob Roy*, Scott's *Lady of the Lake*, Byron's *Corsair*, Wordsworth's *Rob Roy's Grave*, Stevenson's *Macaire*, Mr Max Pemberton's *Iron Pirate*, and a thousand more works distributed systematically as prizes and Christmas presents. Nobody imagines that an admiration of Locksley in *Ivanhoe* will lead a boy to shoot Japanese arrows at the deer in Richmond Park; no one thinks that the incautious opening of Wordsworth at the poem on Rob Roy will set him up for life as a blackmailer. In the case of our own class, we recognize that this wild life is contemplated with pleasure by the young, not bceause it is like their own life, but because it is different from it. It might at least cross our minds that, for whatever other reason the errand-boy reads *The Red Revenge*, it really is not because he is dripping with the gore of his own friends and relatives.

In this matter, as in all such matters, we lose our bearings entirely by speaking of the 'lower classes' when we mean humanity minus ourselves. This trivial romantic literature is not specially plebeian: it is simply human. The philanthropist can never forget classes and callings. He says, with a modest swagger, 'I have invited twenty-five factory hands to tea.' If he said, 'I have invited twenty-five chartered accountants to tea', every one would see the humour of so simple a classification. But this is what we have done with this lumberland of foolish writing: we have probed, as if it were some monstrous new disease, what is, in fact, nothing but the foolish and valiant heart of man. Ordinary men will always be sentimentalists: for a sentimentalist is simply a man who has feelings and does not trouble to invent a new way of expressing them. These common and current publications have nothing essentially evil about them. They express the sanguine and heroic truisms on which civilization is built; for it is clear that unless civilization is built on truisms, it is not built at all. Clearly, there could be no safety for a society in which the remark by the Chief Justice that murder was wrong was regarded as an original and dazzling epigram.

If the authors and publishers of *Dick Deadshot*, and such remarkable works, were suddenly to make a raid upon the educated class, were to take down the name of every man, however distinguished, who was caught at a University Extension Lecture, were to confiscate all our novels and warn us all to correct our lives, we should be seriously annoyed. Yet they have far more right to do so than we; for they, with all their idiocy, are normal and we are abnormal. It is the modern literature of the educated, not of the uneducated, which is avowedly and aggressively criminal. Books recommending profligacy and pessimism, at which the high-souled errand-boy would shudder, lie upon all our drawing-room tables. If the dirtiest old owner of the dirtiest old bookstall in Whitechapel dared to display works really recommending polygamy or suicide, his stock would be seized by the police. These things are our luxuries. And with a hypocrisy so ludicrous as to be almost unparalleled in history, we rate the gutter-boys for their immorality at the very time that we are discussing (with equivocal German professors) whether morality is valid at all. At the very instant that we curse the Penny Dreadful for encouraging thefts upon property, we can-

vass the proposition that all property is theft. At the very instant
we accuse it (quite unjustly) of lubricity and indecency, we are
cheerfully reading philosophies which glory in lubricity and
indecency. At the very instant that we charge it with encouraging the young to destroy life, we are placidly discussing whether
life is worth preserving.

But it is we who are the morbid exceptions; it is we who are
the criminal class. This should be our great comfort. The vast
mass of humanity, with their vast mass of idle books and idle
words, have never doubted and never will doubt that courage is
splendid, that fidelity is noble, that distressed ladies should be
rescued, and vanquished enemies spared. There are a large number of cultivated persons who doubt these maxims of daily life,
just as there are a large number of persons who believe they are
the Prince of Wales; and I am told that both classes of people
are entertaining conversationalists. But the average man or boy
writes daily in these great gaudy diaries of his soul, which we call
Penny Dreadfuls, a plainer and better gospel than any of those
iridescent ethical paradoxes that the fashionable change as often
as their bonnets. It may be a very limited aim in morality to
shoot a 'many-faced and fickle traitor', but at least it is a better
aim than to be a many-faced and fickle traitor, which is a simple
summary of a good many modern systems from Mr d'Annunzio's
downwards. So long as the coarse and thin texture of mere current popular romance is not touched by a paltry culture it will
never be vitally immoral. It is always on the side of life. The
poor—the slaves who really stoop under the burden of life—
have often been mad, scatter-brained, and cruel, but never hopeless. That is a class privilege, like cigars. Their drivelling literature will always be a 'blood and thunder' literature, as simple as
the thunder of heaven and the blood of men.

1901

On Sandals and Simplicity

T HE great misfortune of the modern English is not at all that they are more boastful than other people (they are not); it is that they are boastful about those particular things which nobody can boast of without losing them. A Frenchman can be proud of being bold and logical, and still remain bold and logical. A German can be proud of being reflective and orderly, and still remain reflective and orderly. But an Englishman cannot be proud of being simple and direct, and still remain simple and direct. In the matter of these strange virtues, to know them is to kill them. A man may be conscious of being heroic or conscious of being divine, but he cannot (in spite of all the Anglo-Saxon poets) be conscious of being unconscious.

Now, I do not think that it can be honestly denied that some portion of this impossibility attaches to a class very different in their own opinion, at least, to the school of Anglo-Saxonism. I mean that school of the simple life, commonly associated with Tolstoy. If a perpetual talk about one's own robustness leads to being less robust, it is even more true that a perpetual talking about one's own simplicity leads to being less simple. One great complaint, I think, must stand against the modern upholders of the simple life—the simple life in all its varied forms, from vegetarianism to the honourable consistency of the Doukhobors. This complaint against them stands, that they would make us simple in the unimportant things, but complex in the important things. They would make us simple in the things that do not matter—that is, in diet, in costume, in etiquette, in economic system. But they would make us complex in the things that do matter—in philosophy, in loyalty, in spiritual acceptance, and spiritual rejection. It does not so very much matter whether a man eats a grilled tomato or a plain tomato; it does very much matter whether he eats a plain tomato with a grilled mind. The only kind of simplicity worth preserving is the simplicity of the heart, the simplicity which accepts and enjoys. There may be a reasonable doubt as to what system preserves this; there can surely be no doubt that a system of simplicity destroys it. There is more simplicity in the man who eats caviar on impulse than in the man who eats grape-nuts on principle.

The chief error of these people is to be found in the very phrase to which they are most attached—'plain living and high thinking.' These people do not stand in need of, will not be improved by, plain living and high thinking. They stand in need of the contrary. They would be improved by high living and plain thinking. A little high living (I say, having a full sense of responsibility, a little high living) would teach them the force and meaning of the human festivities, of the banquet that has gone on from the beginning of the world. It would teach them the historic fact that the artificial is, if anything, older than the natural. It would teach them that the loving-cup is as old as any hunger. It would teach them that ritualism is older than any religion. And a little plain thinking would teach them how harsh and fanciful are the mass of their own ethics, how very civilized and very complicated must be the brain of the Tolstoyan who really believes it to be evil to love one's country and wicked to strike a blow.

A man approaches, wearing sandals and simple raiment, a raw tomato held firmly in his right hand, and says, 'The affections of family and country alike are hindrances to the fuller development of human love'; but the plain thinker will only answer him, with a wonder not untinged with admiration, 'What a great deal of trouble you must have taken in order to feel like that.' High living will reject the tomato. Plain thinking will equally decisively reject the idea of the invariable sinfulness of war. High living will convince us that nothing is more materialistic than to despise a pleasure as purely material. And plain thinking will convince us that nothing is more materialistic than to reserve our horror chiefly for material wounds.

The only simplicity that matters is the simplicity of the heart. If that be gone, it can be brought back by no turnips or cellular clothing; but only by tears and terror and the fires that are not quenched. If that remain, it matters very little if a few Early Victorian armchairs remain along with it. Let us put a complex *entrée* into a simple old gentleman; let us not put a simple *entrée* into a complex old gentleman. So long as human society will leave my spiritual inside alone, I will allow it, with a comparative submission, to work its wild will with my physical interior. I will submit to cigars. I will meekly embrace a bottle of Burgundy. I will humble myself to a hansom cab. If only by this means I may preserve to myself the virginity of the spirit, which enjoys with astonishment and fear. I do not say that these are the only methods of preserving it. I incline to the belief that there are others. But I will have nothing to do with simplicity which lacks the fear, the astonishment, and the joy alike. I will have nothing to do with the devilish vision of a child who is too simple to like toys.

The child is, indeed, in these, and many other matters, the best guide. And in nothing is the child so righteously childlike, in nothing does he exhibit more accurately the sounder order of simplicity, than in the fact that he sees everything with a simple pleasure, even the complex things. The false type of naturalness harps always on the distinction between the natural and the artificial. The higher kind of naturalness ignores that distinction. To the child the tree and the lamp-post are as natural and as artificial as each other; or rather, neither of them are natural but both supernatural. For both are splendid and unexplained. The flower with which God crowns the one, and the flame with which Sam the lamplighter crowns the other, are equally of the gold of fairy-tales. In the middle of the wildest fields the most rustic child is, ten to one, playing at steam-engines. And the only spiritual or philosophical objection to steam-engines is not that men pay for them or work at them, or make them very ugly, or even that men are killed by them; but merely that men do not play at them. The evil is that the childish poetry of clockwork does not remain. The wrong is not that engines are too much admired, but that they are not admired enough. The sin is not that engines are mechanical, but that men are mechanical.

In this matter, then, our main conclusion is that it is a fundamental point of view, a philosophy or religion which is needed, and not any change in habit or social routine. The things we need most for immediate practical purposes are all abstractions. We need a right view of the human lot, a right view of the human society; and if we were living eagerly and angrily in the enthusiasm of those things, we should, *ipso facto*, be living simply in the genuine and spiritual sense. Desire and danger make every one simple. And to those who talk to us with interfering eloquence about Jaeger and the pores of the skin, and about Plasmon and

the coats of the stomach, at them shall only be hurled the words
that are hurled at fops and gluttons, 'Take no thought what ye
shall eat or what ye shall drink, or wherewithal ye shall be
clothed. For after all these things do the Gentiles seek. But seek
first the kingdom of God and His righteousness, and all these
things shall be added unto you.' Those amazing words are not
only extraordinarily good, practical politics; they are also
superlatively good hygiene. The one supreme way of making all
those processes go right, the processes of health, and strength,
and grace, and beauty, the one and only way of making certain of
their accuracy, is to think about something else. If a man is bent
on climbing into the seventh heaven, he may be quite easy about
the pores of his skin. If he harnesses his waggon to a star, the
process will have a most satisfactory effect upon the coats of his
stomach. For the thing called 'taking thought,' the thing for
which the best modern word is 'rationalizing,' is in its nature,
inapplicable to all plain and urgent things. Men take thought and
ponder rationalistically, touching remote things—things that
only theoretically matter, such as the transit of Venus. But only
at their peril can men rationalize about so practical a matter as
health.

1905

Invective

THE late Lord Morley, when he was editing the *Pall Mall*, was amused by a young journalist who, when asked his particular line, replied 'Invective.' 'Invective? May I ask against what?' 'Oh ... anything—general invective.' One recognizes that impartial faculty for getting angry. It can produce sneers, tropes, tremendous metaphors; out of it some pages of memorable prose have been written. Such anger is delicious to experience, for it is accompanied by a glowing sense of superiority, and it can be an immense stimulus to composition. But it can only be cultivated at the sacrifice of some spiritual honesty: that is the price which must be paid. Success depends upon rapidly draining into general channels the contents of your private reservoir of resentments, vainglory, thwarted ambitions, wrongs and grudges. Such emotions are ductile. It is particularly easy to make, for instance, a little current of envy turn furiously the mills of righteous indignation. But then the writer must be unconscious of the sources of this energy. Hence the necessity of a certain dishonesty or lack of self-awareness which, whether inborn or acquired, may sooner or later make a fool of the specialist in invective.

Again, invective which has become a habit is apt, like charm, to lose its virtue, for both depend for their effectiveness on spontaneity. Personal charm which has been extravagantly used for personal ends, from winning hearts to securing corner seats in railway carriages, in time grows blowsy. It gradually loses the brave delicacy and sweet candour proper to it, though its possessor may be quite unaware that this is happening. In the same way the adept at invective does not notice when something has crept into his style which makes it ineffective. His attack may still amuse, even impress the detached reader, but it has become incapable of giving pain to the victim, which is its proper end. A

self-delighting exuberance in animosity, a too obvious contentment in the sleek sarcastic phrase, actually bring balm to the
wounds which deadly statements ought to inflict. The victim is
relieved by observing that the writer is licking the chops of his
own malice, and executing a war-dance instead of thrusting at
the vitals of his enemy.

Swinburne was master of a glorious exuberant invective, but I
doubt if his fiercest tirades gave much pain even to Dr Furnivall
or Robert Buchanan. The first effect of his 'Under the Microscope' is to convince the reader that it must have been immense
fun to write it; he is sure that the author, after giving rein to his
emotions, must have enjoyed a sunset-calm of mind. This, of
course, is fatal to the proper purpose of invective.

1935

My Own Centenary

(From *The Times* of AD 2027)

It is a hundred years ago to-day since Forster died; we celebrate his centenary indeed within a few months of the bicentenary of Beethoven, within a few weeks of that of Blake. What special tribute shall we bring him? The question is not easy to answer, and were he himself still alive he would no doubt reply, 'My work is my truest memorial.' It is the reply that a great artist can always be trusted to make. Conscious of his lofty mission, endowed with the divine gift of self-expression, he may rest content, he is at peace, doubly at peace. But we, we who are not great artists, only the recipients of their bounty—what shall we say about Forster? What can we say that has not already been said about Beethoven, about Blake? Whatever shall we say?

The Dean of Dulborough, preaching last Sunday in his own beautiful cathedral, struck perhaps the truest note. Taking as his text that profound verse in Ecclesiasticus, 'Let us now praise famous men,' he took it word by word, paused when he came to the word 'famous,' and, slowly raising his voice, said: 'He whose hundredth anniversary we celebrate on Thursday next is famous, and why?' No answer was needed, none came. The lofty Gothic nave, the great western windows, the silent congregation—they gave answer sufficient, and passing on to the final word of his text, 'men,' the Dean expatiated upon what is perhaps the most mysterious characteristic of genius, its tendency to appear among members of the human race. Why this is, why, since it is, it is not accompanied by some definite outward sign through which it might be recognized easily, are questions not lightly to be raised. There can be no doubt that his contemporaries did not recognize the greatness of Forster. Immersed in their own little affairs, they either ignored him, or forgot him, or confused him, or, strangest of all, discussed him as if he was their equal. We may smile at their blindness, but for him it can have been no laughing matter, he must have had much to bear, and indeed he could scarcely have endured to put forth masterpiece after masterpiece had he not felt assured of the verdict of posterity.

Sir Vincent Edwards, when broadcasting last night, voiced that verdict not uncertainly, and was fortunately able to employ more wealth of illustration than had been appropriate in Dulborough Minster for the Dean. The point he very properly stressed was our writer's loftiness of aim. 'It would be impossible,' he said, 'to quote a single sentence that was not written from the very loftiest motive,' and he drew from this a sharp and salutary lesson for the so-called writers of to-day. As permanent head of the Ministry of Edification, Sir Vincent has, we believe, frequently come into contact with the younger generation, and has checked with the kindliness of which he is a past-master their self-styled individualism—an individualism which is the precise antithesis of true genius. They confuse violence with strength, cynicism with open-mindedness, frivolity with joyousness—mistakes never made by Forster who was never gay until he had earned the right to be so, and only criticized the religious and social institutions of his time because they were notoriously corrupt. We know what the twentieth century was. We know the sort of men who were in power under George V. We know what the State was, what were the churches. We can as easily conceive of Beethoven as a Privy Councillor or of Blake as, forsooth, an Archbishop as of this burning and sensitive soul acquiescing in the deadening conditions of his age. What he worked for—what all great men work for—was for a New Jerusalem, a vitalized State, a purified Church; and the offertory at Dulborough last Sunday, like the success of Sir Vincent's appeal for voluntary workers under the Ministry, shows that he did not labour in vain.

The official ceremony is for this morning. This afternoon Lady Turton will unveil Mr Boston Jack's charming statue in Kensington Gardens, and so illustrate another aspect of our national hero: his love of little children. It had originally been Mr Boston Jack's intention to represent him as pursuing an ideal. Since, however, the Gardens are largely frequented by the young and their immediate supervisors, it was felt that something more whimsical would be in place, and a butterfly was substituted.

The change is certainly for the better. It is true that we cannot
have too many ideals. On the other hand, we must not have too
much of them too soon, nor, attached as it will be to a long cop-
per wire, can the butterfly be confused with any existing species
and regarded as an incentive to immature collectors. Lady
Turton will couple her remarks with an appeal for the Imperial
Daisy Chain, of which she is the energetic Vice-President, and
simultaneously there will be a flag collection throughout the
provinces.

Dulborough, the Ministry of Edification, the official ceremony, Kensington Gardens! What more could be said? Not a
little. Yet enough has been said to remind the public of its heritage, and to emphasize and define the central essence of these
immortal works. And what is that essence? Need we say? Not
their greatness—they are obviously great. Not their profundity—they are admittedly profound. It is something more precious than either: their nobility. Noble works, nobly conceived,
nobly executed, nobler than the Ninth Symphony or the Songs
of Innocence. Here is no small praise, yet it can be given, we are
in the presence of the very loftiest, we need not spare or mince
our words, nay, we will add one more word, a word that has
been implicit in all that have gone before: like Beethoven, like
Blake, Forster was essentially English, and in commemorating
him we can yet again celebrate what is best and most permanent
in ourselves.

                                                      1927

Creighton

THE Church of England is one of the most extraordinary of institutions. An incredible concoction of Queen Elizabeth's, it still flourishes, apparently, and for three hundred years has remained true to type. Or perhaps, in reality, Queen Elizabeth had not very much to do with it; perhaps she only gave, with her long, strong fingers, the final twist to a stem that had been growing for ages, deep-rooted in the national life. Certainly our cathedrals—so careful and so unaesthetic, so class-conscious and so competent—suggest that view of the case. English Gothic seems to show that England was Anglican long before the Reformation—as soon as she ceased to be Norman, in fact. Pure piety, it cannot be denied, has never been her Church's strong point. Anglicanism has never produced—never could produce —a St Teresa. The characteristic great men of the institution— Whitgift, Hooker, Laud, Butler, Jowett—have always been remarkable for virtues of a more secular kind: those of scholarship or of administrative energy. Mandell Creighton was (perhaps) the last of the long line. Perhaps; for who can tell? It is difficult to believe that a man of Creighton's attainments will ever again be Bishop of London. That particular concatenation seems to have required a set of causes to bring it into existence—a state of society, a habit of mind—which have become obsolete. But the whirligigs of time are, indeed, unpredictable; and England, some day or other, may well be blessed with another Victorian Age.

In Creighton *both* the great qualities of Anglican tradition were present to a remarkable degree. It would be hard to say whether he were more distinguished as a scholar or a man of affairs; but—such is the rather unfair persistence of the written word—there can be little doubt that he will be remembered chiefly as the historian of the Papacy. Born when the world was

becoming extremely scientific, he belonged to the post-Carlyle-
and-Macaulay generation—the school of Oxford and Cambridge
inquirers, who sought to reconstruct the past solidly and patiently, with nothing but facts to assist them—pure facts, un-
twisted by political or metaphysical bias and uncoloured by
romance. In this attempt Creighton succeeded admirably. He was
industrious, exact, clear-headed, and possessed of a command
over words that was quite sufficient for his purposes. He succeeded more completely than Professor Samuel Gardiner, whose
history of the Early Stuarts and the Civil Wars was a contemporary work. Gardiner did his best, but he was not an absolute
master of the method. Strive as he would, he could not prevent
himself, now and then, from being a little sympathetic to one or
other of his personages; sometimes he positively alluded to a
physical circumstance; in short, humanity would come creeping
in. A mistake! For Professor Gardiner's feelings about mankind
are not illuminating; and the result is a slight blur. Creighton
was made of sterner stuff. In his work a perfectly grey light
prevails everywhere; there is not a single lapse into psychological
profundity; every trace of local colour, every suggestion of personal passion, has been studiously removed. In many ways all
this is a great comfort. One is not worried by moral lectures or
purple patches, and the field is kept clear for what Creighton
really excelled in—the lucid exposition of complicated political
transactions, and the intricate movements of thought with which
they were accompanied. The biscuit is certainly exceedingly dry;
but at any rate there are no weevils in it. As one reads, one gets
to relish, with a sober satisfaction, this plumless fare. It begins to
be very nearly a pleasure to follow the intrigues of the great
Councils, or to tread the labyrinth of the theological theory of
indulgences. It is a curious cross-section of history that Creighton offers to the view. He has cut the great tree so near to the
ground that leaf and flower have vanished; but he has worked
his saw with such steadiness and precision that every grain in the
wood is visible, and one can look *down* at the mighty structure,
revealed in all its complex solidity like a map to the mind's eye.

Charming, indeed, are the ironies of history; and not the least
charming those that involve the historian. It was very natural
that Creighton, a clever and studious clergyman of the Church
of England, should choose as the subject of his investigations

that group of events which, centring round the Italian popes, produced at last the Reformation. The ironical fact was that those events happened to take place in a world where no clever and studious clergyman of the Church of England had any business to be. 'Sobriety,' as he himself said, was his aim; but what could sobriety do when faced with such figures as Savonarola, Caesar Borgia, Julius II, and Luther? It could only look somewhere else. It is pleasant to witness the high-minded husband and father, the clever talker at Cambridge dinner tables, the industrious diocesan administrator, picking his way with an air of calm detachment amid the recklessness, the brutality, the fanaticism, the cynicism, the lasciviousness, of those Renaissance spirits. 'In his private life,' Creighton says of Alexander VI, 'it is sufficiently clear that he was at little pains to repress a strongly sensual nature. . . . We may hesitate to believe the worst charges brought against him; but the evidence is too strong to enable us to admit that even after his accession to the papal office he discontinued the irregularities of his previous life.' There is high comedy in such a tone on such a topic. One can imagine the father of the Borgias, if he could have read that sentence, throwing up his hands in delighted amazement, and roaring out the obscene blasphemy of his favourite oath.

The truth was that, in spite of his wits and his Oxford training, the admirable north-country middle-class stock, from which Creighton came, dominated his nature. His paradoxes might astound academical circles, his free speech might agitate the lesser clergy, but at heart he was absolutely sound. Even a friendship with that daemonic imp, Samuel Butler, left him uncorroded. He believed in the Real Presence. He was opposed to Home Rule. He read with grave attention the novels of Mrs Humphry Ward. The emancipation of a Victorian bishop could never be as that of other men. The string that tied him to the peg of tradition might be quite a long one; but it was always there. Creighton enjoyed his little runs with the gusto and vitality that were invariably his. The sharp aquiline face, with the grizzled beard, the bald forehead, and the gold spectacles, gleamed and glistened, the long, slim form, so dapper in its episcopal gaiters, preened itself delightedly, as an epigram—a devastating epigram—shot off and exploded, and the Fulham teacups tinkled as they had never tinkled before. Then, a moment later, the guests gone, the firm mouth closed in severe determination; work was resumed. The duties of the day were dispatched swiftly; the vast and stormy diocese of London was controlled with extraordinary efficiency; while a punctual calmness reigned, for, however pressed and pestered, the Bishop was never known to fuss. Only once on a railway journey, when he believed that some valuable papers had gone astray, did his equanimity desert him. 'Where's my black bag?' was his repeated inquiry. His mischievous children treasured up this single lapse; and, ever afterwards, 'Where's my black bag?' was thrown across the table at the good-humoured prelate when his family was in a teasing mood.

When the fourth volume of the *History of the Papacy* appeared there was a curious little controversy, which illustrated Creighton's attitude to history and, indeed, to life. 'It seems to me,' he wrote in the preface, 'neither necessary to moralise at every turn in historical writing, nor becoming to adopt an attitude of lofty superiority over any one who ever played a prominent part in European affairs, nor charitable to lavish undiscriminating censure on any man.' The wrath of Lord Acton was roused. He wrote a violent letter of protest. The learning of the eminent Catholic was at least equal to Creighton's, but he made no complaint upon matters of erudition; it was his moral sense that was outraged. Creighton, it seemed to him, had passed over, with inexcusable indifference, the persecution and intolerance of the mediaeval Church. The popes of the thirteenth and fourteenth centuries, he wrote, '... instituted a system of persecution.... It is the most conspicuous fact in the history of the mediaeval Papacy.... But what amazes and disables me is that you speak of the Papacy not as exercising a just severity, but as not exercising any severity. You ignore, you even deny, at least implicitly, the existence of the torture chamber and the stake.... Now the Liberals think persecution a crime of a worse order than adultery, and the acts done by Ximenes considerably worse than the entertainment of Roman courtesans by Alexander VI. The responsibility exists whether the thing permitted be good or bad. If the thing be criminal, then the authority permitting it bears the guilt.... You say that people in authority are not to be snubbed or sneered at from our pinnacle of conscious rectitude. I really don't know whether you exempt them because of their rank, or of their success and power, or of their date.... Historic responsibility has to make up for the want of legal responsibility. Power tends to corrupt, and absolute power corrupts absolutely. Great men are almost always bad.' These words, surely, are magnificent. One sees with surprise and exhilaration the roles reversed—the uncompromising fervour of Catholicism calling down fire from Heaven upon its own abominable popes and the worldly Protestantism that excused them. Creighton's reply was as Anglican as might have been expected. He hedged. One day, he wrote, John Bright had said, 'If the people knew what sort of men statesmen were, they would rise and hang the whole lot of them.' Next day Gladstone had said 'Statesmanship is the noblest way to serve mankind.' 'I am sufficient of a Hegelian to be able to combine both judgments; but the results of my combination cannot be expressed in the terms of the logic of Aristotle.... Society is an organism,' etc. It is clear enough that his real difference with Lord Acton was not so much over the place of morals in history as over the nature of the historical acts upon which moral judgments are to be passed. The Bishop's imagination was not deeply stirred by the atrocities of the Inquisition; what interested him, what appealed to him, what he really understood, were the difficulties and the expedients of a man of affairs who found himself at the head of a great administration. He knew too well, with ritualists on one side and Kensitites on the other, the trials and troubles from which a clerical ruler had to extricate himself as best he could, not to sympathise (in his heart of hearts) with the clerical rulers of another age who had been clever enough to devise regulations for the elimination of heresy and schism, and strong enough to put those regulations into force.

He himself, however, was never a persecutor; his great practical intelligence prevented that. Firmly fixed in the English tradition of common sense, compromise and comprehension, he held on his way amid the shrieking of extremists with imperturbable moderation. One of his very last acts was to refuse to prosecute two recalcitrant clergymen who had persisted in burning incense in a forbidden manner. He knew that, in England at any rate, persecution did not work. Elsewhere, perhaps, it might be different; in Russia, for instance.... There was an exciting moment in Creighton's life when he was sent to Moscow to represent the Church of England at the Coronation of the Emperor

Nicholas; and his comments on that occasion were significant. Clad in a gorgeous cope of red and gold, with mitre and crozier, the English prelate attracted every eye. He thoroughly relished the fact; he tasted, too, to the full, the splendour of the great ceremonies and the extraordinary display of autocratic power. That there might have been some degree of spiritual squalor mixed with those magnificent appearances never seemed to occur to him. He was fascinated by the apparatus of a mighty organisation, and, with unerring instinct, made straight for the prime mover of it, the Chief Procurator of the Holy Synod, the sinister Pobiedonostzeff, with whom he struck up a warm friendship. He was presented to the Emperor and Empress, and found them charming. 'I was treated with great distinction, as I was called in first. The Empress looked very nice, dressed in white silk.' The aristocratic Acton would, no doubt, have viewed things in a different light. 'Absolute power corrupts absolutely'—so he had said; but Creighton had forgotten the remark. He was no Daniel. He saw no Writing on the Wall.

The Bishop died in his prime, at the height of his success and energy, and was buried in St Paul's Cathedral. Not far from his tomb, which a Victorian sculptor did his best to beautify, stands the strange effigy of John Donne, preaching, in his shroud, an incredible sermon upon mortality. Lingering in that corner, one's mind flashes oddly to other scenes and other persons. One passes down the mouldering street of Ferrara, and reaches an obscure church. In the half-light, from an inner door, an elderly humble nun approaches, indicating with her patois a marble slab in the pavement—a Latin inscription—the grave of Lucrezia Borgia. Mystery and oblivion were never united more pathetically. But there is another flash, and one is on a railway platform under the grey sky of England. A tall figure hurries by, spectacled and bearded, with swift clerical legs, and a voice—a competent, commanding, yet slightly agitated voice—says sharply: 'Where's my black bag?'

1925

H. L. MENCKEN

The Libido for the Ugly

ON a Winter day, not long ago, coming out of Pittsburgh on one of the swift, luxurious expresses of the eminent Pennsylvania Railroad, I rolled eastward for an hour through the coal and steel towns of Westmoreland county. It was familiar ground; boy and man, I had been through it often before. But somehow I had never quite sensed its appalling desolation. Here was the very heart of industrial America, the center of its most lucrative and characteristic activity, the boast and pride of the richest and grandest nation ever seen on earth—and here was a scene so dreadfully hideous, so intolerably bleak and forlorn that it reduced the whole aspiration of man to a macabre and depressing joke. Here was wealth beyond computation, almost beyond imagination—and here were human habitations so abominable that they would have disgraced a race of alley cats.

I am not speaking of mere filth. One expects steel towns to be dirty. What I allude to is the unbroken and agonizing ugliness, the sheer revolting monstrousness, of every house in sight. From East Liberty to Greensburg, a distance of twenty-five miles, there was not one in sight from the train that did not insult and lacerate the eye. Some were so bad, and they were among the most pretentious—churches, stores, warehouses, and the like— that they were downright startling: one blinked before them as one blinks before a man with his face shot away. It was as if all the more advanced Expressionist architects of Berlin had been got drunk on *Schnapps*, and put to matching aberrations. A few masterpieces linger in memory, horrible even there: a crazy little church just west of Jeannette, set like a dormer-window on the side of a bare, leprous hill; the headquarters of the Veterans of Foreign Wars at Irwin; a steel stadium like a huge rat-trap somewhere further down the line. But most of all I recall the general effect—of hideousness without a break. There was not a single decent house within eye-range from the Pittsburgh suburbs to the Greensburg yards. There was not one that was not mis-shapen, and there was not one that was not shabby.

The country itself is not uncomely, despite the grime of the endless mills. It is, in form, a narrow river valley, with deep gullies running up into the hills. It is thickly settled, but not noticeably overcrowded. There is still plenty of room for building, even in the larger towns, and there are very few solid blocks. Nearly every house, big and little, has space on all four sides. Obviously, if there were architects of any professional sense or dignity in the region, they would have perfected a châlet to hug the hillsides—a châlet with a high-pitched roof, to throw off the heavy Winter snows, but still essentially a low and clinging building, wider than it was tall. But what have they done? They have taken as their model a brick set on end. This they have converted into a thing of dingy clapboards, with a narrow, low-pitched roof. And the whole they have set upon thin, preposterous brick piers. What could be more appalling? By the hundreds and thousands these abominable houses cover the bare hillsides, like gravestones in some gigantic and decaying cemetery. On their deep sides they are three, four and even five stories high; on their low sides they bury themselves swinishly in the mud. Not a fifth of them are perpendicular. They lean this way and that, hanging on to their bases precariously. And one and all they are streaked in grime, with dead and eczematous patches of paint peeping through the streaks.

Now and then there is a house of brick. But what brick! When it is new it is the color of a fried egg. When it has taken on the patina of the mills it is the color of an egg long past all hope or caring. Was it necessary to adopt that shocking color? No more than it was necessary to set all of the houses on end. Red brick, even in a steel town, ages with some dignity. Let it become downright black, and it is still sightly, especially if its trimmings are of white stone, with soot in the depths and the high spots washed by the rain. But in Westmoreland they prefer that uremic yellow, and so they have the most loathsome towns and villages ever seen by mortal eye.

I award this championship only after laborious research and incessant prayer. I have seen, I believe, all of the most unlovely towns of the world; they are all to be found in the United States.

I have seen the mill towns of decomposing New England and the desert towns of Utah, Arizona and Texas. I am familiar with the back streets of Newark, Brooklyn, Chicago and Pittsburgh, and have made bold scientific explorations to Camden, N.J. and Newport News, Va. Safe in a Pullman, I have whirled through the gloomy, God-forsaken villages of Iowa and Kansas, and the malarious tide-water hamlets of Georgia. I have been to Bridgeport, Conn., and to Los Angeles. But nowhere on this earth, at home or abroad, have I seen anything to compare to the villages that huddle along the line of the Pennsylvania from the Pittsburgh yards to Greensburg. They are incomparable in color, and they are incomparable in design. It is as if some titanic and aberrant genius, uncompromisingly inimical to man, had devoted all the ingenuity of Hell to the making of them. They show grotesqueries of ugliness that, in retrospect, become almost diabolical. One cannot imagine mere human beings concocting such dreadful things, and one can scarcely imagine human begins bearing life in them.

Are they so frightful because the valley is full of foreigners—dull, insensate brutes, with no love of beauty in them? Then why didn't these foreigners set up similar abominations in the countries that they came from? You will, in fact, find nothing of the sort in Europe—save perhaps in a few putrefying parts of England. There is scarcely an ugly village on the whole Continent. The peasants, however poor, somehow manage to make themselves graceful and charming habitations, even in Italy and Spain. But in the American village and small town the pull is always toward ugliness, and in that Westmoreland valley it has been yielded to with an eagerness bordering upon passion. It is incredible that mere ignorance should have achieved such masterpieces of horror. There is a voluptuous quality in them—the same quality that one finds in a Methodist sermon or an editorial in the New York *Herald-Tribune*. They look deliberate.

On certain levels of the human race, indeed, there seems to be a positive libido for the ugly, as on other and less Christian levels there is a libido for the beautiful. It is impossible to put down the wallpaper that defaces the average American home of the lower middle class to mere inadvertence, or to the obscene humor of the manufacturers. Such ghastly designs, it must be obvious, give a genuine delight to a certain type of mind. They meet, in some unfathomable way, its obscure and unintelligible demands. They caress it as 'The Palms' caresses it, or the art of Landseer, or the ecclesiastical architecture of the United Brethren. The taste for them is as enigmatical and yet as common as the taste for vaudeville, dogmatic theology, sentimental movies, and the poetry of Edgar A. Guest. Or for the metaphysical speculations of Arthur Brisbane. Thus I suspect (though confessedly without knowing) that the vast majority of the honest folk of Westmoreland county, and especially the 100% Americans among them, actually admire the houses they live in, and are proud of them. For the same money they could get vastly better ones, but they prefer what they have got. Certainly there was no pressure upon the Veterans of Foreign Wars at Irwin to choose the dreadful edifice that bears their banner, for there are plenty of vacant buildings along the trackside, and some of them are appreciably better. They might, indeed, have built a better one of their own. But they chose that clapboarded horror with their eyes open, and having chosen it, they let it mellow into its present shocking depravity. They like it as it is: beside it, the Parthenon would no doubt offend them. In precisely the same way the authors of the rat-trap stadium that I have mentioned made a deliberate choice. After painfully designing and erecting it, they made it perfect in their own sight by putting a completely impossible pent-house, painted a staring yellow, on top of it. The effect is truly appalling. It is that of a fat woman with a black eye. It is that of a Presbyterian grinning. But they like it.

Here is something that the psychologists have so far neglected: the love of ugliness for its own sake, the lust to make the world intolerable. Its habitat is the United States. Out of the melting pot emerges a race which hates beauty as it hates truth. The etiology of this madness deserves a great deal more study than it has got. There must be causes behind it; it arises and flourishes in obedience to biological laws, and not as a mere act of God. What, precisely, are the terms of those laws? And why do they run stronger in America than elsewhere? Let some honest *Privat Dozent* apply himself to the problem.

1928

Funeral March

WHERE is the graveyard of dead gods? What lingering mourner waters their mounds? There was a day when Jupiter was the king of the gods, and any man who doubted his puissance was *ipso facto* a barbarian and an ignoramus. But where in all the world is there a man who worships Jupiter to-day? And what of Huitzilopochtli? In one year—and it is no more than five hundred years ago—50,000 youths and maidens were slain in sacrifice to him. To-day, if he is remembered at all, it is only by some vagrant savage in the depth of the Mexican forest. Huitzilopochtli, like many other gods, had no human father; his mother was a virtuous widow; he was born of an apparently innocent flirtation that she carried on with the sun. When he frowned, his father, the sun, stood still. When he roared with rage, earthquakes engulfed whole cities. When he thirsted he was watered with 10,000 gallons of human blood. But to-day Huitzilopochtli is as magnificently forgotten as Marie Corelli. Once the peer of Allah, Buddha and Wotan, he is now the peer of Father Rasputin, J. B. Planché, Sadi Carnot, General Boulanger, Lottie Collins, and Little Tich.

Speaking of Huitzilopochtli recalls his brother, Tezcatilpoca. Tezcatilpoca was almost as powerful: he consumed 25,000 virgins a year. Lead me to his tomb: I would weep, and hang a *couronne des perles*. But who knows where it is? Or where the grave of Quitzalcoatl is? Or Tialoc? Or Chalchihuitlicue? Or Xiehtecutli? Or Centeotl, that sweet one? Or Tlazolteotl, the goddess of love? Or Mictlan? Or Ixtlilton? Or Omacatl? Or Yacatecutli? Or Mixcoatl? Or Xipe? Or all the host of Tzitzimitles? Where are their bones? Where is the willow on which they hung their harps? In what forlorn and unheard-of hell do they await the resurrection morn? Who enjoys their residuary estates? Or that of Dis, whom Caesar found to be the chief god of the Celts? Or that of Tarves, the bull? Or that of Moccos, the pig? Or that of Epona, the mare? Or that of Mullo, the celestial ass? There was a time when the Irish revered all these gods as violently as they now revere the Pope. But to-day even the drunkest Irishman laughs at them.

But they have company in oblivion: the hell of dead gods is as crowded as the Presbyterian hell for babies. Damona is there, and Esus, and Drunemeton, and Silvana, and Dervones, and Adsalluta, and Deva, and Belisama, and Axona, and Vintios, and Taranuous, and Sulis, and Cocidius, and Adsmerius, and Dumiatis, and Caletos, and Moccus, and Ollovidius, and Albiorix, and Leucitius, and Vitucadrus, and Ogmios, and Uxellimus, and Borvo, and Grannos, and Mogons. All mighty gods in their day, worshipped by millions, full of demands and impositions, able to bind and loose—all gods of the first class, not dilettanti. Men laboured for generations to build vast temples to them—temples with stones as large as motor-lorries. The business of interpreting their whims occupied thousands of priests, wizards, archdeacons, canons, deans, bishops, archbishops. To doubt them was to die, usually at the stake. Armies took to the field to defend them against infidels: villages were burned, women and children were butchered, cattle were driven off. Yet in the end they all withered and died, and to-day there is none so poor to do them reverence. Worse, the very tombs in which they lie are lost, and so even a respectful stranger is debarred from paying them the slightest and politest homage.

What has become of Sutehl, once the high god of the whole Nile Valley? What has become of:

Resheph	Baal
Anath	Astarte
Ashtoreth	Hadad
El	Addu
Nergal	Shalem
Nebo	Dagon
Ninib	Sharrab
Melek	Yau
Ahijah	Amon-Re
Isis	Osiris
Ptah	Sebek
Anubis	Moloch?

All these were once gods of the highest eminence. Many of them are mentioned with fear and trembling in the Old Testament. They ranked, five or six thousand years ago, with Jahveh himself; the worst of them stood far higher than Thor. Yet they have all gone down the chute, and with them the following:

Bilé	Gwydion
Lêr	Manawyddan
Arianrod	Nuada Argetlam
Morrigu	Tagd
Govannon	Goibniu
Gunfled	Odin
Sokk-mimi	Llaw Gyffes
Memetona	Lleu
Dagda	Ogma
Kerridwen	Mider
Pwyll	Rigantona
Ogyrvan	Marzin
Dea Dia	Mars
Ceros	Jupiter
Vaticanus	Cunina
Edulia	Potina
Adeona	Statilinus
Iuno Lucina	Diana of Ephesus
Saturn	Robigus
Furrina	Pluto
Vediovis	Ops
Consus	Meditrina
Cronos	Vesta
Enki	Tilmun
Engurra	Zer-panitu
Belus	Merodach
Dimmer	U-ki
Mu-ul-lil	Dauke
Ubargisi	Gasan-abzu
Ubilulu	Elum
Gasan-lil	U-Tin-dir-ki
U-dimmer-an-kia	Marduk
Enurestu	Nin-lil-la
U-sab-sib	Nin
U-Mersi	Persephone
Tammuz	Istar
Venus	Lagas
Bau	U-urugal
Mulu-hursang	Sirtumu
Anu	Ea
Beltis	Nirig
Nusku	Nebo
Ni-zu	Samas

Sahi	Ma-banba-anna
Aa	En-Mersi
Allatu	Amurru
Sin	Assur
Abil Addu	Aku
Apsu	Beltu
Dagan	Dumu-zi-abzu
Elali	Kuski-banda
Isum	Kaawanu
Mami	Nin-azu
Nin-man	Lugal-Amarada
Zaraqu	Qarradu
Suqamunu	Ura-gala
Zagaga	Ueras

You may think I spoof. That I invent the names. I do not. Ask the rector to lend you any good treatise on comparative religion: you will find them all listed. They were gods of the highest standing and dignity—gods of civilized peoples—worshipped and believed in by millions. All were theoretically omnipotent, omniscient, and immortal. And all are dead.

1922

Evening Parties

Human beings are curious creatures, and in nothing more curious than in the forms of diversion which they devise for themselves. Some of these are quite comprehensible; they give physical or mental pleasure. Bathing in the sea, for instance; or watching a play; or visiting the Zoo; or eating agreeable food at someone else's expense, or even at one's own; or playing some game with a ball. It is easy to understand that having one's person surrounded by water, in which one floats and swims, or watching human life enacted improbably by others on a stage, or seeing strange beasts in cages, or rolling elegant foods about the palate, or chasing after a ball, is pleasing. But, besides these simple pleasures, humanity has devised some so-called amusements which seem to depend for their reputations as entertainments less on pleasing sensations inflicted on the participants than on some convention which has ordained that these pursuits shall be held agreeable. It speaks well, perhaps, for the kindliness and amiability of the human race that most such pursuits are of a gregarious nature. Assembling together; dearly we love to do this. 'Neglect not the assembling of yourselves together,' says (I think) St Paul somewhere; and it was a superfluous piece of admonition. Neglect of this will never be numbered among the many omissions of mankind. Seeing one another; meeting the others of our race; exchanging remarks; or merely observing in what particular garments they have elected to clothe themselves to-day; this is so nearly universal a custom that it has become dignified into an entertainment, and we issue to one another invitations to attend such gatherings.

We issue them and we accept them, and, when the appointed date arrives, we assume such of our clothes as we believe to be suitable to the gathering, and sally forth to the party of pleasure. Often, indeed usually, it is in the evening. Therefore we clothe ourselves in such garb as men and women have agreed, in their strange symbolism, to consider appropriate to the hours after eight o'clock or so. And perhaps—who knows?—it is in the exercise of this savage and primitive conventionalism that a large part of the pleasure of an evening gathering consists. We are very primitive creatures, and the mere satisfaction of self-adornment, and of assuming for a particular occasion a particular set of clothes, may well tickle our sensibilities. Be that as it may, we arrive at our party dolled, so to speak, up, and find ourselves in a crowd of our fellow-creatures, all dolled up too. Now we are off. The party of pleasure has begun. We see friends and talk to them. But this we could do with greater comfort at our own homes or in theirs; this cannot, surely, be the promised pleasure. As a matter of fact, if you succeed in getting into a corner with a friend and talking, be sure you will be very soon torn asunder by an energetic hostess, whose motto is 'Keep them moving.' We are introduced to new acquaintances. This may, no doubt, be very agreeable. They may be persons you are glad to know. But it is doubtful whether your acquaintanceship will prosper very much to-night. It may well be that no topics suitable for discussion will present themselves to either of you at the moment of introduction. I know someone who says that she never can think of anything to say to persons introduced to her at a party except 'Do you like parties?' And that is too crude; it simply cannot be said. You must think of some more sophisticated remark. Having thought of it, you must launch it, in the peculiarly resonant pitch necessary to carry it above the clamour (for this clamour, which somewhat resembles the shrieking of a jazz band, is an essential accompaniment to a party, and part of the entertainment provided). A conversation will then ensue, and must be carried on until one or other of you either flags or breaks away, or until someone intervenes between you. One way and another, a very great deal gets said at a party. Let us hope that this is a good thing. It is apparent, anyhow, that the mere use of the tongue, quite apart from the words it utters, gives pleasure to many. If it gives you no pleasure, and if, further, you derive none from listening to the remarks of others, there is no need to converse. You had better then take up a position in a solitary corner (if possible on a chair, but this is a rare treat) and merely listen to the noise as to a concert, not endeavouring to form out of it sentences. As a matter of fact, if thus listened to, the noise of a party will be found a very interesting noise, containing a great variety of different sounds. If you are of those who like also to look at the clothes of others, you will, from this point of vantage, have a good view of these.

It is very possible, however, that you have only come to the party on the chance of obtaining something good to eat. This is, after all, as good a reason as another. You will, with any luck, be offered some comestible—a sandwich, or a chocolate, or some kind of a drink, or, if you are very fortunate, an ice. With a view to this, you cannot do better than to stand solitary, so that your host or hostess may, in despair of making you talk, give you to eat. If you have eaten or drunk, you have anyhow got something out of the party; you can say, in recalling it, 'I ate two chocolates, and that sandwich pleased me,' or, better still, 'I drank.' Words spoken are empty air, and drift windily into oblivion; and, anyhow, there are greatly too many of these; but about food and drink there is something solid and consoling. An hour in which you have consumed nourishment is seldom an hour spent in vain.

But far be it from me to suggest that we should, or do, take such pains over attiring ourselves, and go to so much trouble, and possibly expense, in travelling from one house to another, merely for the sake of some foolish edible trifle which could be procured and consumed with greater ease in the home. I am convinced that the majority of human creatures do not go to parties for the sake of any food, or even drink, that they may get there. No; the reason (if reason indeed there is beyond blind habit) is, fundamentally, that primitive instinct to take any chance of herding together which led our earliest forefathers to form tribes, village communities, and cities. It is the same reason for which great spaces of the countryside in all lands stand empty, while those who might live there herd, instead, in hideous, shrieking and dreadful cities. It is, in short, the gregarious instinct, based on fear of solitude, on terror of such dangers and uncanny visitants as may, we feel, attack us unless we hide within the crowd. We are a haunted race, fleeing from silence and great spaces, feeling safe only when surrounded by warm, comprehensible, chattering humanity like ourselves. So, when there comes for us a little pasteboard card inscribed with an

address where, and a date when, we may thus surround ourselves, under some hospitable roof, we may say with our minds
and lips, 'Shall I go to this!' casually, as if it mattered not at
all; but deep down in our hidden souls the primal whisper
sounds—'There will be people there. There is safety in a crowd.
Go!'

This is, at least, what I presume occurs in that buried self of
which we know so little. Anyhow, for one reason or another, go
we do, quite often. And if anyone knows of any other reason
why, I should be glad to hear it.

Not that, personally, I do not enjoy parties . . .

                                                        1926

Harriette Wilson

ACROSS the broad continent of a woman's life falls the shadow of a sword. On one side all is correct, definite, orderly; the paths are strait, the trees regular, the sun shaded; escorted by gentlemen, protected by policemen, wedded and buried by clergymen, she has only to walk demurely from cradle to grave and no one will touch a hair of her head. But on the other side all is confusion. Nothing follows a regular course. The paths wind between bogs and precipices. The trees roar and rock and fall in ruin. There, too, what strange company is to be met—in what bewildering variety! Stonemasons hobnob with Dukes of the blood royal—Mr Blore treads on the heels of His Grace the Duke of Argyll. Byron rambles through, the Duke of Wellington marches in with all his orders on him. For in that strange land gentlemen are immune; any being of the male sex can cross from sun to shade with perfect safety. In that strange land money is poured out lavishly; bank-notes drop on to breakfast plates; pearl rings are found beneath pillows; champagne flows in fountains; but over it all broods the fever of a nightmare and the transiency of a dream. The brilliant fade; the great mysteriously disappear; the diamonds turn to cinders, and the Queens are left sitting on three-legged stools shivering in the cold. That great Princess, Harriette Wilson, with her box at the Opera and the Peerage at her feet, found herself before she was fifty reduced to solitude, to poverty, to life in foreign parts, to marriage with a Colonel, to scribbling for cash whatever she could remember or invent of her past.

Nevertheless it would be a grave mistake to think that Harriette repented her ways or would have chosen another career had she had the chance. She was a girl of fifteen when she stepped across the sword and became, for reasons which she will not specify, the mistress of the Earl of Craven. A few facts leak out later. She was educated at a convent and shocked the nuns. Her parents had fifteen children; their home was 'truly uncomfortable'; her father was a Swiss with a passion for mathematics, always on the point of solving a problem, and furious if interrupted; while the unhappiness of her parents' married life had decided Harriette before she was ten 'to live free as air from any restraint but that of my own conscience'. So she stepped across. And at once, the instant her foot touched those shifting sands, everything wobbled; her character, her principles, the world itself—all suffered a sea change. For ever after (it is one of the curiosities of her memoirs—one of the obstacles to any certain knowledge of her character) she is outside the pale of ordinary values and must protest till she is black in the face, and run up a whole fabric of lies into the bargain, before she can make good her claim to a share in the emotions of human kind. Could a mere prostitute grieve genuinely for a mother's death? Mr Thomas Seccombe, in the *Dictionary of National Biography*, had his doubts. Harriette Wilson, he said, described her sister's death 'with an appearance of feeling', whereas to Mr Seccombe Lord Hertford's kindness in soothing the same creature's last hours was indisputably genuine.

Outcast as she was, her position had another and an incongruous result. She was impelled, though nothing was further from her liking than serious thought, to speculate a little curiously about the law of the society, to consult, with odd results, the verdict of 'my own conscience'. For example, the marriage law—was that as impeccably moral as people made out? 'I cannot for the life of me divest myself of the idea that if all were alike honourable and true, as I wish to be, it would be unnecessary to bind men and women together by law, since two persons who may have chosen each other from affection, possessing heart and honour, could not part, and where there is neither the one nor the other, even marriage does not bind. My idea may be wicked or erroneous', she adds hastily, for what could be more absurd than that Harriette Wilson should set herself up as a judge of morality—Harry, as the gentlemen called her, whose only rule of conduct was 'One wants a little variety in life', who left one man because he bored her, and another because he drew pictures of cocoa-trees on vellum paper, and seduced poor young Lord Worcester, and went off to Melton Mowbray with Mr

Meyler, and, in short, was the mistress of any man who had money and rank and a person that took her fancy? No, Harriette was not moral, nor refined, nor, it appears, very beautiful, but merely a bustling bouncing vivacious creature with good eyes and dark hair and 'the manners of a wild schoolboy', said Sir Walter Scott, who had dined in her presence. But it cannot be doubted—otherwise her triumph is inexplicable—that gifts she had, gifts of dash and go and enthusiasm, which still stir among the dead leaves of her memoirs and impart even to their rambling verbosity and archness and vulgarity some thrill of that old impetuosity, some flash of those fine dark eyes, some fling of those wild schoolboy manners which, when furbished up in plumes and red plush and diamonds, held our ancestors enthralled.

She was, of course, always falling in love. She saw a stranger riding with a Newfoundland dog in Knightsbridge and lost her heart to his 'pale expressive beauty' at once. She venerated his door-knocker even, and when Lord Ponsonby—for Lord Ponsonby it was—deserted her, she flung herself sobbing on a doorstep in Half Moon Street and was carried, raving and almost dying, back to bed. Large and voluptuous herself, she loved for the most part little men with small hands and feet, and, like Mr Meyler, skins of remarkable transparency, 'churchyard skins', foreboding perhaps an early death; 'yet it would be hard to die, in the bloom of youth and beauty, beloved by everybody, and with thirty thousand a year'. She loved, too, the Apollo Belvedere, and sat entranced at the Louvre, exclaiming in ecstasy at the 'quivering lips—the throat!', till it seemed as if she must share the fate of another lady who sat by Apollo, 'whom she could not warm, till she went raving mad, and in that state died'. But it is not her loves that distinguish her; her passions tend to become perfunctory; her young men with fine skins and large fortunes innumerable; her rhapsodies and recriminations monotonous. It is when off duty, released from the necessity of painting the usual picture in the usual way, that she becomes capable of drawing one of those pictures which only seem to await some final stroke to become a page in *Vanity Fair* or a sketch by Hogarth. All the materials of comedy seem heaped in disorder before us as she, the most notorious woman in London, retires to Charmouth to await the return of her lover, Lord Worcester, from the Spanish wars, trots to church on the arm of the curate's aged father, or peeps from her window at the rustic beauties of Lyme Regis tripping down to the sixpenny Assembly Rooms with 'turbans or artificial flowers twined around their wigs' to dance at five in the evening on the shores of the innocent sea. So a famous prima donna, hidden behind a curtain in strict incognito, might listen to country girls singing a rustic ballad with contempt and amusement, and a dash of envy too, for how simply the good people accepted her. Harriette could not help reflecting how kindly they sympathized with her anxiety about her husband at the wars, and sat up with her to watch for the light of the postwoman's lanthorn as she came late at night over the hill from Lyme Regis with letters from Mr Wilson in Spain! All she could do to show her gratitude was to pay twice what they asked her, to shower clothes upon ragged children, to mend a poor country-woman's roof, and then, tired of the role of Lady Bountiful, she was off to join Lord Worcester in Spain.

Now, for a moment, before the old story is resumed, sketched with a stump of rapid charcoal, springs into existence, to fade for ever after, the figure of Miss Martha Edmonds, her landlady's sister. 'I am old enough,' exclaimed the gallant old maid, 'and thank God I am no beauty.... I have never yet been ten miles from my native place, and I want to see the world.' She declared her intention of escorting Mrs Wilson to Falmouth; she had her ancient habit made up for the purpose. Off they started, the old maid and the famous courtesan, to starve and freeze in an upper room of a crowded Falmouth inn, the winds being adverse, until in some mysterious way Mrs Wilson got into touch first with the Consul and then with the Captain, who were so hospitable, so generous, so kind, that Aunt Martha bought a red rose for her cap, drank champagne, took a hand at cards, and was taught to waltz by Mr Brown. Their gaieties were cut short, however; a letter demanded Mrs Wilson's instant presence in London, and Aunt Martha, deposited in Charmouth, could only regret that she had not seen something of life a little sooner, and declare that there 'was a boldness and grandeur about the views in Cornwall which far exceeded anything she had seen in Devonshire'.

Involved once more with Meylers, Lornes, Lambtons, Berkeleys, Leicesters, gossiping as usual in her box at the Opera about this lady and that gentleman, letting young noblemen pull her

hair, tapping late at night at Lord Hertford's little private gate in
Park Lane, Harriette's life wound in and out among the bogs
and precipices of the shadowy underworld which lies on the far
side of the sword. Occasionally the jingling and junketing was
interrupted by a military figure; the great Duke himself, very like
a rat-catcher in his red ribbon, marched in; asked questions; left
money; said he remembered her; had dreamed of her in Spain. 'I
dreamed you came out on my staff,' he said. Or there was Lord
Byron sitting entirely alone, dressed in brown flowing robes at a
masquerade, 'bright, severe, beautiful', demanding 'in a tone of
wild and thrilling despondency "Who shall console us for acute
bodily anguish?"' Or again the spangled curtain goes up and we
see those famous entertainers the sisters Wilson sitting at home
at their ease, sparring and squabbling and joking about their
lovers; Amy, who adored black puddings; good-natured Fanny,
who doted upon donkey-riding; foolish Sophie, who was made a
Peeress by Lord Berwick and dropped her sisters; Moll Raffles,
Julia, niece to Lord Carysfort and daughter to a maid of honour
with the finest legs in Europe—there they sit gossiping profanely
and larding their chatter with quotations from Shakespeare and
Sterne. Some died prematurely; some married and turned virtuous; some became villains, sorceresses, serpents, and had best be
forgotten; while as for Harriette herself, she was scandalously
treated by the Beauforts, had to retire to France with her
Colonel, would continue to tell the truth about her fine friends
so long as they treated her as they did, and grew, we cannot
doubt, into a fat good-humoured disreputable old woman who
never doubted the goodness of God or denied that the world
had treated her well, or regretted, even when the darkness of
obscurity and poverty blotted her entirely from view, that she
had lived her life on the shady side of the sword.

1925

The Death of the Moth

MOTHS that fly by day are not properly to be called moths; they do not excite that pleasant sense of dark autumn nights and ivy-blossom which the commonest yellow-underwing asleep in the shadow of the curtain never fails to rouse in us. They are hybrid creatures, neither gay like butterflies nor sombre like their own species. Nevertheless the present specimen, with his narrow hay-coloured wings, fringed with a tassel of the same colour, seemed to be content with life. It was a pleasant morning, mid-September, mild, benignant, yet with a keener breath than that of the summer months. The plough was already scoring the field opposite the window, and where the share had been, the earth was pressed flat and gleamed with moisture. Such vigour came rolling in from the fields and the down beyond that it was difficult to keep the eyes strictly turned upon the book. The rooks too were keeping one of their annual festivities; soaring round the tree tops until it looked as if a vast net with thousands of black knots in it had been cast up into the air; which, after a few moments sank slowly down upon the trees until every twig seemed to have a knot at the end of it. Then, suddenly, the net would be thrown into the air again in a wider circle this time, with the utmost clamour and vociferation, as though to be thrown into the air and settle slowly down upon the tree tops were a tremendously exciting experience.

The same energy which inspired the rooks, the ploughmen, the horses, and even, it seemed, the lean bare-backed downs, sent the moth fluttering from side to side of his square of the window-pane. One could not help watching him. One was, indeed, conscious of a queer feeling of pity for him. The possibilities of pleasure seemed that morning so enormous and so various that to have only a moth's part in life, and a day moth's at that, appeared a hard fate, and his zest in enjoying his meagre opportunities to the full, pathetic. He flew vigorously to one corner of his compartment, and, after waiting there a second, flew across to the other. What remained for him but to fly to a third corner and then to a fourth? That was all he could do, in spite of the size of the downs, the width of the sky, the far-off smoke of houses, and the romantic voice, now and then, of a steamer out at sea. What he could do he did. Watching him, it seemed as if a fibre, very thin but pure, of the enormous energy of the world had been thrust into his frail and diminutive body. As often as he crossed the pane, I could fancy that a thread of vital light became visible. He was little or nothing but life.

Yet, because he was so small, and so simple a form of the energy that was rolling in at the open window and driving its way through so many narrow and intricate corridors in my own brain and in those of other human beings, there was something marvellous as well as pathetic about him. It was as if someone had taken a tiny bead of pure life and decking it as lightly as possible with down and feathers, had set it dancing and zigzagging to show us the true nature of life. Thus displayed one could not get over the strangeness of it. One is apt to forget all about life, seeing it humped and bossed and garnished and cumbered so that it has to move with the greatest circumspection and dignity. Again, the thought of all that life might have been had he been born in any other shape caused one to view his simple activities with a kind of pity.

After a time, tired by his dancing apparently, he settled on the window ledge in the sun, and, the queer spectacle being at an end, I forgot about him. Then, looking up, my eye was caught by him. He was trying to resume his dancing, but seemed either so stiff or so awkward that he could only flutter to the bottom of the window-pane; and when he tried to fly across it he failed. Being intent on other matters I watched these futile attempts for a time without thinking, unconsciously waiting for him to resume his flight, as one waits for a machine, that has stopped momentarily, to start again without considering the reason of its failure. After perhaps a seventh attempt he slipped from the wooden ledge and fell, fluttering his wings, on to his back on the window sill. The helplessness of his attitude roused me. It flashed upon me that he was in difficulties; he could no longer raise himself; his legs struggled vainly. But, as I stretched out a pencil, meaning to help him to right himself, it came over me that the failure and awkwardness were the approach of death. I laid the pencil down again.

The legs agitated themselves once more. I looked as if for the enemy against which he struggled. I looked out of doors. What had happened there? Presumably it was midday, and work in the fields had stopped. Stillness and quiet had replaced the previous animation. The birds had taken themselves off to feed in the brooks. The horses stood still. Yet the power was there all the same, massed outside indifferent, impersonal, not attending to anything in particular. Somehow it was opposed to the little hay-coloured moth. It was useless to try to do anything. One could only watch the extraordinary efforts made by those tiny legs against an oncoming doom which could, had it chosen, have submerged an entire city, not merely a city, but masses of human beings; nothing, I knew, had any chance against death. Nevertheless after a pause of exhaustion the legs fluttered again. It was superb this last protest, and so frantic that he succeeded at last in righting himself. One's sympathies, of course, were all on the side of life. Also, when there was nobody to care or to know, this gigantic effort on the part of an insignificant little moth, against a power of such magnitude, to retain what no one else valued or desired to keep, moved one strangely. Again, somehow, one saw life, a pure bead. I lifted the pencil again, useless though I knew it to be. But even as I did so, the unmistakable tokens of death showed themselves. The body relaxed, and instantly grew stiff. The struggle was over. The insignificant little creature now knew death. As I looked at the dead moth, this minute wayside triumph of so great a force over so mean an antagonist filled me with wonder. Just as life had been strange a few minutes before, so death was now as strange. The moth having righted himself now lay most decently and uncomplainingly composed. O yes, he seemed to say, death is stronger than I am.

Published posthumously, 1942

JAMES STEPHENS

Finnegans Wake

I WOULD call *Finnegans Wake* Joyce's autobiography: factual, imaginative, spiritual, and all curiously disguised; for Joyce was a secretive man, as we all are, and we all tell what we do tell with some precaution.

Sometimes I think that when you are discussing a book you had better get rid of the author. Where would Hamlet be if Shakespeare was hanging around? Or if you read *Pickwick* and have Dickens in mind, you are at home with neither of these curious extremes of each other. Every soul in that book is Dickens and Every Man.

There are a number of things to say about Joyce, for he was always a little different from what he was—that is the definition of an author. One says of him that he was an Irishman, a writer, a Catholic and a linguist; but these permit of a narrower definition in each case. He was a writer; that is, he was a prose-writer. He was a Catholic; that is, he was trained in the Jesuit System. He was an Irishman; that is, he was a Leinster man—which is, of course, the top of the world and the only animal worth being. And he was a linguist; that is, he preferred City of Dublin English to any other lingo or pigeon that ever came his way. He loved the City of Dublin with a passion which was both innocent and wicked, the way that every passion is; for if it isn't both of these, it is just animal and stupid; and that Dublin love-affair is responsible for this book.

And then there is a curious fact about *Finnegans Wake*. Every other prose book is written in prose. This book is written in speech.

Speech and prose are not the same thing. They have different wave-lengths, for speech moves at the speed of light, where prose moves at the speed of the alphabet, and must be consecut-

ive and grammatical and word-perfect. Prose cannot gesticulate.
Speech can sometimes do nothing else.

Finnegans Wake is all speech. Now it is soliloquy; now it is dialogue; it becomes at times oration and tittle-tattle and scandal,
but it is always a speech, and however it be punned upon by all
the European and a few of the Asiatic tongues, it is fundamentally the speech that used to be Dublin-English.

Consecutiveness and such-like doesn't quite matter to this
speech; it hops and skips and jumps at its own sweet will; it is
extraordinarily varied and sportive, and even when it is serious it
isn't as serious as all that, for it easily makes up in abundance and
exuberance for all that it lacks in meaning. The meaning isn't
lacking, but it isn't meaning as the crow flies; 'tis, rather, meaning as the bee bumbles: honey here and honey there and
heather-honey on the mountain.

Where he liked he disliked a bit; where he disliked he liked
also. He rather dislikingly liked everything that happened. This
extraordinary prose-poet, Leinster Jesuit, and Liffey linguist
loved Joyce and Dublin. He was so almost a pessimist by limitation, but the English language doesn't permit more than a
spoonful of pessimism—to be well shaken before taken—for
it isn't built that way; and we can only be pessimists before
breakfast.

1947

G. M. YOUNG

The Greatest Victorian

THOSE who have had occasion to adjudicate at some country festival are aware of an embarrassing difficulty which comes upon them half-way through. The first arrangement of red, white and blue strikes one as very pretty and apt to the occasion. But is the fiftieth better or worse than the tenth? Are there no colours in the world but red, white and blue? And, by the hundredth, one is in a mood to award the prize, with a hearty ejaculation of relief, to anyone who had the originality to display the Swastika or the Hammer and Sickle. I have been turning over in my mind the names of some who might be candidates for the title of the Greatest Victorian. One needs a man, or woman, who is typical of a large and important class: rich in the abilities which the age fostered: one who made a difference, and under whose influence or direction we are still living. These being the notes by which posterity, looking back, recognizes the really great men of a former time, in whom, among the remembered figures of the Victorian Age are they best exemplified? And I wondered if I might include Karl Marx. He had two of the qualifications which, perhaps unfairly, one associates with eminence in that time. He once had to pawn his spoons, and he was buried in Highgate Cemetery.

Fifty years ago a large body of intelligent persons, headed by Lord Acton, would have adjudged the place to George Eliot. She had genius of the blend, humorous, observant, didactic, which that age most appreciated: she had raised herself to the head of the literary profession: her gnomic wisdom was a light to thousands who had learnt from her that, in the darkness deepening over all ancient faiths, the star of Duty shone clearer than ever. Grave and wise men thought that George Eliot had, single-handed, by her ethical teaching, saved us from the moral catastrophe which might have been expected to follow upon the waning of religious conviction. They were not altogether wrong. George Eliot did give body, and expression, to a great volume of moral thought necessary to her time. In so doing she shaped a generation, and through that generation something of her influence is still at work in an age which knows as much of her writings as it may catch sight of displayed on the Wayside Pulpit.

Tennyson, 'illustrious and consummate', is a strong candidate. We have outgrown the days when it was possible to pretend that he was not a great poet: but what strikes me now, and what explains the hold he had upon his age, is the dexterity, the almost journalistic address, with which his poetry follows and records its intellectual moment, putting all the questions which the advance of science was forcing it to ask, and indicating answers with which, in the general confusion of faith, it might, with some allowance, be contented. But on the test of lasting direction, the claim must be disallowed.

To be typically great, a man must be, as Tennyson was, profoundly in sympathy with the chief preoccupations of his time; and the preoccupations of the Victorian mind, the points to which it swung most constantly and anxiously, were on the one side theological and moral, on the other social. Something therefore might be said for Matthew Arnold, in whose admirably clear intelligence both were in due subordination to the higher and more permanent rights of culture: much more for Ruskin, and on Ruskin's claims one must pause carefully and long. If the test of influence were solely to be applied, then the title would go beyond doubt to Darwin. His work, however, belongs to the isolated and timeless world of pure scientific speculation; he only happened to be a subject of Victoria as Pasteur happened to be a subject of Napoleon III. But what other age or country could have fostered the genius of Ruskin or given it such a field to work in? And of Ruskin it may, I think, with truth be said that, using no doubt the reflections of other men and their experience, absorbing for example all of Carlyle that is really Victorian, and taking not a little of Maurice and Kingsley, he evolved, and forced his world to accept, a new set of axioms as the basis of all future political science in England.

If anyone reckons this claim too bold, I would ask him to consider it thus. Let him first call up the world of political and social thought in 1837; the atomism, the individualism, the economic determinism from which the young intelligence of Disraeli so violently recoiled. Let him compare it with the common assumptions of our own time as they disclose themselves in our legislation and administration. Then, taking his stand almost midway, let him read the address of the economists to Ruskin in 1885: and their acknowledgement that the world of thought in which they now moved, 'where Political Economy can furnish sound laws of life and work only when it respects the dignity and moral destiny of man: and the wise use of wealth, in developing a complete human life, is of incomparably greater moment both to men and actions than its production or accumulation, and can alone give these any vital significance', was a world of his making.

But Ruskin is too fantastic, too childlike, too incoherent, to be typical of an age which loved solidity and efficiency—in politics, for example, the efficiency of Peel and Gladstone, in science of Faraday, in controversy of Huxley: and was always a little dubious and distrustful of genius, like that of Browning, or Newman, not precisely to be specified. And I am not sure that the great and world-known figures are before us to be judged: they are at the Coronation, we are holding our village feast. We are looking for a man who was in and of his age, and who could have been of no other: a man with sympathy to share, and genius to judge, its sentiments and movements: a man not too illustrious or too consummate to be companionable, but one, nevertheless, whose ideas took root and are still bearing; whose influence, passing from one fit mind to another, could transmit, and can still impart, the most precious element in Victorian civilization, its robust and masculine sanity. Such a man there was: and I award the place to Walter Bagehot.

I do not assert that Bagehot was the greatest man alive and working between 1837 and 1901: I am not sure that the statement would mean anything: and I agree that the landscape of that age is a range of varied eminences with no dominating peak. Indeed, in a footnote to my *Portrait*, which somehow got lost in the proofs, I suggested that anyone who wished to understand the Victorian mind should turn away from the remembered names and survey the careers of three men: Whitwell Elwin, Alderman Thomasson of Bolton, and Charles Adderley, first Lord Norton: reflecting, as he went, on the breadth of their interests, from sound prose to sound religion, and from town planning to Imperial policy, and the quiet and substantial permanence of what they did. It is along this level that we must look, to find 'if not the greatest, at least the truest' Victorian. As I looked, my eye fell on Walter Bagehot and there it has stayed. *Victorianorum maximus*, no. But *Victorianum maxime* I still aver him to be.

Of the Victorian mind, by which I mean the kind of intelligence that one learns to look for and recognize in the years of his maturity, say, from 1846 when he was twenty to 1877 when he died, the characteristics that most impress me are capaciousness and energy. It had room for so many ideas, and it threw them about as lustily as a giant baby playing skittles. The breadth and vigour of Bagehot's mind appear on every page he has left, and they were, we know, not less conspicuous in his conversation and the conduct of affairs. But what was peculiarly his own was the perfect management of all this energy and all these resources. He was as well aware of his superiority in intelligence as Matthew Arnold of his superiority in culture. But he carried it with such genial and ironic delight, that his influence—and he was through the *Economist* and the Reviews a very influential man—encountered no resistance. His paradoxes became axioms: and there are thousands of people thinking and even speaking Bagehot to-day, who might be hard put to it to say when exactly he lived and what exactly he did. Let me give an illustration:

If one makes a close study of a society different from one's own, one finds that institutions the very opposite of one's own are defended by the people to whom they belong with as much fervour as that with which we defend ours. They do not seek to be delivered from them and endowed with something better. Self-government, in fact, does not mean responsible government: it means government by the authority you have been brought up to respect, whom you obey readily because you as well as he take the obedience for granted, who is hallowed by all the dignity of tradition and religious belief and is a symbol of national pride and achievement. Above all in a period of rapid change such as is confronting men to-day, the preservation of such continuity with the past, with the standards they are used to, and the social world where they can find their way about, is essential if the transition is to be effected without producing mere confusion and chaos.

That is pure Bagehot. Observe the psychological realism

which is concerned only to discover how men in societies actually do behave, and the unpretentious colloquialism of the style.
No one ever thought or wrote quite like that before: and it
contains the gist of the famous doctrine, which he first propounded, with much youthful flippancy, after observing the *coup
d'état* of 1851, and restated more gravely in *Physics and Politics*,
that the surest guarantee of stability and freedom in a State is
'stupidity', or the general habit of identical response. And to-
day, will anyone deny it? But the odd thing, to use a common
phrase of his, is that the passage I have quoted is not Bagehot at
all. It is Dr Lucy Mair, speaking in the later Thirties on the
administration of Tanganyika.

But Bagehot was no lonely thinker, anticipating the commonplaces of another age. He was as thoroughly immersed in the
Victorian matter as the most pugnacious, self-satisfied, dogmatic
business man of his day. In his profession as banker, economist
and editor he was highly successful, his word carried equal
weight in Threadneedle Street and Downing Street. He could
even write verses, beginning (or ending, I forget which)

Thou Church of Rome!

and it was his affectionate and humorous interest in all the
doings of his time that furnished him with the material of his
philosophy. Of Macaulay he acutely says that he lacked 'the
experiencing mind'. Bagehot's mind was always experiencing,
and always working its observation into pattern, into system,
but—and here we touch on his central excellence or virtue—into
a system open towards the future. He distrusted swift, unreflect-
ing action. Equally he distrusted all closed, dogmatic combinations: here picking up the true English tradition which the
Radicals had done their best to sever, the tradition of Burke:

When he forewarns, denounces, launches forth
Against all systems built on abstract rights
Keen ridicule: the majesty proclaims
Of Institutes and Laws, hallowed by time:
Declares the vital power of social ties
Endeared by Custom; and with high disdain
Exploding upstart Theory, insists
Upon the allegiance to which men are born,

—which laborious flight of Wordsworthian eloquence Bagehot would probably have countered with his favourite 'How much?' Uncorrected, this insistence on habit leads to an unthinking Liverpudlian conservatism, and Bagehot was a Liberal. What, then, is the correction? In his answer, I confess I see no flaw, and I think that the experience of sixty years has established its truth and disclosed its profundity. People do like splendour, distinction, and authority in their rulers. This is their natural allegiance. Very well; then see to it that the allegiance of the rulers themselves is rightly directed. And to what? You will find the answer in a brief paper published in 1871, and called 'The Emotion of Conviction'. And if there be in English a more 'wholesome doctrine or necessary for these times' than is contained in the last pages of that essay, I must own it has escaped me.

1938

Insouciance

My balcony is on the east side of the hotel, and my neighbours on the right are a Frenchman, white-haired, and his white-haired wife; my neighbours on the left are two little white-haired English ladies. And we are all mortally shy of one another.

When I peep out of my room in the morning and see the matronly French lady in a purple silk wrapper, standing like the captain on the bridge surveying the morning, I pop in again before she can see me. And whenever I emerge during the day, I am aware of the two little white-haired ladies popping back like two white rabbits, so that literally I only see the whisk of their skirt-hems.

This afternoon being hot and thundery, I woke up suddenly and went out on the balcony barefoot. There I sat serenely contemplating the world, and ignoring the two bundles of feet of the two little ladies which protruded from their open doorways, upon the end of the two *chaises longues*. A hot, still afternoon! the lake shining rather glassy away below, the mountains rather sulky, the greenness very green, all a little silent and lurid, and two mowers mowing with scythes, downhill just near: *slush! slush!* sound the scythe-strokes.

The two little ladies become aware of my presence. I become aware of a certain agitation in the two bundles of feet wrapped in two discreet steamer rugs and protruding on the end of two *chaises longues* from the pair of doorways upon the balcony next me. One bundle of feet suddenly disappears; so does the other. Silence!

Then lo! with odd sliding suddenness a little white-haired lady in grey silk, with round blue eyes, emerges and looks straight at me, and remarks that it is pleasant now. A little cooler, say I, with false amiability. She quite agrees, and we speak of the men mowing; how plainly one hears the long breaths of the scythes!

By now we are *tête-à-tête*. We speak of cherries, strawberries, and the promise of the vine crop. This somehow leads ·to Italy, and to Signor Mussolini. Before I know where I am, the little white-haired lady has swept me off my balcony, away from the glassy lake, the veiled mountains, the two men mowing, and the cherry trees, away into the troubled ether of international politics.

I am not allowed to sit like a dandelion on my own stem. The little lady in a breath blows me abroad. And I was so pleasantly musing over the two men mowing: the young one, with long legs in bright blue cotton trousers, and with bare black head, swinging so lightly downhill, and the other, in black trousers, rather stout in front, and wearing a new straw hat of the boater variety, coming rather stiffly after, crunching the end of his stroke with a certain violent effort.

I was watching the curiously different motions of the two men, the young thin one in bright blue trousers, the elderly fat one in shabby black trousers that stick out in front, the different amount of effort in their mowing, the lack of grace in the elderly one, his jerky advance, the unpleasant effect of the new 'boater' on his head—and I tried to interest the little lady.

But it meant nothing to her. The mowers, the mountains, the cherry trees, the lake, all the things that were *actually* there, she didn't care about. They even seemed to scare her off the balcony. But she held her ground, and instead of herself being scared away, she snatched me up like some ogress, and swept me off into the empty desert spaces of right and wrong, politics, Fascism and the rest.

The worst ogress couldn't have treated me more villainously. I don't care about right and wrong, politics, Fascism, abstract liberty or anything else of the sort. I want to look at the mowers, and wonder why fatness, elderliness and black trousers should inevitably wear a new straw hat of the boater variety, move in stiff jerks, shove the end of the scythe-stroke with a certain violence, and win my hearty disapproval, as contrasted with young long thinness, bright blue cotton trousers, a bare black head, and a pretty lifting movement at the end of the scythe-stroke.

Why do modern people almost invariably ignore the things that are actually present to them? Why, having come out from

England to find mountains, lakes, scythe-mowers and cherry trees, does the little blue-eyed lady resolutely close her blue eyes to them all, now she's got them, and gaze away to Signor Mussolini, whom she hasn't got, and to Fascism, which is invisible anyhow? Why isn't she content to be where she is? Why can't she be happy with what she's got? Why must she *care*?

I see now why her round blue eyes are so round, so noticeably round. It is because she 'cares.' She is haunted by that mysterious bugbear of 'caring.' For everything on earth that doesn't concern her she 'cares.' She cares terribly because far-off, invisible, hypothetical Italians wear black shirts, but she doesn't care a rap that one elderly mower whose stroke she can hear, wears black trousers instead of bright blue cotton ones. Now if she would descend from the balcony and climb the grassy slope and say to the fat mower: '*Cher monsieur, pourquoi portez-vous les pantalons noirs*? Why, oh, why do you wear black trousers?'—then I should say: What an on-the-spot little lady!—But since she only torments me with international politics, I can only remark: What a tiresome off-the-spot old woman!

They care! They simply are eaten up with caring. They are so busy caring about Fascism or Leagues of Nations or whether France is right or whether Marriage is threatened, that they never know where they are. They certainly never live on the spot where they are. They inhabit a abstract space, the desert void of politics, principles, right and wrong, and so forth. They are doomed to be abstract. Talking to them is like trying to have a human relationship with the letter *x* in algebra.

There simply is a deadly breach between actual living and this abstract caring. What is actual living? It is a question mostly of direct contact. There was a direct sensuous contact between me, the lake, mountains, cherry trees, mowers, and a certain invisible but noisy chaffinch in a clipped lime tree. All this was cut off by the fatal shears of that abstract word Fascism, and the little old lady next door was the Atropos who cut the thread of my actual life this afternoon. She beheaded me, and flung my head into abstract space. Then we are supposed to love our neighbours!

When it comes to living, we live through our instincts and our intuitions. Instinct makes me run from little over-earnest ladies; instinct makes me sniff the lime blossom and reach for the darkest cherry. But it is intuition which makes me feel the uncanny

glassiness of the lake this afternoon, the sulkiness of the mountains, the vividness of near green in thunder-sun, the young man
in bright blue trousers lightly tossing the grass from the scythe,
the elderly man in a boater stiffly shoving his scythe-strokes,
both of them sweating in the silence of the intense light.

1928

What There is to See at the Zoo

THE peacock spreads his tail, and the nearly circular eyes at regular intervals in the fan are a sight at which to marvel—forming a lacework of white on more delicate white if the peacock is a white one; of indigo, lighter blue, emerald and fawn if the peacock is blue and green.

Look at a tiger. The light and dark of his stripes and the black edge encircling the white patch on his ear help him to look like the jungle with flecks of sun on it. In the way of color, we rarely see a blacker black than tiger stripes, unless it is the black body down of the blue bird of paradise.

Tiger stripes have a merely comparative symmetry beside the almost exact symmetry of a Grevy's zebra. The small lines on one side of the zebra's face precisely match those on the other side, and the small sock stripes on one front leg are an exact duplicate of those on the other front leg.

Although a young giraffe is also an example of 'marking,' it is even more impressive as a study in harmony and of similarities that are not monotony—of sycamore-tree white, beside amber and topaz yellow fading into cream. The giraffe's togue is violet; his eyes are a glossy cider brown. No wonder Thomas Bewick (pronounced Buick, like the car), whose woodcuts of birds and animals are among the best we have, said, 'If I were a painter, I would go to nature for all my patterns.'

Such colors and contrasts educate the eye and stir the imagination. They also demonstrate something of man's and the animals' power of adaptation to environment, since differing surroundings result in differences of appearance and behavior.

The giraffe grows to the height of certain trees that it may reach its leafy food. David Fleay, an authority on Australian wild life, tells us that the lyrebird 'has a very large eye that it may see [grubs] in the dim light of the tree-fern gullies in which it lives.'

Certain chameleons have an eye that revolves in its socket, as some searchlights turn on a revolving swivel, in order to look forward and back.

The bodies of sea lions, frogs, and eels are streamlined so that they can slip through the water with the least possible effort. Living almost entirely in the water, an alligator is shaped like a boat and propels itself by its tail as if it were feathering a sculling oar.

The elephant has an inconsequential tail, but its long nose, or trunk, has the uses of a hand as well as the power of a battering-ram. It can pull down branches for food or push flat the trees that block its progress through the jungle. Helen Fischer, in her photo series 'The Educated Elephants of Thailand,' shows how 'up and onto the waiting truck, an elephant maneuvers a heavy log as easily as we would a piece of kindling.' Then 'after work, it wades and splashes in a cool stream.'

An elephant can use its trunk to draw up water and shower its back or to hose an intruder. With the finger at the end of its trunk, the elephant can pluck grass that has overgrown a paved walk, leaving a line as even as if sheared by man. It can pick up a coin and reach it up to the rider on its back—its mahout (ma-howt', as he is called in India). What prettier sight is there than the parabola described by an elephant's trunk as it spirals a banana into its mouth?

A certain gorilla at the Central Park Zoo in New York sometimes takes a standing leap to her broad trapeze. She sits there, swinging violently for a time, and then suddenly drops without a jar—indeed, descends as lightly as a feather might float to the ground. Walking through the monkey house at the Bronx Zoo, we stop before the cage of an orangutan as he jumps to his lead-pipe trapeze with half an orange in one hand and a handful of straw in the other. He tucks the wisp of hay under his neck and, lying on his back as contentedly as if at rest in a hammock, sucks at the orange from time to time—an exhibition of equilibrium that is difficult to account for.

The gorilla's master feat—the standing leap to a swing the height of her head—is matched by the pigeon when it flies at full speed, stops short, pauses and without a detour flies back in the direction from which it came. At dusk, four or five impalas will timidly emerge from their shelter, then bound through the air, in a succession of twenty-foot leaps, to the end of their runway. Perhaps Clement Moore had seen or heard of impalas and was thinking of them when, in *A Visit from St Nicholas*, he wrote of Santa Claus' reindeer skimming the housetops.

The swimmer has a valuable lesson in muscular control as he watches a sea lion round the curve of its pool, corkscrewing in a spiral as it changes from the usual position to swim upside down. Hardening-up exercises in military training, with obstacles to surmount and ditches to clear, involve skills neatly mastered by animals. In the wilds, bands of gibbons swing from tree to tree as army trainees swing by ropes or work along the bars of a jungle-gym.

Animals are 'propelled by muscles that move their bones as levers, up and down or from side to side.' The ways in which the movements of their muscles vary provide an ever fascinating sight. The motions of animals are so rapid that we really need the aid of an expert such as James Gray to analyze them for us. In his book *The Motions of Animals*, Mr Gray says that the bear—a browser, not a runner—rests on the entire foot when walking. The horse and the deer—built for speed—rest on tiptoe (the hoof); the hock never touches the ground.

An essential rule of safe living is well illustrated by animals: work when you work, play when you play, and rest when you rest. Watch two young bears wrestling, rolling, pushing and attacking. One tires, climbs to a broad rock and stretches out full length on its paws. The other stands up, strains forward till it can reach with its mouth the ear of the bear on the rock and keeps tugging at the ear as though dragging a hassock forward by the ear. The rester gets up, comes down and once more both are tumbling, capsized and capsizing.

There is nothing more concentrated than the perseverance with which a duck preens its feathers or a cat washes its fur. The duck spreads oil on its feathers with its beak from a small sac above the tail. The feathers then lie smooth and waterproof, reminding us that we too must take time to care for our bodies and equipment. For as much as fifteen minutes at a time, a leopard will, without digressing to another area, wash a small patch of fur that is not sleek enough to satisfy it. It may then leap to its shelf, a board suspended by rods from the ceiling of the cage. Dangling a foreleg and a hindleg on either side of the shelf, its tail hanging motionless, the leopard will close its eyes and rest.

Patience on the part of animals is self-evident. In studying, photographing or rearing young animals, human beings also need patience. We have in Helen Martini a thrilling example of what may be done for young animals by a human being. Mrs Martini has reared two sets of tiger cubs, a lion cub and various other baby animals for the Bronx Zoo.

The zoo shows us that privacy is a fundamental need of all animals. For considerable periods, animals in the zoo will remain out of sight in the quiet of their dens or houses. Glass, recently installed in certain parts of the snake house at the Bronx Zoo makes it possible to see in from the outside, but not out from the inside.

We are the guests of science when we enter a zoo; and, in accepting privileges, we incur obligations. Animals are masters of earth, air and water, brought from their natural surroundings to benefit us. It is short-sighted, as well as ungrateful, to frighten them or to feed them if we are told that feeding will harm them. If we stop to think, we will always respect chains, gates, wires or barriers of any kind that are installed to protect the animals and to keep the zoo a museum of living marvels for our pleasure and instruction.

1955

Marie Lloyd

IT requires some effort to understand why one person, among many who do a thing with accomplished skill, should be greater than the others; and it is not always easy to distinguish superiority from great popularity, when the two go together. Although I have always admired the genius of Marie Lloyd I do not think that I always appreciated its uniqueness; I certainly did not realize that her death would strike me as the important event that it was. Marie Lloyd was the greatest music-hall artist of her time in England: she was also the most popular. And popularity in her case was not merely evidence of her accomplishment; it was something more than success. It is evidence of the extent to which she represented and expressed that part of the English nation which has perhaps the greatest vitality and interest.

Among all of that small number of music-hall performers, whose names are familiar to what is called the lower class, Marie Lloyd had far the strongest hold on popular affection. The attitude of audiences toward Marie Lloyd was different from their attitude toward any other of their favourites of that day, and this difference represents the difference in her art. Marie Lloyd's audiences were invariably sympathetic, and it was through this sympathy that she controlled them. Among living music-hall artists none can better control an audience than Nellie Wallace. I have seen Nellie Wallace interrupted by jeering or hostile comment from a boxful of Eastenders; I have seen her, hardly pausing in her act, make some quick retort that silenced her tormentors for the rest of the evening. But I have never known Marie Lloyd to be confronted by this kind of hostility; in any case, the feeling of the vast majority of the audience was so manifestly on her side, that no objector would have dared to lift his voice. And the difference is this: that whereas other comedians amuse their audiences as much and sometimes more than Marie Lloyd, no other comedian succeeded so well in giving expression to the life of that audience, in raising it to a kind of art. It was, I think, this capacity for expressing the soul of the people that made Marie Lloyd unique, and that made her audiences, even when they joined in the chorus, not so much hilarious as happy.

In the details of acting Marie Lloyd was perhaps the most perfect, in her own style, of British actresses. There are no cinema records of her; she never descended to this form of moneymaking; it is to be regretted, however, that there is no film of her to preserve for the recollection of her admirers the perfect expressiveness of her smallest gestures. But it is less in the accomplishment of her act than in what she made it, that she differed from other comedians. There was nothing about her of the grotesque; none of her comic appeal was due to exaggeration; it was all a matter of selection and concentration. The most remarkable of the survivors of the music-hall stage, to my mind, are Nellie Wallace and Little Tich; but each of these is a kind of grotesque; their acts are an orgy of parody of the human race. For this reason, the appreciation of these artists requires less knowledge of the environment. To appreciate, for instance, the last turn in which Marie Lloyd appeared, one ought to know what objects a middle-aged woman of the charwoman class would carry in her bag; exactly how she would go through her bag in search of something; and exactly the tone of voice in which she would enumerate the objects she found in it. This was only part of the acting in Marie Lloyd's last song, 'One of the Ruins that Cromwell Knocked Abaht a Bit'.

Marie Lloyd's art will, I hope, be discussed by more competent critics of the theatre than I. My own chief point is that I consider her superiority over other performers to be in a way a moral superiority: it was her understanding of the people and sympathy with them, and the people's recognition of the fact that she embodied the virtues which they genuinely most respected in private life, that raised her to the position she occupied at her death. And her death is itself a significant moment in English history. I have called her the expressive figure of the lower classes. There is no such expressive figure for any other class. The middle classes have no such idol: the middle classes

are morally corrupt. That is to say, their own life fails to find a
Marie Lloyd to express it; nor have they any independent virtues
which might give them as a conscious class any dignity. The
middle classes, in England as elsewhere, under democracy, are
morally dependent upon the aristocracy, and the aristocracy are
subordinate to the middle class, which is gradually absorbing
and destroying them. The lower class still exists; but perhaps it
will not exist for long. In the music-hall comedians they find the
expression and dignity of their own lives; and this is not found in
the most elaborate and expensive revue. In England, at any rate,
the revue expresses almost nothing. With the decay of the
music-hall, with the encroachment of the cheap and rapid-breed-
ing cinema, the lower classes will tend to drop into the same
state of protoplasm as the bourgeoisie. The working man who
went to the music-hall and saw Marie Lloyd and joined in the
chorus was himself performing part of the act; he was engaged in
that collaboration of the audience with the artist which is necessary in all art and most obviously in dramatic art. He will now
go to the cinema, where his mind is lulled by continuous sense-
less music and continuous action too rapid for the brain to act
upon, and will receive, without giving, in that same listless ap-
athy with which the middle and upper classes regard any entertainment of the nature of art. He will also have lost some of his
interest in life. Perhaps this will be the only solution. In an
interesting essay in the volume of *Essays on the Depopulation of
Melanesia*, the psychologist W. H. R. Rivers adduced evidence
which has led him to believe that the natives of that unfortunate
archipelago are dying out principally for the reason that the
'Civilization' forced upon them has deprived them of all interest
in life. They are dying from pure boredom. When every theatre
has been replaced by 100 cinemas, when every musical instrument has been replaced by 100 gramophones, when every horse
has been replaced by 100 cheap motor-cars, when electrical
ingenuity has made it possible for every child to hear its bedtime
stories from a loudspeaker, when applied science has done everything possible with the materials on this earth to make life as
interesting as possible, it will not be surprising if the population
of the entire civilized world rapidly follows the fate of the
Melanesians.

                                                                    1923

Symmetry and Repetition

THE effort which people put up to avoid thinking might almost enable them to think and to have some new ideas. But having ideas produces anxiety and *malaise* and runs counter to the deepest instincts of human nature, which loves symmetry, repetition, and routine. Mine certainly does, and to such a degree that I get sick of them, and then notice that proclivity in others and criticise it.

When an East European peasant sits to a photographer, he places his hands symmetrically on his knees, like the statues of the Pharaohs: obviously a primeval instinct. Italian peasants and *petit bourgeois* cannot stand asymmetry in the distribution of windows, and paint them on the wall if it is impossible to have real ones in their place. German faces are marvellously symmetrical—look, for instance, at that of Hindenburg. No other European nation ever attains that square, stolid facial symmetry. The love which the Germans have for symmetry, repetition, and routine helps to make them great organisers.

One would expect people to remember the past and to imagine the future. But in fact, when discoursing or writing about history, they imagine it in terms of their own experience, and when trying to gauge the future they cite supposed analogies from the past: till, by a double process of repetition, they imagine the past and remember the future.

There are fetishes of dates, places, and methods. One of the reasons of the Austrian disaster in Serbia, in August 1914, was that the Austrian commander made haste to have a victory for the Emperor Francis Joseph's birthday, which fell on the 18th of that month. The date and region of Sedan were not without influence on the German operations which preceded the first Battle of the Marne; nor again was the name in 1940. On the other hand, the memory of defeats turned into victories cheered the

British public in May 1940, especially as the retreating armies approached the Marne. When the possibility of successful resistance vanished in France one weighty reason against her continuing it from the colonies was that the French had never done so before. The editor of *La France Libre* writes in a brilliant article on 'La Capitulation': 'L'idée de défendre la France de l'extérieur restait abstraite, parce qu'aucun souvenir, aucune tradition ne l'animait'. Our own memories of the last war are of the battles of Ypres, the Somme, Vimy Ridge, and Passchendaele, and we therefore do not relish the idea of fighting in Flanders and France, but should cheerfully take to action in Spain, for then we should have a Peninsular War of blessed memory.

Among the German troops now in Poland the best-behaved are those who had been there with the German army of occupation in the last war, partly because these are older men who have not been in the Hitler Youth, but partly because they remember how, after having conquered country after country, Germany collapsed. 'Last time,' said such a soldier to a Pole a year ago, 'I was disarmed by a washerwoman. I wonder who will disarm me this time?' The memories of 1918 weigh on the Germans. On the other hand, the Italians are fortified in their misfortunes by remembering that they never won a battle (unless both sides were Italian), but invariably managed to profit by somebody else's victory (the fame of Garibaldi as a soldier rests on his having fought only Italians). 'When the Cardinal of Rouen said to me that the Italians had no aptitude for war,' wrote Machiavelli four centuries ago, 'I answered him that the French had no aptitude for politics.' In another passage, while giving copious excuses, Machiavelli admits that 'in all the many wars of the last twenty years, whenever an army was wholly Italian it failed to stand the test'.

A book could be written on the plagiarism of revolutions. The imaginative and emotional element is strong in them, and while objective observation and thought draws on the infinite variety of nature, human imagination and feelings are restricted and stereotyped. Most novelists or dramatists have only one or two plots, which they adorn with artificial variations, and Napoleon fought all his battles on two variations of one single plan, confessing at St Helena that in his last battle he did not know more than in his first. Revolutions have their tradition, ritual, and magic tricks; moreover, it is easy to acquire the habit of revolutions: revolution breeds revolution. To quote Machiavelli once more: 'Perchè sempre una mutazione lascia lo addentellato per la edificazione dell' altra' ('For one change always leaves an indent for the next').

Continuity is a compromise between novelty and repetition. 'The English angel of progress moves from precedent to precedent', and that is why we are invariably well prepared to fight the previous war. In 1914 we had the equipment and training which would have served us well in the Boer War, and in 1939 we had all that was needed in 1914. The French held in 1914 ideas about offensive action inappropriate to trench warfare, and in 1939 ideas based on trench warfare irrelevant to a war of movement. The position of the Poles was even worse. They had, for a century, been without State and army, and when they had left off, the cavalry of Jan Sobieski and of the Napoleonic Legions had still a *raison d'être*; so they prepared an excellent and numerous cavalry for a war of tanks and aeroplanes. I asked a Polish officer why they had done so. He replied that they had prepared for war with Russia rather than with Germany. 'But Russia, too, has a mechanised army,' I remarked. He answered that this was so.

It is a mistake to suppose that people think: they wobble with the brain, and sometimes the brain does not wobble.

1941

The Necessary Enemy

SHE is a frank, charming, fresh-hearted young woman who married for love. She and her husband are one of those gay, good-looking young pairs who ornament this modern scene rather more in profusion perhaps than ever before in our history. They are handsome, with a talent for finding their way in their world, they work at things that interest them, their tastes agree and their hopes. They intend in all good faith to spend their lives together, to have children and do well by them and each other—to be happy, in fact, which for them is the whole point of their marriage. And all in stride, keeping their wits about them. Nothing romantic, mind you; their feet are on the ground.

Unless they were this sort of person, there would be not much point to what I wish to say; for they would seem to be an example of the high-spirited, right-minded young whom the critics are always invoking to come forth and do their duty and practice all those sterling old-fashioned virtues which in every generation seem to be falling into disrepair. As for virtues, these young people are more or less on their own, like most of their kind; they get very little moral or other aid from their society; but after three years of marriage this very contemporary young woman finds herself facing the oldest and ugliest dilemma of marriage.

She is dismayed, horrified, full of guilt and forebodings because she is finding out little by little that she is capable of hating her husband, whom she loves faithfully. She can hate him at times as fiercely and mysteriously, indeed in terribly much the same way, as often she hated her parents, her brothers and sisters, whom she loves, when she was a child. Even then it had seemed to her a kind of black treacherousness in her, her private wickedness that, just the same, gave her her only private life. That was one thing her parents never knew about her, never seemed to suspect. For it was never given a name. They did and said hateful things to her and to each other as if by right, as if in them it was a kind of virtue. But when they said to her, 'Control your feelings,' it was never when she was amiable and obedient, only in the black times of her hate. So it was her secret, a shameful one. When they punished her, sometimes for the strangest reasons, it was, they said, only because they loved her—it was for her good. She did not believe this, but she thought herself guilty of something worse than ever they had punished her for. None of this really frightened her: the real fright came when she discovered that at times her father and mother hated each other; this was like standing on the doorsill of a familiar room and seeing in a lightning flash that the floor was gone, you were on the edge of a bottomless pit. Sometimes she felt that both of them hated her, but that passed, it was simply not a thing to be thought of, much less believed. She thought she had outgrown all this, but here it was again, an element in her own nature she could not control, or feared she could not. She would have to hide from her husband, if she could, the same spot in her feelings she had hidden from her parents, and for the same no doubt disreputable, selfish reason: she wants to keep his love.

Above all, she wants him to be absolutely confident that she loves him, for that is the real truth, no matter how unreasonable it sounds, and no matter how her own feelings betray them both at times. She depends recklessly on his love; yet while she is hating him, he might very well be hating her as much or even more, and it would serve her right. But she does not want to be served right, she wants to be loved and forgiven—that is, to be sure he would forgive her anything, if he had any notion of what she had done. But best of all she would like not to have anything in her love that should ask for forgiveness. She doesn't mean about their quarrels—they are not so bad. Her feelings are out of proportion, perhaps. She knows it is perfectly natural for people to disagree, have fits of temper, fight it out; they learn quite a lot about each other that way, and not all of it disappointing either. When it passes, her hatred seems quite unreal. It always did.

Love. We are early taught to say it. I love you. We are trained to the thought of it as if there were nothing else, or nothing else worth having without it, or nothing worth having which it could not bring with it. Love is taught, always by precept, sometimes by example. Then hate, which no one meant to teach us, comes of itself. It is true that if we say I love you, it may be received with doubt, for there are times when it is hard to believe. Say I hate you, and the one spoken to believes it instantly, once for all.

Say I love you a thousand times to that person afterward and mean it every time, and still it does not change the fact that once we said I hate you, and meant that too. It leaves a mark on that surface love had worn so smooth with its eternal caresses. Love must be learned, and learned again and again; there is no end to it. Hate needs no instruction, but waits only to be provoked . . . hate, the unspoken word, the unacknowledged presence in the house, that faint smell of brimstone among the roses, that invisible tongue-tripper, that unkempt finger in every pie, that sudden oh-so-curiously *chilling* look—could it be boredom?—on your dear one's features, making them quite ugly. Be careful: love, perfect love, is in danger.

If it is not perfect, it is not love, and if it is not love, it is bound to be hate sooner or later. This is perhaps a not too exaggerated statement of the extreme position of Romantic Love, more especially in America, where we are all brought up on it, whether we know it or not. Romantic Love is changeless, faithful, passionate, and its sole end is to render the two lovers happy. It has no obstacles save those provided by the hazards of fate (that is to say, society), and such sufferings as the lovers may cause each other are only another word for delight: exciting jealousies, thrilling uncertainties, the ritual dance of courtship within the charmed closed circle of their secret alliance; all *real* troubles come from without, they face them unitedly in perfect confidence. Marriage is not the end but only the beginning of true happiness, cloudless, changeless to the end. That the candidates for this blissful condition have never seen an example of it, nor ever knew anyone who had, makes no difference. That is the ideal and they will achieve it.

How did Romantic Love manage to get into marriage at last, where it was most certainly never intended to be? At its highest it was tragic: the love of Héloïse and Abélard. At its most graceful, it was the homage of the trouvère for his lady. In its most popular form, the adulterous strayings of solidly married couples who meant to stray for their own good reasons, but at the same time do nothing to upset the property settlements or the line of legitimacy; at its most trivial, the pretty trifling of shepherd and shepherdess.

This was generally condemned by church and state and a word of fear to honest wives whose mortal enemy it was. Love within the sober, sacred realities of marriage was a matter of personal luck, but in any case, private feelings were strictly a private affair having, at least in theory, no bearing whatever on the fixed practice of the rules of an institution never intended as a recreation ground for either sex. If the couple discharged their religious and social obligations, furnished forth a copious progeny, kept their troubles to themselves, maintained public civility and died under the same roof, even if not always on speaking terms, it was rightly regarded as a successful marriage. Apparently this testing ground was too severe for all but the stoutest spirits; it too was based on an ideal, as impossible in its way as the ideal Romantic Love. One good thing to be said for it is that society took responsibility for the conditions of marriage, and the sufferers within its bonds could always blame the system, not themselves. But Romantic Love crept into the marriage bed, very stealthily, by centuries, bringing its absurd notions about love as eternal springtime and marriage as a personal adventure meant to provide personal happiness. To a Western romantic such as I, though my views have been much modified by painful experience, it still seems to me a charming work of the human imagination, and it is a pity its central notion has been taken too literally and has hardened into a convention as cramping and enslaving as the older one. The refusal to acknowledge the evils in ourselves which therefore are implicit in any human situation is as extreme and unworkable a proposition as the doctrine of total depravity; but somewhere between them, or maybe beyond them, there does exist a possibility for reconciliation between our desires for impossible satisfactions and the simple unalterable fact that we also desire to be unhappy and that we create our own sufferings; and out of these sufferings we salvage our fragments of happiness.

Our young woman who has been taught that an important part of her human nature is not real because it makes trouble and interferes with her peace of mind and shakes her self-love, has been very badly taught; but she has arrived at a most important stage of her re-education. She is afraid her marriage is going to fail because she has not love enough to face its difficulties; and this because at times she feels a painful hostility toward her husband, and cannot admit its reality because such an admission would damage in her own eyes her view of what love should be, an absurd view, based on her vanity of power. Her hatred is real as her love is real, but her hatred has the advantage at present because it works on a blind instinctual level, it is lawless; and her love is subjected to a code of ideal conditions, impossible by their very nature of fulfillment, which prevents its free growth and deprives it of its right to recognize its human limitations and come to grips with them. Hatred is natural in a sense that love, as she conceives it, a young person brought up in the tradition of Romantic Love, is not natural at all. Yet it did not come by hazard, it is the very imperfect expression of the need of the human imagination to create beauty and harmony out of chaos, no matter how mistaken its notion of these things may be, nor how clumsy its methods. It has conjured love out of the air, and seeks to preserve it by incantations; when she spoke a vow to love and honor her husband until death, she did a very reckless thing, for it is not possible by an act of the will to fulfill such an engagement. But it was the necessary act of faith performed in defense of a mode of feeling, the statement of honorable intention to practice as well as she is able the noble, acquired faculty of love, that very mysterious overtone to sex which is the best thing in it. Her hatred is part of it, the necessary enemy and ally.

1948

The Sterner Sex

WHITEHALL AND PIMLICO

THE other day a cousin of mine was married. Though what God or myself had to do with it I do not know, but I was obliged to go to church. I object to going to church except to hear the Athanasian Creed, which stretches the power of belief and leaves one with the same pleasant, warm, tired feeling as an attack of yawning. All the same, I found myself one day last week in the porch of a sombre building adorned with stained-glass windows which proved that the Apostles liked aniline dyes and everything handsome about them: I faced an usher whose eyes dropped doubtfully on the books I held in my arms. I had caught them up carelessly as I left home to read in the train, and it was an unfortunate coincidence that they happened to be the latest batch of literature issued by the Divorce Law Reform Union. With an air of resource he conducted me to a pew in the transept where any disturbance I might make would be quite ineffective.... The organ began to play a Nocturne that Chopin wrote for George Sand when he loved her; their shades were drawn down by the music from the skies and looked in at the church windows. 'I wish they wouldn't bring us to such ugly places!' they grumbled. Perhaps, forgetting their last discords, they looked at each other and smiled a little. 'Things were prettier with us in Venice, weren't they?'

That horrid interior, distempered a deep drab picked out with a lemon-coloured dado, turned the sweetness of the music sour; and I opened 'Marriage and Divorce', a pamphlet complied by the secretary of the Divorce Law Reform Union, which is largely concerned with the findings of the Royal Commission on Matrimonial Causes. That is the body which sat for three years and issued a Majority Report containing recommendations of the utmost importance to the working classes. Because it deals with sex and shocks the Archbishop of Canterbury neither the Liberals nor the Conservatives will ever have the pluck to carry it into law. Thus do men perform the task of government. My eye fell on an extraordinary conversation between Lady Frances Balfour and the Bishop of Birmingham:

'We have had evidence put before us,' said Lady Frances, 'that there are men who live on the prostitution of their wives. Is such a wife, being a Christian, to stick to her husband and to do as he commands her?'

'Yes,' replied the Bishop, probably plucking at the edge of his apron, 'I am afraid so.'

I felt sorry that I had come to church. It was true that Norwood is not in the diocese of Birmingham, but I did not wish to encourage the Church of England at all after this. The lack of pity for hurt flesh, the tolerance of squalid sin!

But a shadow fell across my book and I looked up into the candid gaze of two blue eyes. The bridegroom stood beside me, extending a slender hand and smiling shyly and frankly, but quite without coquetry. The sun lay on his fair head like a benediction, and no fretful or angry passion had ever lined that boyish brow: only the gentlest words had shaped that mouth. Personal grace of a high order was accentuated by dainty dressing; his frock-coat fitted him like the pelt of a young antelope, and his trousers had a silvery gleam like willows seen at twilight. He was like a pure white rose.

How can I tell you how this flower-like radiance of untainted youth affected me, who had just come from Fleet Street? Suffice it to say that hastily I closed my book and put it from me, lest his eyes should fall on the—to his young mind—strange and disturbing title. I had an impulse to raise his gloved hands to my lips: only by some such old-fashioned courtesy could I express the million mingled feelings of love and pity, foreboding and strong hope, that rose in my heart as I saw this young creature, so grave, yet so unsuspecting of life's darker side, standing on the threshold of his new life. I did not do it. People in Norwood are so ununderstanding. He murmured a few graceful words, his delicate skin suffused with a flush, and flitted on.

An hour or so afterwards I threaded my way to my aunt through a number of people who were eating oyster patties and meringues for the same excellent reason that they had previously gone to church: because my cousin was being married. My aunt shook my hand warmly, looked out of the window into the garden, and said with deep feeling: 'I shall never see geraniums and calceolarias again without remembering how I lost my little Rachel.' Our conversation was at first disconnected, for it seemed that I had been dropping pamphlets on cheap divorce all round the wedding breakfast, and people were constantly retrieving them and bringing them to me. Then, casting a glance to where a trembling but composed figure stood beside my cousin's athletic form, I asked:

'What does this young man that Rachel's married do?'

'Nothing. He's in the War Office,' replied my aunt, and her attention wandered again. 'I always think there's something depressing about a bridesmaid if she's over twenty. Poor things, they can't help feeling it ... Yes, Cyril is in the War Office. Your uncle is very pleased about that. He said to me last night, quite impressively, that a government worker is in as secure a position as any member of the royal family.'

I was glad that my cousin had chosen her husband from that sheltered parterre, the Civil Service. When I go down Whitehall and meet all the cool and uncreased young men emerging from the cloister of their offices, I feel as one does when visiting an Irish convent boarding-school where girls of good family learn to paint on satin and play the harp; one feels one's voice a little loud, one's boots a little large. There is something appealing about the turning of the middle-class man to the refuge of unresponsible administrative work: the working man knows peril in his various occupations, and all women may die through motherhood; but he alone walks always in the sunlight, almost as happy as the capitalist. And so England would have her manhood. Do not the Insurance Commissioners congratulate themselves on the fact that they are spending thousands not on sick or maternity benefits, but on the employment of unskilled clerk labour? I see a day when the manhood of England will spring like a lily from an office-stool. All the same, I felt that there was a certain exuberance about the the statement that a government worker is in as secure a position as any member of the royal family.

For Whitehall leads to Pimlico. And in Pimlico there are a

number of government workers who seem in an extraordinarily insecure position. It is true that they are not decorative persons like Cyril, whose personal appearance is, as I have explained, such that no one could raise a hand against him save in the way of kindness: still, their humanity entitles them to a certain measure of justice. On 1 July a notice was put up in the yard which announced that the rate of pay for making soldiers' drab jackets was altered from 2s. 11½d. to 2s. 6d.: which meant a reduction of from 1s. 6d. to 2s. on weekly earnings which now average 17s. 4d. It was a reduction that one cannot ascribe either to the law of supply or demand or too much gold. It was simply an act of pure devilish thrift on the part of the War Office: the Army Clothing Employees' Union has just forced them to raise the rate of pay for making scarlet uniforms by 3½d., and it is now trying to make the other workers pay for this generous rise. So six hundred women marched up Whitehall to the War Office. Cyril, I believe, watched them from behind a blind. Cyril's superiors received them with politeness and promised that, pending the arrangement for a round-table conference, the reductions should not be carried into effect. Meantime, the authorities are hunting for some justification of the prejudice that equates in the official mind like a blind-eyed toad: that some women of exceptional skill make 25s. a week at this work, and it is not the will of God that a woman should earn more than a pound a week. And the need for keeping down expenses is so great . . .

Good God enlighten us! Which of these two belongs to the sterner sex—the man who sits in Whitehall all his life on a comfortable salary, or the woman who has to keep her teeth bared lest she has her meatless bone of 17s. 4d. a week snatched away from her and who has to produce the next generation on her off-days? I remembered a member of the Army Clothing Employees' Union that I met a week or two ago: her brown hands were strong, every sentence she spoke bit into the truth, she had the faculty of deep and rowdy laughter. I looked at Cyril. He had sat down now and, with a solemnity that lay prettily on his Dresden-like charm, was toying with a vanilla ice. I had a vision of the world fifty years hence, when we have simply had to take over the dangerous adventures of the earth. I saw some bronzed and travel-scarred pioneer returning from the Wild

West with hard-earned treasure, buying a fresh and unspoiled bridegroom who had hardly ever stirred from the office of, let us say, the Director of Public Prosecutions. I saw a world of women struggling as the American capitalist men of today struggle, to maintain a parasitic sex that is at once its tyrant and its delight. . . . We must keep men up to the mark.

1913

The Colloid and the Crystal

THE first real snow was soon followed by a second. Over the radio the weatherman talked lengthily about cold masses and warm masses, about what was moving out to sea and what wasn't. Did Benjamin Franklin, I wondered, know what he was starting when it first occurred to him to trace by correspondence the course of storms? From my stationary position the most reasonable explanation seemed to be simply that winter had not quite liked the looks of the landscape as she first made it up. She was changing her sheets.

Another forty-eight hours brought one of those nights ideal for frosting the panes. When I came down to breakfast, two of the windows were almost opaque and the others were etched with graceful, fernlike sprays of ice which looked rather like the impressions left in rocks by some of the antediluvian plants, and they were almost as beautiful as anything which the living can achieve. Nothing else which has never lived looks so much as though it were actually informed with life.

I resisted, I am proud to say, the almost universal impulse to scratch my initials into one of the surfaces. The effect, I knew, would not be an improvement. But so, of course, do those less virtuous than I. That indeed is precisely why they scratch. The impulse to mar and to destroy is as ancient and almost as nearly universal as the impulse to create. The one is an easier way than the other of demonstrating power. Why else should anyone not hungry prefer a dead rabbit to a live one? Not even those horrible Dutch painters of bloody still—or shall we say stilled? —lifes can have really believed that their subjects were more beautiful dead.

Indoors it so happened that a Christmas cactus had chosen this moment to bloom. Its lush blossoms, fuchsia-shaped but pure red rather than magenta, hung at the drooping ends of strange, thick stems and outlined themselves in blood against the glistening background of the frosty pane—jungle flower against frostflower; the warm beauty that breathes and lives and dies competing with the cold beauty that burgeons, not because it wants to, but merely because it is obeying the laws of physics which require that crystals shall take the shape they have always taken since the world began. The effect of red flower against white tracery was almost too theatrical, not quite in good taste perhaps. My eye recoiled in shock and sought through a clear area of the glass the more normal out-of-doors.

On the snow-capped summit of my bird-feeder a chickadee pecked at the new-fallen snow and swallowed and a few of the flakes which serve him in lieu of the water he sometimes sadly lacks when there is nothing except ice too solid to be picked at. A downy woodpecker was hammering at a lump of suet and at the coconut full of peanut butter. One nuthatch was dining while the mate waited his—or was it her?—turn. The woodpecker announces the fact that he is a male by the bright red spot on the back of his neck, but to me, at least, the sexes of the nuthatch are indistinguishable. I shall never know whether it is the male or the female who eats first. And that is a pity. If I knew, I could say, like the Ugly Duchess, 'and the moral of that is . . .'

But I soon realized that at the moment the frosted windows were what interested me most—especially the fact that there is no other natural pheonomenon in which the lifeless mocks so closely the living. One might almost think that the frostflower had got the idea from the leaf and the branch if one did not know how inconceivably more ancient the first is. No wonder that enthusiastic biologists in the nineteenth century, anxious to conclude that there was no qualitative difference between life and chemical processes, tried to believe that the crystal furnished the link, that its growth was actually the same as the growth of a living organism. But excusable though the fancy was, no one, I think, believes anything of the sort today. Protoplasm is a colloid and the colloids are fundamentally different from the crystalline substances. Instead of crystallizing they jell, and life in its simplest known form is a shapeless blob of rebellious jelly rather than a crystal eternally obeying the most ancient law.

No man ever saw a dinosaur. The last of these giant reptiles was dead eons before the most dubious halfman surveyed the world about him. Not even the dinosaurs ever cast their dim eyes upon many of the still earlier creatures which preceded them. Life changes so rapidly that its later phases know nothing of those which preceded them. But the frostflower is older than the dinosaur, older than the protozoan, older no doubt than the enzyme or the ferment. Yet it is precisely what it has always been. Millions of years before there were any eyes to see it, millions of years before any life existed, it grew in its own special way, crystallized along its preordained lines of cleavage, stretched out its pseudo-branches and pseudo-leaves. It was beautiful before beauty itself existed.

We find it difficult to conceive a world except in terms of purpose, of will, or of intention. At the thought of the something without beginning and presumably without end, of something which is, nevertheless, regular though blind, and organized without any end in view, the mind reels. Constituted as we are it is easier to conceive how the slime floating upon the waters might become in time Homo sapiens than it is to imagine how so complex a thing as a crystal could have always been and can always remain just what it is—complicated and perfect but without any meaning, even for itself. How can the lifeless even obey a law?

To a mathematical physicist I once confessed somewhat shamefacedly that I had never been able to understand how inanimate nature managed to follow so invariably and so promptly her own laws. If I flip a coin across a table, it will come to rest at a certain point. But before it stops at just that point, many factors must be taken into consideration. There is the question of the strength of the initial impulse, of the exact amount of resistance offered by the friction of that particular table top, and of the density of the air at the moment. It would take a physicist a long time to work out the problem and he could achieve only an approximation at that. Yet presumably the coin will stop exactly where it should. Some very rapid calculations have to be made before it can do so, and they are, presumably, always accurate.

And then, just as I was blushing at what I supposed he must regard as my folly, the mathematician came to my rescue by informing me that Laplace had been puzzled by exactly the same fact. 'Nature laughs at the difficulties of integration,' he remarked—and by 'integration' he meant, of course, the mathematician's word for the process involved when a man solves one of the differential equations to which he has reduced the laws of motion.

When my Christmas cactus blooms so theatrically a few inches in front of the frost-covered pane, it also is obeying laws but obeying them much less rigidly and in a different way. It blooms at about Christmastime because it has got into the habit of doing so, because, one is tempted to say, it wants to. As a matter of fact it was, this year, not a Christmas cactus but a New Year's cactus, and because of this unpredictability I would like to call it 'he,' not 'it.' His flowers assume their accustomed shape and take on their accustomed color. But not as the frostflowers follow their predestined pattern. Like me, the cactus has a history which stretches back over a long past full of changes and developments. He has not always been merely obeying fixed laws. He has resisted and rebelled; he has attempted novelties, passed through many phases. Like all living things he has had a will of his own. He has made laws, not merely obeyed them.

'Life,' so the platitudinarian is fond of saying, 'is strange.' But from our standpoint it is not really so strange as those things which have no life and yet nevertheless move in their predestined orbits and 'act' though they do not 'behave.' At the very least one ought to say that if life is strange there is nothing about it more strange than the fact that it has its being in a universe so astonishingly shared on the one hand by 'things' and on the other by 'creatures,' that man himself is both a 'thing' which obeys the laws of chemistry or physics and a 'creature' who to some extent defies them. No other contrast, certainly not the contrast between the human being and the animal, or the animal and the plant, or even the spirit and the body, is so tremendous as this contrast between what lives and what does not.

To think of the lifeless as merely inert, to make the contrast merely in terms of a negative, is to miss the real strangeness. Not the shapeless stone which seems to be merely waiting to be acted upon but the snowflake or the frostflower is the true representative of the lifeless universe as opposed to ours. They represent plainly, as the stone does not, the fixed and perfect system of organization which includes the sun and its planets, includes therefore this earth itself, but against which life has set up its seemingly puny opposition. Order and obedience are the primary characteristics of that which is not alive. The snowflake eternally obeys its one and only law: 'Be thou six pointed'; the planets their one and only: 'Travel thou in an ellipse.' The astronomer can tell where the North Star will be ten thousand years hence; the botanist cannot tell where the dandelion will bloom tomorrow.

Life is rebellious and anarchial, always testing the supposed immutability of the rules which the nonliving changelessly accepts. Because the snowflake goes on doing as it was told, its story up to the end of time was finished when it first assumed the form which it has kept ever since. But the story of every living thing is still in the telling. It may hope and it may try. Moreover, though it may succeed or fail, it will certainly change. No form of frostflower ever became extinct. Such, if you like, is its glory. But such also is the fact which makes it alien. It may melt but it cannot die.

If I wanted to contemplate what is to me the deepest of all mysteries, I should choose as my object lesson a snowflake under a lens and an amoeba under the microscope. To a detached observer—if one can possibly imagine any observer who *could* be detached when faced with such an ultimate choice—the snowflake would certainly seem the 'higher' of the two. Against its intricate glistening perfection one would have to place a shapeless, slightly turbid glob, perpetually oozing out in this direction or that but not suggesting so strongly as the snowflake does, intelligence and plan. Crystal and colloid, the chemist would call them, but what an inconceivable contrast those neutral terms imply! Like the star, the snowflake seems to declare the glory of God, while the promise of the amoeba, given only perhaps to itself, seems only contemptible. But its jelly holds, nevertheless, not only its promise but ours also, while the snowflake represents some achievement which we cannot possibly share. After the passage of billions of years, one can see and be aware of the other, but the relationship can never be reciprocal. Even after these billions of years no aggregate of colloids can be as beautiful as the crystal always was, but it can know, as the crystal cannot, what beauty is.

Even to admire too much or too exclusively the alien kind of beauty is dangerous. Much as I love and am moved by the grand, inanimate forms of nature, I am always shocked and a little frightened by those of her professed lovers to whom landscape is the most important thing, and to whom landscape is merely a matter of forms and colors. If they see or are moved by an animal or flower, it is to them merely a matter of a picturesque completion and their fellow creatures are no more than decorative details. But without some continuous awareness of the two great realms of the inanimate and the animate there can be no love of nature as I understand it, and what is worse, there must be a sort of disloyalty to our cause, to us who are colloid, not crystal. The pantheist who feels the oneness of all living things, I can understand; perhaps indeed he and I are in essential agreement. But the ultimate All is not one thing, but two. And because the alien half is in its way as proud and confident and successful as our half, its fundamental difference may not be disregarded with impunity. Of us and all we stand for, the enemy is not so much death as the not-living, or rather that great system which succeeds without ever having had the need to be alive. The frostflower is not merely a wonder; it is also a threat and a warning. How admirable, it seems to say, not living can be! What triumphs mere immutable law can achieve!

Some of Charles Peirce's strange speculations about the possibility that 'natural law' is not law at all but merely a set of habits fixed more firmly than any habits we know anything about in ourselves or in the animals suggest the possibility that the snowflake was not, after all, always inanimate, that it merely surrendered at some time impossibly remote the life which once achieved its perfect organization. Yet even if we can imagine such a thing to be true, it serves only to warn us all the more strongly against the possibility that what we call the living might in the end succumb also to the seduction of the immutably fixed.

No student of the anthill has ever failed to be astonished either into admiration or horror by what is sometimes called the perfection of its society. Though even the anthill can change its ways, though even ant individuals—ridiculous as the conjunction of the two words may seem—can sometimes make choices, the perfection of the techniques, the regularity of the habits almost suggest the possibility that the insect is on its way back to inanition, that, vast as the difference still is, an anthill crystallizes somewhat as a snowflake does. But not even the anthill, nothing else indeed in the whole known universe is so perfectly planned as one of these same snowflakes. Would, then, the ultimately planned society be, like the anthill, one in which no one makes plans, any more than a snowflake does? From the cradle in which it is not really born to the grave where it is only a little deader than it always was, the ant-citizen follows a plan to the making of which he no longer contributes anything.

Perhaps we men represent the ultimate to which the rebellion, begun so long ago in some amoeba-like jelly, can go. And perhaps the inanimate is beginning the slow process of subduing us again. Certainly the psychologist and the philosopher are tending more and more to think of us as creatures who obey laws rather than as creatures of will and responsibility. We are, they say, 'conditioned' by this or by that. Even the greatest heroes are studied on the assumption that they can be 'accounted for' by something outside themselves. They are, it is explained, 'the product of forces.' All the emphasis is placed, not upon that power to resist and rebel which we were once supposed to have, but upon the 'influences' which 'formed us.' Men are made by society, not society by men. History as well as character 'obeys laws.' In their view, we crystallize in obedience to some dictate from without instead of moving in conformity with something within.

And so my eye goes questioningly back to the frosted pane. While I slept the graceful pseudo-fronds crept across the glass, assuming, as life itself does, an intricate organization. 'Why live,' they seem to say, 'when we can be beautiful, complicated, and orderly without the uncertainty and effort required of a living thing? Once we were all that was. Perhaps some day we shall be all that is. Why not join us?'

Last summer no clod or no stone would have been heard if it had asked such a question. The hundreds of things which walked and sang, the millions which crawled and twined were all having their day. What was dead seemed to exist only in order that the living might live upon it. The plants were busy turning the inorganic into green life and the animals were busy turning that green into red. When we moved, we walked mostly upon grass. Our pre-eminence was unchallenged.

On this winter day nothing seems so successful as the frostflower. It thrives on the very thing which has driven some of us indoors or underground and which has been fatal to many. It is having now its hour of triumph, as we before had ours. Like the cactus flower itself, I am a hothouse plant. Even my cats gaze dreamily out of the window at a universe which is no longer theirs.

How are we to resist, if resist we can? This house into which I have withdrawn is merely an expedient and it serves only my mere physical existence. What mental or spiritual convictions, what will to maintain to my own kind of existence can I assert? For me it is not enough merely to say, as I do say, that I shall resist the invitation to submerge myself into a crystalline society and to stop planning in order that I may be planned for. Neither is it enough to go further, as I do go, and to insist that the most important thing about a man is not that part of him which is 'the product of forces' but that part, however small it may be, which enables him to become something other than what the most accomplished sociologist, working in conjunction with the most accomplished psychologist, could predict that he would be.

I need, so I am told, a faith, something outside myself to which I can be loyal. And with that I agree, in my own way. I am on what I call 'our side,' and I know, though vaguely, what I think that is. Wordsworth's God had his dwelling in the light of setting suns. But the God who dwells there seems to me most probably the God of the atom, the star, and the crystal. Mine, if I have one, reveals Himself in another class of phenomena. He makes the grass green and the blood red.

1950

On Being the Right Size

T HE most obvious differences between different animals are differences of size, but for some reason the zoologists have paid singularly little attention to them. In a large textbook of zoology before me I find no indication that the eagle is larger than the sparrow, or the hippopotamus bigger than the hare, though some grudging admissions are made in the case of the mouse and the whale. But yet it is easy to show that a hare could not be as large as a hippopotamus, or a whale as small as a herring. For every type of animal there is a most convenient size, and a large change in size inevitably carries with it a change of form.

Let us take the most obvious of possible cases, and consider a giant man sixty feet high—about the height of Giant Pope and Giant Pagan in the illustrated *Pilgrim's Progress* of my childhood. These monsters were not only ten times as high as Christian, but ten times as wide and ten times as thick, so that their total weight was a thousand times his, or about eighty to ninety tons. Unfortunately the cross sections of their bones were only a hundred times those of Christian, so that every square inch of giant bone had to support ten times the weight borne by a square inch of human bone. As the human thigh-bone breaks under about ten times the human weight, Pope and Pagan would have broken their thighs every time they took a step. This was doubtless why they were sitting down in the picture I remember. But it lessens one's respect for Christian and Jack the Giant Killer.

To turn to zoology, suppose that a gazelle, a graceful little creature with long thin legs, is to become large, it will break its bones unless it does one of two things. It may make its legs short and thick, like the rhinoceros, so that every pound of weight has still about the same area of bone to support it. Or it can compress its body and stretch out its legs obliquely to gain stability, like the giraffe. I mention these two beasts because they happen to belong to the same order as the gazelle, and both are quite successful mechanically, being remarkably fast runners.

Gravity, a mere nuisance to Christian, was a terror to Pope, Pagan, and Despair. To the mouse and any smaller animal it presents practically no dangers. You can drop a mouse down a thousand-yard mine shaft; and, on arriving at the bottom, it gets a slight shock and walks away, so long as the ground is fairly soft. A rat is killed, a man is broken, a horse splashes. For the resistance presented to movement by the air is proportional to the surface of the moving object. Divide an animal's length, breadth, and height each by ten; its weight is reduced to a thousandth, but its surface only to a hundredth. So the resistance to falling in the case of the small animal is relatively ten times greater than the driving force.

An insect, therefore, is not afraid of gravity; it can fall without danger, and can cling to the ceiling with remarkably little trouble. It can go in for elegant and fantastic forms of support like that of the daddy-long-legs. But there is a force which is as formidable to an insect as gravitation to a mammal. This is surface tension. A man coming out of a bath carries with him a film of water of about one-fiftieth of an inch in thickness. This weighs roughly a pound. A wet mouse has to carry about its own weight of water. A wet fly has to lift many times its own weight and, as every one knows, a fly once wetted by water or any other liquid is in a very serious position indeed. An insect going for a drink is in as great danger as a man leaning out over a precipice in search of food. If it once falls into the grip of the surface tension of the water—that is to say, gets wet—it is likely to remain so until it drowns. A few insects, such as water-beetles, contrive to be unwettable, the majority keep well away from their drink by means of a long proboscis.

Of course tall land animals have other difficulties. They have to pump their blood to greater heights than a man and, therefore, require a larger blood pressure and tougher blood-vessels. A great many men die from burst arteries, especially in the brain, and this danger is presumably still greater for an elephant or a giraffe. But animals of all kinds find difficulties in size for the following reason. A typical small animal, say a microscopic worm or rotifer, has a smooth skin through which all the oxygen it requires can soak in, a straight gut with sufficient surface to absorb its food, and a simple kidney. Increase its dimensions ten-fold in every direction, and its weight is increased a thousand times, so that if it is to use its muscles as efficiently as its minia-ture counterpart, it will need a thousand times as much food and oxygen per day and will excrete a thousand times as much of waste products.

Now if its shape is unaltered its surface will be increased only a hundredfold, and ten times as much oxygen must enter per minute through each square millimetre of skin, ten times as much food through each square millimetre of intestine. When a limit is reached to their absorptive powers their surface has to be increased by some special device. For example, a part of the skin may be drawn out into tufts to make gills or pushed in to make lungs, thus increasing the oxygen-absorbing surface in pro-portion to the animal's bulk. A man, for example, has a hundred square yards of lung. Similarly, the gut, instead of being smooth and straight, becomes coiled and develops a velvety surface, and other organs increase in complication. The higher animals are not larger than the lower because they are more complicated. They are more complicated because they are larger. Just the same is true of plants. The simplest plants, such as the green algae growing in stagnant water or on the bark of trees, are mere round cells. The higher plants increase their surface by putting out leaves and roots. Comparative anatomy is largely the story of the struggle to increase surface in proportion to volume.

Some of the methods of increasing the surface are useful up to a point, but not capable of a very wide adaptation. For example, while vertebrates carry the oxygen from the gills or lungs all over the body in the blood, insects take air directly to every part of their body by tiny blind tubes called tracheae which open to the surface at many different points. Now, although by their breathing movements they can renew the air in the outer part of the tracheal system, the oxygen has to penetrate the finer branches by means of diffusion. Gases can diffuse easily through very small distances, not many times larger than the average length travelled by a gas molecule between collisions with other molecules. But when such vast journeys—from the point of view of a molecule—as a quarter of an inch have to be made, the pro-cess becomes slow. So the portions of an insect's body more than a quarter of an inch from the air would always be short of oxygen. In consequence hardly any insects are much more than half an inch thick. Land crabs are built on the same general plan as insects, but are much clumsier. Yet like ourselves they carry oxygen around in their blood, and are therefore able to grow far larger than any insects. If the insects had hit on a plan for driving air through their tissues instead of letting it soak in, they might well have become as large as lobsters, though other considerations would have prevented them from becoming as large as man.

Exactly the same difficulties attach to flying. It is an elementary principle of aeronautics that the minimum speed needed to keep an aeroplane of a given shape in the air varies as the square root of its length. If its linear dimensions are increased four times, it must fly twice as fast. Now the power needed for the minimum speed increases more rapidly than the weight of the machine. So the larger aeroplane, which weighs sixty-four times as much as the smaller, needs one hundred and twenty-eight times its horse-power to keep up. Applying the same principles to the birds, we find that the limit to their size is soon reached. An angel whose muscles developed no more power weight for weight than those of an eagle or a pigeon would require a breast projecting for about four feet to house the muscles engaged in working its wings, while to economize in weight, its legs would have to be reduced to mere stilts. Actually a large bird such as an eagle or kite does not keep in the air mainly by moving its wings. It is generally to be seen soaring, that is to say balanced on a rising column of air. And even soaring becomes more and more difficult with increasing size. Were this not the case eagles might be as large as tigers and as formidable to man as hostile aeroplanes.

But it is time that we passed to some of the advantages of size. One of the most obvious is that it enables one to keep warm. All warm-blooded animals at rest lose the same amount of heat from a unit area of skin, for which purpose they need a food-supply proportional to their surface and not to their weight. Five thousand mice weigh as much as a man. Their combined surface and food or oxygen consumption are about seventeen times a man's. In fact a mouse eats about one quarter its own weight of food every day, which is mainly used in keeping it warm. For the same reason small animals cannot live in cold countries. In the arctic regions there are no reptiles or amphibians, and no small mammals. The smallest mammal in Spitzbergen is the fox. The small birds fly away in the winter, while the insects die, though their eggs can survive six months or more of frost. The most successful mammals are bears, seals, and walruses.

Similarly, the eye is a rather inefficient organ until it reaches a large size. The back of the human eye on which an image of the outside world is thrown, and which corresponds to the film of a camera, is composed of a mosaic of 'rods and cones' whose diameter is little more than a length of an average light wave. Each eye has about half a million, and for two objects to be distinguishable their images must fall on separate rods or cones. It is obvious that with fewer but larger rods and cones we should see less distinctly. If they were twice as broad two points would have to be twice as far apart before we could distinguish them at a given distance. But if their size were diminished and their number increased we should see no better. For it is impossible to form a definite image smaller than a wave-length of light. Hence a mouse's eye is not a small-scale model of a human eye. Its rods and cones are not much smaller than ours, and therefore there are far fewer of them. A mouse could not distinguish one human face from another six feet away. In order that they should be of any use at all the eyes of small animals have to be much larger in proportion to their bodies than our own. Large animals on the other hand only require relatively small eyes, and those of the whale and elephant are little larger than our own.

For rather more recondite reasons the same general principle holds true of the brain. If we compare the brain-weights of a set of very similar animals such as the cat, cheetah, leopard, and tiger, we find that as we quadruple the body-weight the brain-weight is only doubled. The larger animal with proportionately larger bones can economize on brain, eyes, and certain other organs.

Such are a very few of the considerations which show that for every type of animal there is an optimum size. Yet although Galileo demonstrated the contrary more than three hundred years ago, people still believe that if a flea were as large as a man it could jump a thousand feet into the air. As a matter of fact the height to which an animal can jump is more nearly independent of its size than proportional to it. A flea can jump about two feet, a man about five. To jump a given height, if we neglect the resistance of the air, requires an expenditure of energy proportional to the jumper's weight. But if the jumping muscles form a constant fraction of the animal's body, the energy developed per ounce of muscle is independent of the size, provided it can be developed quickly enough in the small animal. As a matter of fact an insect's muscles, although they can contract more quickly than our own, appear to be less efficient; as otherwise a flea or grasshopper could rise six feet into the air.

And just as there is a best size for every animal, so the same is true for every human institution. In the Greek type of democracy all the citizens could listen to a series of orators and vote directly on questions of legislation. Hence their philosophers held that a small city was the largest possible democratic state. The English invention of representative government made a democratic nation possible, and the possibility was first realized in the United States, and later elsewhere. With the development of broadcasting it has once more become possible for every citizen to listen to the political views of representative orators, and the future may perhaps see the return of the national state to the Greek form of democracy. Even the referendum has been made possible only by the institution of daily newspapers.

To the biologist the problem of socialism appears largely as a problem of size. The extreme socialists desire to run every nation as a single business concern. I do not suppose that Henry Ford would find much difficulty in running Andorra or Luxembourg on a socialistic basis. He has already more men on his pay-roll than their population. It is conceivable that a syndicate of Fords, if we could find them, would make Belgium Ltd or Denmark Inc. pay their way. But while nationalization of certain industries is an obvious possibility in the largest of states, I find it no easier to picture a completely socialized British Empire or United States than an elephant turning somersaults or a hippopotamus jumping a hedge.

1927

Meditation on the Moon

MATERIALISM and mentalism—the philosophies of 'nothing but'. How wearily familiar we have become with that 'nothing but space, time, matter and motion', that 'nothing but sex', that 'nothing but economics'! And the no less intolerant 'nothing but spirit', 'nothing but consciousness', 'nothing but psychology'— how boring and tiresome they also are! 'Nothing but' is mean as well as stupid. It lacks generosity. Enough of 'nothing but'. It is time to say again, with primitive common sense (but for better reasons), 'not only, but also'.

Outside my window the night is struggling to wake; in the moonlight, the blinded garden dreams so vividly of its lost colours that the black roses are almost crimson, the trees stand expectantly on the verge of living greenness. The white-washed parapet of the terrace is brilliant against the dark-blue sky. (Does the oasis lie there below, and, beyond the last of the palm trees, is that the desert?) The white walls of the house coldly reverberate the lunar radiance. (Shall I turn to look at the Dolomites rising naked out of the long slopes of snow?) The moon is full. And not only full, but also beautiful. And not only beautiful, but also . . .

Socrates was accused by his enemies of having affirmed, heretically, that the moon was a stone. He denied the accusation. All men, said he, know that the moon is a god, and he agreed with all men. As an answer to the materialistic philosophy of 'nothing but' his retort was sensible and even scientific. More sensible and scientific, for instance, than the retort invented by D. H. Lawrence in that strange book, so true in its psychological substance, so preposterous, very often, in its pseudo-scientific form, *Fantasia of the Unconscious*. 'The moon,' writes Lawrence, 'certainly isn't a snowy cold world, like a world of our own gone cold. Nonsense. It is a globe of dynamic substance, like radium, or phosphorus, coagulated upon a vivid pole of energy.' The defect of this statement is that it happens to be demonstrably untrue. The moon is quite certainly not made of radium or phosphorus. The moon is, materially, 'a stone'. Lawrence was angry (and he did well to be angry) with the nothing-but philosophers who insist that the moon is only a stone. He knew that it was something more; he had the empirical certainty of its deep significance and importance. But he tried to explain this empirically established fact of its significance in the wrong terms—in terms of matter and not of spirit. To say that the moon is made of radium is nonsense. But to say, with Socrates, that it is made of god-stuff is strictly accurate. For there is nothing, of course, to prevent the moon from being both a stone and a god. The evidence for its stoniness and against its radiuminess may be found in any children's encyclopaedia. It carries an absolute conviction. No less convincing, however, is the evidence for the moon's divinity. It may be extracted from our own experiences, from the writings of the poets, and, in fragments, even from certain textbooks of physiology and medicine.

But what is this 'divinity'? How shall we define a 'god'? Expressed in psychological terms (which are primary—there is no getting behind them), a god is something that gives us the peculiar kind of feeling which Professor Otto has called 'numinous' (from the Latin *numen*, a supernatural being). Numinous feelings are the original god-stuff, from which the theory-making mind extracts the individualized gods of the pantheons, the various attributes of the One. Once formulated, a theology evokes in its turn numinous feelings. Thus, men's terrors in face of the enigmatically dangerous universe led them to postulate the existence of angry gods; and, later, thinking about angry gods made them feel terror, even when the universe was giving them, for the moment, no cause of alarm. Emotion, rationalization, emotion—the process is circular and continuous. Man's religious life works on the principle of a hot-water system.

The moon is a stone; but it is a highly numinous stone. Or, to be more precise, it is a stone about which and because of which men and women have numinous feelings. Thus, there is a soft moonlight that can give us the peace that passes understanding. There is a moonlight that inspires a kind of awe. There is a cold and austere moonlight that tells the soul of its loneliness and desperate isolation, its insignificance or its uncleanness. There is an amorous moonlight prompting to love—to love not only for an individual but sometimes even for the whole universe. But the moon shines on the body as well as, through the windows of the eyes, within the mind. It affects the soul directly; but it can affect it also by obscure and circuitous ways—through the blood. Half the human race lives in manifest obedience to the lunar rhythm; and there is evidence to show that the physiological and therefore the spiritual life, not only of women, but of men too, mysteriously ebbs and flows with the changes of the moon. There are unreasoned joys, inexplicable miseries, laughters and remorses without a cause. Their sudden and fantastic alternations constitute the ordinary weather of our minds. These moods, of which the more gravely numinous may be hypostasized as gods, the lighter, if we will, as hobgoblins and fairies, are the children of the blood and humours. But the blood and humours obey, among many other masters, the changing moon. Touching the soul directly through the eyes and, indirectly, along the dark channels of the blood, the moon is doubly a divinity. Even dogs and wolves, to judge at least by their nocturnal howlings, seem to feel in some dim bestial fashion a kind of numinous emotion about the full moon. Artemis, the goddess of wild things, is identified in the later mythology with Selene.

Even if we think of the moon as only a stone, we shall find its very stoniness potentially a *numen*. A stone gone cold. An airless, waterless stone and the prophetic image of our own earth when, some few million years from now, the senescent sun shall have lost its present fostering power.... And so on. This passage could easily be prolonged—a Study in Purple. But I forbear. Let every reader lay on as much of the royal rhetorical colour as he finds to his taste. Anyhow, purple or no purple, there the stone is—stony. You cannot think about it for long without finding yourself invaded by one or other of several essentially numinous sentiments. These sentiments belong to one or other of two contrasted and complementary groups. The name of the first family is Sentiments of Human Insignificance, of the second, Sentiments of Human Greatness. Meditating on that derelict stone afloat there in the abyss, you may feel most numinously a worm, abject and futile in the face of wholly incomprehensible immensities. 'The silence of those infinite spaces frightens me.'

You may feel as Pascal felt. Or, alternatively, you may feel as M. Paul Valéry has said that he feels. 'The silence of those infinite spaces does *not* frighten me.' For the spectacle of that stony astronomical moon need not necessarily make you feel like a worm. It may, on the contrary, cause you to rejoice exultantly in your manhood. There floats the stone, the nearest and most familiar symbol of all the astronomical horrors; but the astronomers who discovered those horrors of space and time were men. The universe throws down a challenge to the human spirit; in spite of his insignificance and abjection, man has taken it up. The stone glares down at us out of the black boundlessness, a *memento mori*. But the fact that we know it for a *memento mori* justifies us in feeling a certain human pride. We have a right to our moods of sober exultation.

1931

JAMES THURBER

My Own Ten Rules for a Happy Marriage

Nobody, I hasten to announce, has asked me to formulate a set of rules for the perpetuation of marital bliss and the preservation of the tranquil American boudoir and inglenook. The idea just came to me one day, when I watched a couple in an apartment across the court from mine gesturing and banging tables and throwing *objets d'art* at each other. I couldn't hear what they were saying, but it was obvious, as the shot-put followed the hammer throw, that he and/or she (as the lawyers would put it) had deeply offended her and/or him.

Their apartment, before they began to take it apart, had been quietly and tastefully arranged, but it was a little hard to believe this now, as he stood there by the fireplace, using an andiron to bat back the Royal Doulton figurines she was curving at him from her strongly entrenched position behind the davenport. I wondered what had started the exciting but costly battle, and, brooding on the general subject of Husbands and Wives, I found myself compiling my own Ten Rules for a Happy Marriage.

I have avoided the timeworn admonitions, such as 'Praise her new hat', 'Share his hobbies', 'Be a sweetheart as well as a wife', and 'Don't keep a blonde in the guest room', not only because they are threadbare from repetition, but also because they don't seem to have accomplished their purpose. Maybe what we need is a brand-new set of rules. Anyway, ready or not, here they come, the result of fifty years (I began as a little boy) spent in studying the nature and behaviour, mistakes and misunderstandings, of the American Male (*homo Americansis*) and his Mate.

RULE ONE: Neither party to a sacred union should run down, disparage or badmouth the other's former girls or beaux, as the case may be. The tendency to attack the character, looks, intelligence, capability, and achievements of one's mate's former friends of the opposite sex is a common cause of domestic discontent. Sweetheart-slurring, as we will call this deplorable practice, is encouraged by a long spell of gloomy weather, too many highballs, hang-overs, and the suspicion that one's spouse is hiding, and finding, letters in a hollow tree, or is intercepting the postman, or putting in secret phone calls from the corner drugstore. These fears almost always turn out to be unfounded, but the unfounded fear, as we all know, is worse than the founded.

Aspersions, insinuations, reflections or just plain cracks about old boy friends and girl friends should be avoided at all times. Here are some of the expressions that should be especially eschewed: 'That waffle-fingered, minor-league third baseman you latched on to at Cornell'; 'You know the girl I mean—the one with the hips who couldn't read'; 'That old flame of yours with the vocabulary of a hoot owl'; and 'You remember her—that old bat who chewed gum and dressed like Daniel Boone.'

This kind of derogatory remark, if persisted in by one or both parties to a marriage, will surely lead to divorce or, at best, a blow on the head with a glass ash tray.

RULE TWO: A man should make an honest effort to get the names of his wife's friends right. This is not easy. The average wife who was graduated from college at any time during the past thirty years keeps in close touch with at least seven old classmates. These ladies, known as 'the girls', are named, respectively: Mary, Marian, Melissa, Marjorie, Maribel, Madeleine, and Miriam; and all of them are called Myrtle by the careless husband we are talking about. Furthermore, he gets their nicknames wrong. This, to be sure, is understandable, since their nicknames are, respectively: Molly, Muffy, Missy, Midge, Mabby, Maddy, and Mims. The careless husband, out of thoughtlessness or pure cussedness, calls them all Mugs, or, when he is feeling particularly brutal, Mucky.

All the girls are married, one of them to a Ben Tompkins, and as this is the only one he can remember, our hero calls all the husbands Ben, or Tompkins, adding to the general annoyance and confusion.

If you are married to a college graduate, then, try to get the names of her girl friends and their husbands straight. This will prevent some of those interminable arguments that begin after Midge and Harry (not Mucky and Ben) have said a stiff goodnight and gone home.

RULE THREE: A husband should not insult his wife publicly, at parties. He should insult her in the privacy of the home. Thus, if a man thinks the soufflés his wife makes are as tough as an outfielder's glove, he should tell her so when they are at home, not when they are out at a formal dinner party where a perfect soufflé has just been served. The same rule applies to the wife. She should not regale his men friends, or women friends, with hilarious accounts of her husband's clumsiness, remarking that he dances like a 1907 Pope Hartford, or that he locked himself in the children's rabbit pen and couldn't get out. All parties must end finally, and the husband or wife who has revealed all may find that there is hell to pay in the taxi going home.

RULE FOUR: The wife who keeps saying, 'Isn't that just like a man?' and the husband who keeps saying, 'Oh, well, you know how women are,' are likely to grow farther and farther apart through the years. These famous generalizations have the effect of reducing an individual to the anonymous status of a mere unit in a mass. The wife who, just in time, comes upon her husband about to fry an egg in a dry skillet should not classify him with all other males but should give him the accolade of a special distinction. She might say, for example, 'George, no other man in the world would try to do a thing like that.' Similarly, a husband watching his wife labouring to start the car without turning on the ignition should not say to the gardener or a passer-by, 'Oh, well, you know, etc.' Instead, he should remark to his wife, 'I've seen a lot of women in my life, Nellie, but I've never seen one who could touch you.'

Certain critics of this rule will point out that the specific comments I would substitute for the old familiar generalities do not solve the problem. They will maintain that the husband and wife will be sore and sulky for several days, no matter what is said. One wife, reading Rule Four over my shoulder, exclaimed, 'Isn't that just like a man?' This brings us right back where we started. Oh, well, you know how women are!

RULE FIVE: When a husband is reading aloud, a wife should sit quietly in her chair, relaxed but attentive. If he has decided to read the Republican platform, an article on elm blight, or a blow-by-blow account of a prize fight, it is not going to be easy, but she should at least pretend to be interested. She should not keep swinging one foot, start to wind her wrist watch, file her fingernails, or clap her hands in an effort to catch a mosquito. The good wife allows the mosquito to bite her when her husband is reading aloud.

She should not break in to correct her husband's pronunciation, or to tell him one of his socks is wrong side out. When the husband has finished, the wife should not lunge instantly into some irrelevant subject. It's wiser to exclaim, 'How intersting!' or, at the very least, 'Well, well!' She might even compliment him on his diction and his grasp of politics, elm blight or boxing. If he should ask some shrewd question to test her attention, she can cry, 'Good heavens!' leap up, and rush out to the kitchen on some urgent fictitious errand. This may fool him, or it may not. I hope, for her sake—and his—that it does.

RULE SIX: A husband should try to remember where things are around the house so that he does not have to wait for his wife to get home from the hairdresser's before he can put his hands on what he wants. Among the things a husband is usually unable to locate are the iodine, the aspirin, the nail file, the French vermouth, his cuff links, studs, black silk socks and evening shirts, the snapshots taken at Nantucket last summer, his favourite record of 'Kentucky Babe', the borrowed copy of *My Cousin Rachel*, the garage key, his own towel, the last bill from Brooks Bros, his pipe cleaners, the poker chips, crackers, cheese, the whetstone, his new raincoat, and the screens for the upstairs windows.

I don't really know the solution to this problem, but one should be found. Perhaps every wife should draw for her husband a detailed map of the house, showing clearly the location of everything he might need. Trouble is, I suppose, he would lay the map down somewhere and not be able to find it until his wife got home.

RULE SEVEN: If a husband is not listening to what his wife is saying, he should not grunt, 'Okay' or 'Yeah, sure', or make little affirmative noises. A husband lost in thought or worry is likely not to take in the sense of such a statement as this: 'We're going to the Gordons for dinner tonight, John, so I'm letting the servants off. Don't come home from the office first. Remember, we both have to be at the dentist's at five, and I'll pick you up there with the car.' Now, an 'Okay' or a 'Yeah, sure' at this point can raise havoc if the husband hasn't really been listening. As usual, he goes all the way out to his home in Glenville—thirteen miles from the dentist's office and seventeen miles from the Gordons' house—and he can't find his wife. He can't find the servants. His wife can't get him on the phone because all she gets is the busy buzz. John is calling everybody he can think of except, of course, in his characteristic way, the dentist and the Gordons. At last he hangs up, exhausted and enraged. Then the phone rings. It is his wife. And here let us leave them.

RULE EIGHT: If your husband ceases to call you 'Sugar-foot' or 'Candy Eyes' or 'Cutie Fudge Pie' during the first year of your marriage, it is not neccessarily a sign that he has come to take you for granted or that he no longer cares. It is probaby an indication that he has recovered his normal perspective. Many a young husband who once called his wife 'Tender Mittens' or 'Taffy Ears' or 'Rose Lips' has become austere or important, like a Common Pleas Judge, and he wouldn't want reports of his youthful frivolity to get around. If he doesn't call you Dagmar when your name is Daisy, you are sitting pretty.

RULE NINE: For those whose husbands insist on pitching for the Married Men against the Single Men at the Fourth-of-July picnic of the First M. E. Church, I have the following suggestion: don't sit on the sidelines and watch him. Get lost. George is sure to be struck out by a fourteen-year-old boy, pull up with a Charley horse running to first, and get his teeth knocked out by an easy grounder to the mound. When you see him after the game, tell him everybody knew the little boy was throwing illegal spitballs, everybody saw the first baseman spike George, and everybody said that grounder took such a nasty bounce even Phil Rizzuto couldn't have fielded it. Remember, most middle-aged husbands get to sleep at night by imagining they are striking out the entire batting order of the Yankees.

RULE TEN: A wife's dressing table should be inviolable. It is the one place in the house a husband should get away from and stay away from. And yet, the average husband is drawn to it as by a magnet, especially when he is carrying something wet, oily, greasy or sticky, such as a universal joint, a hub cap, or the blades of a lawn mower. His excuse for bringing these alien objects into his wife's bedroom in the first place is that he is looking for 'an old rag' with which to wipe them off. There are no old rags in a lady's boudoir, but husbands never seem to learn this. They search hampers, closets, and bureau drawers, expecting to find a suitable piece of cloth, but first they set the greasy object on the dressing table. The aggrieved wife may be tempted, following this kind of vandalism, to lock her bedroom door and kick her husband out for good. I suggest, however, a less stringent punishment. Put a turtle in his bed. The wife who is afraid to pick up a turtle should ask Junior to help her. Junior will love it.

Now I realize, in glancing back over these rules, that some of my solutions to marital problems may seem a little untidy; that I have, indeed, left a number of loose ends here and there. For example, if the husbands are going to mislay their detailed maps of household objects, I have accomplished nothing except to add one item for the distraught gentleman to lose.

Then, there is that turtle. Captious critics will point out that a turtle in a husband's bed is not a valid solution to anything, but merely a further provocation. The outraged husband will deliberately trip his wife during their next mixed-doubles match. She will thereupon retaliate by putting salt in his breakfast coffee ...

Let somebody else try to figure out what to do about the Running Feud in marriage. The Williamses are coming to dinner tonight, and I promised to put the white wine on the ice at three o'clock. It is now six-thirty. After all, I have my own problems.

1953

The Toy Farm

ANGELA, at the house where I am staying, has just celebrated a
birthday, her seventh, and is now the breathless mistress of a toy
farm. You never saw such a farm. It has barns, haystacks, sties,
hurdles, gates, trees (which must be looked at only from the
front), and a yellow tumbril with scarlet wheels. There are fat
brown horses, cows that stand up and cows that sit down, black
pigs and pink pigs, sheep with their lambs, a goat, two dogs (one
staring fiercely out of a kennel), and a coloured host of turkeys,
ducks, hens and chicks. There are even people on this farm, five
of them, and very fine people they are too. A man in his shirt-
sleeves perpetually pushes a crimson wheelbarrow; and two
carters, wearing white smocks, brown gaiters, red scarves, and
little round hats, for ever stride forward, whips in hand, whistling tunes that we shall never catch. Then there is the farmer
himself, bluff, whiskered, in all his bravery of scarlet waistcoat,
white cravat, and green breeches, who grasps his stout stick and
stares at things from under his hard brown hat. His wife, neat
and buxom in a blue bonnet, a pink gown, and snowy apron,
with a basket in one hand and a large green umbrella in the
other, is setting out upon some never-to-be-accomplished
errand. All these people, labourers, master, mistress, though not
more than two inches high and only made of painted tin, stand
there for ever confident, ruddy, smiling in perpetual sunshine:
they seem to stare at us out of a lost Arcadia.

Perhaps that is why poor Angela has not so far had that farm
to herself, being compelled to share it with a number of shameless adults. It is, of course, an engaging toy, and there is not
one of us here, I am thankful to say, so old and wicked, so
desiccated, as to have lost all delight in toys, particularly those
that present something huge and elaborate, such as a fort
crammed with soldiers, a battleship, a railway station, a farm, on

a tiny scale and in brighter hues than Nature ever knew. These toys transform you at a stroke into a god, and a happier god than any who look down upon our sad muddle. It is, of course, the more poetical of our activities that are chosen as subjects for these bright miniatures of the nursery, yet there is so much poetry in the toys themselves that even if they mirrored in little even the most prosaic things, they would still be satisfying. I remember that when I was a child, the boy next door was given a tiny printing machine, a gasping, wheezing affair that would print nothing but the blurred image of three ducks. He and I, however, collecting all the paper we could lay our hands on, would spend hours, hours full to the brim, printing ducks, thousands and thousands of ducks, and while we were engaged in producing this monotonous sequence of dim fowls we asked nothing more from life beyond the promise of suety meals at odd intervals.

Yet so far, nobody, not even in America, I imagine, has produced a toy miniature of business life, the Limited Company complete in box from ten shillings upwards. What Angela and her like would think of such a toy I do not know, though their sense of wonder is sufficiently strong for them to find entertainment in anything; but I do know that I should be tempted to buy one this very morning. You would have a building, with the front wall removed as it is in the best doll's-houses, so that you could arrange the people and the furniture just as you pleased. There would be tiny stenographers and clerks and cashiers; typewriters, calculating machines, ledgers and files no bigger than your finger-nail; telephones that you could just see and never hear; and all manner of things, chairs and tables and desks, to be shifted from one room to another, from the Counting House to the Foreign Department, and so forth. There would be a Board Room with four or five directors, fat little chaps in shiny black, with the neatest, tiniest spats imaginable, all sitting round a table some six inches long. In the best sets you would be given a Chairman, quarter of an inch taller than the others and costing perhaps a penny more, who might be so contrived that he stood perpetually at the head of the table addressing his fellow directors. If I had him I should call him Sir Glossy Tinman. Then, if you wanted to do the thing properly, you would be able to buy Debenture Holders at two shillings or half a crown the dozen, complete with an interrupter who was rising to his feet and holding up an arm, the very image, in tin and varnish, of a retired Colonel of the Indian Army. Nor would you stop there; the possibilities are almost endless; and I promise to outline some of them to any enterprising manufacturer of toys who should consider putting the complete Limited Company on the market.

It may be, though, that there are special reasons why we should all be finding the toy farm so enchanting. Its little people, as I have said, seem to stare at us out of a lost Arcadia. Behind them, and their bright paraphernalia of beasts and belongings, is the Idea, dominating the imagination. This farmer and his wife are the happy epitome of all farmers and their wives, but they are unmistakably idealised. These white-smocked carters, for ever soundlessly whistling among the clover, are not the countrymen we know in miniature, but are images from an old dream of the countryside. Looking at these trees, or at least looking at them from the front, we might cry with Keats:

> Ah, happy, happy boughs! that cannot shed
> Your leaves, nor ever bid the Spring adieu.

Here is the bright epitome, not of the country we can find where the tram-lines come to an end and the street lamps fade out, but of the country that has always existed in our imagination, so clean, trim, lavishly coloured. None of us here, I venture to say, has any passion for agriculture as a pursuit, for real farms, with their actual lumbering beasts, their mud and manure, their clumsy and endless obstetrics, their mortgages and loans and market prices, their long days of wet fields and dirty straw. We may regard the farmer as an excellent solid fellow or as a grasping ruffian, but certainly he never seems to us a poetical figure whose existence is passed in a golden atmosphere. Yet there is such a farmer somewhere at the back of our minds, a farmer in a picture-book, and this piece of painted, moulded tin is his portrait. If we could only find him in this actual life, not all the pleasures of the town would keep us from living in his shadow all the rest of our days, for we know that his world is one long dreamed of, that countryside where there are no ugly downpours, no sodden fields and lanes choked with mud, where only the gentlest shower of rain breaks through the sunshine, where

everything is as clean as a new pin and fresh from the paint-box,
where men and women are innocent and gay and the very beasts
are old friends, where sin and suffering and death are not even a
distant rumour. Is not this the Arcadia that men lost long ago
and have never found again?

How long this dream has lasted no man can say. It shines
through all literature, from the poets and novelists of yesterday
to Virgil and Theocritus. It is the burden of more than one half
of our old songs, with all their 'Hawthorne buds, and sweet
Eglantine, and girlonds of roses, and Sopps in wine', their
Corydons and neat-handed Phyllises, their haymakers, rakers,
reapers, and mowers waiting on their Summer Queen, their
hey-down-derry or shepherd lovers in the shade. And always this
lovely time

> When Tom came home from labour,
>   Or Cis to milking rose,
> Then merrily went their tabor,
>   And nimbly went their toes

had just passed away. Nobody ever saw this contryside, but it
was always somewhere round the corner; a turn at the end of a
long road, a descent from some strange hill, and there it might
be, shining in the sun. It is not the perfervid vision of townsmen,
longing for the fields in their wilderness of bricks and mortar, a
revolt against the ugly mechanical things of today, but a dream
that would appear to be as old as civilised man himself, touching
men's imagination when towns were little more than specks
in the green countryside. Poets who lived in the country, who
passed all their days among real shepherds and dairymaids, could
sing of this other country where there was nothing ugly nor any
pain and sorrow, knowing full well that this was not the land
that stretched itself beyond their gates but a land they had never
seen. It is one of the more homely manifestations of that ideal of
unchanging beauty which haunts the mind of man everywhere
and in every age, and from which there is no escape except into
brutishness. Its shadow can fall even upon a number of little
pieces of painted tin newly come from the toyshop.

1927

The Case for Xanthippe

T HOUGH I rely on intuition for the writing of poems and for the general management of my life, intuition must obviously be checked by reason whenever possible. Poets are (or ought to be) reasonable people; poems, though born of intuition, are (or ought to be) reasonable entities, and make perfect sense in their unique way. I should not, however, describe either poems or poets as 'rational'. 'Reasonable' has warm human connotations; 'rational' has coldly inhuman ones. Examine the abstract nouns that both adjectives yield. The stock epithet for 'reasonableness', first used by Matthew Arnold, is 'sweet'. The usual epithets for 'rationality' are not at all affectionate; and those for 'rationalization' are often positively crude.

Dear, useful Reason! The technique of isolating hard facts from a sea of guess, or hearsay, or legend; and of building them, when checked and counter-checked, into an orderly system of cause and effect! But too much power and glory can be claimed for this technique. Though helpful in a number of routine tasks, Reason has its limitations. It fails, for example, to prompt the writing of original poems, or the painting of original pictures, or the composing of original music; and shows no spark of humour or religious feeling.

Reason was still warmly reasonable three thousand years ago. Even geometry ('the measurement of land') began as a practical means of redistributing the cornlands of Egypt, which were left without landmarks after each annual Nile flood. But Greek philosophers could not keep their fingers off geometry. To convince doubters that the lands had been justly divided, they took it upon themselves to prove the measurements rational as well as reasonable, by abstract argument. Hence the theorems of Pythagoras and Euclid. Reason gets out of hand once it deals with abstractions.

Abstract reasoning under the name of 'philosophy' became a new sport for the leisured classes of Greece, and was applied not only to mathematics and physics, but to metaphysics. Abstract reason soon ranked higher than practical reason, as seeming more remote and godlike, and as further distinguishing mankind from the beasts. Indeed, philosophers belittled poetic myth; and their habit of substituting rational abstractions for gods and goddesses conceived in the human image caused a religious malaise from which Greece never recovered. When old gods lost their hold, the sanctity of oaths and treaties dimmed. The infection spread through neighbouring countries. Jesus' warning 'Do not put new wine into old bottles!' seems to have been directed against the Grecians—Jewish-Egyptian followers of the philosopher Philo, who had given Jehovah's practical laws of 'thou shalt!' and 'thou shalt not!', promulgated by Moses, an exciting new Platonic interpretation.

Philosophy is antipoetic. Philosophize about mankind and you brush aside individual uniqueness, which a poet cannot do without self-damage. Unless, for a start, he has a strong personal rhythm to vary his metrics, he is nothing. Poets mistrust philosophy. They know that once heads are counted, each owner of a head loses his personal identify and becomes a number in some government scheme: if not as a slave or serf, at least as a party to the device of majority voting, which smothers personal views. In either case, to use an old-fashioned phrase, he will be a 'mere cipher'. An ominous count of man-power always precedes its translation into sword- or cannon-fodder. Shortly before our Christian era began, the Romans took to philosophy and rationalized their politics by killing more than three million people in a few years of civil and foreign warfare. Politics were further rationalized when Julius Caesar, after gaining supreme power, established a semblance of order by making godhead the superman's privilege.

Women and poets are natural allies. Greek women had opposed the philosophers' free exercise of abstract reasoning—which was then, as now, a predominantly male field of thought. They considered it a threat to themselves; and they were right. Many important discoveries had been made in bygone centuries: such as plough-agriculture, the potter's wheel, the alphabet, weights and measures, navigation by the stars. But each of these had been reasonably absorbed into the corpus of poetic myth, by attributing its discovery to some god, goddess or hero, and hallowing its use with religious rites. Poetic myths gave city-states their charters and kept society on an even keel—until the metaphysicians proved myth irrational and chose Socrates as their revolutionary hero and master. Sweet reasonableness was wanting in Socrates: 'So long as I breathe,' he declared, 'I will never stop philosophizing!' His homosexual leanings, his absent-minded behaviour, his idleness, and his love of proving everyone wrong, would have endeared him to no wife of mettle. Yet Xanthippe is still pilloried as a shrew who could not understand her husband's spiritual greatness; and Socrates is still regarded as a saint because he patiently bore with her reproaches.

Let me break a lance for Xanthippe. Her intuitions were sound. She foresaw that his metaphysical theories would bring the family into public disgrace and endanger the equipoise of the world she knew. Whenever the rational male intellect asserts itself at the expense of simple faith, natural feeling and sweet reasonableness, there follows a decline in the status of women—who then figure in statistics merely as child-bearers and sexual conveniences to men; and a decline in the status of poets—who cannot be given any effective social recognition; also an immediate increase in wars, crime, mental ill-health and physical excess.

With Christianity, the pendulum at first swung back towards personal religion. But fourth century Roman bishops, by offering a simple faith to every class in the Empire and winning over a large part of the Army, rose to be the Emperor Constantius's State priests. They soon closed the pagan universities and took control of education. All reputable branches of learning and the arts were brought into the Christian fold, poets ceased to exist, and all schools became Church schools. Yet the Church, while basing her doctrines on primitive Jewish beliefs, and the hope of salvation, had not dared to ban philosophy and thus admit herself irrational. Theologians married personal faith with abstract reason, taking immense pains to make Christian doctrine logically unassailable. And though Jesus had warned his disciples not to prepare their arguments beforehand, but to extemporize them intuitively, missionaries and apologists were trained as rhetoricians. Nor did rhetoric any longer mean the practical art of reasoning from evidence: it had been rationalized as the science of dazzling with irrelevances, and of misleading by ingenious twists of argument. Worse: abstract theological speculations about the Otherworld, which now engrossed intellectuals, bred an unreasonable disregard for the practical problems of living. Gregory Nazianzenus' fourth-century metrical homilies were considered enough for the Greek-speaking world: the works of famous pagan poets went to light fires—which accounts for our loss of Sappho's poems, except in quotation. Rome decayed, fresh barbarian hordes descended from the North, the Dark Ages ensued.

The Church's close monopoly of all mental exercises did not weaken until the Crusaders, coming into contact with the Orient, tapped new sources of knowledge; and until vernacular poetry sprang up again in the ruins of the Empire. This monopoly weakened still more after the Reformation, when the pursuit·of experimental science was revived in territory outside the Pope's control. The Church, having frozen her dogma a thousand years before, failed to check, or even keep track of, new scientific heresies. Nevertheless, she still applied a severe censorship within her own realm. As late as 1632, Galileo was imprisoned by Papal authority for endorsing a heretical German view that the Earth goes round the Sun, and not contrariwise: as though it made a pennyworth of difference to agriculture, navigation, medicine, industry or morals which went round which! Galileo's misfortune set religious dogma and anti-religious science at loggerheads: a struggle culminating in the bloody French Revolution, when the Paris mob disavowed its Catholicism by enthroning Reason as a goddess.

Since those unhappy days, tacit agreements of co-existence have been reached between rival systems. of abstract thought. Scientists abstain from attacks on theology; theologians tolerate the free pursuit of science. But co-existence is not integration. The belief in miracles, when first preached by the Church, was reconcilable with contemporary Greek thought—parthenogenesis, resurrections from the dead, and ascensions to Heaven having been widely accepted as rare yet authenticated occurrences. If, some centuries later, miracles seemed anti-scientific, the

Church's belief could be framed in more cautious terms: 'From time to time, during the Biblical epoch, certain Laws of Nature which had hitherto always held good, and still hold good, were briefly suspended by God within a chosen geographical area.' This doctrine is logically defensible, given All-Powerful God as an axiom, and still prevails among the elder Churches. Even Catholic priests now give scientific technology their blessing, though aware that scientists pursue knowledge for the sake of knowledge alone; remain officially divorced from religion; and, as scientists, have a very limited moral code. The code prevents them from faking the results of their experiments, from withholding due credit to fellow-scientists, and from suppressing newly discovered knowledge; but that is all.

Abstract reason, formerly the servant of practical human reason, has everywhere become its master, and denies poetry any excuse for existence. What were the Nazi surgeons who conducted 'devilish' experiments on 'non-Aryan' prisoners—if not dedicated scientists taking full advantage of the unusual opportunities offered them by Hitler's irreligion? The first essay in nuclear warfare, however, originated in a Christian country. And as for the more recent tests that scatter long-lived, man-eating isotopes throughout the biosphere—any woman of healthy intuitive powers could have told the scientists long ago that they were playing with worse than fire.

But who cares for female intuition? Most modern Xanthippes behave reasonably: rather than scold, they shrug, and leave Socrates to theorize and experiment at his pleasure—though, of course, reserving the right to blame him afterwards when things go wrong. And who cares for poetic intuition, except perhaps women? Certainly neither priests nor scientists do. A sentimental pagan glory may attach to the name of 'poet', despite the Church, and despite scientists; but what he says carries no weight. He is not rational, but intuitive.

The word 'intuition' must be used with extreme care. Intuition, like instinct, is a natural faculty shared by both sexes—instinct being the feeling which prompts habitual actions. You see a wriggling in the grass; instinct tells you (foolishly perhaps) to retreat. You burn your finger; instinct tells you (also foolishly perhaps) to dip it in cold water. Intuition, however, has no concern with habits: it is the mind working in a trance at problems which offer only meagre data for their rational solution. Male intellectuals therefore tend to despise it as an irrational female way of thinking. Granted, fewer women than men have their intuitive powers blunted by formal schooling; yet, oddly enough, only men who have preserved them unspoilt can hope to earn the name of 'genius'.

Genius has been mistaken for the obsessional industry that often goes with it; but to be a true genius, whether as a poet or scientist, implies thought on a profound intuitive level—the drawing upon an inexhaustible store of miscellaneous experience absorbed and filed away in subterranean cellars of memory, and then making a mental leap across the dark void of ignorance. The nucleus of every true poem is a single phrase which (the poet's intuition tells him) provides a key to its eventual form. But this nucleus is as much as he has to work upon consciously.

All scientists ratiocinate; few have the intuition that will carry them safely across the dark void into some new field of research. If it were not for occasional geniuses—minds which first think intuitively and then rationalize their findings—science would still be back in the Dark Ages. Kekulé the chemist, who discovered the aniline complex, and Rowan Hamilton the mathematician, who invented quaternions, did so by sudden visionary flashes—Kekulé in a waking dream; Hamilton while walking idly one morning across a Dublin park. Later, these inspirations were built into the scientific system, and acclaimed as triumphs of reason. Unfortunately Kekulé, Rowan Hamilton, and the rest applied their intuitive powers to abstract rather than human problems. The exploitation of their findings in the chemical and electronic industries has therefore made life more rational than reasonable.

Rational schooling on the shallow level (as a discipline imposed on all students, whatever their bent) discourages intuitive thought. Our civilization is geared to mass-demand, statistically determinable, for commodities which everyone should either make, market, or consume. Good citizens eat what others eat, wear what others wear, behave as others behave, read what others read, think as others think. This system has its obvious economic advantages, and supports vast populations; but crime and sickness due to maladjustment are rationally glossed over as being inseparable from a pursuit of the majority good.

Our technologists can do little better than provide the wretched, maladjusted minority with tranquillizers, drink, prisons, mental hospitals, and teams of social workers.

Maladjustment is due, largely, to conflicts between rationalization and human instinct. City dwellers are forced to live by artifice, rather than by natural appetite. And every year the urban dragon swallows up more small towns; every year wider agricultural areas are industrialized. Not that old-fashioned country-folk are demonstrably happier than well-adjusted city-dwellers—who have grown so used to their surroundings that they find the taste of fresh milk and farmhouse bread positively repulsive, and feel ill at ease in antiquated crooked houses. But humane, creative thought, which depends on intuition, withers under the abstract rule of urban reason.

Myself, I left the city long ago and, whenever I return on a visit, find medicine, art, literature and entertainment still further rationalized. Too few physicians practise the intuitive diagnosis that used to be expected from every medical man while medicine was still a calling rather than a business. Too many young painters have decided that there is no escape from commercial art but to go non-representational; and non-representational art, once the prerogative of wild men, has therefore turned academic. Literature is a trade: 'creative writing' courses supply the know-how. Organized entertainment rests on a pseudo-science of audience-reaction—an axiom of which is that the public should have its drama, sentiment and humour as hygienically and economically processed as its food.

Our predicament is technological maturity linked with emotional immaturity. While the scientific world needs intuitive thinking for further progress, the world of politics demands a purely rational approach to life, and rejects all reasonable intuitions of a humaner sort. Efforts to save mankind from near-suicide made by the United Nations and lesser philanthropic groups are too impersonally expressed to be of much avail; nor have women been given an opportunity, even at this late stage, to exert an influence commensurate with their numbers and wealth. The truth is that politicians, salesmen, priests, teachers and scientists are (often against their private conscience) forcibly banded together in the public interest against all who demand personal liberty of thought. Thus, if catastrophe finally halts the blind progress of rationalization ... But this is a favourite field of science-fiction, and better left untrod.

Meanwhile the few remaining poets are pledged, by the age-old loyalty which they owe their Muse-goddess, to resist all pressure from mechanistic philosophy. I should not claim, as Shelley did, that poets are the 'law-givers of mankind'; but some still uphold the principle of imaginative freedom. Socrates was perhaps aware of this peculiarity when he banned poets from his humourless Republic—I even wonder whether it may not have been the cause of Xanthippe's quarrel with him. Did a dialogue, unrecorded by Plato, take place perhaps between Socrates and an angry, unyielding poet, Xanthippe's lover—in which the honours went elsewhere?

I am no outlaw by temperament; I simply suffer from a poetic obsession which an increasing number of reasonable people share with me. Poetry, for us, means not merely poems but a peculiar attitude to life. Though philosophers like to define poetry as irrational fancy, for us it is a practical, humorous, reasonable way of being ourselves. Of never acquiescing in a fraud; of never accepting the second-rate in poetry, painting, music, love, friends. Of safeguarding our poetic institutions against the encroachments of mechanized, insensate, inhumane, abstract rationality.

1960

A Preface to Persius

MAUDLIN MEDITATIONS IN A SPEAKEASY

THE other evening I had to dine out alone, and, stopping in on my way at a book store, bought a little eighteenth-century edition of Persius, with notes and a translation. The editor and translator was a man named William Drummond, who had also been a member of Parliament. The attractive duodecimo was bound in green morocco and stamped in gold, and, inside the cover, had gray marbled paper with green and yellow runnings. There was a medallion of Aulus Persius Flaccus, with crisp metallic curls on the title-page, and the whole volume, with the edges of its pages gilded on all three sides and its well-spaced and small clear type, had the aspect of a little casket in which something precious was kept. I went to an Italian restaurant and, while I was waiting for the antipasto, I began to read the preface. 'In offering to the public,' it ran, 'a new English version of Persius, my object has rather been to express his meaning clearly, than either to translate his words literally, or to copy his manner servilely. The sentiments of this satirist are indeed admirable, and deserve to be better known than they are; but his poetry cannot be praised for its elegance, nor his language for its urbanity.'

The plate arrived, with a glistening sardine, little purple olives, two pearly leeks, bronze anchovies and bright red pimiento slices that looked like little tongues, all against a lining of pale lettuce. There were also a bottle of yellow wine and a goblet of pale green glass. I wondered what had become of the sort of thing that this editor of Persius represented. In the middle of the room where I sat was a party of men and women, all pink and of huge size, who were uncorking loud sour laughter; across the room was a quite pretty young girl, of an obviously simple nature, who had some sort of keen professional interest in pleasing a rather defective-looking half-aquiline man, whose eyes one couldn't see through his eyeglasses. Craning around behind me, however, I caught sight of E. E. Cummings. Cummings, I reflected, was a cultivated fellow and a good writer and came from Boston, but was not a bit like William Drummond. For a point of view like that of Drummond, who would have reproached the inelegance of Persius's poetry yet applauded his admirable sentiments, one would have to go to our own eighteenth-century literature—to Joel Barlow or Philip Freneau. But Drummond was a fancier of letters and a political figure at the same time—Jefferson might have thought and written so. Some of the logic, some of the elegance, some of the moderate and equable opinion which seemed to be the qualities—here found, as it were, in their pure state—of this preface to an ancient classic had gone to the announcement of our national policies and the construction of our constitution. But new interests had taken over the government; and I had been reading in the paper that day of a typical example of their methods—an assault by the state constabulary on a meeting of Italian miners; men clubbed insensible, children gassed, old people badly beaten, a nursing woman knocked down. With the exception, I reflected, of Cummings, and possibly the Italian waiters, there was probably not a person in that room who would not either approve this action or refuse to believe that it had happened; and Cummings was as powerless to prevent its recurrence as the illiterate waiters would be. Perhaps the only element in sight which had anything in common with Drummond on Persius was the Italian dinner itself, of which a bowl of minestrone, with its cabbage, big brown beans and round noodles, had just reached me as the second course, and in the richness and balance of whose composition I could still see the standards of a civilization based on something more comfortably human than commercial and industrial interests. Yet how few generations of Italians speaking English and competing with the natives would it take for them to forget their cooking and their ideal of a good dinner and to go in, as one already saw them beginning to do, for expensive à la carte restaurants with heterogeneous and uninteresting menus?

'The defects of Persius, considered with respect to composition, cannot perhaps be easily defended. Even Casaubon, his fondest admirer, and most successful interpreter, admits that his style is obscure. If, however, any apology can be made for this

first sin against good writing, it is in the case of a satirist, and
above all, of a satirist who dared to reprobate the crimes, and to
ridicule the follies of a tyrant. If Persius be obscure, let it be
remembered, he lived in the time of Nero. But it has been
remarked, that this author is not obscure, only when he lashes
and exposes the Roman emperor. It was very well, it has been
said, to employ hints, and to speak in half sentences, while he
censured the vices of a cruel and luxurious despot; but there
could be no occasion for enveloping himself in obscurity, while
he expounded the doctrines of the Stoics to his friend Cornutus,
or expatiated to the poet Bassus on the true use of riches.'—
I looked up, as the chicken and greens were being set down
before me, and saw Cummings, who had finished and was leaving. If they had felt that way about Persius, what would they
have thought of Cummings? And what was the use, in the eighteenth century, of the critics' having cultivated those standards, if
Cummings was what the future had in store? He stopped at my
table, and I asked him where he had spent the summer. 'I
thought of going to Boston,' he ejaculated, 'to see the machine-
guns!'—but we've all seen plenty of machine-guns!—commonest
thing in the world!—so I walked around New York, expecting
to be blown up any moment—be a fine thing to blow the
subways up!—of course, my attitude toward this whole thing—I
mean, it's just unfortunate—it's a bore, like somebody losing his
pants—it's embarrassing, but it oughtn't to be a surprise to anybody—what surprises me is that they managed to stay alive for
seven years!—why, I've seen them shoot people first and search
them afterwards—and if they've got any bullets in them, they
arrest them for carrying concealed weapons!' He slipped away
with his spirited crest of hair and his narrow self-regarding eyes.
I addressed myself to the salad. So Persius, in another age that
combined moral anarchy with harsh repression, had, it seemed,
expressed himself confusedly, inelegantly and obscenely. And it
is Persius who is the writer and not the complacent Drummond,
as it is Cummings and not the persons who publish books on
American poetry. Where life is disorderly, the poets will express

---

' Sacco and Vanzetti had been executed in Boston on August 23, 1927, and the
demonstrations in protest against this had alarmed the city authorities to the point of
having the State House guarded with tommy-guns.

themselves in nonsense. I had looked at the beginning of Persius—'*O curas hominum! O quantum est in rebus inane!*'—it seemed to me entirely in the modern spirit.

I went on reading the preface: 'While, therefore, I fully admit the charge against Persius, I cannot allow to it that weight, which it would have in most other cases. Indeed, we may as well complain of the rust on an ancient coin, as of the obscurity of an ancient satire. Nature, it is true, always holds up the same mirror, but prejudice, habit, and education, are continually changing the appearance of the objects seen in it. The objections which have been made to my author in some other respects, are more difficult to answer. His unpolished verses, his coarse comparisons, and his ungraceful transitions from one subject to another, manifest, it is said, either his contempt or his ignorance of elegant composition. It cannot, indeed, be contended, that Persius displays the politeness of Horace, or that he shows himself an adept in the *callida junctura*. His poetry is a strong and rapid torrent, which pours in its infracted course over the rocks and precipices, and which occasionally, like the waters of the Rhone, disappears from view, and loses itself underground.' Yes: like Cummings's poetry and conversation. Yet Drummond had his poetry, too: 'the rust on an ancient coin,' 'a strong and rapid torrent, which pours in its infracted course'—and it appeared that he was, after all, sympathetic with the unpolished Persius and had earnestly undertaken to defend him against the taste of the time. That was the paradox of literature: provoked only by the anomalies of reality, by its discord, its chaos, its pain, it attempted, from poetry to metaphysics, to impose on that chaos some order, to find some resolution for that discord, to render that pain acceptable—to strike some permanent mark of the mind on the mysterious flux of experience which escapes beneath our hand. So with Persius, poised, as it were, on the edge of the collapse of the Empire, for whom the criticism of the satirist, the philosophy of the Stoic—at the least, the hexameter itself—were all ways, for even so 'inelegant' a poet, of introducing some logic and some meaning into the ceaseless struggles of men to make themselves at home in the universe. Then, as it were, relieving the poet, the critic who studies him, in turn, must stand firm against these miseries and horrors, these disquieting shocks of reality—he must pick up the poet's verses, all twisted where disaster has struck them, and he must carry them further, like Drummond, to where there is tranquillity and leisure enough for him to point out what form and what sense the poet had tried to give them, to supply by his own judicial comments the straightness and the soundness they lack. Yet, even beneath the shelter of that firmament of eighteenth-century order, he, too, has felt the shock of reality—the dullness of a rusted coin, the turbulence of a river. For without the impulse from reality, neither criticism nor poetry nor any other human work can be valid.

I had finished the apple, the Brie cheese and the little black demi-tasse, and I turned to the book again: 'I cannot conclude this Preface, without lamenting that an early and untimely death should have prevented the Poet, whom I have translated, from giving a more finished appearance to his works.' How extraordinary that William Drummond, almost two thousand years after Christ, should have felt this solidarity with Persius, that, bridging the ruin of Rome, bridging the confusion of the Middle Ages, we should find him lamenting this early death as if it were that of some able young man whom he had known in his time in London, some young man who had been educated at the same institutions and shared with him the same values. The discord and chaos of reality! From the point of view of civilization, the whole of the West had caved in. The geographical void of Europe had been too big for Rome to fill; and then later—to change the metaphor, as my wine made it easy to do—when plantations that had been ploughed under had scattered their seed abroad, and at last there had been bred all through Europe such a race as had formerly flourished only on the Mediterranean, a new race to whom Persius could speak as men of his own education—when this had been achieved, there opened, as it were in another dimension, a new void, the social void, below the class of educated people to which Persius and Drummond belonged, and into that yawning gulf of illiteracy and mean ambitions, even while Drummond wrote—the book was dated 1797—Europe heavily and dully sank, not without some loud crackings of her structure. America, in a sense, was the gulf.

I had finished the bottle of wine, which was certainly better than they had had last year—they were really making an effort, I thought, to improve the quality of their wine. How much, I wondered, was it due to the wine that I now myself felt so warmed by this sense of continuity with the past, with Persius and William Drummond, by this spirit of stubborn endurance? Suppose that the federal agents should succeed in suppressing these restaurants where pretty good wine was still served. These restaurants were run by Italians, and it had lately been against Italians that the machine-guns of the State had been trained and the police incited to butchery. In the meantime, there was nothing to do save to work with the dead for allies, and at odds with the ignorance of most of the living, that that edifice, so many times begun, so discouragingly reduced to ruins, might yet stand as the headquarters of humanity!

I left the restaurant in meditation, and, on my way out, had a collision that jarred me with a couple of those bulky pink people who had stopped laughing and were dancing to the radio.

1927

About Myself

I AM A man of medium height. I keep my records in a Weis Folder Re-order Number 8003. The unpaid balance of my estimated tax for the year 1945 is item 3 less the sum of items 4 and 5. My eyes are gray. My Selective Service order number is 10789. The serial number is T1654. I am in Class IV-A, and have been variously in Class 3-A, Class I-A (H), and Class 4-H. My social security number is 067-01-9841. I am married to US Woman Number 067-01-9807. Her eyes are gray. This is not a joint declaration, nor is it made by an agent; therefore it need be signed only by me—and, as I said, I am a man of medium height.

I am the holder of a quit-claim deed recorded in Book 682, Page 501, in the county where I live. I hold Fire Insurance Policy Number 424747, continuing until the 23 day of October in the year nineteen hundred forty-five, at noon, and it is important that the written portions of all policies covering the same property read exactly alike. My cervical spine shows relatively good alignment with evidence of proliferative changes about the bodies consistent with early arthritis. (Essential clinical data: pain in neck radiating to mastoids and occipito-temporal region, not constant, moderately severe; patient in good general health and working.) My operator's licence is Number 16200. It expired December 31, 1943, more than a year ago, but I am still carrying it and it appears to be serving the purpose. I shall renew it when I get time. I have made, published, and declared my last will and testament, and it thereby revokes all other wills and codicils at any time heretofore made by me. I hold Basic A Mileage Ration 108950, OPA Form R-525-C. The number of my car is 18-388. Tickets A-14 are valid through March 21st.

I was born in District Number 5903, New York State. My birth is registered in Volume 3/58 of the Department of Health.

My father was a man of medium height. His telephone number was 484. My mother was a housewife. Her eyes were blue. Neither parent had a social security number and neither was secure socially. They drove to the depot behind an unnumbered horse.

I hold Individual Certificate Number 4320–209 with the Equitable Life Assurance Society, in which a corporation hereinafter called the employer has contracted to insure my life for the sum of two thousand dollars. My left front tire is Number 48KE8846, my right front tire is Number 63T6895. My rear tires are, from left to right, Number 6N4M5384 and Number A26E5806D. I brush my hair with Whiting-Adams Brush Number 010 and comb my hair with Pro-Phy-Lac-Tic Comb Number 1201. My shaving brush is sterilized. I take Pill Number 43934 after each meal and I can get more of them by calling ELdorado 5-6770. I spray my nose with De Vilbiss Atomizer Number 14. Sometimes I stop the pain with Squibb Pill, Control Number 3K49979 (aspirin). My wife (Number 067-01-9807) takes Pill Number 49345.

I hold War Ration Book 40289EW, from which have been torn Airplane Stamps Numbers 1, 2, and 3. I also hold Book 159378CD, from which have been torn Spare Number 2, Spare Number 37, and certain other coupons. My wife holds Book 40288EW and Book 159374CD. In accepting them, she recognized that they remained the property of the United States Government.

I have a black dog with cheeks of tan. Her number is 11032. It is an old number. I shall renew it when I get time. The analysis of her prepared food is guaranteed and is Case Number 1312. The ingredients are: Cereal Flaked feeds (from Corn, Rice, Bran, and Wheat), Meat Meal, Fish Liver and Glandular Meal, Soybean Oil Meal, Wheat Bran, Corn Germ Meal, 5% Kel-Centrate [containing Dried Skim Milk, Dehydrated Cheese, Vitamin B_1 (Thiamin), Flavin Concentrate, Carotene, Yeast, Vitamin A and D Feeding Oil (containing 3,000 USP units Vitamin A and 400 USP units Vitamin D per gram), Diastase (Enzyme), Wheat Germ Meal, Rice Polish Extract], $1\frac{1}{2}$% Calcium Carbonate, .00037% Potassium Iodide, and $\frac{1}{4}$% Salt. She prefers offal.

When I finish what I am now writing it will be late in the day. It will be about half past five. I will then take up Purchase Order

Number 245-9077-B-Final, which I received this morning from
the Office of War Information and which covers the use of certain material they want to translate into a foreign language. Attached to the order are Standard Form Number 1034 (white) and
three copies of Standard Form Number 1034a (yellow), also
'Instructions for Preparation of Voucher by Vendor and Example of Prepared Voucher.' The Appropriation Symbol of the
Purchase Order is 1153700.001-501. The requisition number is
B-827. The allotment is X5-207.1-R2-11. Voucher shall be prepared in ink, indelible pencil, or typewriter. For a while I will be
vendor preparing voucher. Later on, when my head gets bad
and the pain radiates, I will be voucher preparing vendor. I see
that there is a list of twenty-one instructions which I will be following. Number One on the list is: 'Name of payor agency as
shown in the block "appropriation symbol and title" in the
upper left-hand corner of the Purchase Order.' Number Five on
the list is: 'Vendor's personal account or invoice number,' but
whether that means Order Number 245-9077-B-Final, or Requisition B-827, or Allotment X5-207.1-R2-11, or Appropriation
Symbol 1153700.001-501, I do not know, nor will I know later
on in the evening after several hours of meditation, nor will I be
able to find out by consulting Woman 067-01-9807, who is no
better at filling out forms than I am, nor after taking Pill Number 43934, which tends merely to make me drowsy.

I owe a letter to Corporal 32413654, Hq and Hq Sq., VII
AAF SC, APO 953, c/o PM San Francisco, Calif., thanking him
for the necktie he sent me at Christmas. In 1918 I was a private
in the Army. My number was 4,345,016. I was a boy of medium
height. I had light hair. I had no absences from duty under GO
31,1912, or GO 45, 1914. The number of that war was Number
One.

1945

Our Half-Hogarth

THE English humorists! Through a fog compounded of tobacco smoke, the stink of spirits and the breath of bailiffs, we see their melancholy faces. Look at Thomas Hood, his eyes swollen with the cardiac's solemnity, his mouth pouting after tears. There is a terrible account of his last days in Canon Ainger's *Memoir*, where we see the poet famous, forty-six, bankrupt and dying of heart disease, writing farewells to his friends and unable to stop making puns. They beset him like a St Vitus' dance. They come off his lips in an obsessional patter as if his tongue had become a cuckoo clock and his mind a lunatic asylum of double meanings. And around him his doting family and his friends are weeping, 'Poor Tom Hood.' This is, alas, one of the too many crying scenes of Victorian biography. It brims with that homemade beverage of laughter and tears which is handed round like a negus from the chiffonier of the lighter Victorian literature. The savage and vital indignation of the eighteenth century, its moral dogmatism, its body full of laughter and its roars of pain, have gone; melodrama replaces morality, a sprite-like pathos, all grace and weeping, and inked by fear of life, steps in where Caliban groaned and blubbered. Charles Lamb called Thomas Hood 'our half-Hogarth,' and that is the measure of the difference between the two periods.

Hood marks the difference well. Only in Goldsmith do we find a tenderness comparable to his. We look at the eighteenth century and, when all is said, we can hardly deny that it had a coherent and integrated mind, a mind not deeply divided against itself. The proper study of mankind is man, who is very corrupt, but presently Divine Reason will teach him to cast off his chains and he will become a free child of nature. By the end of the century the chains are removed. And what is the result? Hood's early nineteenth century shows us. Man has not become free; he has vanished. Or rather, that humane abstraction called Man has been succeeded by two warring groups. Man has degenerated and has become the middle classes and the poor. No longer, like Swift, do the Victorians feel horror of mankind; on the contrary, looking at the little circle of mankind in which they live, they find the species has very much improved. At Clapham, at Wanstead Flats, even in Russell Square and Fleet Street, he is kindly, charitable and good. Their horror moves from man as a whole to a section of men. They are horrified, they are frightened—philanthropical and well policed though they are—by the poor. For generations now they will not stop talking about the poor. Did they pull down the blinds and turn to conceits and fancies because this fear is outside the window after dark? The feeling is that outside the sitting room is an undefined world of wickedness, hunger, catastrophe and crime. Pickpockets are nabbed, poachers are imprisoned, desperate laborers threaten arson, and children go to the mills and up the chimneys; the press gang and transportation are living memories, and sailors drown—oh, how many sailors drown!—in calamitous storms. These terrible things happen—to the poor. There we have Hood's background. There is his material. But writers are urged and taught to write not by society only but by other writers whose background and intention make them utterly different from their pupils. It is a strange fact that the England of Hood is not delineated by revolutionary realists, but has come down to us in the fantastic dress of German Gothic. The Cruikshank who frightens us; Mr Punch, with his pot belly, his fairy legs and the arching nose like some cathedral fragment, who squats on Dicky Doyle's cover, are part of the Gothic colony that settle like a migration of gargoyles among the English chimneys and their myth-creating smoke.

Hood, who was a Cockney of Scottish parentage, writes very early in his career of 'doing something in the German manner.' In his serious verses he is a Romantic, with his eye on Shakespeare, Scott and Keats. But this is the less readable part of Hood. His serious verses, if one excepts pieces of singular purity like 'I remember, I remember,' hardly amount to more than poetic dilutions for the family album, though contemporaries like Lamb, Southey and Byron had a higher opinion of them. Hood's best work is inflected by the basic early Victorian fear and the fancies to which it led. He is on the side of the poor, of course, and wrote for the early, unsuccessful Radical *Punch*: but the Hood of *The Song of the Shirt*—which trebled the circulation of *Punch*—*The Lay of the Labourer* and *The Bridge of Sighs* is the dying Hood who is touched by the indignation of the hungry Forties. The earlier Hood thinks the poor are quaint and that their crimes can be sardonically disinfected. The result is a vein of fanciful horror which fathered a whole school of ballad writing:

> The body-snatchers they have come,
> And made a snatch at me;
> It's very hard them kind of men
> Won't let a body be!
>
> You thought that I was buried deep
> Quite decent-like and chary,
> But from her grave in Mary-bone
> They've come and boned your Mary.

That is from *Mary's Ghost*. I could have quoted from *The Volunteer* or *Death's Ramble*. There is *The Careless Nurse Mayd*:

> I saw a Mayd sitte on a Bank
> Beguiled by Wooer fayne and fond;
> And whiles his flatteryinge Vowes she drank,
> Her Nurselynge slipt within a Pond!
>
> All Even Tide they Talkde and Kist
> For She was Fayre and He was Kinde;
> The Sunne went down before she wist
> Another Sonne had sett behinde!

Or from *Sally Simpkin's Lament*:

> Oh! What is that comes gliding in
> And quite in middling haste?
> It is the picture of my Jones,
> And painted to the waist.
>
> Oh Sally dear, it is too true—
> The half that you remark
> Is come to say my other half
> Is bit off by a shark.

Gilbert, Lear, Carroll, Thackeray, the authors of *Struwwelpeter* and the cautionary tales continue this comic macabre tradition,

which today appears to be exhausted. There is Mr Belloc, who
digressed intellectually; and there are the sardonic ballads of Mr
William Plomer. He has added brilliantly the horrors of vulgarity
to the horrors of crime and accident.

Hood's special idiosyncrasy is to turn the screw of verbal conceit upon his subject. In *Eugene Aram* alone he cut out these
tricks, even forbearing in the last verse when his temptation was
always strongest. (How was it Hood failed to ruin what are,
surely, the most frightening dramatic lines in English narrative
verse?) But if Hood's puns are often disastrous, they do frequently show, as Walter Jerrold (his biographer) has said, a kind
of second sight. They are like the cackle out of the grave in
Hamlet. They add malice to the knife and give the macabre its
own morbid whimsicalities. Take that terrible poem, *The Last
Man*. The earth has been desolated by plague and only two men
are left alive. They meet at a gallows and one, out of jealousy,
decides to hang the other. He does so and is left, wracked by
conscience, to lament that he cannot now hang himself:

> For there is not another man alive,
> In the world to pull my legs.

The wit in *Death's Ramble* shocks one first of all and then freezes
the blood one degree colder. Death sees two duelists:

> He saw two duellists going to fight,
> In fear they could not smother;
> And he shot one through at once—for he knew
> They never would shoot each other.

And the comic funk of *The Volunteer* gets a grotesque double
meaning. He fears the alarm:

> My jaws with utter dread, enclos'd
> The morsel I was munching,
> And terror lock'd them up too tight,
> My very teeth went crunching
> All through my bread and tongue at once
> Like sandwich made at lunching.

To the poor, Hood draws our attention by shuddering and
laughing with them at the same time. His detachment, when he

is writing about crime and catastrophe, is dropped when he is putting the case of the poor. Then he writes with something like the garrulous, flat statement of the broadsheets. These odes and poems lumber along. There is the washerwoman's attack on the new steam laundry which has taken her living. There is the chimney boy's lament that the law against street cries forbids him to cry 'Sweep' in the streets. Drapers' assistants plead politely with people to shop early. These are pieces of topical journalism which time has blunted, and Hood's pen dipped deeply into that sentimentality which the philanthropical outlook of the period demanded. He was a prolific writer, and knew how to turn out his stuff. Like Dickens he was a sentimental Radical who hoped, as Dickens also hoped, that the problem of the poor could be solved by kindness; but the abiding note is that unpleasant one of Uriah Heep's: 'Me and mother is very humble.'

Hood prefers to let the poor or oppressed describe their lives uncouthly, rather than to attack the rich. The grotesque poem called *Miss Kilmansegg and her Precious Leg* is an exception. This poem startles because it is the first documented account of the upbringing of the perfect middle-class young lady whose parents are rising in the world. She is brought up to be a proud heiress, and the wonderful picture of arrogant surfeit recalls the awful overfed daughter of the mine owner in Zola's *Germinal*. Money is the only subject of conversation. Then one day Miss Kilmansegg has an accident, her leg is amputated and is replaced by a golden one. A wooden one would not be good enough. Far from spoiling her chances, the golden leg doubles the number of her suitors. Her parents select the most plausible and least trustworthy one who is an alleged aristocrat. He turns out to be a bankrupt gambler who, very soon after the wedding night, gives a knowing look at the leg and

> The Countess heard in language low
> That her Precious leg was precious slow,
> A good 'un to look at, but bad to go
> And kept quite a sum lying idle.

She refuses to sell it. But unhappily she is in the habit of taking it off at night, and the Count sees his chance. Using the leg as a cudgel he bashes her brains out and absconds.

This long poem is like a grotesque novel, something of de la Mare's, perhaps, packed with realistic description, and if its plot groans the lines scamper along as fast as Browning's dramatic narratives and are delighted with their own wit. And here the puns give the poem a kind of jeering muttered undertone. Hood had a great gift for domestic realism and the conversational phrase. In *Miss Kilmansegg* he is not half a Hogarth, but Hogarth whole. Or ought one to say, half a Hogarth and the other half that fanciful melodramatic sermonizer—as Dickens was in the *Christmas Carol*—which the nineteenth century loved? The poem is labored but it is alive.

Hood's wit quietened and compassion melted him in his last years. *The Song of the Shirt* and *The Lay of the Labourer* last very well in their genre, because of their metrical brilliance and because they are taken directly from life. One would want to remove only two or three lines of self-parody from *The Bridge of Sighs*. Hood is as well documented as the realistic novelists were to become. *The Lay of the Labourer* is based on a true incident. An agricultural laborer was convicted for threatening arson because he could not get work or food, and Hood kept the newspaper cutting about the event on his mantelpiece until he wrote the poem. The sentiment is bearable, the rant is bearable, because the facts cry out and are so tellingly reported. One must regret that his feeling for narrative, his instinct for the right tune to put it in and his kind of conscience too, died out of verse with the Victorians. In the higher regions where Hardy lived, as in the lower regions of the music hall, the art of writing dramatic stories in verse seems to have gone for good.

1947

The Ant-Lion

THE Maures are my favourite mountains, a range of old rounded mammalian granite which rise three thousand feet above the coast of Provence. In summer they are covered by dark forests of cork and pine, with paler interludes on the northern slopes of bright splay-trunked chestnut, and an undergrowth of arbutus and bracken. There is always water in the Maures, and the mountains are green throughout the summer, never baked like the limestone, or like the Southern Alps a slagheap of gritty oyster-shell. They swim in a golden light in which the radiant ebony green of their vegetation stands out against the sky, a region hardly inhabited, yet friendly as those dazzling landscapes of Claude and Poussin, in which shepherds and sailors from antique ships meander under incongruous elms. Harmonies of light and colour, drip of water over fern; they inculcate in those who stay long in the Midi, and whose brains are addled by iodine, a habit of moralizing, a brooding about causes. What makes men divide up into nations and go to war? Why do they live in cities? And what is the true relationship between Nature and Man?

The beaches of the Maures are of white sand, wide, with a ribbon of umbrella-pines, below which juicy mesembrianthemum and dry flowers of the sand stretch to within a yard of the sea. Lying there amid the pacific blues and greens one shuts the eyes and opens them on the white surface: the vague blurred philosophizing continues. Animism, pantheism, images of the earth soaring through space with the swerve of a ping-pong ball circulate in the head; the woolly brain meddles with ethics. No more power, no aggression, no intolerance. All must be free. Then whizz! A disturbance. Under the eye the soil is pitted into a conical depression, about the size of a candle extinguisher, down whose walls the sand trickles gently, moved by a suspicion of wind. Whizz, and a clot is hurled to the top again, the bottom of the funnel cleared, in disobedience to the natural law! As the funnel silts up it is cleared by another whirr, and there appears, at the nadir of the cone, a brown pair of curved earwig horns, antlers of a giant earwig that churn the sand upwards like a steam shovel.

Now an ant is traversing the dangerous *arête*. He sidles, slithers, and goes fumbling down the Wall of Death to the waiting chopper. Snap! He struggles up, mounting the steep banking grain by grain as it shelves beneath him, till a new eruption is engineered by his waiting enemy. Sand belches out, the avalanche engulfs him, the horny sickles contract and disappear with their beady victim under the whiteness. Mystery, frustration, tragedy, death are then at large in this peaceful wilderness! Can the aggressive instinct be analysed out of those clippers? Or its lethal headpiece be removed by a more equitable distribution of raw materials? The funnels, I observe, are all round me. The sand is pockmarked with these geometrical death-traps, engineering triumphs of insect art. And this horsefly might be used for an experiment. I shove it downwards. The Claws seize on a wing, and the struggle is on. The fight proceeds like an atrocity of chemical warfare. The great fly threshes the soil with its wings, it buzzes and drones while the sand heaves round its propellers and the facets of its giant projectors glitter with light. But the clippers do not relax, and disappear tugging the fly beneath the surface. The threshing countinues, a faint buzzing comes from the invisible horsefly, and its undercarriage appears, with legs waving. Will it take off? The wings of the insect bomber pound the air, the fly starts forward and upwards, and hauls after it—O fiend, embodiment of evil! A creature whose clippers are joined to a muscle-bound thorax and a vile yellow armour-plated body, squat and powerful, with a beetle set of legs to manoeuvre this engine of destruction. The Tank with a Mind now scuttles backwards in reverse, the stern, then the legs disappear, then the jaws which drag its prey. Legs beat the ground. A fainter wheeze and whirr, no hope now, the last wing-tip vanished, the air colder, the pines greener, the cone empty except for the trickle, the sifting and silting down the funnel of the grains of pearl-coloured sand.

Nature arranged this; bestowed on the Ant-Lion its dredging

skill and its cannon-ball service. How can it tell, buried except
for the striking choppers, that the pebble which rolls down has
to be volleyed out of the death-trap, while the approaching ant
must be collected by gentle eruptions, dismayed by a perpetual
sandy shower? And, answer as usual, we do not know.

Yet the relationship between the Ant-Lion and the curving
beaches of Pampelone suggests a parallel. This time at Albi. Here
Art and Nature have formed one of the most harmonious scenes
in Europe. The fortress cathedral, the Bishop's Palace with its
hanging gardens, and the old bridge, all of ancient brick, blend
into the tawny landscape through which the emancipated Tarn
flows from its gorges to the Garonne. Here again one wanders
through this dream of the Middle Ages, by precincts of the rosy
cathedral where the pious buzz like cockchafers, to be brought
up by a notice on the portcullis of the Bishop's Palace. 'Musée
Toulouse-Lautrec.' Tucked in the conventional Gothic of the
fortress is a suite of long rooms in which the mother of the artist,
using all her feudal powers, forced the municipal authorities to
hang the pictures of her son. Less fortunate than those of Aix,
who refused Cézanne's request to leave his pictures to the city,
the fathers were intimidated by the Countess into placing them
in this most sacred corner, lighted and hung in salons whose
decoration has concealed all traces of the unsightly past.

The concierge turns proudly to the Early Work—pastoral
scenes and sentimental evocations of Millert—these he likes best;
they are what the Count was doing before he left his home and
was corrupted by the Capital. Then come the drawings, in which
emerges the fine savage line of the mature artist, that bold, but
not (as in some of the paintings) vulgar stroke, which hits off the
brutaliy of his subjects, or the beauty of those young girls
doomed to such an inevitable end. In the large room beyond are
the paintings, a morgue of End of Century vice, a succession of
canvases in which there is hardly daylight, and where the only
creature who lives by day is the wizened little Irish jockey. The
world of the hunchback Count is nocturnal, gas-lit, racy, depraved and vicious; the shocked Albigeois who pass through the
gallery are riveted by the extraordinary picture of the laundress
who checks over with the *sous-maîtresse* the linen from her
Maison. As one goes from picture to picture the atmosphere
intensifies, Valentin le Désossé amd La Goulue become familiars,

and the lovely girls blur into the dark of the Moulin Rouge, where one distinguishes a favourite figure, the long, sad, nocturnal, utterly empty but doggedly boring face of 'L'Anglais,' —some English habitué to whom constant all-night attendance has given the polish of a sentry at his post.

At the end of the gallery is a door before which the concierge smiles mysteriously, as if to prepare us for Pompeian revelations. He opens it, and we emerge on a small terrace. The sun is shining, the sky is blue, the Tarn ripples underneath. Beyond the ancient brick of the bishop's citadel and the arches of the bridge stretches the landscape of the Albigeois—foothills of green corn delicately crowned by pink hill villages, which merge into the brown of the distant Cevennes under the pale penetrating light of the near-south, the transitional-Mediterranean. A lovely and healthy prospect, in which fields and cities of men blend everywhere into the earth and the sunshine. One takes a deep breath, when obstinately, from behind the closed door, one feels a suction; attraction fights repulsion as in the cold wavering opposition between the like poles of a magnet. Deep in his lair the Ant-Lion is at work; the hunchback Count recalls us; the world of poverty, greed, bad air, consumption, and of those who never go to bed awaits, but there awaits also an artist's integration of it, a world in which all trace of sentiment or decadence is excluded by the realism of the painter, and the vitality of his line. In the sunlight on the terrace we are given the choice between the world of Nature and the world of Art. Nature seems to win, but at the moment of victory there is something lacking, and it is that lack which only the unnatural world inside can supply—progress, for example, for the view from the Palace has not altered, except slightly to deteriorate, for several hundred years. The enjoyment of it requires no more perception than had Erasmus, while the art of Lautrec is modern, and can be appreciated only by those who combine a certain kind of aristocratic satisfaction at human beings acting in character, and in gross character, with the love of fine drawing and colour.

Not that Lautrec was a great artist; he is to Degas what Maupassant is to Flaubert, one who extended the noble conception of realism by which a great master accepts the world as it is for the sake of its dynamism, and for the passive, extraordinarily responsive quality of that world to the artist who has learnt how to impose his will on it. The world of Lautrec is artificial because it excludes goodness and beauty as carefully as it excludes the sun. But it is an arranged world, a world of melancholy and ignorance (figures melancholy because ignorant, patient in the treadmill of pleasure), and so the artist drags us in from the terrace because force and intelligence dominate that arrangement. And once back, we are back in his dream, in a hunchback's dream of the world; the sunlight seems tawdry, the red brick vulgar, the palace ornate; the crowd who stand in their tall hats gaping at the blossoming Can-Can dancers are in the only place worth being.

Now I understand the Ant-Lion. It is in Nature and with a natural right to its existence. There is no conflict between them; it is an advanced gadget in the scheme which includes the peaceful hills and the beach with its reedy pools of brackish water. Nor is there any opposition between Lautrec and the landscape of Albi. Albi was the oyster, and the contents of the museum are the Pearl. The irritant? The action of a physical deformity on an aristocratic, artistic but unoriginal mind which was happiest in the company of its inferiors, and which liked to be surrounded by the opposite sex in places where the deformity could be concealed by potency, or by the distribution of money. The result, a highly specialized painter, one of Nature's very latest experiments. And yet even that peaceful landscape was the home in the Middle Ages of a subversive, doctrine, the Albigensian heresy; a primitive anarchism which taught that men were equal and free, which disbelieved in violence and believed in a chosen priesthood, in the Cathari who attained purity by abstinence, while they encouraged the Count's royal ancestors to come through excess and indulgence to heavenly wisdom. It was they who believed that the human race should cease to procreate, and so solve the problem of evil, who were massacred at Muret and Lavaur, and whom Simon de Montfort slaughtered with the remark, 'The Lord will know his own.' And the Heretics were right. Had a revolt against procreation spread outwards from Albi the world would have become an empty place, nor would such obstinate human beings who survived have been driven to kill each other for living-room, victims, for all we may know, of some deeper instinct of self-destruction which bids them make way for a new experiment, the civilization of the termite or the rat.

Much has happened since the summer. To-day the Maures are out of bounds, the Museum closed, and many generalizations based on incorrect assessment of the facts fallen to pieces, but (since the operations of the Ant-Lion have now been extended) it seems worth while to recall that the statements on the life of pleasure which Lautrec took from his witnesses at the Tabarin and the Moulin de la Galette, and which he so vigorously recorded on canvas, are still available to the traveller of the future, and assert their truth.

1939

GEORGE ORWELL

Reflections on Gandhi

Saints should always be judged guilty until they are proved innocent, but the tests that have to be applied to them are not, of course, the same in all cases. In Gandhi's case the questions one feels inclined to ask are: to what extent was Gandhi moved by vanity—by the consciousness of himself as a humble, naked old man, sitting on a praying mat and shaking empires by sheer spiritual power—and to what extent did he compromise his own principles by entering politics, which of their nature are inseparable from coercion and fraud? To give a definite answer one would have to study Gandhi's acts and writings in immense detail, for his whole life was a sort of pilgrimage in which every act was significant. But this partial autobiography, which ends in the nineteen-twenties, is strong evidence in his favour, all the more because it covers what he would have called the unregenerate part of his life and reminds one that inside the saint, or near-saint, there was a very shrewd, able person who could, if he had chosen, have been a brilliant success as a lawyer, an administrator or perhaps even a business man.

At about the time when the autobiography first appeared I remember reading its opening chapters in the ill-printed pages of some Indian newspaper. They made a good impression on me, which Gandhi himslef at that time, did not. The things that one associated with him—home-spun cloth, 'soul forces' and vegetarianism—were unappealing, and his medievalist programme was obviously not viable in a backward, starving, over-populated country. It was also apparent that the British were making use of him, or thought they were making use of him. Strictly speaking, as a Nationalist, he was an enemy, but since in every crisis he would exert himself to prevent violence—which, from the British point of view, meant preventing any effective action whatever—he could be regarded as 'our man'. In private this was sometimes cynically admitted. The attitude of the Indian millionaires was similar. Gandhi called upon them to repent, and naturally they preferred him to the Socialists and Communists who, given the chance, would actually have taken their money away. How reliable such calculations are in the long run is doubtful; as Gandhi himself says, 'in the end deceivers deceive only themselves'; but at any rate the gentleness with which he was nearly always handled was due partly to the feeling that he was useful. The British Conservatives only became really angry with him when, as in 1942, he was in effect turning his non-violence against a different conqueror.

But I could see even then that the British officials who spoke of him with a mixutre of amusement and disapproval also genuinely liked and admired him, after a fashion. Nobody ever suggested that he was corrupt, or ambitious in any vulgar way, or that anything he did was actuated by fear or malice. In judging a man like Gandhi one seems instinctively to apply high standards, so that some of his virtues have passed almost unnoticed. For instance, it is clear even from the autobiography that his natural physical courage was quite outstanding: the manner of his death was a later illustration of this, for a public man who attached any value to his own skin would have been more adequately guarded. Again, he seems to have been quite free from that maniacal suspiciousness which, as E. M. Forster rightly says in *A Passage to India*, is the besetting Indian vice, as hypocrisy is the British vice. Although no doubt he was shrewd enough in detecting dishonesty, he seems wherever possible to have believed that other people were acting in good faith and had a better nature through which they could be approached. And though he came of a poor middle-class family, started life rather unfavourably, and was probably of unimpressive physical appearance, he was not afflicted by envy or by the feeling of inferiority. Colour feeling when he first met it in its worst form in South Africa, seems rather to have astonished him. Even when he was fighting what was in effect a colour war, he did not think of people in terms of race or status. The governor of a province, a cotton millionaire, a half-starved Dravidian coolie, a British private soldier were all equally human beings, to be approached in much the same way. It is noticeable that even in the worst possible circumstances, as in South Africa when he was making himself unpopular as the champion of the Indian community, he did not lack European friends.

Written in short lengths for newspaper serialization, the auto-biography is not a literary masterpiece, but it is the more impressive because of the commonplaceness of much of its material. It is well to be reminded that Gandhi started out with the normal ambitions of a young Indian student and only adopted his extremist opinions by degrees and, in some cases, rather unwillingly. There was a time, it is interesting to learn, when he wore a top hat, took dancing lessons, studied French and Latin, went up the Eiffel Tower and even tried to learn the violin—all this with the idea of assimilating European civiliza-tion as thoroughly as possible. He was not one of those saints who are marked out by their phenomenal piety from childhood onwards, nor one of the other kind who forsake the world after sensational debaucheries. He makes full confession of the misdeeds of his youth, but in fact there is not much to confess. As a frontispiece to the book there is a photograph of Gandhi's possessions at the time of his death. The whole outfit could be purchased for about £5, and Gandhi's sins, at least his fleshly sins, would make the same sort of appearance if placed all in one heap. A few cigarettes, a few mouthfuls of meat, a few annas pilfered in childhood from the maidservant, two visits to a brothel (on each occasion he got away without 'doing any-thing'), one narrowly escaped lapse with his landlady in Plymouth, one outburst of temper—that is about the whole col-lection. Almost from childhood onwards he had a deep earnest-ness, an attitude ethical rather than religious, but, until he was about thirty, no very definite sense of direction. His first entry into anything describable as public life was made by way of veget-arianism. Underneath his less ordinary qualities one feels all the time the solid middle-class businessmen who were his ancestors. One feels that even after he had abandoned personal ambition he must have been a resourceful, energetic lawyer and a hard-headed political organizer, careful in keeping down expenses, an adroit handler of committees and an indefatigable chaser of subscriptions. His character was an extraordinarily mixed one, but there was almost nothing in it that you can put your finger on and call bad, and I believe that even Gandhi's worst enemies would admit that he was an interesting and unusual man who enriched the world simply by being alive. Whether he was also a lovable man, and whether his teachings can have much value for those who do not accept the religious beliefs on which they are founded, I have never felt fully certain.

Of late years it has been the fashion to talk about Gandhi as though he were not only sympathetic to the Western Left-wing movement, but were integrally part of it. Anarchists and pacifists, in particular, have claimed him for their own, noticing only that he was opposed to centralism and State violence and ignoring the other-worldly, anti-humanist tendency of his doctrines. But one should, I think, realize that Gandhi's teachings cannot be squared with the belief that Man is the measure of all things and that our job is to make life worth living on this earth, which is the only earth we have. They make sense only on the assumption that God exists and that the world of solid objects is an illusion to be escaped from. It is worth considering the disciplines which Gandhi imposed on himself and which—though he might not insist on every one of his followers observing every detail—he considered indispensable if one wanted to serve either God or humanity. First of all, no meat-eating, and if possible no animal food in any form. (Gandhi himself, for the sake of his health, had to compromise on milk, but seems to have felt this to be a backsliding.) No alcohol or tobacco, and no spices or condiments even of a vegetable kind, since food should be taken not for its own sake but solely in order to preserve one's strength. Secondly, if possible, no sexual intercourse. If sexual intercourse must happen, then it should be for the sole purpose of begetting children and presumably at long intervals. Gandhi himself, in his middle thirties, took the vow of *bramahcharya*, which means not only complete chastity but the elimination of sexual desire. This condition, it seems, is difficult to attain without a special diet and frequent fasting. One of the dangers of milk-drinking is that it is apt to arouse sexual desire. And finally—this is the cardinal point—for the seeker after goodness there must be no close friendships and no exclusive loves whatever.

Close friendships, Gandhi says, are dangerous, because 'friends react on one another' and through loyalty to a friend one can be led into wrong-doing. This is unquestionably true. Moreover, if one is to love God, or to love humanity as a whole, one cannot give one's preference to any individual person. This again is true, and it marks the point at which the humanistic and the religious attitude cease to be reconcilable. To an ordinary human being, love means nothing if it does not mean loving some people more than others. The autobiography leaves it uncertain whether Gandhi behaved in an inconsiderate way to his wife and children, but at any rate it makes clear that on three occasions he was willing to let his wife or a child die rather than administer the animal food prescribed by the doctor. It is true that the threatened death never actually occurred, and also that Gandhi—with, one gathers, a good deal of moral pressure in the opposite direction—always gave the patient the choice of staying alive at the price of committing a sin: still, if the decision had been solely his own, he would have forbidden the animal food, whatever the risks might be. There must, he says, be some limit to what we will do in order to remain alive, and the limit is well on this side of chicken broth. This attitude is perhaps a noble one, but, in the sense which—I think—most people would give to the word, it is inhuman. The essence of being human is that one does not seek perfection, that one *is* sometimes willing to commit sins for the sake of loyalty, that one does not push asceticism to the point where it makes friendly intercourse impossible, and that one is prepared in the end to be defeated and broken up by life, which is the inevitable price of fastening one's love upon other human individuals. No doubt alcohol, tobacco and so forth are things that a saint must avoid, but sainthood is also a thing that human beings must avoid. There is an obvious retort to this, but one should be wary about making it. In this yogi-ridden age, it is too readily assumed that 'nonattachment' is not only better than a full acceptance of earthly life, but that the ordinary man only rejects it because it is too difficult: in other words, that the average human being is a failed saint. It is doubtful whether this is true. Many people genuinely do not wish to be saints, and it is probable that some who achieve or aspire to sainthood have never felt much temptation to be human beings. If one could follow it to its psychological roots, one would, I believe, find that the main motive for 'nonattachment' is a desire to escape from the pain of living, and above all from love, which, sexual or non-sexual, is hard work.

But it is not necessary here to argue whether the other-worldly or the humanistic ideal is 'higher'. The point is that they are incompatible. One must choose between God and Man, and all 'radicals' and 'progressives', from the mildest Liberal to the most extreme Anarchist, have in effect chosen Man.

However, Gandhi's pacifism can be separated to some extent from his other teachings. Its motive was religious, but he claimed also for it that it was a definite technique, a method, capable of producing desired political results. Gandhi's attitude was not that of most Western pacifists. *Satyagraha*, first evolved in South Africa, was a sort of non-violent warfare, a way of defeating the enemy without hurting him and without feeling or arousing hatred. It entailed such things as civil disobedience, strikes, lying down in front of railway trains, enduring police charges without running away and without hitting back, and the like. Gandhi objected to 'passive resistance' as a translation of *Satyagraha*: in Gujarati, it seems, the word means 'firmness in the truth'. In his early days Gandhi served as a stretcher-bearer on the British side in the Boer War, and he was prepared to do the same again in the war of 1914–18. Even after he had completely abjured violence he was honest enough to see that in war it is usually necessary to take sides. He did not—indeed, since his whole political life centred round a struggle for national independence, he could not—take the sterile and dishonest line of pretending that in every war both sides are exactly the same and it makes no difference who wins. Nor did he, like most Western pacifists, specialize in avoiding awkward questions. In relation to the late war, one question that every pacifist had a clear obligation to answer was: 'What about the Jews? Are you prepared to see them exterminated? If not, how do you propose to save them without resorting to war?' I must say that I have never heard, from any Western pacifist, an honest answer to this question, though I have heard plenty of evasions, usually of the 'you're another' type. But it so happens that Gandhi was asked a somewhat similar question in 1938 and that his answer is on record in Mr Louis Fischer's *Gandhi and Stalin*. According to Mr Fischer, Gandhi's view was that the German Jews ought to commit collective suicide, which 'would have aroused the world and the people of Germany to Hitler's violence'. After the war he justified himself: the Jews had been killed anyway, and might as

well have died significantly. One has the impression that this attitude staggered even so warm an admirer as Mr Fischer, but
Gandhi was merely being honest. If you are not prepared to take
life, you must often be prepared for lives to be lost in some other
way. When, in 1942, he urged non-violent resistance against
a Japanese invasion, he was ready to admit that it might cost
several million deaths.

At the same time there is reason to think that Gandhi who
after all was born in 1869, did not understand the nature of
totalitarianism and saw everything in terms of his own struggle
against the British government. The important point here is not
so much that the British treated him forbearingly as that he was
always able to command publicity. As can be seen from the
phrase quoted above, he believed in 'arousing the world', which
is only possible if the world gets a chance to hear what you are
doing. It is difficult to see how Gandhi's methods could be
applied in a country where opponents of the régime disappear in
the middle of the night and are never heard of again. Without a
free Press and the right of assembly, it is impossible not merely
to appeal to outside opinion, but to bring a mass movement into
being, or even to make your intentions known to your adversary. Is there a Gandhi in Russia at this moment? And if there is,
what is he accomplishing? The Russian masses could only practise civil disobedience if the same idea happened to occur to all of
them simultaneously, and even then, to judge by the history of
the Ukraine famine, it would make no difference. But let it be
granted that non-violent resistance can be effective against one's
own government, or against an occupying power: even so, how
does one put it into practice internationally? Gandhi's various
conflicting statements on the late war seem to show that he felt
the difficulty of this. Applied to foreign politics, pacifism either
stops being pacifist or becomes appeasement. Moreover the
assumption, which served Gandhi so well in dealing with individuals, that all human beings are more or less approachable and
will respond to a generous gesture, needs to be seriously ques-
ioned. It is not necessarily true, for example, when you are dealing with lunatics. Then the question becomes: Who is sane? Was
Hitler sane? And is it not possible for one whole culture to be
insane by the standards of another? And, so far as one can gauge
the feelings of whole nations, is there any apparent connection

between a generous deed and a friendly response? Is gratitude a factor in international politics?

These and kindred questions need discussion, and need it urgently, in the few years left to us before somebody presses the button and the rockets begin to fly. It seems doubtful whether civilization can stand another major war, and it is at least thinkable that the way out lies through non-violence. It is Gandhi's virtue that he would have been ready to give honest consideration to the kind of question that I have raised above; and, indeed, he probably did discuss most of these questions somewhere or other in his innumerable newspaper articles. One feels of him that there was much that he did not understand, but not that there was anything that he was frightened of saying or thinking. I have never been able to feel much liking for Gandhi, but I do not feel sure that as a political thinker he was wrong in the main, nor do I believe that his life was a failure. It is curious that when he was assassinated, many of his warmest admirers exclaimed sorrowfully that he had lived just long enough to see his life work in ruins, because India was engaged in a civil war which had always been foreseen as one of the by-products of the transfer of power. But it was not in trying to smooth down Hindu–Moslem rivalry that Gandhi had spent his life. His main political objective, the peaceful ending of British rule, had after all been attained. As usual the relevant facts cut across one another. On the one hand, the British did get out of India without fighting, an event which very few observers indeed would have predicted until about a year before it happened. On the other hand, this was done by a Labour government, and it is certain that a Conservative government, especially a government headed by Churchill, would have acted differently. But if, by 1945, there had grown up in Britain a large body of opinion sympathetic to Indian independence, how far was this due to Gandhi's personal influence? And if, as may happen, India and Britain finally settle down into a decent and friendly relationship, will this be partly because Gandhi, by keeping up his struggle obstinately and without hatred, disinfected the political air? That one even thinks of asking such questions indicates his stature. One may feel, as I do, a sort of aesthetic distaste for Gandhi, one may reject the claims of sainthood made on his behalf (he never made any such claim himself, by the way), one may also reject sainthood as an ideal and therefore feel that Gandhi's basic aims were anti-human and reactionary: but regarded simply as a politician, and compared with the other leading political figures of our time, how clean a smell he has managed to leave behind!

1949

EVELYN WAUGH

Well-Informed Circles ... and How to Move in Them

In the vocabulary of the daily press 'well-informed circles' have by now been relegated to a place of secondary importance. On those very frequent days when foreign and diplomatic correspondents find themselves without any credible information to report, it is their custom to appease their editors with modest forecasts of their own. On these occasions, it is usual to evoke as authority some anonymous source. If, for instance, they have heard something from the postman, they attribute it to 'a semi-official statement'; if they have fallen into conversation with a stranger at a bar, they can conscientiously describe him as 'a source that has hitherto proved umimpeachable'. It is only when the journalist is reporting a whim of his own, and one to which he attaches minor importance, that he defines it as the opinion of 'well-informed circles'.

At home, however, in ordinary social intercourse, 'well-informed circles' still retain their prestige. It is significant of the diffidence with which we, as a nation, hold our opinions that the English for '*on dit*' is 'They say'. The Parisian reports what is being said at the café by his cronies and by himself; the Londoner pays homage to the enigmatic They—the people in the know. To be well-informed in England does not mean—as it used to in Germany and still does in the United States—to have studied the subject, written a thesis and earned a diploma. It means to be in constant, intimate association with the Great. Nothing is more helpful to a shy young man than to get this reputation. It is by no means difficult. Like all arts, it is simply a matter of the proper use of raw material.

The difficulty is not in meeting the Great—most of us from time to time find ourselves within measurable distance of them—but in making the proper use of our meetings. Suppose, for instance, you are asked to luncheon at the last moment by a harassed hostess and, on arrival, see in the distance a face which has long been familiar to you in newsreels and caricatures. You are introduced and drift away to a more obscure part of the room. At the table, he sits six places from you. You are dimly aware of his expressing a liking for porridge. He leaves immediately after luncheon. You wait until he is clear of the hall and then go, too. Nothing much there, you think, to qualify you for a member of the 'well-informed circles'.

'Mary had the Prime Minister to lunch today,' you report.

'Oh, how exciting. What did he say about Palestine?'

'I don't think he mentioned it.'

'Oh.'

No ice cut. But try it this way: 'I had luncheon with the Prime Minister today.'

'Oh, how exciting. What did he say about Palestine?'

'Mary was there, and, as you know, the PM never talks in front of her. He won't be saying anything about Palestine this week anyway. Ask me next Thursday, and I may be able to tell you something rather interesting.'

When celebrities fail, it is always possible to introduce quite unknown names with such an air of authority that no one dares challenge you. This is particularly useful when you meet a rival Well-Informed Man and are getting the worst of the encounter. You have asserted, for example, that the Paraguayan government is in the hands of a military clique; you have been caught on this by a sudden disclosure of superior knowledge. 'What about Hernandes, Cervantes and Alvarez?' you are suddenly asked. You have never heard of them; nor, in all probability, has your rival. Counter smartly with, 'You need not worry about them. Perhaps I ought not to say that. I got it only this morning from Henry Scudamore himself.' There is only one answer to this particular gambit. 'Ah yes. I suppose Scudamore was cutting your hair at the time?' It is conclusive, but it makes a lifelong enemy. Generally speaking, a certain reciprocal loyalty should be observed by 'well-informed men'.

Of these, there are two distinct schools, both of which enjoy wide popularity at the moment. Anyone who wishes to make a social career on these lines should decide early what school he wishes to belong to and follow it without deviation. His temperament must be the deciding factor.

The simpler, perhaps, is the Pseudo-Secret-Service. Those who seek admission to this honourable corps must have travelled a little in the Near East and, if possible, beyond. They must exhibit an interest in languages—a different and vastly easier thing than a knowledge of them. If, for instance, you are caught out by the menu, say blandly, 'I've never been able to pay much attention to Latin languages,' or, better still, 'the Romance Group'; and to such direct questions as, 'Do you speak Magyar?' answer, 'Not nearly as well as I ought.' It is good policy to introduce linguistic questions whenever possible; for instance, if someone says he has spent three weeks in Cairo, instead of asking about the hotels, say, 'Tell me, is much demotic Armenian spoken there now?', and if big-game hunting in Kenya is mentioned, say, 'I suppose one can muddle along with Swahili, Arabic and Kikuyu?' You must also be an expert on accents —'. . . she spoke Catalan with a strong Cretan accent . . .'

In appearance, the Pseudo-Secret-Service are conventional. From time to time, they must be seen in public with very queer company and, when asked about it, reply, 'Well in a way it's more or less my job.' They must have a keen memory for diplomatic appointments; not only our own, but the whole boiling. '. . . Going to Warsaw? Let's see, who have the Siamese got there now? . . .' You can also flatter your friends and enhance your own prestige by giving them little commissions to execute for you: 'Going to Paris? I wonder if you could find out something for me. I should very much like to know who owns a little weekly called *Le Faux Bonhomme* . . .' Or '. . . I wonder if you'd mind posting a letter for me in Budapest. I'd rather prefer the government not to have it through their hands . . .'

Above all, you must assume a mysterious compulsion behind all your movements. 'I may have to go abroad next week. Where? Well, I shan't really know until I reach Paris. It depends on what I hear when I get there. What shall I do? Well,' (with a knowing smile) 'I expect I shall play a little golf. I find it is a very good tip to take golf-clubs about with me abroad. They save one a lot of awkward questions.'

The strength of this school is that, as one of its prime objects is evasion, it is almost impossible to be shown up; the weakness is that it is very easy, in a confidential or convivial moment, to show oneself up. It also imposes restraints that often become irksome. For more boisterous and expansive spirits, the Bluff-and-Glory school is recommended.

Personal appearance counts for a lot here: an opulent and in-artistic Bohemianism is the effect aimed at. The Pseudo-Secret-Service have affiliations with the Russian Ballet, Wiltshire and the fashionable weeklies; the Bluff-and-Glory boys move about the Stock Exchange, Fleet Street and the House of Commons smoking-room. They have definite traces of City soot behind the ears, and they are usually too busy to visit the barber. They have hoarse and rather hectoring voices, a gangster vocabulary, effusive geniality. They eschew moderation and either drink to excess or not at all; they are boastful in love and pursue rather access-ible quarry. They know the names of everyone with more than twenty thousand pounds a year and can furnish, unasked, exact details of the dispositions of their fortunes. 'Old So-and-so moved back one hundred thousand in Commodities,' they say, or, 'I will hand it to So-and-so, he made a very pretty clean-up last week in Oxides.' In Parliament, they know all the gossip from the lobbies and the Whips' offices. Cabinet secrets are no secrets to them, particularly in regard to personal dissensions. In spite of their ruggedness of appearance, they have a keen regard for personal comfort, and few of them have travelled further than Los Angeles and the Lido.

An essential quality is resilience in face of exposure. For example, you have been dominating the table for some time about the character of Catalan nationalism, and, towards the end of your discourse, you reveal the fact that you thought Bilbao and Guernica were in Catalonia. Do not be put out. Either say offensively, 'It's no use trying to talk reason to Communists'—or 'Fascists', at will; or shout, 'I'm not talking about Catalonia; I'm talking about the Basques,' or 'Who said anything about Bilbao? I'm talking about Barcelona,' or 'My dear fellow, look at the map. I'm not here to teach you elementary geography.' Any of these replies, or all of them in one fine peroration, should suffice to clear your reputation. Treat all discussion as though you were being heckled in a tough ward at an election. Rely on the impromptu statistic; e.g., someone says, 'All ships' engineers seem to be Scotsmen'; reply, 'The latest Mercantile Marine figures give the percentage at 78.4 recurring.' Attribute all facts of common knowledge to personal information; for instance, do not say, 'What a wet week it has been,' but, 'They tell me at Greenwich they have registered the highest rainfall for six weeks.' Instead of 'I see there have been a lot of jewel robberies lately,' say, 'The Chief Commissioner tells me that Scotland Yard is up against it.' Always refer to big-business concerns by the name of their chief magnate. 'Ashfield is making a new station,' 'Mond is putting up the price of pills,' 'Write to Astor about it.'

By following these simple instructions and studying the methods of those who have already made good in the job, you can assure yourself a glamorous youth, prosperous middle age, the title of Grand Old Man, and finally some laudatory obituaries.

1939

The Lost Childhood

PERHAPS it is only in childhood that books have any deep influence on our lives. In later life we admire, we are entertained, we may modify some views we already hold, but we are more likely to find in books merely a confirmation of what is in our minds already: as in a love affair it is our own features that we see reflected flatteringly back.

But in childhood all books are books of divination, telling us about the future, and like the fortune-teller who sees a long journey in the cards or death by water they influence the future. I suppose that is why books excited us so much. What do we ever get nowadays from reading to equal the excitement and the revelation in those first fourteen years? Of course I should be interested to hear that a new novel by Mr E. M. Forster was going to appear this spring, but I could never compare that mild expectation of civilized pleasure with the missed heartbeat, the appalled glee I felt when I found on a library shelf a novel by Rider Haggard, Percy Westerman, Captain Brereton or Stanley Weyman which I had not read before. No, it is in those early years that I would look for the crisis, the moment when life took a new slant in its journey towards death.

I remember distinctly the suddenness with which a key turned in a lock and I found I could read—not just the sentences in a reading book with the syllables coupled like railway carriages, but a real book. It was paper-covered with the picture of a boy, bound and gagged, dangling at the end of a rope inside a well with the water rising above his waist—an adventure of Dixon Brett, detective. All a long summer holiday I kept my secret, as I believed: I did not want anybody to know that I could read. I suppose I half consciously realized even then that this was the dangerous moment. I was safe so long as I could not read—the wheels had not begun to turn, but now the future stood around on bookshelves everywhere waiting for the child to choose—the life of a chartered accountant perhaps, a colonial civil servant, a planter in China, a steady job in a bank, happiness and misery, eventually one particular form of death, for surely we choose our death much as we choose our job. It grows out of our acts and our evasions, out of our fears and out of our moments of courage. I suppose my mother must have discovered my secret, for on the journey home I was presented for the train with another real book, a copy of Ballantyne's *Coral Island* with only a single picture to look at, a coloured frontispiece. But I would admit nothing. All the long journey I stared at the one picture and never opened the book.

But there on the shelves at home (so many shelves for we were a large family) the books waited—one book in particular, but before I reach that one down let me take a few others at random from the shelf. Each was a crystal in which the child dreamed that he saw life moving. Here in a cover stamped dramatically in several colours was Captain Gilson's *The Pirate Aeroplane*. I must have read that book six times at least—the story of a lost civilization in the Sahara and of a villainous Yankee pirate with an aeroplane like a box kite and bombs the size of tennis balls who held the golden city to ransom. It was saved by the hero, a young subaltern who crept up to the pirate camp to put the aeroplane out of action. He was captured and watched his enemies dig his grave. He was to be shot at dawn, and to pass the time and keep his mind from uncomfortable thoughts the amiable Yankee pirate played cards with him—the mild nursery game of Kuhn Kan. The memory of that nocturnal game on the edge of life haunted me for years, until I set it to rest at last in one of my own novels with a game of poker played in remotely similar circumstances.

And here is *Sophy of Kravonia* by Anthony Hope—the story of a kitchen-maid who became a queen. One of the first films I ever saw, about 1911, was made from that book, and I can hear still the rumble of the Queen's guns crossing the high Kravonian pass beaten hollowly out on a single piano. Then there was Stanley Weyman's *The Story of Francis Cludde,* and above all other books at that time of my life *King Solomon's Mines*.

This book did not perhaps provide the crisis, but it certainly influenced the future. If it had not been for that romantic tale of

Allan Quatermain, Sir Henry Curtis, Captain Good, and, above all, the ancient witch Gagool, would I at nineteen have studied the appointments list of the Colonial Office and very nearly picked on the Nigerian Navy for a career? And later, when surely I ought to have known better, the odd African fixation remained. In 1935 I found myself sick with fever on a camp bed in a Liberian native's hut with a candle going out in an empty whisky bottle and a rat moving in the shadows. Wasn't it the incurable fascination of Gagool with her bare yellow skull, the wrinkled scalp that moved and contracted like the hood of a cobra, that led me to work all through 1942 in a little stuffy office in Freetown, Sierra Leone? There is not much in common between the land of the Kukuanas, behind the desert and the mountain range of Sheba's Breast, and a tin-roofed house on a bit of swamp where the vultures moved like domestic turkeys and the pi-dogs kept me awake on moonlit nights with their wailing, and the white women yellowed by atebrin drove by to the club; but the two belonged at any rate to the same continent, and, however distantly, to the same region of the imagination—the region of uncertainty, of not knowing the way about. Once I came a little nearer to Gagool and her witch-hunters, one night in Zigita on the Liberian side of the French Guinea border, when my servants sat in their shuttered hut with their hands over their eyes and someone beat a drum and a whole town stayed behind closed doors while the big bush devil—whom it would mean blindness to see—moved between the huts.

But *King Solomon's Mines* could not finally satisfy. It was not the right answer. The key did not quite fit. Gagool I could recognize—didn't she wait for me in dreams every night, in the passage by the linen cupboard, near the nursery door? and she continues to wait, when the mind is sick or tired, though now she is dressed in the theological garments of Despair and speaks in Spenser's accents:

> The longer life, I wote the greater sin,
> The greater sin, the greater punishment.

Yes, Gagool has remained a permanent part of the imagination, but Quatermain and Curtis—weren't they, even when I was only ten years old, a little too good to be true? They were men of such unyielding integrity (they would only admit to a fault in order to show how it might be overcome) that the wavering personality of a child could not rest for long against those monumental shoulders. A child, after all, knows most of the game—it is only an attitude to it that he lacks. He is quite well aware of cowardice, shame, deception, disappointment. Sir Henry Curtis perched upon a rock bleeding from a dozen wounds but fighting on with the remnant of the Greys against the hordes of Twala was too heroic. These men were like Platonic ideas: they were not life as one had already begun to know it.

But when—perhaps I was fourteen by that time—I took Miss Marjorie Bowen's *The Viper of Milan* from the library shelf, the future for better or worse really struck. From that moment I began to write. All the other possible futures slid away: the potential civil servant, the don, the clerk had to look for other incarnations. Imitation after imitation of Miss Bowen's magnificent novel went into exercise-books—stories of sixteenth-century Italy or twelfth-century England marked with enormous brutality and a despairing romanticism. It was as if I had been supplied once and for all with a subject.

Why? On the surface *The Viper of Milan* is only the story of a war between Gian Galeazzo Visconti, Duke of Milan, and Mastino della Scala, Duke of Verona, told with zest and cunning and an amazing pictorial sense. Why did it creep in and colour and explain the terrible living world of the stone stairs and the never quiet dormitory? It was no good in that real world to dream that one would ever be a Sir Henry Curtis, but della Scala who at last turned from an honesty that never paid and betrayed his friends and died dishonoured and a failure even at treachery—it was easier for a child to escape behind his mask. As for Visconti, with his beauty, his patience, and his genius for evil, I had watched him pass by many a time in his black Sunday suit smelling of mothballs. His name was Carter. He exercised terror from a distance like a snowcloud over the young fields. Goodness has only once found a perfect incarnation in a human body and never will again, but evil can always find a home there. Human nature is not black and white but black and grey. I read all that in *The Viper of Milan* and I looked round and I saw that it was so.

There was another theme I found there. At the end of *The Viper of Milan*—you will remember if you have once read it—

comes the great scene of complete success—della Scala is dead, Ferrara, Verona, Novara, Mantua have all fallen, the messengers pour in with news of fresh victories, the whole world outside is cracking up, and Visconti sits and jokes in the wine light. I was not on the classical side or I would have discovered, I suppose, in Greek literature instead of in Miss Bowen's novel the sense of doom that lies over success—the feeling that the pendulum is about to swing. That too made sense; one looked around and saw the doomed everywhere—the champion runner who one day would sag over the tape; the head of the school who would atone, poor devil, during forty dreary undistinguished years; the scholar . . . and when success began to touch oneself too, however mildly, one could only pray that failure would not be held off for too long.

One had lived for fourteen years in a wild jungle country without a map, but now the paths had been traced and naturally one had to follow them. But I think it was Miss Bowen's apparent zest that made me want to write. One could not read her without believing that to write was to live and to enjoy, and before one had discovered one's mistake it was too late—the first book one does enjoy. Anyway she had given me my pattern—religion might later explain it to me in other terms, but the pattern was already there—perfect evil walking the world where perfect good can never walk again, and only the pendulum ensures that after all in the end justice is done. Man is never satisfied, and often I have wished that my hand had not moved further than *King Solomon's Mines*, and that the future I had taken down from the nursery shelf had been a district office in Sierra Leone and twelve tours of malarial duty and a finishing dose of blackwater fever when the danger of retirement approached. What is the good of wishing? The books are always there, the moment of crisis waits, and now our children in their turn are taking down the future and opening the pages. In his poem 'Germinal' A.E. wrote:

> In ancient shadows and twilights
> Where childhood had strayed,
> The world's great sorrows were born
> And its heroes were made.
> In the lost boyhood of Judas
> Christ was betrayed.

1947

Adams at Ease

I T is impossible to be consistent in the feelings with which we respond to Henry Adams. Sometimes he is irresistible, as in his memories of his boyhood, or in the exercise or expression of friendship, or in some fleeting reference to his dead wife. Sometimes he is hateful, as in his anti-Jewish utterances, or in the queer malice with which he infused the visions of doom of his later life. There are times when he is supreme in manly delicacy, and times when he seems feline, or trifling and shallow. It often occurs to us to believe that his is the finest American intelligence we can possibly know, while again it sometimes seems that his mind is so special, and so refined in specialness, as to be beside any possible point.

It would of course be easier for us to settle our personal accounts with him if only his personality were a private one. We might then choose simply to conclude that the flaws of his temper are of a kind that prevents us from giving him full credence. Or we might feel, with more charity and worldliness, that this was a man who lived to be eighty years old, who was articulate for more than sixty of those years, and who, one way or another, made himself the subject of all that he said, who, although not a confessional writer as we nowadays understand that term, was not bound by conventions of reticence other than those he created for himself—we might well feel of such a man that it would be strange indeed if he did not exhibit a good many of the inadequacies that the human spirit is heir to.

But it is not easy to come to a final settlement with Adams's personality either in the way of condemnation or in the way of tolerance because, as I say, it isn't a private personality that we are dealing with: it is a public issue. And it is not an issue that in the course of our lifetime we are likely to take a fixed position on. Once we involve ourselves with Adams, we are fated to be back and forth with him, now on one side of the issue, now on the other, as the necessities of our mood and circumstance dictate. We are at one with Adams whenever our sense of the American loneliness and isolation becomes especially strong, whenever we feel that our culture belongs to everyone except ourselves and our friends, whenever we believe that our talents and our devotion are not being sufficiently used. At such moments we have scarcely any fault to find with Adams the man. His temperamental failings sink out of sight beneath his large and noble significance.

Yet it isn't possible to identify ourselves with Adams for very long. One's parsnips must be already buttered, as Adams's were, before one can despair as wholeheartedly as he. One needs what he had, a certain elegance of *décor,* something of an almost princely style of life, a close and intimate view of the actuality of the power one undertakes to despise, and freedom to travel and observe, and leisure to pursue the studies by which one fleshes the anatomy of one's dark beliefs. And even apart from the economic considerations, we can't long afford the identification with Adams. We come to see, as William James saw, that there is a kind of corruption and corruptingness in the perfect plenitude of his despair. With James we understand that Adams's despair is a chief condition of its own existence, and that the right to hope is earned by our courage in hoping. And when we see this, we turn on Adams, using against him every weapon on which we can lay our hands. We look for the weaknesses in his theory of history and of society (it is not hard to find them), we question his understanding of science, we seek out the rifts in his logic—and we insist, of course, on the faults of his personal temper, we permit his irony to irritate us, we call him snobbish, and over-fastidious, and *fainéant*.

But we shall be wrong, we shall do ourselves a great disservice, if ever we try to read Adams permanently out of our intellectual life. I have called him an issue—he is even more than that, he is an indispensable element of our thought, he is an instrument of our intelligence. To succeed in getting rid of Adams would be to diminish materially the seriousness of our thought. In the intellectual life there ought to be frequent occasions for the exercise of ambivalence, and nothing can be more salutary for the American intelligence than to remain aware of Adams and to maintain toward him a strict ambivalence, to weigh our admiration and affection for him against our impatience and suspicion.

Two recent Adams items seem to me peculiarly useful in helping us keep the right balance of our emotions toward their author. Newton Arvin has made a selection of Adams's letters in the Great Letters Series, the publishers of which, Farrar, Straus, and Young, have met the occasion of the presidential convention season with a new printing of Adams's Washington novel, *Democracy*. If, as I think, the scales have tipped rather against Adams in the last few years—is this because some of his worst predictions have come dismally true?—if at the moment he is more out of favor than in, these two books will do much to restore the equipoise of our judgment.

Adams, as we all remember, thought of himself as a child of the eighteenth century, and his belief in his anachronism is substantiated by nothing so much as his letters. His family was formidable in the epistolary art, his grandmother Abigail being something of a genius in it, and Henry practiced it with an appropriate seriousness. Early in his career he speaks half-jokingly of a desire to emulate Horace Walpole in the representation of the manners and habits of his time, and he made bold to hope that his letters would be remembered when much in the historical scene was forgotten. There is a common belief that those who write letters to posterity as well as to their friends are bound to write dully, but this is not true in general and it certainly is not true of Adams. Since personal letters were first valued and published, no public or quasi-public person can write them without the awareness of posterity. And of course for Adams archives were no great thing, and the notion that he was adding to their number did not make him awkward or, in the bad sense, self-conscious. The historical past, the historical present, the historical future, were the stuff of his existence, the accepted circumstance of his most private thoughts and most intimate friendships.

Adams's capacity for friendship was one of the most notable things about him, and it is of course a decisive element of the greatness of his letters. In this he is peculiarly a man of the nineteenth century, which was the great age of friendship. Men then felt that the sharing of experience with certain chosen spirits was one of the essential pleasures of life, and even writers and revolutionaries found it possible to have close, continuing communication with each other. Adams, we almost come to believe as we read his letters, was the last man, or perhaps the last American, to have had actual friendships. He mistrusted much in the world, but he trusted his friends, and he so far developed his great civilized talent for connection that he could be in lively communicative relationship even with his family, and even with women.

Mr Arvin has selected the letters with his usual tact and perceptiveness, and with the awareness of how much they add to, and deduct from, and in general qualify the image of Adams's mind which we get from his two most famous books, *Mont-Saint-Michel and Chartres* and *The Education*. In the letters it was possible for Adams's mind to work without the excessive elaborations of irony which are characteristic of his published late writings. This irony is no doubt always very brilliant, but as Mr Arvin observes, it is all too obviously less a function of the author's intelligence than of his personal uneasiness in relation to the unknown reader, of that excess of delicacy and self-regard which led him into the irritating high-jinks of flirtatious hesitation about publishing his two great works. But in the letters there is no embarrassment and there is no irony beyond what normally and naturally goes with the exercise of a complex intelligence.

Then the letters, as Mr Arvin remarks, give us an Adams who loved the world in its manifold variety much more than we might ever conclude from the books. He delighted in the pleasures that the world could offer, in what might be observed of the world for the joy of observation rather than for the support of a theory of the world's uninhabitability. Further, it is from the letters that we get a notion of the development of Adams's mind which is more accurate than his own formal account of it in *The Education*. 'Only the reader of the letters,' as Mr Arvin says, 'has a full sense of the delicacy with which Adams's mind was for many years balanced between the poles of hopefulness and despair, affirmation and denial, belief and skepticism.'

Certainly the reader of the letters gets what the reader of *The Education* does not get, the awareness of how late in coming was

Adams's disillusionment with democracy. *The Degradation of the Democratic Dogma* was not the title that Adams himself gave to the oddly contrived posthumous volume that his brother edited and published under that name, but it is a phrase that accurately suggests Adams's attitude in his last years, and *The Education* would lead us to believe that this attitude was established with him upon the defeat of his youthful political expectations with the publication of the list of Grant's cabinet. Yet the letters of the middle years, even after the process of disillusionment had begun, are full of references to his continuing faith in democracy. In 1877 he wrote to his English friend, Charles Milnes Gaskell: 'As I belong to the class of people who have great faith in this country and who believe that in another century it will be saying in its turn the last word of civilisation, I enjoy the expectation of the coming day.' And in 1881 he could write to Wayne MacVeagh upon the occasion of Garfield's assassination: 'Luckily we are a democracy and a sound one. Nothing can shake society with us, now that slavery is gone.'

And the same essential faith in the American democratic ideal is implicit in Adams's novel, *Democracy,* despite its satiric rejection of the actualities of American government in 1879. There is no touch of irony in the speech which Adams puts into the mouth of his questing heroine, Mrs Lightfoot Lee, when she takes it upon herself to chasten a young Italian Secretary of Legation who too easily accepts the idea that 'there was no society except in the old world':

'Society in America? Indeed there is society in America and very good society too; but it has a code of its own, and newcomers seldom understand it. I will tell you what it is, Mr Orsini, and you will never be in danger of making any mistake. "Society" in America means all the honest, kindly mannered, pleasant-voiced women, and all the good, brave, unassuming men, between the Atlantic and the Pacific. Each of these has a free pass in every city and village, "good for this generation only," and it depends on each to make use of this pass or not as it may happen to suit his or her fancy. To this rule there are *no* exceptions, and those who say "Abraham is our father" will surely furnish food for that humour which is the staple product of our country.'

It rings, does it not, with the passionate naivety of an old-fashioned high-school oration, yet such sentiments about their country once served to aerate and brighten the minds of even the most sophisticated and critical of Americans—one finds them being uttered at a certain period by Henry James and with a passion of optimism no less naive than that of his friend Adams.

The awareness of this very attractive naivety of idealism is essential for the understanding of Adams's ultimate development in pessimism. It is endemic in *Democracy,* appearing in the use which is made of the simplicity and plainness of General Lee's house at Arlington, and in the elaborate discussion of Mount Vernon and General Washington as representing the republican virtues which, although receding into the past, are still part of the American dream.

In his introduction to the *Selected Letters,* Mr Arvin refers to Adams's two anonymously published novels, *Democracy* and *Esther,* as 'remarkable books, more remarkable than they have usually been recognized as being.' And so they are. Of the two, *Democracy* is, I think, the more attractive. *Esther,* which stands in the same interrogative relation to religion that *Democracy* stands to politics, is full of velleities of thought and feeling about its subject, and these, while possibly they make for an interesting darkness in the work, also make for uncertainty and irresolution. But *Democracy* is all clarity and brightness, and entirely satisfying so far as it goes. It does not, as a novel, go very far—does not pretend to go very far. It is brief and witty and schematic; everything in it is contrived and controlled by the author's intelligence and his gaiety. It can claim a degree of cousinage with Peacock's novels of intellectual humors; it shares the light speed of this form and it has more than a few Peacockian moments, such as that in which the bright young American heiress explains to a British visitor the trouble that Americans have with their sundials:

'Look at that one! they all behave like that. The wear and tear of our sun is too much for them; they don't last. My uncle, who has a place at Long Branch, had five sun-dials in ten years.'

'How very odd! But really now, Miss Dare, I don't see how a sun-dial could wear out.'

'Don't you? How strange! Don't you see, they get soaked with sunshine so that they can't hold shadow.'

But although its humor is frequent and its wit pervasive, *Democracy* does not move among ideas with the Peacockian lack of commitment to anything but common sense. It is concerned to ask a question which was of the greatest importance to Adams himself, and to his countrymen: The nature of American political life being what it is, is it possible for a person of moral sensibility to participate in it?

The person upon whom the test is made is the attractive and intelligent young widow, Madeleine Lee—Adams is already, in middle life, making woman the touchstone and center of civilization. Bereaved of a husband and baby in a single year, Mrs Lee has tried to fill her life with civilized interests, and, having found philosophy and philanthropy of no avail, has established herself in Washington intent on trying political activity as a last resort. 'She wanted to see with her own eyes the action of primary forces; to touch with her own hand the massive machinery of society; to measure with her own mind the capacity of the motive power. She was bent upon getting to the heart of the great American mystery of democracy and government.'

She wanted, in short, the experience of power, as did Adams himself. It was not merely her being a woman that brought it about that 'the force of the engine was a little confused in her mind with that of the engineer, the power with the men who wielded it'; the confusion had existed in Adams's mind when he had decided that there was no possible place for him in American political life; the confusion is no confusion at all but an accurate statement of the fact.

Madeleine's experience of the men who wield the power, or try to wield it, is the substance of her sad education. Presidents of the United States, she learns, are likely to be foolish, vulgar, bedeviled men, who, with their impossible wives, lead the most hideous lives of public ceremony. Reformers maintain their equanimity only by a bland ignorance of the nature of what they are trying to change. Most senators are nonentities, and the one senator who is more than that, who does have the strength and craft to control the great machine, is, as poor Madeleine finds, venal—corrupt not merely through personal motives but by the acknowledged terms of his profession, by his devotion to party. And since it is this Senator Ratcliffe whom Madeleine is drawn to by reason of his power and even thinks of marrying, his acceptance and rationalization of his immorality is decisive with her. She surrenders Washington and withdraws from the political life, and the posed question is conclusively answered: No, it is not possible for a person of moral sensibility to take part in American politics.

Very likely the means by which the answer is made will seem too simple to us nowadays and not quite relevant to our situation—not that the moral probity of senators is now an article of our political faith, but that we do not take senatorial corruption for granted as it was taken for granted in the Seventies when Mark Twain instituted his famous comparison between the moral character of senators and that of hogs. But the question is still a valid one, and so, in some important part, is Adams's answer.

1952

A New Westminster

WHEN I first learned that Westminster Abbey was to be demolished in the foreseeable future I was as dumbfounded as, no doubt, will be the readers of these words. I sought permission on the very highest level to present the case for demolition to an intelligent public in the favourable light in which I now myself see it and received an express intimation from my Minister himself via the deputy comptroller that I could do so. Neither my Minister nor the London County Council Planning Committee nor the works and buildings committee of the Westminster City Council, all of whom are of course directly and indirectly concerned with the proposed demolition, wished the matter to be discussed yet in the national press. It was considered that what my Minister calls 'a feeler' might be put out in the *Spectator* or the *Manchester Guardian,* to test the more enlightened reaction of an exclusive and cultivated public to a scheme the benefits of which might not at first seem to outweigh the somewhat sentimental losses. Both my Minister, the LCC and any local planning committees have always found that in practice it is best to present the general public with a *fait accompli* when a scheme is ultimately for its own good. I must, therefore, ask my readers not to pass on the information they read here to their lady helps, domestic science assistants, public cleansing officers, etc., but to confine their information to administrative grades.

For some time now the Minister of Transport has been concerned by the increase of traffic between Victoria and Parliament Square and notably by the bottleneck caused by the projection of the western towers of the Abbey into the roadway. In the near future it is proposed to erect on the site of the old Westminster Hospital a much-needed block of government offices, with the result that the bottleneck will be further intensified. My Minister was reluctant to take the drastic course of demolishing the

Abbey without first examining all possible alternatives. The most obvious of these was the setting back of the proposed new government offices so as to secure a consistent width of roadway the whole way down Victoria Street to Parliament Square. To this there were insuperable objections: the plans for the new offices were already in an advanced state and could not be altered except at prohibitive cost to the public funds; the roadway itself would make an unnecessary curve to avoid the Abbey and interrupt a fine vista the LCC planning authorities had, with imaginative foresight, arranged whereby Big Ben would be visible with the Houses of Parliament from as far away down Victoria Street as the Army and Navy Stores. Another course to be taken was that of leaving things as they are, which *prima facie,* is impossible.

My Minister had then to consider the pros and cons of the Abbey itself. It has undoubted historic associations going back as far as Saxon times, though the vestiges of these interesting days are so slight as to cause very little trouble in their preservation, if it is envisaged, in the new scheme for developing the site. Then there are the memorials of eminent persons in the political, scientific, economic and artistic worlds whose bones are interred in the Abbey. By arrangement with the development company which is to erect the fine new building on the site, my Minister has arranged that these shall be moved and re-erected in a suitable cloister or close at Brookwood Cemetery, where they will be open to those members of the public who still enjoy the rather morbid occupation of examining gravestones.

Next my Minister was faced with finding alternative accommodation for the purposes for which the Abbey is used at present. There is still a certain amount of religious services carried on there, though, we may confidently expect, as material progress continues, rapidly diminishing numbers. By arrangement with neighbouring vicars of the Church of England and with the full assurance of the authorities of the adjacent Roman Catholic Cathedral at Westminster that they will receive any members of the existing congregation of the Abbey who may care to join them, it should be possible to cater for these persons without undue inconvenience. Finally, there are the rare occasions when the building is used for Coronations, and we must assume, for the present, that the monarchy will continue to exist. All will agree that the present building is too small, too inconvenient and too ill-planned to enable those many thousands who may wish to witness this quaint and historic ceremony to see it. It is suggested that a place with better visibility, say, the Festival Hall or Wembley Stadium, be used for future Coronations. This will have the additional advantage of being non-sectarian.

As to the fabric itself, my Minister has given this careful consideration. He has consulted acknowledged experts and learns that the building, though ancient in origin, was not all built at one period and therefore lacks the consistency of a single unit of architecture such as is envisaged on its site. The controversial western towers are indeed a fake, having been ascribed to Sir Christopher Wren and being in a Gothic which, if my readers will pardon the phrase, can only be called 'bastard'. The exterior was largely refaced by the Victorians, who notably lacked artistic taste. The only feature which all are agreed as being of exceptional merit is the Henry VII chapel, which, though very late and decadent Gothic, has a certain charm. The developers have expressed themselves as willing to retain a portion of this, if possible, in their new building, since they maintain that as it is the best the Middle Ages could do in the way of glass and stone (stainless steel not then having been discovered) it can be made to harmonise with the simpler and more honest expression of our own age in steel and glass which they are proposing to erect. But if they keep a part of this chapel they will have, for economic reasons, to develop on the site of the somewhat redundant church of St Margaret, Westminster.

Finally, there come the advantages of the proposed scheme, which may be summarised under the following heads.

Practical: London's traffic problem will be materially eased by a free passage of transport between the busy stations of Victoria and Waterloo and buses and cars will be able to travel much faster from the South-West to Whitehall. A more suitable building will be provided elsewhere for Coronations. Much needed government and commercial offices will be provided in Westminster which, in the neighbourhood of the existing Abbey, lags far behind the City of London in commercial development.

Economic: The development company is willing to pay a high enough sum for this key site to offset the cost of the road improvements and gain in public parking space which will result, thus putting no burden on the ratepayers.

Artistic: The very best architects are to be employed by the company and the design will of course be submitted to the LCC, the Westminster Corporation and possibly even to the Royal Fine Arts Commission. The resulting achievement, to be in the form of a glass and steel tower hung with specially designed curtain walling and three hundred feet high with subsidiary light and airy blocks rising to not more than one hundred feet, will challenge, as our own age should if we have any faith in it, the Houses of Parliament to which it will act as a vast foil. A new vista will be opened from Victoria Street. A worthy contribution to a famous skyline will at last be added in a part too long dominated by the obsolescent buildings of past eras.

As a Government servant and Public Relations Officer I cannot, for obvious reasons, subscribe my name to this article, but have paid a journalist to do so who has pleasure in signing himself

J. BETJEMAN

1958

SIR WILLIAM EMPSON

The Faces of Buddha

THERE is room for an amateur to say something about Buddha faces, because the experts tend rather to avoid so indefinite a topic, while there are two likely misunderstandings for a man in the street: that the Buddhas have no expression at all, an idea set on foot by Lafcadio Hearn, who had a genuine feeling for the East but was almost blind; or else that they all sneer, a thing G. K. Chesterton, for instance, often says, which is less easy to answer. Certainly in each Buddhist country, after a few centuries, the type becomes conventional and is liable to be complacent; also one thinks first of the Buddhas of China, and as soon as the Buddha arrived in China he was given something of the polite irony of a social superior. There was some real falsity when they came to treat the Goddess of Mercy as a fashion plate of the court lady. Yet before merely disliking that look it is only fair to see where it comes in the system. The Buddha has delivered himself from the world and may well look superior to it, but he is telling you that you can do the same; also he could not achieve this apparently selfish aim without first learning complete unselfishness. The Ajanta caves occasionally give him the face of a typical Italian Christ, but only in previous lives, while he was dealing with that aspect (giving his body to a hungry tiger and so on). As to the after-dinner look of many Buddhas, and the rings of fat on the neck, a puzzle of the translators seems to show the point; one expert gives a remark of the Buddha as 'While I live thus, after having felt the extreme sensations, I am pure,' and another as 'after having felt my last sensation.' An idea that you must be somehow satisfied as well as mortified before entering repose goes deep into the system, and perhaps into human life. However, what you are meant to feel in a Bodhisattva (which is roughly any 'Buddha' with a headdress, shown not as a monk but a king) has escaped these doctrinal puzzles and become clearly sacrificial. They are saints who have given up their Nirvana, their heaven, till they have helped their last fellow-creature into heaven before them, and the face is meant to show it. In a sense they have given up their deaths, not their lives, but the conception appeared in the first centuries after Christ and along the caravan routes to Europe; the two religions may very well be connected. The drooping eyelids of the great creatures are heavy with patience and suffering, and the subtle irony which offends us in their raised eyebrows (it is quite a common expression in Europeans, though curiously avoided in our portraits) is in effect an appeal to us to feel, as they do, that it is odd that we let our desires subject us to so much torment in the world. The first thing to say about the Buddha face, granted that many later ones are complacent, is that the smile of superiority can mean and be felt to mean simply the power to help.

The next thing, I think, about the stock type, is that it is the simplest conception of high divinity the human race has devised; people say it is monotonous, but there is a sort of democracy about its repetition. In a way Europe has agreed on the face of Christ, but you have to be a good artist to do it. Anyone who cares about the Lord Buddha can do his face in a few ignorant strokes on sand or blotting paper, and among all the crude versions I have walked past I do not remember one that failed to give him his effect of eternity. It is done by the high brow, soaring outwards; by the long slit eye, almost shut in meditation, with a suggestion of a squint, that would be a frighteningly large eye if opened; and by a suggestion of the calm of childhood in the smooth lines of the mature face—a certain puppy quality in the long ear helps to bring this out. If you get these they carry the main thought of the religion; for one thing the face is at once blind and all-seeing ('he knows no more than a Buddha,' they say of a deceived husband in the Far East), so at once sufficient to itself and of universal charity. This essential formula for the face allows of great variety and is hardly more than a blank cheque, but one on a strong bank, so to speak. To my feeling a quite unrealistic Buddha is far more ready than a European head of Christ to be conceived as a real person in the room; as you sink into it you seem to know what it would feel like to have those extraordinary hands.

It is a mistake to explain this type as merely racial, though it was exaggerated in the Far East and somehow fits in with their normal outlook. Greco-Roman artists in the northwest, about the first century AD, seem to have broken the Indian convention that the Buddha must not be portrayed, and the calm of their Apollo made a conflict with the human and muscular earth-god tradition of Mathura. Then by the Indian Gupta period (fifth century) the Buddha has settled down to a high brow and half-shut eye; they are not a Far Eastern invention. The eye had to follow the brow; a wide open eye under the high brow would be in great danger of the coy surprise of George Robey, or anyway of an unquiet sort of surprise, which is not meant. Of course a good enough artist can avoid the obvious; it is terrible when the Buddha in the Ajanta caves once fully opens his eyes, as he takes his last look at his wife—a picture, by the way, which has been destroyed by varnish, and can now be seen only in photographs. But this, I think, gave a main reason for closing it. The photograph here from the great Bodhisattva of Cave 1 will serve to show how the type was going, though not to show the Titian richness of the flesh-painting and the Tintoretto glitter of the crown. Not that the Far East was afraid of Robey; there is a further threat of him when the brows curl down again on the outside, and this was used mainly for the late Vairocana Bodhisattva, who stands for the energy behind the universe (or thereabouts). This strange conception tends to particularly puppy ears and a certain winning bounce in the raised finger; the type can aim at something near Robey and be still a god. For that matter both Kwannon and Maitreya have a version as a great fat laughing sprawler, which helps to show that this is not a misunderstanding. The merely racial difficulty in understanding the faces is indeed smaller than you would expect, and the artists at Angkor no less than Ajanta seem to have amused themselves by putting the same face on to all the races of mankind.

The formula leaves much of the face free. The nose can do what it likes, and is used for anything between childishness, sensuality, and administrative power. The mouth can do what it likes, and varies from a rich sensual repose to the strained tight-lipped alert smile seen on flying aces and archaic Greek sculpture. This of course is not borrowed from Greece; the Greek influence was not archaistic, and anyway the typical thing about an archaic Apollo is not simply the mouth but a peculiar half-baked look about the jowl. The point about the archaic fixed smile, on Buddhas or elsewhere, is that it would be made by a pull on the main zygomatic, the muscle most under conscious control, leaving the others at rest; thus it is an easy way to make a statue look socially conscious, wilful, alert. Many of the Chinese Buddhas from the Yun-kang caves, the earliest period, get a strong effect from using this quite flatly (e.g., the fine one that dominates Room 2 of the Chinese Exhibition). But you have only to sink the ends into the cheeks to give it an ironical or complacent character, and my example from Yun-kang, almost winking as it is, gets, I think, with these simple means, an extraordinary effect both of secure hold on strength and peace and of the humorous goodwill of complete understanding. The Koriuji example is traditionally a gift from Korea and can stand for the second main influence on early Japan; its very subtle mouth is not at all of this type, and the future Buddha has a plaintive and somewhat foxy elegance not yet developed as an active force in the world. In the Chuguji one, who will also when he is born bring a new revelation, it is rather the older convention for the mouth, toned down and with a couple of ripples in the smooth wood, that gives all that lightness and tenderness which will at any moment brush away the present universe as an unwise dream. The Horiuji Goddess of Mercy, though not very clearly on her copy in the British Museum, uses it for a rueful puggy puzzled expression, faintly suggesting the White Queen, that needs for its interpretation the great sweep of the flamelike draperies, stretching as far as earth, and the jerk of the extended arm, like a stalk, exhausted but still offering. Noble stupid creature; at least no one can say that she is sneering.

1936

LOREN EISELEY

The Snout

I HAVE long been an admirer of the octopus. The cephalopods
are very old, and they have slipped, protean, through many
shapes. They are the wisest of the mollusks, and I have always
felt it to be just as well for us that they never came ashore,
but—there are other things that have.

There is no need to be frightened. It is true some of the
creatures are odd, but I find the situation rather heartening than
otherwise. It gives one a feeling of confidence to see nature still
busy with experiments, still dynamic, and not through nor satisfied because a Devonian fish managed to end as a two-legged
character with a straw hat. There are other things brewing and
growing in the oceanic vat. It pays to know this. It pays to know
there is just as much future as there is past. The only thing that
doesn't pay is to be sure of man's own part in it.

There are things down there still coming ashore. Never make
the mistake of thinking life is now adjusted for eternity. It gets
into your head—the certainty, I mean—the human certainty, and
then you miss it all: the things on the tide flats and what they
mean, and why, as my wife says, 'they ought to be watched.'

The trouble is we don't know what to watch for. I have a
friend, one of these Explorers Club people, who drops in now
and then between trips to tell me about the size of crocodile jaws
in Uganda, or what happened on some back beach in Arnhem
Land.

'They fell out of the trees,' he said. 'Like rain. And into the
boat.'

'Uh?' I said, noncommittally.

'They did *so*,' he protested, 'and they were hard to catch.'

'Really—' I said.

'We were pushing a dugout up one of the tidal creeks in
northern Australia and going fast when *smacko* we jam this man-
grove bush and the things come tumbling down.

'What were they doing sitting up there in bunches? I ask you. It's no place for a fish. Besides that they had a way of sidling off with those popeyes trained on you. I never liked it. Somebody ought to keep an eye on them.'

'Why?' I asked.

'I don't know why,' he said impatiently, running a rough, square hand through his hair and wrinkling his forehead. 'I just mean they make you feel that way, is all. A fish belongs in the water. It ought to stay there—just as we live on land in houses. Things ought to know their place and stay in it, but those fish have got a way of sidling off. As though they had mental reservations and weren't keeping any contracts. See what I mean?'

'I see what you mean,' I said gravely. 'They ought to be watched. My wife thinks so too. About a lot of things.'

'She does?' He brightened. 'Then that's two of us. I don't know why, but they give you that feeling.'

He didn't know why, but I thought that I did.

It began as such things always begin—in the ooze of unnoticed swamps, in the darkness of eclipsed moons. It began with a strangled gasping for air.

The pond was a place of reek and corruption, of fetid smells and of oxygen-starved fish breathing through laboring gills. At times the slowly contracting circle of the water left little windrows of minnows who skittered desperately to escape the sun, but who died, nevertheless, in the fat, warm mud. It was a place of low life. In it the human brain began.

There were strange snouts in those waters, strange barbels nuzzling the bottom ooze, and there was time—three hundred million years of it—but mostly, I think, it was the ooze. By day the temperature in the world outside the pond rose to a frightful intensity; at night the sun went down in smoking red. Dust storms marched in incessant progression across a wilderness whose plants were the plants of long ago. Leafless and weird and stiff they lingered by the water, while over vast areas of grassless uplands the winds blew until red stones took on the polish of reflecting mirrors. There was nothing to hold the land in place. Winds howled, dust clouds rolled, and brief erratic torrents choked with silt ran down to the sea. It was a time of dizzying contrasts, a time of change.

On the oily surface of the pond, from time to time a snout thrust upward, took in air with a queer grunting inspiration, and swirled back to the bottom. The pond was doomed, the water was foul, and the oxygen almost gone, but the creature would not die. It could breathe air direct through a little accessory lung, and it could walk. In all that weird and lifeless landscape, it was the only thing that could. It walked rarely and under protest, but that was not surprising. The creature was a fish.

In the passage of days the pond became a puddle, but the Snout survived. There was dew one dark night and a coolness in the empty stream bed. When the sun rose next morning the pond was an empty place of cracked mud, but the Snout did not lie there. He had gone. Down stream there were other ponds. He breathed air for a few hours and hobbled slowly along on the stumps of heavy fins.

It was an uncanny business if there had been anyone there to see. It was a journey best not observed in daylight, it was something that needed swamps and shadows and the touch of the night dew. It was a monstrous penetration of a forbidden element, and the Snout kept his face from the light. It was just as well, though the face should not be mocked. In three hundred million years it would be our own.

There was something fermenting in the brain of the Snout. He was no longer entirely a fish. The ooze had marked him. It takes a swamp-and-tide-flat zoologist to tell you about life; it is in this domain that the living suffer great extremes, it is here that the water-failures, driven to desperation, make starts in a new element. It is here that strange compromises are made and new senses are born. The Snout was no exception. Though he breathed and walked primarily in order to stay in the water, he was coming ashore.

He was not really a successful fish except that he was managing to stay alive in a noisome, uncomfortable, oxygen-starved environment. In fact the time was coming when the last of his kind, harried by more ferocious and speedier fishes, would slip off the edge of the continental shelf, to seek safety in the sunless abysses of the deep sea. But the Snout was a fresh-water Crossopterygian, to give him his true name, and cumbersome and plodding though he was, something had happened back of his eyes. The ooze had gotten in its work.

It is interesting to consider what sort of creatures we, the remote descendants of the Snout, might be, except for that green quagmire out of which he came. Mammalian insects perhaps we should have been—solid-brained, our neurones wired for mechanical responses, our lives running out with the perfection of beautiful, intricate, and mindless clocks. More likely we should never have existed at all. It was the Snout and the ooze that did it. Perhaps there also, among rotting fish heads and blue, night-burning bog lights, moved the eternal mystery, the careful finger of God. The increase was not much. It was two bubbles, two thin-walled little balloons at the end of the Snout's small brain. The cerebral hemispheres had appeared.

Among all the experiments in that dripping, ooze-filled world, one was vital: the brain had to be fed. The nerve tissues are insatiable devourers of oxygen. If they do not get it, life is gone. In stagnant swamp waters, only the development of a highly efficient blood supply to the brain can prevent disaster. And among those gasping, dying creatures, whose small brains winked out forever in the long Silurian drought, the Snout and his brethren survived.

Over the exterior surface of the Snout's tiny brain ran the myriad blood vessels that served it; through the greatly enlarged choroid plexuses, other vessels pumped oxygen into the spinal fluid. The brain was a thin-walled tube fed from both surfaces. It could only exist as a thing of thin walls permeated with oxygen. To thicken, to lay down solid masses of nervous tissue such as exist among the fishes in oxygenated waters was to invite disaster. The Snout lived on a bubble, two bubbles in his brain.

It was not that his thinking was deep; it was only that it had to be thin. The little bubbles of the hemispheres helped to spread the area upon which higher correlation centers could be built, and yet preserve those areas from the disastrous thickenings which meant oxygen death to the swamp dweller. There is a mystery about those thickenings which culminate in the so-called solid brain. It is the brain of insects, of the modern fishes, of some reptiles and all birds. Always it marks the appearance of elaborate patterns of instinct and the end of thought. A road has been taken which, anatomically, is well-nigh irretraceable; it does not lead in the direction of a high order of consciousness.

Wherever, instead, the thin sheets of gray matter expand

upward into the enormous hemispheres of the human brain, laughter, or it may be sorrow, enters in. Out of the choked Devonian waters emerged sight and sound and the music that rolls invisible through the composer's brain. They are there still in the ooze along the tideline, though no one notices. The world is fixed, we say: fish in the sea, birds in the air. But in the man-grove swamps by the Niger, fish climb trees and ogle uneasy naturalists who try unsuccessfully to chase them back to the water. There are things still coming ashore.

The door to the past is a strange door. It swings open and things pass through it, but they pass in one direction only. No man can return across that threshold, though he can look down still and see the green light waver in the water weeds.

There are two ways to seek the doorway: in the swamps of the inland waterways and along the tide flats of the estuaries where rivers come to the sea. By those two pathways life came ashore. It was not the magnificent march through the breakers and up the cliffs that we fondly imagine. It was a stealthy advance made in suffocation and terror, amidst the leaching bite of chemical discomfort. It was made by the failures of the sea.

Some creatures have slipped through the invisible chemical barrier between salt and fresh water into the tidal rivers, and later come ashore; some have crept upward from the salt. In all cases, however, the first adventure into the dreaded atmosphere seems to have been largely determined by the inexorable crowding of enemies and by the retreat further and further into marginal situations where the oxygen supply was depleted. Finally, in the ruthless selection of the swamp margins, or in the the scramble for food on the tide flats, the land becomes home.

Not the least interesting feature of some of the tide-flat emergents is their definite antipathy for the full tide. It obstructs their food-collecting on the mud banks and brings their enemies. Only extremes of fright will drive them into the water for any period.

I think it was the great nineteenth-century paleontologist Cope who first clearly enunciated what he called the 'law of the unspecialized,' the contention that it was not from the most highly organized and dominant forms of a given geological era that the master type of a succeeding period evolved, but that instead the dominant forms tended to arise from more lowly and generalized animals which were capable of making new adaptations, and which were not narrowly restricted to a given environment.

There is considerable truth to this observation, but, for all that, the idea is not simple. Who is to say without foreknowledge of the future which animal is specialized and which is not? We have only to consider our remote ancestor, the Snout, to see the intricacies into which the law of the unspecialized may lead us.

If we had been making zoological observations in the Paleozoic Age, with no knowledge of the strange realms life was to penetrate in the future, we would probably have regarded the Snout as specialized. We would have seen his air-bladder lung, his stubby, sluggish fins, and his odd ability to wriggle overland as specialized adaptations to a peculiarly restricted environmental niche in stagnant continental waters. We would have thought in water terms and we would have dismissed the Snout as an interesting failure off the main line of progressive evolution, escaping from his enemies and surviving successfully only in the dreary and marginal surroundings scorned by the swift-finned teleost fishes who were destined to dominate the seas and all quick waters.

Yet it was this poor specialization—this bog-trapped failure—whose descendants, in three great movements, were to dominate the earth. It is only now, looking backward, that we dare to regard him as 'generalized.' The Snout was the first vertebrate to pop completely through the water membrane into a new dimension. His very specializations and failures, in a water sense, had preadapted him for a world he scarcely knew existed.

The day of the Snout was over three hundred million years ago. Not long since I read a book in which a prominent scientist spoke cheerfully of some ten billion years of future time remaining to us. He pointed out happily the things that man might do throughout that period. Fish in the sea, I thought again, birds in the air. The climb all far behind us, the species fixed and sure. No wonder my explorer friend had had a momentary qualm when he met the mudskippers with their mental reservations and lack of promises. There is something wrong with our world view. It is still Ptolemaic, though the sun is no longer believed to revolve around the earth.

We teach the past, we see farther backward into time than any race before us, but we stop at the present, or, at best, we project far into the future idealized versions of ourselves. All that long way behind us we see, perhaps inevitably, through human eyes alone. We see ourselves as the culmination and the end, and if we do indeed consider our passing, we think that sunlight will go with us and the earth be dark. We are the end. For us continents rose and fell, for us the waters and the air were mastered, for us the great living web has pulsated and grown more intricate.

To deny this, a man once told me, is to deny God. This puzzled me. I went back along the pathway to the marsh. I went, not in the past, not by the bones of dead things, not down the lost roadway of the Snout. I went instead in daylight, in the Now, to see if the door was still there, and to see what things passed through.

I found that the same experiments were brewing, that up out of that ancient well, fins were still scrambling toward the sunlight. They were small things, and which of them presaged the future I could not say. I saw only that they were many and that they had solved the oxygen death in many marvelous ways, not always ours.

I found that there were modern fishes who breathed air, not through a lung but through their stomachs or through strange chambers where their gills should be, or breathing as the Snout once breathed. I found that some crawled in the fields at nightfall pursuing insects, or slept on the grass by pond sides and who drowned, if kept under water, as men themselves might drown.

Of all these fishes the mudskipper *Periophythalmus* is perhaps the strangest. He climbs trees with his fins and pursues insects; he snaps worms like a robin on the tide flats; he sees as land things see, and above all he dodges and evades with a curious popeyed insolence more suggestive of the land than of the sea. Of a different tribe and a different time he is, nevertheless, oddly reminiscent of the Snout.

But not the same. There lies the hope of life. The old ways are exploited and remain, but new things come, new senses try the unfamiliar air. There are small scuttlings and splashings in the dark, and out of it come the first croaking, illiterate voices of the things to be, just as man once croaked and dreamed darkly in that tiny vesicular forebrain.

Perpetually, now, we search and bicker and disagree. The eternal form eludes us—the shape we conceive as ours. Perhaps the old road through the marsh should tell us. We are one of many appearances of the thing called Life; we are not its perfect image, for it has no image except Life, and life is multitudinous and emergent in the stream of time.

1957

What If—? English versus German and French

I AM asked what I think would have happened if our national language were German instead of English. My first impulse is to retort: 'Why, *isn't* it German?' I think of the thick layers of abstract jargon we carry on top of our heads, of the incessant urge to rename everything in roundabout phrases (Personal Armor System = the new army helmet), of the piling up of modifiers before the noun (easy-to-store safety folding ironing board), of the evil passion for agglutinating half-baked ideas into single terms (*surprizathon* = advertising goods by lottery) and I can only grudgingly concede: 'True, it isn't German, but some of it is more German than English.'

Had the Pilger Fathers brought with them the pure Plattdeutsch of their time, all might have been well. After separation from its source and under stress of the hard frontier life, the language would have melted and clarified like butter, lost its twisted shapes and hard corners, and become a model of lucidity and force. What only the greatest German writers—Goethe, Schopenhauer, Nietzsche, and a few others—managed to do by main strength in their prose would have been done anonymously by everybody in Massachusetts and in the wagons crossing the plains. Tough characters like Thoreau, Lincoln, Mark Twain, and Ambrose Bierce would not have tolerated the stacking of clause within clause of yard-long words, uncaring whether meaning comes out at the other end. They were articulate beings and they articulated their thoughts—as we are doing less and less every day.

For on our former, flexible and clear Anglo-Latin-French, which we call American English, the überwältigend academic fog has descended and we grope about, our minds damp and moving in circles. Similar forms of the blight have struck the other languages of Western civilization, with the inevitable result of a growing inability to think sharp and straight about anything—whence half our 'prahblems'.

Had the good forthright people who built this country in the last century met this verbal miasma on landing here, they would have either perished soon from suffocation or made tracks for the open air of Canada, which would now number 210 million. Make no mistake: syntax can change the course of history.

English has a great advantage over German, on the one hand, French and the rest of the Romance languages, on the other, in that it possesses two vocabularies, nearly parallel, which carry the respective suggestions of abstract and concrete, formal and vernacular. A writer can say *concede* or *give in*; *assume* or *take up*; *deliver* or *hand over*; *insert* or *put in*; *retreat* or *fall back*; a shop in New York can even call inself 'Motherhood Maternity'. The two series of terms are not complete, and the connotations of a word in either set must be heeded before it can be used as a substitute for its first cousin, but the existence of the quasi duplicate makes for a wide range of coloring in style and nuances in thought. Only a mechanical mind believes that the so-called Anglo-Saxon derivatives should always be preferred, and only the starched and stilted will persistently fall into the Latinate.

In contrast, the corresponding words in German always show their concrete origins: *Empfindung* means *perception*, but whereas the English word conceals the Latin *take* (capere), the German keeps in plain sight the *find* (come upon). Similarly, *Gelegenheit* (occasion) has *lie* in it; *abrichten* (adjust) has *straight*; *Verhältnis* (proportion) has *hold*; *Entwurf* (project) has *throw*, and so on. All the everyday words reappear in the compounds. Not merely the associations of these words but their uses and contexts are influenced by this 'open plumbing': the abstract idea has not been fully abstracted away.

French, having lost much of its brisk medieval vocabulary during the Latinizing vogue of the Renaissance, has been left with very formal-sounding words for everyday use—for example *comestible* and *consommation* for cases in which we would say *food* and *drink*. The reason why American and English tourists think that French hotel porters are highly educated is that they say such things as: Monsieur est *matinal*; vous allez au *spectacle*; il serait *prudent* de prendre un *imperméable*; c'est un *indigène*; oui, la *représentation* est *intégrale*—and so on. The truth is, no other words are available (except slang), and all these 'learned' terms are the familiar ones, just as the highfalutin *emergency* in English is the only way to refer to a very commonplace event.

The results of these contrasting developments in the leading languages of the West go beyond differences of style; they may plausibly be held responsible for tendencies of thought. Thus, when philosophy stopped being written in Latin, the English school that arose was the Empiricist—thinkers who believed in the primacy of *things*: ideas were viewed as coming from objects in the world concretely felt. In French philosophy, *notions* came first: abstract words breed generalities at once, and the realm of thought is then seen as cut off from the world of things, the mind from the body. See M. Descartes. The historians Tocqueville and Taine thought that some of the greatest errors of the French Revolution were due to unconscious and misplaced abstraction.

By the same token, the French language has a reputation— wholly undeserved—for being the most logical of all. For three hundred years French writers have repeated this myth in good faith, because the act of fitting together abstract, generalizing terms lends a geometrical aspect to the product. But French grammar and usage and spelling are full of illogicalities—like those of other languages.

As for German, its lumpy compounds and awkward syntax present a paradox. There is a sense in which a formal German sentence delivers its core meaning three times over—once in the root of the verb, again in the noun, and finally in the adjectives or adverbs almost always tacked on to those other terms. One might therefore have expected that German thought would be peculiarly down-to-earth; yet everybody knows that is is has been peculiarly cloudy. The probable explanation is that the words that have to be used for abstract ideas (like *Vorstellung*—'put before') acquire the abstract quality while keeping visible their original concreteness. This double aspect makes the user confident that he is on solid ground. The upshot is the German academic prose that made Kierkegaard, Nietzsche, and William James tear their hair (*Eigeneshaarsıchauszupflückenplage*). If anybody is inclined to belittle English for its mongrel character and its 'illogicalities', let him remember the limitations of its rivals. We are lucky to have, in James's words, a language 'with all the modern improvements'.

1984

MAURICE RICHARDSON

In Search of Nib-Joy

THE first fountain pen in my life was Nanny's stylograph, the Dwarf; a stumpy terracotta word tool with a fine point. At the nursery table with the blue cloth with bobbles round the edge, Nanny wrote letters to her sister with it. She wrote in neat mousey handwriting that I longed to imitate. My governess decreed that anything except a plain steel nib—but not the decadent sophisticated Relief with its oblique point—was bad for forming the hand.

In the early part of the Kaiser's war, England was still penholder-minded. Arriving at Brighton station en route for my prep school, I was greeted by a huge poster:

THEY COME AS A BOON AND A BLESSING TO MEN
THE PICKWICK, THE OWL, AND THE WAVERLEY PEN.

The inkpot was unsuited to trench life. The demand for fountain pens increased. A popluar model was the Blackbird. It was advertised by a picture of a soldier writing home: 'Dear Mum, I hope this finds you as it leaves me, in the pink.' Shells were bursting overhead. I bought a Blackbird and was happy with it for a time. To fill it, you unscrewed the nib and its holder and injected ink into the barrel with a fountain pen filler. Apt to be messy.

More desirable, and more symbolically potent, were the self-filling makes; Swan, Waterman and Onoto were the best known of many. The Onoto was the first pen to give me real nib-joy. It had the advantage of a screw by which you could control the flow—always a problem. Its barrel was long, slender and beautifully balanced.

The perfectionist is never satisfied. A friend pointed out that the great penmen had written with quills. I began to experiment. The quill was capricious: ecstasy one moment, despair the next, when you failed to cut a new point.

About this time I acquired a *Manuel de Graphologie*. Being French gave it extra esoteric significance. I studied it with concentration. My aim, since handwriting was indicative of one's character, was to change my character, which was getting me into trouble just then, by changing my handwriting. It was simply a question of finding a suitable model. But the hand of those illustrated which seemed to suit me best was *Une Écriture Extrêmement Bizarre*; its writer, I gathered, had ended in a Maison des Fous.

Pen-fetishism lay dormant until I was sixteen. Then I went one winter with my sister to Gstaad, not so fashionable as it has become but even then sporting a crowned head or two. I collided with one, a plump olive-skinned little fellow, on my *luge* in the drive of the Palace Hotel. He was the Shah of Persia, soon to be deposed by the present Shah's father and retire to Paris where he kept a scent shop. In my hotel was a gambling machine, a sort of simplified roulette wheel. I discovered that if you banged the machine against the wall at the critical moment it would pay out whatever the pointer had stopped at. I came down very early one morning and won over five pounds before I was warned off.

In the window of a stationer's shop in the town was a fountain pen that had already aroused my passion. It was the largest pen I'd ever seen, a Continental Waterman, the kind of fountain pen you might expect to find one of the great villains of crime fiction signing his false name with, and now I could afford it. Its nib was retractile, worked by a screw at the hinder end. Capping and uncapping were major operations. If you didn't screw everything up tight you were in danger of leaking half a pint of blue-black.

'I can understand you getting ink all over yourself only too well,' said my sister. 'But how did you manage to get it all over that unfortunate girl you were dancing with?'

Time passed. Pens came and went and my passion ebbed and flowed with them. The ballpoints, like Nanny's stylo, returned on a higher plane of history's spiral, were wildly exciting at first but somehow too inflexible, too impersonal. The Parker 51 with its half concealed nib, which many an amateur psychoanalyst has compared to the uncircumcised penis, seemed at first to have enormous promise. But I could never, in spite of the patience of the girls at the nib-changing counter, get a nib that really suited me. One day I heard a girl say: 'Oh, Christ! Here's Old Nibby again!' It was time to pack it in.

But this year, something is stirring. Judging by the advertisements, there seems to be a renewed interest in fountain pens. Old Nibby prowls again from 'Pencraft' to 'Penfriends', trying them all: Parkers, Sheaffers, Watermans, and the German makes, Mont Blanc, and the Lamy. Which will he get first: nib-joy or the bum's rush?

*c.*1970

Young Hunger

IT is very hard for people who have passed the age of, say, fifty to remember with any charity the hunger of their own puberty and adolescence when they are dealing with the young human animals who may be frolicking about them. Too often I have seen good people helpless with exasperation and real anger upon finding in the morning that cupboards and iceboxes have been stripped of their supplies by two or three youths—or even *one*—who apparently could have eaten four times their planned share at the dinner table the night before.

Such avidity is revolting, once past. But I can recall its intensity still; I am not yet too far from it to understand its ferocious demands when I see a fifteen-year-old boy wince and whiten at the prospect of waiting politely a few more hours for food, when his guts are howling for meat-bread-candy-fruit-cheese-milkmilk milk—ANYTHING IN THE WORLD TO EAT.

I can still remember my almost insane desperation when I was about eighteen and was staying overnight with my comparatively aged godparents. I had come home alone from France in a bad continuous storm and was literally concave with solitude and hunger. The one night on the train seemed even rougher than those on board ship, and by the time I reached my godparents' home I was almost lightheaded.

I got there just in time for lunch. It is clear as ice in my mind: a little cup of very weak chicken broth, one salted cracker, one-half piece of thinly sliced toast, and then, ah then, a whole waffle, crisp and brown and with a piece of beautiful butter melting in its middle—which the maid deftly cut into four sections! One section she put on my godmother's plate. The next *two*, after a nod of approval from her mistress, she put on mine. My god-father ate the fourth.

There was a tiny pot of honey, and I dutifully put a dab of it on my piggish portion, and we all nibbled away and drank one cup apiece of tea with lemon. Both my godparents left part of their waffles.

It was simply that they were old and sedentary and quite out of the habit of eating amply with younger people: a good thing for them, but pure hell for me. I did not have the sense to explain to them how starved I was—which I would not hesitate to do now. Instead I prowled around my bedroom while the house slumbered through its afternoon siesta, wondering if I dared sneak to the strange kitchen for something, anything, to eat, and knowing I would rather die than meet the silent, stern maid or my nice, gentle little hostess.

Later we walked slowly down to the village, and I was thinking sensuously of double malted ice-cream sodas at the corner drugstore, but there was no possibility of such heaven. When we got back to the quiet house, the maid brought my godfather a tall glass of exquisitely rich milk, with a handful of dried fruit on the saucer under it, because he had been ill; but as we sat and watched him unwillingly down it, his wife said softly that it was such a short time until dinner that she was sure I did not want to spoil my appetite, and I agreed with her because I was young and shy.

When I dressed, I noticed that the front of my pelvic basin jutted out like two bricks under my skirt: I looked like a scarecrow.

Dinner was very long, but all I can remember is that it had, as *pièce de résistance*, half of the tiny chicken previously boiled for broth at luncheon, which my godmother carved carefully so that we should each have a bit of the breast and I, as guest, should have the leg, after a snippet had been sliced from it for her husband, who like dark meat too.

There were hot biscuits, yes, the smallest I have ever seen, two apiece under a napkin on a silver dish. Because of them we had no dessert: it would be too rich, my godmother said.

We drank little cups of decaffeinized coffee on the screened porch in the hot Midwestern night, and when I went up to my room I saw that the maid had left a large glass of rich malted milk beside my poor godfather's bed.

My train would leave before five in the morning, and I slept little and unhappily, dreaming of the breakfast I would order on it. Of course when I finally saw it all before me, twinkling on the Pullman silver dishes, I could eat very little, from too much hunger and a sense of outrage.

I felt that my hosts had been indescribably rude to me, and selfish and conceited and stupid. Now I know that they were none of these things. They had simply forgotten about any but their own dwindling and cautious needs for nourishment. They had forgotten about being hungry, being young, being . . .

In an essay by Max Beerbohm about hosts and guests, the tyrants and the tyrannized, there is a story of what happened to him once when he was a schoolboy and someone sent him a hamper that held, not the usual collection of marmalade, sardines, and potted tongue, but twelve whole sausage-rolls.

'Of sausage-rolls I was particularly fond,' he says. He could have dominated all his friends with them, of course, but 'I carried the box up to my cubicle, and, having eaten two of the sausage-rolls, said nothing that day about the other ten, nor anything about them when, three days later, I had eaten them all—all, up there, alone.'

What strange secret memories such a tale evokes! Is there a grown-up person anywhere who cannot remember some such shameful, almost insane act of greediness of his childhood? In recollection his scalp will prickle, and his palms will sweat, at the thought of the murderous risk he may have run from his outraged companions.

When I was about sixteen, and in boarding-school, we were allowed one bar of chocolate a day, which we were supposed to eat sometime between the sale of them at the little school bookstore at four-thirty and the seven o'clock dinner gong. I felt an almost unbearable hunger for them—not for one, but for three or four or five at a time, so that I should have *enough*, for once, in my yawning stomach.

I hid my own purchases for several days, no mean trick in a school where every drawer and cupboard was inspected, openly and snoopingly too, at least twice a week. I cannot remember now how I managed it, with such lack of privacy and my own almost insurmountable hunger every afternoon, but by Saturday I had probably ten chocolate bars—my own and a few I had bribed my friends who were trying to lose weight to buy for me.

I did not sign up for any of the usual weekend debauchery such as a walk to the village drugstore for a well-chaperoned double butterscotch and pecan sundae. Instead I lay languidly on my bed, trying to look as if I had a headache and pretending to read a very fancy book called, I think, *Martin Pippin in the Apple Orchard*, until the halls quieted.

Then I arranged all my own and my roommate's pillows in a voluptuous pile, placed so that I could see whether a silent housemotherly foot stood outside the swaying monk's-cloth curtain that served as a door (to cut down our libidinous chitchat, the school board believed), and I put my hoard of Hersheys discreetly under a fold of the bedspread.

I unwrapped their rich brown covers and their tinfoil as silently as any prisoner chipping his way through a granite wall, and lay there breaking off the rather warm, rubbery, delicious pieces and feeling them melt down my gullet, and reading the lush symbolism of the book; and all the time I was hot and almost panting with the fear that people would suddenly walk in and see me there. And the strange thing is that nothing would have happened if they had!

It is true that I had more than my allotted share of candy, but that was not a crime. And my friends, full of their Saturday delights, would not have wanted ordinary chocolate. And anyway I had much more than I could eat, and was basically what Beerbohm calls, somewhat scornfully, 'a host' and not 'a guest': I loved to entertain people and dominate them with my generosity.

Then why was I breathless and nervous all during that solitary and not particularly enjoyable orgy? I suppose there is a Freudian explanation for it, or some other kind. Certainly the experience does not make me sound very attractive to myself. Even the certainty of being in good company is no real solace.

1946

Churchill and Roosevelt

(*from* Winston Churchill in 1940)

I T is an error to regard the imagination as a mainly revolutionary force—if it destroys and alters, it also fuses hitherto isolated beliefs, insights, mental habits, into strongly unified systems. These, if they are filled with sufficient energy and force of will—and, it may be added, fantasy, which is less frightened by the facts and creates ideal models in terms of which the facts are ordered in the mind—sometimes transform the outlook of an entire people and generation.

The British statesman most richly endowed with these gifts was Disraeli, who in effect conceived that imperialist mystique, that splendid but most un-English vision which, romantic to the point of exoticism, full of metaphysical emotion, to all appearances utterly opposed to everything most soberly empirical, utilitarian, antisystematic in the British tradition, bound its spell on the mind of England for two generations.

Churchill's political imagination has something of the same magical power to transform. It is a magic which belongs equally to demagogues and great democratic leaders: Franklin Roosevelt, who as much as any man altered his country's inner image of itself and of its character and its history, possessed it in a high degree. But the differences between him and the Prime Minister of Britain are greater than the similarities, and to some degree epitomise the differences of continents and civilisations. The contrast is brought out vividly by the respective parts which they played in the war which drew them so closely together.

The Second World War in some ways gave birth to less novelty and genius than the First. It was, of course, a greater cataclysm, fought over a wider area, and altered the social and political contours of the world at least as radically as its predecessor, perhaps more so. But the break in continuity in 1914 was far more violent. The years before 1914 look to us now, and looked even in the 1920s, as the end of a long period of largely peaceful development, broken suddenly and catastrophically. In Europe, at least, the years before 1914 were viewed with understandable nostalgia by those who after them knew no real peace.

The period between the wars marks a decline in the development of human culture if it is compared with that sustained and fruitful period which makes the nineteenth century seem a unique human achievement, so powerful that it persisted, even during the war which broke it, to a degree which seems astonishing to us now. The quality of literature, for example, which is surely one of the most reliable criteria of intellectual and moral vitality, was incomparably higher during the war of 1914–18 than it has been after 1939. In western Europe alone these four years of slaughter and destruction were also years in which works of genius and talent continued to be produced by such established writers as Shaw and Wells and Kipling, Hauptmann and Gide, Chesterton and Arnold Bennett, Beerbohm and Yeats, as well as such younger writers as Proust and Joyce, Virginia Woolf and E. M. Forster, T. S. Eliot and Alexander Blok, Rilke, Stefan George and Valéry. Nor did natural science, philosophy and history cease to develop fruitfully. What has the recent war to offer by comparison?

Yet perhaps there is one respect in which the Second World War did outshine its predecessor: the leaders of the nations involved in it were, with the significant exception of France, men of greater stature, psychologically more interesting, than their prototypes. It would hardly be disputed that Stalin is a more fascinating figure than the Tsar Nicholas II; Hitler more arresting than the Kaiser; Mussolini than Victor Emmanuel; and, memorable as they were, President Wilson and Lloyd George yield in the attribute of sheer historical magnitude to Franklin Roosevelt and Winston Churchill.

'History', we are told by Aristotle, 'is what Alcibiades did and suffered.' This notion, despite all the efforts of the social sciences to overthrow it, remains a good deal more valid than rival hypotheses, provided that history is defined as that which historians actually do. At any rate Churchill accepts it wholeheartedly, and takes full advantage of his opportunities. And because his narrative deals largely in personalities and gives individual genius its full and sometimes more than its full due, the appearance of the great wartime protagonists in his pages gives his narrative some of the quality of an epic, whose heroes and villains acquire their stature not merely—or indeed at all— from the importance of the events in which they are involved, but from their own intrinsic human size upon the stage of human history; their characteristics, involved as they are in perpetual juxtaposition and occasional collision with one another, set each other off in vast relief.

Comparisons and contrasts are bound to arise in the mind of the reader which sometimes take him beyond Churchill's pages. Thus Roosevelt stands out principally by his astonishing appetite for life and by his apparently complete freedom from fear of the future; as a man who welcomed the future eagerly as such, and conveyed the feeling that whatever the times might bring, all would be grist to his mill, nothing would be too formidable or crushing to be subdued and used and moulded into the pattern of the new and unpredictable form of life, into the building of which he, Roosevelt, and his allies and devoted subordinates would throw themselves with unheard-of energy and gusto. This avid anticipation of the future, the lack of nervous fear that the wave might prove too big or violent to navigate, contrasts most sharply with the uneasy longing to insulate themselves so clear in Stalin or Chamberlain. Hitler, too, in a sense, showed no fear, but his assurance sprang from a lunatic's violent and cunning vision, which distorted the facts too easily in his favour.

So passionate a faith in the future, so untroubled a confidence in one's power to mould it, when it is allied to a capacity for realistic appraisal of its true contours, implies an exceptionally sensitive awareness, conscious or half-conscious, of the tend-encies of one's milieu, of the desires, hopes, fears, loves, hatreds, of the human beings who compose it, of what are impersonally described as social and individual 'trends'. Roosevelt had this sensibility developed to the point of genius. He acquired the symbolic significance which he retained throughout his presid-ency, largely because he sensed the tendencies of his time and their projections into the future to a most uncommon degree. His sense, not only of the movement of American public opinion but of the general direction in which the larger human society of his time was moving, was what is called uncanny. The inner currents, the tremors and complicated convolutions of this movement, seemed to register themselves within his nervous system with a kind of seismographical accuracy. The majority of his fellow-citizens recognised this—some with enthusiasm, others with gloom or bitter indignation. Peoples far beyond the frontiers of the United States rightly looked to him as the most genuine and unswerving spokesman of democracy of his time, the most contemporary, the most outward-looking, the boldest, most imaginative, most large-spirited, free from the obsessions of an inner life, with an unparalleled capacity for creating confidence in the power of his insight, his foresight, and his capacity genuinely to identify himself with the ideals of humble people.

The feeling of being at home not merely in the present but in the future, of knowing where he was going and by what means and why, made him, until his health was finally undermined, buoyant and gay: made him delight in the company of the most varied and opposed individuals, provided that they embodied some specific aspect of the turbulent stream of life, stood actively for the foward movement in their particular world, whatever it might be. And this inner *élan* made up, and more than made up, for faults of intellect or character which his enemies—and his victims—never ceased to point out. He seemed genuinely unaffected by their taunts: what he could not abide was, before all, passivity, stillness, melancholy, fear of life or preoccupation with eternity or death, however great the insight or delicate the sensibility by which they were accompanied.

Churchill stands at almost the opposite pole. He too does not fear the future, and no man has ever loved life more vehemently and infused so much of it into everyone and everything that he has touched. But whereas Roosevelt, like all great innovators, had a half-conscious premonitory awareness of the coming shape of society, not wholly unlike that of an artist, Churchill, for all his extrovert air, looks within, and his strongest sense is the sense of the past.

The clear, brightly coloured vision of history, in terms of which he conceives both the present and the future, is the inexhaustible source from which he draws the primary stuff out of which his universe is so solidly built, so richly and elaborately ornamented. So firm and so embracing an edifice could not be constructed by anyone liable to react and respond like a sensitive instrument to the perpetually changing moods and directions of other persons or institutions or peoples. And, indeed, Churchill's strength (and what is most frightening in him) lies precisely in this: that, unlike Roosevelt, he is not equipped with numberless sensitive antennae which communicate the smallest oscillations of the outer world in all its unstable variety. Unlike Roosevelt (and unlike Gladstone and Lloyd George for that matter) he does not reflect a contemporary social or moral world in an intense and concentrated fashion; rather he creates one of such power and coherence that it becomes a reality and alters the external world by being imposed upon it with irresistible force. As his history of the war shows, he has an immense capacity for absorbing facts, but they emerge transformed by the categories which he powerfully imposes on the raw material into something which he can use to build his own massive, simple, impregnably fortified inner world.

Roosevelt, as a public personality, was a spontaneous, optimistic, pleasure-loving ruler who dismayed his assistants by the gay and apparently heedless abandon with which he seemed to delight in pursuing two or more totally incompatible policies, and astonished them even more by the swiftness and ease with which he managed to throw off the cares of office during the darkest and most dangerous moments. Churchill too loves pleasure, and he too lacks neither gaiety nor a capacity for exuberant self-expression, together with the habit of blithely cutting Gordian knots in a manner which often upset his experts; but he is not a frivolous man. His nature possesses a dimension of depth—and a corresponding sense of tragic possibilities— which Roosevelt's light-hearted genius instinctively passed by.

Roosevelt played the game of politics with virtuosity, and both his successes and his failures were carried off in splendid style; his performance seemed to flow with effortless skill. Churchill is acquainted with darkness as well as light. Like all inhabitants and even transient visitors of inner worlds, he gives evidence of seasons of agonised brooding and slow recovery. Roosevelt might have spoken of sweat and blood, but when Churchill offered his people tears, he spoke a word which might have been uttered by Lincoln or Mazzini or Cromwell, but not by Roosevelt, great-hearted, generous and perceptive as he was.

1949

To Err is Human

Everyone must have had at least one personal experience with a computer error by this time. Bank balances are suddenly reported to have jumped from $379 into the millions, appeals for charitable contributions are mailed over and over to people with crazy-sounding names at your address, department stores send the wrong bills, utility companies write that they're turning everything off, that sort of thing. If you manage to get in touch with someone and complain, you then get instantaneously typed, guilty letters from the same computer, saying, 'Our computer was in error, and an adjustment is being made in your account.'

These are supposed to be the sheerest, blindest accidents. Mistakes are not believed to be part of the normal behavior of a good machine. If things go wrong, it must be a personal, human error, the result of fingering, tampering, a button getting stuck, someone hitting the wrong key. The computer, at its normal best, is infallible.

I wonder whether this can be true. After all, the whole point of computers is that they represent an extension of the human brain, vastly improved upon but nonetheless human, super-human maybe. A good computer can think clearly and quickly enough to beat you at chess, and some of them have even been programmed to write obscure verse. They can do anything we can do, and more besides.

It is not yet known whether a computer has its own con-sciousness, and it would be hard to find out about this. When you walk into one of those great halls now built for the huge machines, and stand listening, it is easy to imagine that the faint, distant noises are the sound of thinking, and the turning of the spools gives them the look of wild creatures rolling their eyes in the effort to concentrate, choking with information. But real thinking, and dreaming, are other matters.

On the other hand, the evidences of something like an *un-*

conscious, equivalent to ours, are all around, in every mail. As extensions of the human brain, they have been constructed with the same property of error, spontaneous, uncontrolled, and rich in possibilities.

Mistakes are at the very base of human thought, embedded there, feeding the structure like root nodules. If we were not provided with the knack of being wrong, we could never get anything useful done. We think our way along by choosing between right and wrong alternatives, and the wrong choices have to be made as frequently as the right ones. We get along in life this way. We are built to make mistakes, coded for error.

We learn, as we say, by 'trial and error'. Why do we always say that? Why not 'trial and rightness' or 'trial and triumph'? The old phrase puts it that way because that is, in real life, the way it is done.

A good laboratory, like a good bank or a corporation or government, has to run like a computer. Almost everything is done flawlessly, by the book, and all the numbers add up to the predicted sums. The days go by. And then, if it is a lucky day, and a lucky laboratory, somebody makes a mistake: the wrong buffer, something in one of the blanks, a decimal misplaced in reading counts, the warm room off by a degree and a half, a mouse out of his box, or just a misreading of the day's protocol. Whatever, when the results come in, something is obviously screwed up, and then the action can begin.

The misreading is not the important error; it opens the way. The next step is the crucial one. If the investigator can bring himself to say, 'But even so, look at that!' then the new finding, whatever it is, is ready for snatching. What is needed, for progress to be made, is the move based on the error.

Whenever new kinds of thinking are about to be accomplished, or new varieties of music, there has to be an argument beforehand. With two sides debating in the same mind, haranguing, there is an amiable understanding that one is right and the other wrong. Sooner or later the thing is settled, but there can be no action at all if there are not the two sides, and the argument. The hope is in the faculty of wrongness, the tendency toward error. The capacity to leap across mountains of information to land lightly on the wrong side represents the highest of human endowments.

It may be that this is a uniquely human gift, perhaps even stipulated in our genetic instructions. Other creatures do not seem to have DNA sequences for making mistakes as a routine part of daily living, certainly not for programmed error as a guide for action.

We are at our human finest, dancing with our minds, when there are more choices than two. Sometimes there are ten, even twenty different ways to go, all but one bound to be wrong, and the richness of selection in such situations can lift us onto totally new ground. This process is called exploration and is based on human fallibility. If we had only a single center in our brains, capable of responding only when a correct decision was to be made, instead of the jumble of different, credulous, easily conned clusters of neurones that provide for being flung off into blind alleys, up trees, down dead ends, out into blue sky, along wrong turnings, around bends, we could only stay the way we are today, stuck fast.

The lower animals do not have this splendid freedom. They are limited, most of them, to absolute infallibility. Cats, for all their good side, never make mistakes. I have never seen a maladroit, clumsy, or blundering cat. Dogs are sometimes fallible, occasionally able to make charming minor mistakes, but they get this way by trying to mimic their masters. Fish are flawless in everything they do. Individual cells in a tissue are mindless machines, perfect in their performance, as absolutely inhuman as bees.

We should have this in mind as we become dependent on more complex computers for the arrangement of our affairs. Give the computers their heads, I say; let them go their way. If we can learn to do this, turning our heads to one side and wincing while the work proceeds, the possibilities for the future of mankind, and computerkind, are limitless. Your average good computer can make calculations in an instant which would take a lifetime of slide rules for any of us. Think of what we could gain from the near infinity of precise, machine-made miscomputation which is now so easily within our grasp. We would begin the solving of some of our hardest problems. How, for instance, should we go about organizing ourselves for social living on a planetary scale, now that we have become, as a plain fact of life, a single community? We can assume, as a working hypothesis, that all the right ways of doing this are unworkable. What we need, then, for moving ahead, is a set of wrong alternatives much longer and more interesting than the short list of mistaken courses that any of us can think up right now. We need, in fact, an infinite list, and when it is printed out we need the computer to turn on itself and select, at random, the next way to go. If it is a big enough mistake, we could find ourselves on a new level, stunned, out in the clear, ready to move again.

1979

Bad Poets

SOMETIMES it is hard to criticize, one wants only to chronicle. The good and mediocre books come in from week to week, and I put them aside and read them and think of what to say; but the 'worthless' books come in day after day, like the cries and truck sounds from the street, and there is nothing that anyone could think of that is good enough for them. In the bad type of the thin pamphlets, in hand-set lines on imported paper, people's hard lives and hopeless ambitions have expressed themselves more directly and heartbreakingly than they have ever been expressed in any work of art: it is as if the writers had sent you their ripped-out arms and legs, with 'This is a poem' scrawled on them in lipstick. After a while one is embarrassed not so much for them as for poetry, which is for these poor poets one more of the openings against which everyone in the end beats his brains out; and one finds it unbearable that poetry should be so hard to write—a game of Pin the Tail on the Donkey in which there is for most of the players no tail, no donkey, not even a booby prize. If there were only some mechanism (like Seurat's proposed system of painting, or the projected Universal Algebra that Gödel believes Leibnitz to have perfected and mislaid) for reasonably and systematically converting into poetry what we see and feel and are! When one reads the verse of people who cannot write poems—people who sometimes have more intelligence, sensibility, and moral discrimination than most of the poets—it is hard not to regard the Muse as a sort of fairy godmother who says to the poet, after her colleagues have showered on him the most disconcerting and ambiguous gifts, 'Well, never mind. You're still the only one that can write poetry.'

It seems a detestable joke that the 'national poet of the Ukraine'—kept a private in the army for ten years, and forbidden by the Czar to read, to draw, or even to write a letter

—should not have for his pain one decent poem. A poor Air Corps sergeant spends two and a half years on Attu and Kiska, and at the end of the time his verse about the war is indistinguishable from Browder's brother's parrot's. How cruel that a cardinal—for one of these books is a cardinal's—should write verses worse than his youngest choir-boy's! But in this universe of bad poetry everyone is compelled by the decrees of an unarguable Necessity to murder his mother and marry his father, to turn somersaults widdershins around his own funeral, to do everything that his worst and most imaginative enemy could wish. It would be a hard heart and a dull head that could condemn, except with a sort of sacred awe, such poets for anything that they have done—or rather, for anything that has been done to them: for they have never *made* anything, they have suffered their poetry as helplessly as they have anything else; so that it is neither the imitation of life nor a slice of life but life itself—beyond good, beyond evil, and certainly beyond reviewing.

                                                                    1953

Thomas Hobbes

WHEN Thomas Hobbes, at the age of eighty-four, looked back on his life, he found the key to it in fear. 'Fear and I were born twins,' he wrote; for his birth had been premature, hastened by the panic of the Spanish Armada. Fear characterised his personal life, making him twice a fugitive. Fear is the basis of his political philosophy, as of all dictatorships; the very word tolls like a minute-bell throughout the *Leviathan*. But philosophical systems do not spring from obsessions only. The mind of a revolutionary thinker is rarely simple, and the extraordinary boldness of Hobbes' intellectual method requires some less facile explanation.

Like many great revolutionaries, Hobbes was a convert. His early studies were desultory. In his youth, he loved music, and the lute. At Oxford, he left his books to snare jackdaws. As tutor to the Cavendish family, he hawked and hunted in Derbyshire, and wrote a poem on the wonders of the Peak. His intellectual interests were with the humanists. He jotted notes for Francis Bacon in the stately gardens of Gorhambury. He read Aristotle and translated Thucydides, corresponded with philosophers and conversed at Great Tew. He had already passed his fortieth year, ingenious but infertile, a witty conversationalist and pleasant companion for his aristocratic friends, before he reached that intellectual crisis, which to most men occurs, if at all, at least ten years earlier. Travelling abroad with his patron, he picked up a text of Euclid, and opened it at the forty-seventh proposition. From that moment he was 'in love with geometry'; it was 'the only science that it hath pleased God hitherto to bestow upon mankind'.

Whether his conversion was really as simple as this, we cannot say. Such incidents are usually the culmination of a long and painful process, not a substitute for it. But from that time

Hobbes gradually turned his back on his intellectual past, and trod a new path, which he never forsook. Aristotle, he now discovered, was no better than a country bumpkin; nothing could be more absurd than his Metaphysics, nor more repugnant to government than his Politics, nor more ignorant than his Ethics. Henceforth logic was the only intellectual method which he allowed. The baggage of the past—experience, tradition, observation—was jettisoned. He read little. Had he read as much as other men, he said, he would know no more than they. Instead of reading, or observing, he thought, logically. He walked in France with a pen and inkhorn in his stick, and a note-book in his pocket, 'and as soon as a thought darted, he presently entered it into his book, or otherwise he might perhaps have lost it'. Thus the *Leviathan* was written.

The axiom, fear; the method, logic; the conclusion, despotism. Such is the argument of that extraordinary book. Man, says Hobbes, is by nature unpolitical and unoriginal, a mechanical creature moved by strings and springs. This was the view of the Benthamites after him, and it is no accident that it was they who, in the nineteenth century, revived and edited his works. But the 'springs of action' which Hobbes postulated were simpler than those of Jeremy Bentham: they were fear and emotions derived from fear. 'The cause that moveth a man to become subject to another is fear of not otherwise preserving himself.' Man does not move towards positive ends, but away from fear. It is fear that urges him to 'a perpetual and restless desire of power after power, that ceaseth only in death'. Of all the horrors of the state of nature, so grimly catalogued, the worst of all is 'continual fear and danger of violent death'. And if a man turn to philosophical speculation, what comfort has he?

As Prometheus (which interpreted is, the Prudent Man) was bound to the hill Caucasus, a place of large prospect, where an eagle, feeding on his liver, devoured in the day as much as was repaired in the night; so that man which looks too far before him, in the care of future time, hath his heart all the day long gnawed on by fear of death, poverty, or other calamity, and has no repose, nor pause of his anxiety, but in sleep.

What is the answer to this terrible, this obsessive problem? One answer is given by the Churches, which exploit fear, and particularly 'the fear of darkness and ghosts, which is greater than other fears', building thereupon a pretentious superstructure of myth and mummery only 'to keep in credit the use of exorcism, of crosses, of holy water, and other such inventions of ghostly men'. This answer Hobbes utterly rejects. Religion is not a safeguard against fear, but a parasite on it. Though his prudence made him an Anglican, and his logic an erastian, Hobbes was, in fact (as his enemies maintained), a complete atheist, regarding all religion as a deliberate fraud invented by priests to fool the people. Ill in France, he was pestered by the clergy of three denominations, begging him to die in their communions. 'Let me alone,' he replied, 'or I will detect all your cheats from Aaron to yourselves.' And he attributed 'all the changes of religion in the world to one and the same cause; and that is, unpleasing priests'.

The trenchancy of Hobbes's anti-clericalism, which makes him so readable, suggests that in this, too, he may have been a convert. It is interesting that his contemporaries believed (perhaps on the evidence of his writings) that he was afraid to be alone in the dark; and though his friends denied this, the vividness and frequency of his allusions to supernatural fears suggest that he may not always have been exempt from them. The man who described Brutus, haunted by the ghost of Caesar—

For sitting in his tent, pensive and troubled with the horror of his rash act, it was not hard for him, slumbering in the cold, to dream of that which most affrighted him—

and who, in a series of contemptuous paragraphs, likened the whole apparatus of the Roman Church to the imaginary world of spooks and hobgoblins, at least knew some sympathy with the emotions he disclaimed.

Hobbes's answer is therefore a purely secular answer. To escape the consequences of his bestial and timid nature, man must erect a civil authority of terrifying completeness: a state based on naked, and wielding absolute power, with no other function than to wield power; whose effectiveness alone is its legitimacy; whose opinions are truth; whose orders are justice; resistance to which is a logical absurdity. This is 'that great Leviathan, or rather (to speak more reverently) that mortal god, to which we owe, under the immortal God, our peace and defence'.

The *Leviathan* is a fantastic monster, such as is sometimes cast up, with other strange births, in political, as in marine, convulsions. It is an isolated phenomenon in English thought, without ancestry or posterity; crude, academic, and wrong. Its axioms are inadequate, its method inapplicable, its conclusions preposterous. How seldom in history has any reality corresponded with it! Hobbes's whole system was based on huge errors, uncorrected, because untested, by observation. He had learnt nothing of experimental methods from Bacon, nothing of historical understanding from Thucydides. A vivid impression of civil strife is perhaps all he preserved from the profound wisdom of that greatest of historians. To compare him with Machiavelli is absurd; for Machiavelli tests and illustrates every thesis by historical analogy. Hobbes despised the evidence of the past. It was no better, he said, than prophecy; 'both being grounded only on experience'. He cannot even be regarded symptomatically, as a commentary on contemporary events; for his fundamental ideas were developed before civil war had broken out in England or France. He is a typical, academic *Gelehrte*; which is perhaps why his most enthusiastic commentators have all come from Germany.

Why then is he important? First, for his style. Hobbes was no spellbinder. A complete nominalist, he used words as tools, not as charms. He was contemptuous of fine, meaningless phrases. St Thomas Aquinas had called eternity '*nunc stans*, an ever-abiding Now'; 'which is easy enough to say,' remarked Hobbes drily,'but though I fain would, yet I never could conceive it. They that can are happier than I'. 'Words', he said elsewhere, 'are wise men's counters, they do but reckon by them; but they are the money of fools.' Nevertheless, though Hobbes could never lose himself in an *O Altitudo*, the stock from which he drew his counters was that wonderful, rich vocabulary of the early seventeenth century, the vocabulary of Milton and Donne and Sir Thomas Browne; and by the boldness with which he used them, by the monolithic temper of his mind, and the formidable logic of his argument, he wrote a book which is as striking in its singleness of purpose, its defiant language, its inspired iconoclasm (and sometimes its dullness), as the poem of Lucretius.

Secondly, he concentrated his doctrines into a single, timely and complete work. Many of those doctrines had already appeared in the works of French lawyers and English pamphleteers, and in Hobbes's other English and Latin works; but in the *Leviathan* they are brought together in a logical system that allows no further development. And this book, by an accident of date, acquired a terrible significance. In 1651 the ingenuous author presented his academic thesis to the exiled Charles II. Two years later, Cromwell seized power in England, and Hobbes's outrageous doctrines suddenly corresponded, or seemed to correspond, with a fearful reality. Already 'the father of atheists', he now appeared as the theorist of the usurper, made yet more dangerous and detestable by the ringing phrases, the exultant nihilism, with which he swept away the tinselled rubbish of traditional thought in order to make room, in the desolate void which he had created, for his grim, impersonal idol. Humane, conventional, practical, religious men—whether Puritan or Anglican—saw what he had done and trembled. No voice was raised in his support. The universal horror which he inspired became yet another argument for conservatism, for a royal restoration.

Nevertheless, his work—at least his work of destruction—could not be undone. The idol might have crumbled, but the great void remained, and pious hands were never again able to reassemble the old intellectual *bric-à-brac* which he had swept away. So, when King and Court returned to authority, the execrated philosopher somehow survived. The royalist clergy might snipe timidly at the old pachyderm as he brushed through their shady preserves; Dr Beale, in his court sermon, might still dispute the old question, whether angels have beards, and decide learnedly that they have; the theorists of Divine right might mumble away about Noah and Nimrod; the bishops might 'make a motion to have the good old gentleman burnt for a heretique'; the Presbyterian Baxter might join with high-flying Anglicans against him: but it was all rather ineffective. Besides the old man was so genial, so witty, so entertaining: one could not really dislike him. And he had royal support: Charles II preferred wit to orthodoxy, and protected his former tutor. As for the philosophers, they might keep his unpopular name from their books, but they could not exclude his achievement from their minds. He had cleared political thought of its ancient, biblical cobwebs, and set it firmly on the secular basis of human psychology. That his psychology was inadequate, elementary, and wrong is an irrelevant objection. The function of genius is not to give new answers, but to pose new questions, which time and mediocrity can resolve. This Hobbes had achieved. By one great thunderstorm he had changed the climate of thought; and his achievement is not the thunderstorm, but the change.

After the storm, the old philosopher enjoyed his ease. He was back in Derbyshire, still with the Cavendishes, the friends and patrons of seventy years. Erect and sprightly, his health improving yearly, he still played tennis at seventy-five; after which a servant would rub him down in bed. Then, in the privacy of his chamber, the old bachelor would lift up his voice and sing prick-song, for the health of his lungs. It would prolong his life, he believed. Certainly he went on living. He seemed immortal, like Satan himself; a genial Satan. At eighty, he wrote *Behemoth*, incorrigibly erroneous. At eighty-six, feeling bored (for conversation at Chatsworth was sometimes thin) he dashed off an English translation of the *Iliad* and the *Odyssey*. At ninety, he was still going. His face was rubicund; his bright hazel eye glowed like an ember; and when he took his pipe from his mouth, he delighted all by his brisk and decisive repartees. Only the flies disconcerted him, settling on his bald head. Of the *Leviathan*, that product of his headstrong sixties, he did not speak. There has never been anything to add to its utter finality.

1945

ELIZABETH HARDWICK

The Apotheosis of Martin Luther King

MEMPHIS, ATLANTA, 1968

T HE decaying, downtown shopping section of Memphis—still another Main Street—lay, the weekend before Martin Luther King's funeral, under a siege. The deranging curfew and that state of civic existence called 'tension' made the town seem to be sinister, again very much like a film set, perhaps for a television drama, of breakdown, catastrophe. Since films and television have staged everything imaginable before it happens, a true event, taking place in the real world, brings to mind the landscape of films. There is no meaning in this beyond description and real life only looks like a fabrication and does not feel so.

The streets are completely empty of traffic and persons and yet the emptiness is the signal of dire and dramatic possibilities. In the silence, the horn of a tug gliding up the dark Mississippi is background. The hotel, downtown, overlooking the city park, is a tomb and perhaps that is usual since it is downtown where nobody wants to go in middle-sized cities. It is a shabby place, poorly staffed by aged persons, not grown old in their duties, but newly hired, untrained, depressed, worn-out old people.

The march was called for the next year, a march originally called by King as a renewal of his efforts in the Memphis garbage strike, efforts interrupted by a riot in the poor, black sections the week before. Now he was murdered and the march was called to honor him. Fear of riots, rage, had brought the curfew and the National Guard. Perhaps there was fear, but in civic crises there is always something exciting and even a sort of humidity of smugness seemed to hang over the town. Children kept home from school, bank and ten-cent store closed. If one was not in clear danger, there seemed to be a complacent pleasure in thinking, We have been brought to this by Them.

Beyond and beneath the glassy beige curtains of the hotel room, the courthouse square was spread out like a target, the destination of the next day's march and ceremony. All night long little hammer blows, a ghostly percussion, rang out as the structure for the 'event' was being put together. The stage, slowly forming, plank by plank, seemed in the deluding curfew emptiness and silence like a scaffolding being prepared for a beheading. These overwrought and exaggerated images came to me from the actual scene and from a crush of childhood memories. Memphis was a Southern town in which a murder had taken place. The killer might be over yonder in that deep blue thicket, or holed up in the woods on the edge of town, ready to come back at night. Of course this was altogether different. The assassin's work was completed. Here in Memphis it was not the killer, whoever he might be, who was feared, but the killed one and what his death might bring.

Not far from the downtown was the leprous little hovel where, from a squalid toilet window, the assassin had been able to look across and target the new and hopeful Lorraine Motel. Now the motel was being visited by mourners. The black people of Memphis, dressed in their best, filed silently up and down the ramp, glancing shyly into the room which King had occupied. At the ramp before the door of the room where he fell there were flowers, glads and potted azaleas.

All over the Negro section, rickety little stores, emptied in the 'consumer rebellion,' were boarded up, burned out, or simply empty, with the windows broken. The stores were for the most part of great modesty. Who owned that one? I asked the taxi driver.

'Well, that happened to be Chinee,' he said.

Shops are a dwelling and their goods and stuffs, counters and cash registers are a form of interior decoration. Sacked and disordered, these Memphis boxes were amazingly small and only an active sense of possibility could conceive of them as the site of commercial enterprise. It did not seem possible that by stocking a few shelves these squares of rotting timber could merit ownership, license, investment and produce a profitable exchange. They are lean-tos, chicken coops—measly and optimistic. Looters had sought the consolations of television sets and whiskey. The intrepid dramas of refrigerators and living room suites, deftly transported from store to home, were beyond the range of this poor section of the romantic city on the river.

The day of the march came to the gray, empty streets. The march was solemn and impressive, but on the other hand perhaps somewhat disappointing. A compulsive exaggeration dogs most of the expectations of ideological gatherings and thereby turns success to failure. The forty to sixty thousand predicted belittled the eighteen thousand present. The National Guard, alert with gun and bayonet as if for some important marine landing, made the quiet, orderly march appear a bit of a sell.

The numbers of the National Guard, the body count, spoke almost of a sort of psychotic imagining. They were on every street, blocking every intersection, cutting off each highway. There, in their large brown trucks, crawling out from under the olive-brown canvas, were men in full battle dress, in helmets and chin straps which concealed most of their pale, red-flecked and rather alarmed Southern faces. They guarded the alleys and the horizon, the river and the muddy playground, thoroughfare and esplanade, newspaper store and bank. It was as if by some cancerous multiplication the sensible and necessary had been turned into a monstrous glut.

The march, after all, was mostly made up of Memphis blacks. Was this a victory or a defect? There were also some local white students from Southwestern, a few young ministers, and from New York members of the teachers' union with a free day off and a lunch box. Mrs King came from Atlanta for the gathering, a tribute to her husband and also a tribute to the poor sanitation workers for whom it had all begun.

The people gathered early and waited long in the streets. They stood in neat lines to indicate the absence of unruly feelings. Part of the ritual of every public show of opinion and solidarity is the presence of a name or, preferably, the body of a Notable ('Notable' for a routine occasion, and 'Dignitary' for a more solemn and affecting event such as the funeral to be held in Atlanta the next day). Notables are often from the entertainment world and the rest are usually to be known for political activities. Like a foundation of stone moved from site to site, only on the notables can the petition for funds be based, the protest developed, the idea constructed.

The marchers waited without restlessness for the chartered

airplane to arrive and to announce that it could then truly begin. A limousine will be waiting to take the noted ones to the front of the line, or to leave them off at the stage door. The motors are kept running. After an appearance, a speech, a mere presence, out they go by the back doors used by the celebrated, out to the waiting limousine, off to the waiting plane, and then off.

These persons are symbols of a larger consensus that can be transferred to the mass of the unknown faithful. They are priests giving sanction to idea, struggle, defiance. It is believed that only the famous, the busy, the talented have the power to solicit funds from the rich, notice from the press, and envy from the opposition. Also they are a sunshine, warming. They have the appeal of the lucky.

The march of Memphis was quiet; it was designed as a silent memorial, like a personal prayer. Hold your head high, the instructions read. No gum chewing. For protection in case of trouble, no smoking, no umbrella, no earrings in pierced ears, no fountain pens in jacket pockets. One woman said, 'If they make me take off my shades, I'm quitting the march.' Among those who had come from some distance a decision had been made in favor of the small gathering in Memphis over the 'national' funeral in Atlanta the following day. 'I feel this is more important,' they would say.

In the march and at the funeral of Martin Luther King, the mood of the earlier Civil Rights days in Alabama and Mississippi returned, a reunion at the grave of squabbling, competitive family members. And no one could doubt that there had been a longing for reunion among the white ministers and students and the liberals from the large cities. The 'love'—locked arms, hymns, good feeling—all of that was remembered with feeling.

This love, if not always refused, was now seldom forthcoming in relations with new black militants, who were set against dependency upon the checkbooks and cooperation of the guilty, longing, loving whites. Everything separated the old Civil Rights people from the new black militants; it could be said, and for once truly, that they did not speak the same language. A harsh, obscene style, unforgiving stares, posturings, insulting accusations and refusal to make distinctions among those of the white world—this was humbling and perplexing. Many of the white people had created their very self-identity out of issues and

distinctions and they felt cast off, ill at ease, with the new street
rhetoric of 'self-defense' and 'self-determination'.

Comradeship, yes, and being in the South again gave one a
remembrance of the meaning of the merely legislative, the newly
visible. Back at the hotel in the late afternoon the marchers were
breaking up. The dining room was suddenly filled with not-too-
pleasing young black boys—not black notables with cameras and
briefcases, or in the company of intimidating, busy-looking
persons from afar—no, just poor boys from Memphis. The aged
waitresses padded about on aching feet and finally approached
with the questions of function. Menu? Yes. Cream in the coffee?
A little.

So, at last business was business, not friendship. The old white
waitresses themselves were deeply wrinkled by the stains of
plebeianism. Manner, accent marked them as 'disadvantaged';
they were diffident, ignorant, and poor and would themselves
cast a blight on the cheerful claims of many dining places. They
seemed to be the enduring remnants of many an old retired
trailer camp couple, the men with tattooed arms and the women
in bright colored stretch pants; those who wander the warm
roads and whose traveling kitchenettes and motorized toilets are
a distress to the genteel and tasteful.

In any case, joy and flush-cheeked nuns were past history, a
folk epic, full of poetry, simplicity and piety. The pastoral period
of the Civil Rights Movement had gone by.

At the funeral in Atlanta, rising above the crowd, the *nez pointu*
of Richard Nixon ... Lester Maddox, short-toothed little mar-
moset, peeking from behind the draperies of the Georgia State
House ... Many Christians have died without the scruples of
Christian principles being to the point. The *belief* of Martin
Luther King—what an unexpected curiosity it was, the strength
of it. His natural mode of address was the sermon. 'So I say to
you, seek God and discover Him and make Him a power in your
life. Without Him all our efforts turn to ashes and our sunrises
into darkest nights.'

At the end of his life, King seemed in some transfigured state,
even though politically he had become more radical and there
were traces of disillusionment—with what? messianic hope per-

haps. He had observed that America was sicker, more intransigent than he had realized when he began his work. The last, ringing, 'I have been to the mountaintop!' gave voice to a transcendent experience. It is this visionary strain that makes him a man elusive in the extreme, difficult to understand as a character.

How was it possible for one so young as King to seem to contain, in himself, so much of the American past? At the very least, the impression he gave was of an experience of life coterminous with the years of his father. The depression, the dust bowl, the sharecropper, the old back-country churches, and even the militance of the earlier IWW—he suggested all of this. He did not appear to belong to the time of Billy Graham (God bless you real good) but to a previous and more spiritual evangelism, to a time of solitude and refined simplicity. In *Adam Bede*, Dinah preaches that Jesus came down from Heaven to tell the good news about God to the poor. 'Why, you and me, dear friends, are poor. We have been brought up in poor cottages, and have been reared on oatcake and lived coarse.... It doesn't cost Him much to give us our little handful of victual and bit of clothing, but how do we know He cares for us any more than we care for the worms ... so long as we rear our carrots and onions?' There remains this old, pure tradition in King. Rare elements of the godly and the political come together, with an affecting naturalness. His political work was indeed a Mission, as well as a political cause.

In spite of the heat of his sermon oratory, King seems lofty and often removed by the singleness of his concentration—an evangelical aristocrat. There is even a coldness in his public character, an impenetrability and solidity often seen in those who have given their entire lives to ideas and causes. The racism in America acts finally as an exhaustion to all except the strongest of black leaders. It leads to the urban, manic frenzy, the sleeplessness, hurry, and edginess that are a contrast to King's steadiness and endurance.

Small-town Christianity, staged in some sense as it was, made King's funeral supremely moving. Its themes were root American, bathed in memory, in forgotten prayers and hymns and dreams. Mule carts, sharecroppers, dusty poverty, sleepy Sunday morning services, and late Wednesday night prayer meetings *after work*. There in the reserved pews it was something else—candidates, former candidates, and hopeful candidates, illuminated, as it were, on prime viewing time, free of charge, you might say, free of past contributions to the collection plate, free of the envelopes of future pledges.

The rare young man was mourned and, without him, the world was fearful indeed. The other side of the funeral, Act Two ready in the wings, was the looting and anger of a black population inconsolable for its many losses.

'Jesus is a trick on niggers,' a character in Flannery O'Connor's *Wise Blood* says. The strength of belief revealed in King and in such associates as the Reverend Abernathy was a chastening irregularity, not a regionalism absorbed. It stands apart from our perfunctory addresses to 'this nation under God.' In a later statement Abernathy has said that God, not Lester Maddox or George Wallace, rules over the South. So, Negro justice is God's work and God's will.

The popular Wesleyan hymns have always urged decent, sober behavior, or that is part of the sense of the urgings. As you sing, 'I can hear my Savior calling,' you are invited to accept the community of the church and also, quite insistently, to behave yourself, stop drinking, gambling and running around. Non-violence of a sort, but personal, thinking of the home and the family, and looking back to an agricultural or small-town life, far from the uprooted, inchoate, *communal* explosions in the ghettos of the cities. The political non-violence of Martin Luther King was an act of brilliant intellectual conviction, very sophisticated and yet perfectly consistent with evangelical religion, but not a necessary condition as we know from the white believers.

One of the cruelties of the South and part of the pathos of Martin Luther King's funeral and the sadness that edges his rhetoric is that the same popular religion is shared by many bellicose white communicants. The religion seems to have sent few peaceful messages to them in so far as their brothers in Christ, the Negroes, are concerned. Experience leads one to suppose there was more respect for King among Jews, atheists, and comfortable Episcopalians, more sympathy and astonishment, than among the white congregations who use, with a different cadence, the same religious tone and the same hymns heard in the Ebenezer Church. Under the robes of the Klan there is an evangelical skin; its dogmatism is touched with the Scriptural, however perverse the reading of the text.

At one of the memorial services in Central Park after the murder, a radical speaker shouted, 'You have killed the last good nigger!' This posturing exclamation was not meant to dishonor King, but to speak of his kind as something gone by, its season over. And perhaps so. The inclination of white leaders to characterize everything unpleasant to themselves in black response to American conditions as a desecration of King's memory was a sordid footnote to what they had named the 'redemptive moment.' But it told in a self-serving way of the peculiarity of the man, of the survival in him of habits of mind from an earlier time.

King's language in the pulpit and in his speeches was effective but not remarkably interesting. His style compares well, however, with the speeches of recent Presidents and even with those of Adlai Stevenson, most of them bland and flat in print. In many ways, King was not Southern and rural in his address, although he had a melting Georgia accent and his discourse was saturated in the Bible. His was a practical, not a frenzied exhortation, inspiring the Southern Negroes to the sacrifices and dangers of protest and yet reassuring them by its clarity and humanity.

His speech was most beautiful in the less oracular cadences, as when he summed up the meaning of the Poor People's March on Washington with, 'We have come for our checks!' The language of the younger generation is another thing altogether. It has the brutality of the city and an assertion of threatening power at hand, not to come. It is military, theatrical, and at its most coherent probably a lasting repudiation of empty courtesy and bureaucratic euphemism.

The murder of Martin Luther King was a 'national disgrace.' That we said again and again and it would be cynical to hint at fraudulent feelings in the scramble for suitable acts of penance. Levittowns would henceforth not abide by local rulings, but would practice open housing; Walter Reuther offered $50,000 to the beleaguered sanitation workers of Memphis; the Field Foundation gave a million to the Southern Christian Leadership movement; Congress acted on the open housing bill. Nevertheless, the mundane continued to nudge the eternal. In 125 cities there was burning and looting. Smoke rose over Washington, DC.

The Reverend Abernathy spoke of a plate of salad shared with Dr King at the Lorraine Motel, creating a grief-laden scenery of the Last Supper. How odd it was after all, this exalted Black Liberation, played out at the holy table and at Gethsemane, 'in the Garden,' as the hymns have it. A moment in history, each instance filled with symbolism and the aura of Christian memory. Perhaps what was celebrated in Atlanta was an end, not a beginning—the waning of the slow, sweet dream of Salvation, through Christ, for the Negro masses.

1968

The Gangster as Tragic Hero

Aᴍᴇʀɪᴄᴀ, as a social and political organization, is committed to a cheerful view of life. It could not be otherwise. The sense of tragedy is a luxury of aristocratic societies, where the fate of the individual is not conceived of as having a direct and legitimate political importance, being determined by a fixed and supra-political—that is, non-controversial—moral order or fate. Modern equalitarian societies, however, whether democratic or authoritarian in their political forms, always base themselves on the claim that they are making life happier; the avowed function of the modern state, at least in its ultimate terms, is not only to regulate social relations, but also to determine the quality and the possibilities of human life in general. Happiness thus becomes the chief political issue—in a sense, the only political issue—and for that reason it can never be treated as an issue at all. If an American or a Russian is unhappy, it implies a certain reprobation of his society, and therefore, by a logic of which we can all recognize the necessity, it becomes an obligation of citizenship to be cheerful; if the authorities find it necessary, the citizen may even be compelled to make a public display of his cheerfulness on important occasions, just as he may be conscripted into the army in time of war.

Naturally, this civic responsibility rests most strongly upon the organs of mass culture. The individual citizen may still be permitted his private unhappiness so long as it does not take on political significance, the extent of this tolerance being determined by how large an area of private life the society can accommodate. But every production of mass culture is a public act and must conform with accepted notions of the public good. Nobody seriously questions the principle that it is the function of mass culture to maintain public morale, and certainly nobody in the mass audience objects to having his morale maintained. At a time when the normal condition of the citizen is a state of anxiety, euphoria spreads over our culture like the broad smile of an idiot. In terms of attitudes towards life, there is very little difference between a 'happy' movie like *Good News*, which ignores death and suffering, and a 'sad' movie like *A Tree Grows in Brooklyn*, which uses death and suffering as incidents in the service of a higher optimism.

But, whatever its effectiveness as a source of consolation and a means of pressure for maintaining 'positive' social attitudes, this optimism is fundamentally satisfying to no one, not even to those who would be most disoriented without its support. Even within the area of mass culture, there always exists a current of opposition, seeking to express by whatever means are available to it that sense of desperation and inevitable failure which optimism itself helps to create. Most often, this opposition is confined to rudimentary or semi-literate forms: in mob politics and journalism, for example, or in certain kinds of religious enthusiasm. When it does enter the field of art, it is likely to be disguised or attenuated: in an unspecific form of expression like jazz, in the basically harmless nihilism of the Marx Brothers, in the continually reasserted strain of hopelessness that often seems to be the real meaning of the soap opera. The gangster film is remarkable in that it fills the need for disguise (though not sufficiently to avoid arousing uneasiness) without requiring any serious distortion. From its beginnings, it has been a consistent and astonishingly complete presentation of the modern sense of tragedy.

In its initial character, the gangster film is simply one example of the movies' constant tendency to create fixed dramatic patterns that can be repeated indefinitely with a reasonable expectation of profit. One gangster film follows another as one musical or one Western follows another. But this rigidity is not necessarily opposed to the requirements of art. There have been very successful types of art in the past which developed such specific and detailed conventions as almost to make individual examples of the type interchangeable. This is true, for example, of Elizabethan revenge tragedy and Restoration comedy.

For such a type to be successful means that its conventions have imposed themselves upon the general consciousness and become the accepted vehicles of a particular set of attitudes and a particular aesthetic effect. One goes to any individual example of the type with very definite expectations, and originality is to be welcomed only in the degree that it intensifies the expected experience without fundamentally altering it. Moreover, the relationship between the conventions which go to make up such a type and the real experience of its audience or the real facts of whatever situation it pretends to describe is of only secondary importance and does not determine its aesthetic force. It is only in an ultimate sense that the type appeals to its audience's experience of reality; much more immediately, it appeals to previous experience of the type itself: it creates its own field of reference.

Thus the importance of the gangster film, and the nature and intensity of its emotional and aesthetic impact, cannot be measured in terms of the place of the gangster himself or the importance of the problem of crime in American life. Those European movie-goers who think there is a gangster on every corner in New York are certainly deceived, but defenders of the 'positive' side of American culture are equally deceived if they think it relevant to point out that most Americans have never seen a gangster. What matters is that the experience of the gangster *as an experience of art* is universal to Americans. There is almost nothing we understand better or react to more readily or with quicker intelligence. The Western film, though it seems never to diminish in popularity, is for most of us no more than the folklore of the past, familiar and understandable only because it has been repeated so often. The gangster film comes much closer. In ways that we do not easily or willingly define, the gangster speaks for us, expressing that part of the American psyche which rejects the qualities and the demands of modern life, which rejects 'Americanism' itself.

The gangster is the man of the city, with the city's language and knowledge, with its queer and dishonest skills and its terrible daring, carrying his life in his hands like a placard, like a club. For everyone else, there is at least the theoretical possibility of another world—in that happier American culture which the gangster denies, the city does not really exist; it is only a more crowded and more brightly lit country—but for the gangster there is only the city; he must inhabit it in order to personify it: not the real city, but that dangerous and sad city of the imagination which is so much more important, which is the modern world. And the gangster—though there are real gangsters—is also, and primarily, a creature of the imagination. The real city, one might say, produces only criminals; the imaginary city produces the gangster: he is what we want to be and what we are afraid we may become.

Thrown into the crowd without background or advantages, with only those ambiguous skills which the rest of us—the real people of the real city—can only pretend to have, the gangster is required to make his way, to make his life and impose it on others. Usually, when we come upon him, he has already made his choice or the choice has already been made for him, it doesn't matter which: we are not permitted to ask whether at some point he could have chosen to be something else than what he is.

The gangster's activity is actually a form of rational enterprise, involving fairly definite goals and various techniques for achieving them. But this rationality is usually no more than a vague background; we know, perhaps, that the gangster sells liquor or that he operates a numbers racket; often we are not given even that much information. So his activity becomes a kind of pure criminality: he hurts people. Certainly our response to the gangster film is most consistently and most universally a response to sadism; we gain the double satisfaction of participating vicariously in the gangster's sadism and then seeing it turned against the gangster himself.

But on another level the quality of irrational brutality and the quality of rational enterprise become one. Since we do not see the rational and routine aspects of the gangster's behavior, the practice of brutality—the quality of unmixed criminality—becomes the totality of his career. At the same time, we are always conscious that the whole meaning of this career is a drive for success: the typical gangster film presents a steady upward progress followed by a very precipitate fall. Thus brutality itself becomes at once the means to success and the content of success—a success that is defined in its most general terms, not as accomplishment or specific gain, but simply as the unlimited possibility of aggression. (In the same way, film presentations of businessmen tend to make it appear that they achieve their success by talking on the telephone and holding conferences and that success *is* talking on the telephone and holding conferences.)

From this point of view, the initial contact between the film and its audience is an agreed conception of human life: that man is a being with the possibilities of success or failure. This principle, too, belongs to the city; one must emerge from the crowd or else one is nothing. On that basis the necessity of the action is established, and it progresses by inalterable paths to the point where the gangster lies dead and the principle has been modified: there is really only one possibility—failure. The final meaning of the city is anonymity and death.

In the opening scene of *Scarface*, we are shown a successful man; we know he is successful because he has just given a party of opulent proportions and because he is called Big Louie. Through some monstrous lack of caution, he permits himself to be alone for a few moments. We understand from this immediately that he is about to be killed. No convention of the gangster film is more strongly established than this: it is dangerous to be alone. And yet the very conditions of success make it impossible not to be alone, for success is always the establishment of an *individual* pre-eminence that must be imposed on others, in whom it automatically arouses hatred; the successful man is an outlaw. The gangster's whole life is an effort to assert himself as an individual, to draw himself out of the crowd, and he always dies *because* he is an individual; the final bullet thrusts him back, makes him, after all, a failure. 'Mother of God,' says the dying Little Caesar, 'is this the end of Rico?'—speaking of himself thus in the third person because what has been brought low is not the undifferentiated *man*, but the individual with a name, the gangster, the success; even to himself he is a creature of the imagination. (T. S. Eliot has pointed out that a number of Shakespeare's tragic heroes have this trick of looking at themselves dramatically; their true identity, the thing that is destroyed when they die, is something outside themselves—not a man, but a style of life, a kind of meaning.)

At bottom, the gangster is doomed because he is under the obligation to succeed, not because the means he employs are unlawful. In the deeper layers of the modern consciousness, *all* means are unlawful, every attempt to succeed is an act of aggression, leaving one alone and guilty and defenseless among enemies: one is *punished* for success. This is our intolerable dilemma: that failure is a kind of death and success is evil and dangerous, is—ultimately—impossible. The effect of the gangster film is to embody this dilemma in the person of the gangster and resolve it by his death. The dilemma is resolved because it is *his* death, not ours. We are safe; for the moment, we can acquiesce in our failure, we can choose to fail.

1948

The Homburg Hat

Even to an unpractised eye, it was apparent that school began at Paddington. For here were all the *signes avant-coureurs* of what was awaiting one, all the untold horrors of the unknown, much further up the line. On the departure platform, there were small groups of twos and threes, keeping their distances from one another, perhaps in a last desperate bid to cling on to the family unit, even reduced to bare essentials, in the absence of sisters and elder or younger brothers, and to hold on to the last tiny particle of home, holiday and privacy. Boys, of vastly different sizes, but all affecting a brave unconcern, almost as if anxious to get it all over and settle down in the compartment; mothers, on the verge of tears, a few actually *over* the verge, fathers sharing with their sons, whether tall or quite tiny, a brave indifference and common stiff upper-lipdom. Some of the boys were standing awkwardly on one leg, others were shifting from leg to leg, as if in need to go to the lavatory, or indeed needing just to do that, but unwilling to admit to its urgency. From inside the compartment, I watched the scene, with some trepidation, but glad at least to have got it all over, as far as I was concerned, without witnesses, on the up platform of Tunbridge Wells Central: once in the train, a welcoming tunnel had at once blotted out the sight of my parents, as they disappeared in a swirl of yellow smoke. It was a station fortunately not suited to prolonged adieux.

I thought that I could distinguish between those who, like myself, were newcomers—later, as I was to learn, their correct designation was New Scum—and the hardend old-timers—two-year-olds, three-year-olds—other expressions of a mysterious vocabulary awaiting to trip me up, like so many other things, at the other end—by the apparently unaffected ease and studied casualness of the latter, some of them standing in slightly drooping postures, as if they were meeting their parents at a social occasion, or as if they were holding invisible sherry glasses, or had come together by chance at a race meeting. One tall boy even had his back to his parents, another could be seen handing his father a silver cigarette-case.

The equine illusion was reflected in the manner, likewise studiedly casual, in which all the boys, both tall and apparently fully grown, and diminutive, were clothed: brown shoes indented with complicated arabesques, or, daringly, suede shoes, grey flannels, hacking jackets with wide vents, or furry sports-coats in brownish, grey or green herring-bone, polka-dot ties and wide-striped Viyella collars, all topped, at the command of a mysterious, unwritten uniformity, by trilbies in brown or dark grey—brown was the majority response, grey suggested that of the vaguely *osé* and raffish—the brim pushed well over the nose: Mark One long and solemn, Mark Two snub, Mark Three bulbous—or worn at a rakish angle. An intent group of tall and tiny punters, dressed for Newmarket, Newbury, or the Eridge point-to-point. Had I been better acquainted with that *milieu*, I would have expected to see brandy-flasks emerging from hip-pockets. As it was, the hip-pockets were, like the rest of the outfit, articles of make-believe. Judged from the studiedly informal scattered groups of normal-sized or miniature racing enthusiasts, the contingent from the South-East must have represented quite a considerable fraction of the school. This I found vaguely reassuring; and I was certainly in dire need of reassurance.

There had been no mention of the trilbies on the lists supplied by Matron. Clearly they belonged to some unspoken tradition, a code that one ignored at one's peril. Certainly all had heard the hidden voice, for there was not a boy to be seen whose head (and sometimes his ears) was not covered, or who was not casually holding his hat by the brim, or twirling it around with abandoned ease, as if he had never set foot out of doors without this sign of middle-class self-advertisement. What strange laws are those of English social conformity! for, as I was later to discover, these hats were only worn six—or, counting half-terms—nine times a year, on the journey in one direction or another, on the Paddington to Birkenhead Express.

It had been a stroke of pure luck that I had been forewarned about the hats, thanks to the fact that two brothers, who had been at the school a year and two years ahead of me, lived in

Tunbridge Wells. Thanks to them, I was saved the awful humiliation of having been seen at Paddington, or at the station of arrival, or on the drive from the station to the school, BAREHEADED. It would have been a very bad start indeed, and one that might have dogged me for years: COBB, THE BOY WHO DID NOT HAVE A HAT.

As it was, I was able to tell my mother that, in addition to the reglementary top-hat for Sunday wear—one was provided, already battered, by my maternal uncle, a country doctor—I absolutely needed an ordinary hat in which to travel. As the hat was only to be in use for travel, my mother did not see the need of providing me with a new one. One of my father's hats, no doubt dating back to official occasions in Cairo, was dug out. It was light grey, with a light grey band, and with an embroidered turned up brim in even lighter grey silk, and a silk lining. Inside it had the name of a Cairo *chapelier*, as well as the Egyptian royal arms, indicating that he was the *fournisseur attitré* to the Court of the Khedive. It was somewhat faded and had certainly seen better days, on no doubt khedival occasions when my father had had to present himself in Cairo. By the Thirties, it was a type of hat already outmoded, though still worn, to go with spats and gloves *gorge de pigeon*, by elderly Dutch politicians of the Anti-Revolutionary Party. Its antique and heavily formal appearance caused me considerable misgiving. If trilbies sat ill on the head of a fourteen-year-old, this stiff Homburg, hinting at a pre-War Karlsbad, could only make me look utterly ridiculous. It was also much too large. But then, I thought, I would only have to wear it for a few minutes, and might even not be noticed in it. I could not hold it, owing to the giveaway upturned brim. And indeed, at Paddington, I had managed to get into the compartment holding the wretched thing against my body, on the train side. I had then hidden it behind my suitcase on the luggage-rack.

It was at Birmingham that the disaster occurred. A grown-up traveller, stretching to put his luggage in the rack, displaced the pearl-grey homburg that, hideously, and as if impelled by familiarity, fell into my lap, witnessed by the whole, observant carriage. There was a dreadful roar of laughter, which even grew in volume when it was discovered that the hat revealed little of me above chin level. The rest of the journey was pure hell, though, by one kind dispensation, none of those who had caught me in this unforgivable social faux-pas was from my House, so that I was never identified there as the owner of the homburg. A few weeks after the beginning of term, profiting from a rainy day, I was able to throw it from the bridge into the Severn. Thus removed from the cause of my social humiliation, I was able to look on the observant world of my fellow-boys with something like self-confidence. Later in the term I profited from a visit from my godmother, who lived at High Ercall, to purchase a new trilby, dark brown in colour. Travelling after that caused no problem, though the approach of my first Speech Day was marred by my trepidation as to my *father's* choice of headgear. In the summer, my father favoured a boater. A panama would have been all right, but not a boater. . . . To my intense relief, he turned up in plus-fours and a cap. What is more, I noticed that the mother of someone in my house was wearing a leopard-skin coat; and no *lady* ever wore one of these. As soon as I left school, I celebrated my freedom by buying myself, in Vienna, a furry black hat with a wide brim.

1985

The People's Victor

> How came it that this prudent, economical man was also
> generous? That this chaste adolescent, this model father, grew
> to be, in his last years, an ageing faun? That this legitimist
> changed, first into a Bonapartist, only, later still, to be hailed
> as the grandfather of the Republic? That this pacifist could
> sing, better than anybody, of the glories of the flags of Wag-
> ram? That this bourgeois in the eyes of other bourgeois came
> to assume the stature of a rebel? These are the questions that
> every biographer of Victor Hugo must answer.
>
> ANDRÉ MAUROIS

Monsieur Maurois, luckily, does not seriously attempt to
answer any of them; his book' is very much better than this little
collection of paradoxes à l'Américaine would suggest. This is a
lucid, well-constructed biography, solidly based on wide and
deep research, and making discriminating and efficient use of
vast materials. M. Maurois's narrative, although fairly long—
about five hundred pages—is compact and extremely readable: it
has the momentum and the sweep necessary for a subject which
demands greater-than-life-size treatment. The work is, the author
tells us, 'the largest in scale and the most difficult that I have
undertaken.' Much as it surpasses his earlier biographies—particularly his rather skimpy exercises on English subjects—it
could only have been written by a man with long experience of
the possibilities and problems of biography—by, in short, a
master craftsman. A craftsman, too, not burdened by excessive
subtlety or overmuch fastidiousness, or irony—and therefore at
home with his subject. Hugo was the concentrated essence of a

---

' *Olympio: The Life of Victor Hugo.*

century of *Français moyens* and it is fitting that he has found a
Français moyen to write his biography.

'We have so rich a native field of romantic poetry,' says the
dust jacket, 'that Hugo's somewhat rhetorical verse leaves us
cold.' No doubt you have; perhaps it does. It could also be that
you know less French than you think you do, and that you have
a taste for misleading comparisons, flattering to yourselves.
There is nothing at all like Hugo in the English field of romantic
poetry. Nor was Hugo's verse just 'somewhat rhetorical,' with
the implication of poor taste and unwarranted excitement that
that conveys. It was a majestic roll of rhetoric sustained for fifty
years, with a marvellous variety of expression and an always
deepening resonance. Hugo was a public man. He felt the events
of his own life—the birth of a child, a bereavement—as public
events, archetypes of human destiny. He felt the great historical
events of his day as events in his personal emotional life. And
always words, millions of words, gushed out of him, scalding
hot and at high pressure, like steam out of his boiling century.
He said so much that in the end he had said something for everybody. A section of his public followed his body from the Étoile
to the Panthéon. There were two million of them.

The corresponding contemporary figure in England was not
Tennyson—nothing is more misleading than to think of Hugo as
in some sense the Poet Laureate of France—and obviously not
Browning or Arnold.

The English Victor Hugo, the prophet and the incarnation
of the century, was not a poet but a politician: Gladstone. It is
not just that both became respected old men who had uttered
more emotive words than any of their contemporary fellow-
countrymen. Nor was it that they shared a power for expressing
and arousing moral indignation, answering to a great need of the
age. It was not even that each was a medium, through whom
inarticulate masses found a voice. The essential was that both
were artists, and artists of the same kind: If love of liberty, that
ambiguous and powerful emotion, was the force that drew their
great audiences to them, those audiences desired that liberty to
appear in an acceptable form, not inchoate and anarchic, but
ordered, rich and beautiful. Gladstone and Hugo had the *souffle*,
the mastery of language, and the legend-focussing personality
that could confer a formal order on a general release of emotion.

In the case of the orator, it has been held that this faculty was daemonic and annunciatory of disasters to come. The road to Nuremberg, on this view, begins at Midlothian. It might be truer to say that it was the nation which had no Midlothians that found itself a voice at Nuremberg. The relevant metaphor, for the age, is still that of 'letting off steam': that was, in part, the function for their nations, and in their very different ways, of Hugo and of Gladstone. It is hardly wise to regard the process with suspicion because another nation, in a later day, blew up the boiler.

Frenchmen, on the very rare occasions when they think of Gladstone at all, think of him as a symbol of English hypocrisy. Did he not veil his face in affected horror at the discovery of Parnell's adultery, about which he had already known for years? And did he not indulge in a life of vice with the pretext that he was reforming prostitutes? (It is of no use to reply that in reality he did neither of these things.) The proper comparison, in the French view, would be with Tartuffe and Félix Faure, certainly not with France's greatest poet: how compare the fustian of a politician with verses that are among the glories of the French language? The argument has weight when one places the works of Hugo beside the collected speeches of Gladstone; it has very much less weight when one thinks of the two men, living, in relation to their communities. It is probable that many of the two million who followed the *corbillard des pauvres* to the Panthéon were there because Hugo had written:

> Je ne fais point fléchir les mots auxquels je crois:
> Raison, progrès, honneur, loyauté, devoirs, droits,
> On ne va point au vrai par une route oblique.
> Sois juste; c'est ainsi qu'on sert la république;
> Le devoir envers elle est l'équité pour tous;
> Pas de colère; et nul n'est juste s'il n'est doux.

It is possible to regard these lines as less successful poetry, but more successful politics than: 'That cloud in the West! That coming storm! God's minister of vengeance upon ancient and inveterate and still but half-atoned injustice!' For Hugo, in pleading in very flat verse for mercy for the Communards, knew that his words would find an echo in hundreds of thousands of French hearts. Gladstone could have felt no corresponding

confidence; the accent of his words, as the first intuition of his
Irish task comes on him, is genuinely tragic, in marked contrast
to the hollow and perfunctory eloquence of Hugo on the Com-
mune. The poet and the tribune are here in the wrong places

If we think of the two men of genius as being of the same
prophetic race, both poet-tribunes, then we may also think it no
accident that in France the prophet turned his face to literature,
in England to politics. Gladstone was an engine in which great
forces at a high temperature were concentrated to make changes
in English life, for good or ill. Hugo changed nothing, except
the personal lives of those around him, and his style. And, since
a human being who becomes an engine becomes less as well as
more than human, Hugo's personality remained the richer and
his life the more exemplary—although certainly not in the vulgar
moral application of the last term. It is not a question of contrasting M. Maurois's ageing faun, seducer of servant-girls, with
the self-dedicated redeemer of fallen women, but of seeing how,
in Gladstone, all emotion tends to turn to controversy, to engage
in public *work*; while with Hugo, a series of magnificent emotions, in a life filled with drama, found, within and without the
limits of his art, free and spectacular *play*. The two men seem
almost like complementary colossal figures—not devoid of
cliché—designed to express the contrasting genius of the two
peoples, art governing and art living. Prodigious creatures,
concentrating and revealing the essential character of the life
around them, one feels, before their force and mystery, something of what Hugo felt, seeing comparable portents, at the zoo:

> Moi, je n'exige pas que Dieu toujours s'observe,
> Il faut bien tolérer quelques excès de verve
> Chez un si grand poète, et ne point se fâcher
> Si Celui qui nuance une fleur de pêcher
> Et courbe l'arc-en-ciel sur l'Océan qu'il dompte,
> Après un colibri, nous donne un mastodonte!
> C'est son humeur à lui d'être de mauvais goût,
> D'ajouter l'hydre au gouffre et le ver à l'égout,
> D'avoir, en toute chose, une stature étrange
> Et d'être un Rabelais d'où sort un Michel-Ange.
> C'est Dieu; moi, je l'accepte . . .

1956

Movies on Television

A FEW years ago, a jet on which I was returning to California after a trip to New York was instructed to delay landing for a half hour. The plane circled above the San Francisco area, and spread out under me were the farm where I was born, the little town where my grandparents were buried, the city where I had gone to school, the cemetery where my parents were, the homes of my brothers and sisters, Berkeley, where I had gone to college, and the house where at that moment, while I hovered high above, my little daughter and my dogs were awaiting my return. It was as though my whole life were suspended in time—as though no matter where you'd gone, what you'd done, the past were all still there, present, if you just got up high enough to attain the proper perspective.

Sometimes I get a comparable sensation when I turn from the news programs or the discussion shows on television to the old movies. So much of what formed our tastes and shaped our experiences, and so much of the garbage of our youth that we never thought we'd see again—preserved and exposed to eyes and minds that might well want not to believe that this was an important part of our past. Now these movies are there for new generations, to whom they cannot possibly have the same impact or meaning, because they are all jumbled together, out of historical sequence. Even what may deserve an honorable position in movie history is somehow dishonored by being so available, so meaninglessly present. Everything is in hopeless disorder, and that is the way new generations experience our movie past. In the other arts, something like natural selection takes place: only the best or the most significant or influential or successful works compete for our attention. Moreover, those from the past are likely to be touched up to accord with the taste of the present. In popular music, old tunes are newly orchestrated. A small repertory of plays is continually reinterpreted for contemporary meanings—the great ones for new relevance, the not so great rewritten, tackily 'brought up to date,' or deliberately treated as period pieces. By contrast, movies, through the accidents of commerce, are sold in blocks or packages to television, the worst with the mediocre and the best, the successes with the failures, the forgotten with the half forgotten, the ones so dreary you don't know whether you ever saw them or just others like them with some so famous you can't be sure whether you actually saw them or only imagined what they were like. A lot of this stuff never really made it with any audience; it played in small towns or it was used to soak up the time just the way TV in bars does.

There are so many things that we, having lived through them, or passed over them, never want to think about again. But in movies nothing is cleaned away, sorted out, purposefully discarded. (The destruction of negatives in studio fires or deliberately, to save space, was as indiscriminate as the preservation and resale.) There's a kind of hopelessness about it: what does not deserve to last lasts, and so it all begins to seem one big pile of junk, and some people say, 'Movies never really were any good—except maybe the Bogarts.' If the same thing had happened in literature or music or painting—if we were constantly surrounded by the piled-up inventory of the past—it's conceivable that modern man's notions of culture and civilization would be very different. Movies, most of them produced as fodder to satisfy the appetite for pleasure and relaxation, turned out to have magical properties—indeed, to *be* magical properties. This fodder can be fed to people over and over again. Yet, not altogether strangely, as the years wear on it doesn't please their palates, though many will go on swallowing it, just because nothing tastier is easily accessible. Watching old movies is like spending an evening with those people next door. They bore us, and we wouldn't go out of our way to see them; we drop in on them because they're so close. If it took some effort to see old movies, we might try to find out which were the good ones, and if people saw only the good ones maybe they would still respect old movies. As it is, people sit and watch movies that audiences walked out on thirty years ago. Like Lot's wife, we are tempted to take another look, attracted not by evil but by something that seems much more shameful—our own innocence. We

don't try to reread the girls' and boys' 'series' books of our adolescence—the very look of them is dismaying. The textbooks we
studied in grammar school are probably more 'dated' than the
movies we saw then, but we never look at the old schoolbooks,
whereas we keep seeing on TV the movies that represent the
same stage in our lives and played much the same part in
them—as things we learned from and, in spite of, went beyond.

Not all old movies look bad now, of course; the good ones are
still good—surprisingly good, often, if you consider how much
of the detail is lost on television. Not only the size but the shape
of the image is changed, and, indeed, almost all the specifically
visual elements are so distorted as to be all but completely
destroyed. On television, a cattle drive or a cavalry charge or a
chase—the climax of so many a big movie—loses the dimensions of space and distance that made it exciting, that sometimes
made it great. And since the structural elements—the rhythm,
the buildup, the suspense—are also partly destroyed by deletions
and commercial breaks and the interruptions incidental to home
viewing, it's amazing that the bare bones of performance, dialogue, story, good directing, and (especially important for close-
range viewing) good editing can still make an old movie more
entertaining than almost anything new on television. (That's
why old movies are taking over television—or, more accurately,
vice versa.) The verbal slapstick of the newspaper-life comedies—*Blessed Event, Roxie Hart, His Girl Friday*—may no longer
be fresh (partly because it has been so widely imitated), but it's
still funny. Movies with good, fast, energetic talk seem better
than ever on television—still not great but, on television, better
than what *is* great. (And as we listen to the tabloid journalists
insulting the corrupt politicians, we respond once again to the
happy effrontery of that period when the targets of popular satire
were still small enough for us to laugh at without choking.) The
wit of dialogue comedies like Preston Sturges's *Unfaithfully Yours*
isn't much diminished, nor does a tight melodrama like *Double
Indemnity* lose a great deal. Movies like Joseph L. Mankiewicz's
A Letter to Three Wives and *All About Eve* look practically the
same on television as in theatres, because they have almost no
visual dimensions to lose. In them the camera serves primarily to
show us the person who is going to speak the next presumably
bright line—a scheme that on television, as in theatres, is accept-

able only when the line *is* bright. Horror and fantasy films like Karl Freund's *The Mummy* or Robert Florey's *The Murders in the Rue Morgue*—even with the loss, through miniaturization, of imaginative special effects—are surprisingly effective, perhaps because they are so primitive in their appeal that the qualities of the imagery matter less than the basic suggestions. Fear counts for more than finesse, and viewing horror films is far more frightening at home than in the shared comfort of an audience that breaks the tension with derision.

Other kinds of movies lose much of what made them worth looking at—the films of von Sternberg, for example, designed in light and shadow, or the subtleties of Max Ophuls, or the lyricism of Satyajit Ray. In the box the work of these men is not as lively or as satisfying as the plain good movies of lesser directors. Reduced to the dead grays of a cheap television print, Orson Welles's *The Magnificent Ambersons*—an uneven work that is nevertheless a triumphant conquest of the movie medium—is as lifelessly dull as a newspaper Wirephoto of a great painting. But when people say of a 'big' movie like *High Noon* that it has dated or that it doesn't hold up, what they are really saying is that their judgment was faulty or has changed. They may have over-responded to its publicity and reputation or to its attempt to deal with a social problem or an idea, and may have ignored the banalities surrounding that attempt; now that the idea doesn't seem so daring, they notice the rest. Perhaps it was a traditional drama that was new to them and that they thought was new to the world; everyone's 'golden age of movies' is the period of his first moviegoing and just before—what he just missed or wasn't allowed to see. (The Bogart films came out just before today's college kids started going.)

Sometimes we suspect, and sometimes rightly, that our memory has improved a picture—that imaginatively we made it what we knew it could have been or should have been—and, fearing this, we may prefer memory to new contact. We'll remember it better if we don't see it again—we'll remember what it meant to us. The nostalgia we may have poured over a performer or over our recollections of a movie has a way of congealing when we try to renew the contact. But sometimes the experience of reseeing is wonderful—a confirmation of the general feeling that was all that remained with us from childhood. And we enjoy the fresh proof of the rightness of our responses that reseeing the film gives us. We re-experience what we once felt, and memories flood back. Then movies seem magical—all those *madeleines* waiting to be dipped in tea. What looks bad in old movies is the culture of which they were part and which they expressed—a tone of American life that we have forgotten. When we see First World War posters, we are far enough away from their patriotic primitivism to be amused at the emotions and sentiments to which they appealed. We can feel charmed but superior. It's not so easy to cut ourselves off from old movies and the old selves who responded to them, because they're not an isolated part of the past held up for derision and amusement and wonder. Although they belong to the same world as stories in *Liberty*, old radio shows, old phonograph records, an America still divided between hayseeds and city slickers, and although they may seem archaic, their pastness isn't so very past. It includes the last decade, last year, yesterday.

Though in advertising movies for TV the recentness is the lure, for many of us what constitutes the attraction is the datedness, and the earlier movies are more compelling than the ones of the fifties or the early sixties. Also, of course, the movies of the thirties and forties look better technically, because, ironically, the competition with television that made movies of the fifites and sixties enlarge their scope and their subject matter has resulted in their looking like a mess in the box—the sides of the image lopped off, the crowds and vistas a boring blur, the color altered, the epic themes incongruous and absurd on the little home screen. In a movie like *The Robe*, the large-scale production values that were depended on to attract TV viewers away from their sets become a negative factor. But even if the quality of the image were improved, these movies are too much like the ones we can see in theatres to be interesting at home. At home, we like to look at those stiff, carefully groomed actors of the thirties, with their clipped, Anglophile stage speech and their regular, clean-cut features—walking profiles, like the figures on Etruscan vases and almost as remote. And there is the faithless wife—how will she decide between her lover and her husband, when they seem as alike as two wax grooms on a wedding cake? For us, all three are doomed not by sin and disgrace but by history. Audiences of the period may have enjoyed these movies for their action, their story, their thrills, their wit, and all this high living. But through our window on the past we see the actors acting out other dramas as well. The Middle European immigrants had children who didn't speak the king's English and, after the Second World War, didn't even respect it so much. A flick of the dial and we are in the fifties amid the slouchers, with their thick lips, shapeless noses, and shaggy haircuts, waiting to say their lines until they think them out, then mumbling something that is barely speech. How long, O Warren Beatty, must we wait before we turn back to beautiful stick figures like Phillips Holmes?

We can take a shortcut through the hell of many lives, turning the dial from the social protest of the thirties to the films of the same writers and directors in the fifties—full of justifications for blabbing, which they shifted onto characters in oddly unrelated situations. We can see in the films of the forties the displaced artists of Europe—the anti-Nazi exiles like Conrad Veidt, the refugees like Peter Lorre, Fritz Kortner, and Alexander Granach. And what are they playing? Nazis, of course, because they have accents and so for Americans—for the whole world—they become images of Nazi brutes. Or we can look at the patriotic sentiments of the Second World War years and those actresses, in their orgies of ersatz nobility, giving their lives—or, at the very least, their bodies—to save their country. It was sickening at the time; it's perversely amusing now—part of the spectacle of our common culture.

Probably in a few years some kid watching *The Sandpiper* on television will say what I recently heard a kid say about *Mrs Miniver*: 'And to think they really believed it in those days.' Of course, we didn't. We didn't accept nearly as much in old movies as we may now fear we did. Many of us went to see big-name pictures just as we went to *The Night of the Iguana*, without believing a minute of it. The James Bond pictures are not to be 'believed,' but they tell us a lot about the conventions that audiences now accept, just as the confessional films of the thirties dealing with sin and illegitimacy and motherhood tell us about the sickly-sentimental tone of American entertainment in the midst of the Depression. Movies indicate what the producers thought people would pay to see—which was not always the same as what they *would* pay to see. Even what they enjoyed seeing does not tell us directly what they believed but only indirectly hints at the tone and style of a culture. There is no reason to assume that people twenty or thirty years ago were stupider than they are now. (Consider how *we* may be judged by people twenty years from now looking at today's movies.) Though it may not seem obvious to us now, part of the original appeal of old movies—which we certainly understood and responded to as children—was that, despite their sentimental tone, they helped to form the liberalized modern consciousness. This trash—and most of it was, and is, trash—probably taught us more about the world, and even about values, than our 'education' did. Movies broke down barriers of all kinds, opened up the world, helped to make us aware. And they were almost always on the side of the mistreated, the socially despised. Almost all drama is. And, because movies were a mass medium, they had to be on the side of the poor.

Nor does it necessarily go without saying that the glimpses of something really good even in mediocre movies—the quickening of excitement at a great performance, the discovery of beauty in a gesture or a phrase or an image—made us understand the meaning of art as our teachers in appreciation courses never could. And—what is more difficult for those who are not movie lovers to grasp—even after this sense of the greater and the higher is developed, we still do not want to live only on the heights. We still want that pleasure of discovering things for ourselves; we need the sustenance of the ordinary, the commonplace, the almost-good as part of the anticipatory atmosphere. And though it all helps us to respond to the moments of greatness, it is not only for this that we want it. The educated person who became interested in cinema as an art form through Bergman or Fellini or Resnais is an alien to me (and my mind goes blank with hostility and indifference when he begins to talk). There isn't much for the art-cinema person on television; to look at a great movie, or even a poor movie carefully designed in terms of textures and contrasts, on television is, in general, maddening, because those movies lose too much. (Educational television, though, persists in this misguided effort to bring the television viewer movie classics.) There are few such movies anyway. But there are all the not-great movies, which we probably wouldn't bother going to see in museums or in theatre revivals—they're just not that important. Seeing them on tele-

vision is a different kind of experience, with different values—
partly because the movie past hasn't been filtered to conform to
anyone's convenient favorite notions of film art. We make our
own, admittedly small, discoveries or rediscoveries. There's Dan
Dailey doing his advertising-wise number in *It's Always Fair
Weather*, or Gene Kelly and Fred Astaire singing and dancing
'The Babbitt and the Bromide' in *Ziegfeld Follies*. And it's like
putting on a record of Ray Charles singing 'Georgia on My
Mind' or Frank Sinatra singing 'Bim Bam Baby' or Elisabeth
Schwarzkopf singing operetta, and feeling again the elation we
felt the first time. Why should we deny these pleasures because
there are other, more complex kinds of pleasure possible? It's
true that these pleasures don't deepen, and that they don't change
us, but maybe that is part of what makes them seem our own—
we realize that we have some emotions and responses that *don't*
change as we get older.

People who see a movie for the first time on television don't
remember it the same way that people do who saw it in a theatre.
Even without the specific visual loss that results from the transfer to another medium, it's doubtful whether a movie could have
as intense an impact as it had in its own time. Probably by definition, works that are not truly great cannot be as compelling
out of their time. Sinclair Lewis's and Hemingway's novels were
becoming archaic while their authors lived. Can *On the Waterfront*
have the impact now that it had in 1954? Not quite. And revivals
in movie theatres don't have the same kind of charge, either.
There's something a little stale in the air, there's a different kind
of audience. At a revival, we must allow for the period, or care
because of the period. Television viewers seeing old movies for
the first time can have very little sense of how and why new stars
moved us when they appeared, of the excitement of new themes,
of what these movies meant to us. They don't even know which
were important in their time, which were 'hits.'

But they can discover *something* in old movies, and there are
few discoveries to be made on dramatic shows produced for
television. In comedies, the nervous tic of canned laughter
neutralizes everything; the laughter is as false for the funny as for
the unfunny and prevents us from responding to either. In general, performances in old movies don't suffer horribly on television except from cuts, and what kindles something like the

early flash fire is the power of personality that comes through in those roles that made a star. Today's high school and college students seeing *East of Eden* and *Rebel Without a Cause* for the first time are almost as caught up in James Dean as the first generation of adolescent viewers was, experiencing that tender, romantic, marvelously masochistic identification with the boy who does everything wrong because he cares so much. And because Dean died young and hard, he is not just another actor who outlived his myth and became ordinary in stale roles—he is the symbol of misunderstood youth. He is inside the skin of moviegoing and television-watching youth—even educated youth—in a way that Keats and Shelley or John Cornford and Julian Bell are not. Youth can respond—though not so strongly —to many of our old heroes and heroines: to Gary Cooper, say, as the elegant, lean, amusingly silent romantic loner of his early Western and aviation films. (And they can more easily ignore the actor who sacrificed that character for blubbering righteous bathos.) Bogart found his myth late, and Dean fulfilled the romantic myth of self-destructiveness, so they look good on television. More often, television, by showing us actors before and after their key starring roles, is a myth-killer. But it keeps acting ability alive.

There is a kind of young television watcher seeing old movies for the first time who is surprisingly sensitive to their values and responds almost with the intensity of a moviegoer. But he's different from the moviegoer. For one thing, he's housebound, inactive, solitary. Unlike a moviegoer, he seems to have no need to discuss what he sees. The kind of television watcher I mean (and the ones I've met are all boys) seems to have extreme empathy with the material in the box (new TV shows as well as old movies, though rarely news), but he may not know how to enter into a conversation, or even how to come into a room or go out of it. He fell in love with his baby-sitter, so he remains a baby. He's unusually polite and intelligent, but in a mechanical way— just going through the motions, without interest. He gives the impression that he wants to withdraw from this human interference and get back to his real life—the box. He is like a prisoner who has everything he wants in prison and is content to stay there. Yet, oddly, he and his fellows seem to be tuned in to each other; just as it sometimes seems that even a teen-ager locked in a closet would pick up the new dance steps at the same moment as other teen-agers, these television watchers react to the same things at the same time. If they can find more intensity in this box than in their own living, then this box can provide *constantly* what we got at the movies only a few times a week. Why should they move away from it, or talk, or go out of the house, when they will only experience that as a loss? Of course, we can see why they should, and their inability to make connections outside is frighteningly suggestive of ways in which we, too, are cut off. It's a matter of degree. If we stay up half the night to watch old movies and can't face the next day, it's partly, at least, because of the fascination of our own movie past; *they* live in a past they never had, like people who become obsessed by places they have only imaginative connections with—Brazil, Venezuela, Arabia Deserta. Either way, there is always something a little shameful about living in the past; we feel guilty, stupid—as if the pleasure we get needed some justification that we can't provide.

For some moviegoers, movies probably contribute to that self-defeating romanticizing of expectations which makes life a series of disappointments. They watch the same movies over and over on television, as if they were constantly returning to the scene of the crime—the life they were so busy dreaming about that they never lived it. They are paralyzed by longing, while those less romantic can leap the hurdle. I heard a story the other day about a man who ever since his school days had been worshipfully 'in love with' a famous movie star, talking about her, fantasizing about her, following her career, with its ups and downs and its stormy romances and marriages to producers and agents and wealthy sportsmen and rich businessmen. Though he became successful himself, it never occurred to him that he could enter her terrain—she was so glamorously above him. Last week, he got a letter from an old classmate, to whom, years before, he had confided his adoration of the star; the classmate—an unattractive guy who had never done anything with his life and had a crummy job in a crummy business—had just married her.

Movies are a combination of art and mass medium, but television is so single in its purpose—selling—that it operates without that painful, poignant mixture of aspiration and effort and compromise. We almost never think of calling a television show 'beautiful,' or even of complaining about the absence of beauty, because we take it for granted that television operates without beauty. When we see on television photographic records of the past, like the pictures of Scott's Antarctic expedition or those series on the First World War, they seem almost too strong for the box, too pure for it. The past has a terror and a fascination and a beauty beyond almost anything else. We are looking at the dead, and they move and grin and wave at us; it's an almost unbearable experience. When our wonder and our grief are interrupted or followed by a commercial, we want to destroy the ugly box. Old movies don't tear us apart like that. They do something else, which we can take more of and take more easily: they give us a sense of the passage of life. Here is Elizabeth Taylor as a plump matron and here, an hour later, as an exquisite child. That charmingly petulant little gigolo with the skinny face and the mustache that seems the most substantial part of him—can he have developed into the great Laurence Olivier? Here is Orson Welles as a young man, playing a handsome old man, and here is Orson Welles as he has really aged. Here are Bette Davis and Charles Boyer traversing the course of their lives from ingenue and juvenile, through major roles, into character parts—back and forth, endlessly, embodying the good and bad characters of many styles, many periods. We see the old character actors put out to pasture in television serials, playing gossipy neighbors or grumpy grandpas, and then we see them in their youth or middle age, in the roles that made them famous—and it's startling to find how good they were, how vital, after we've encountered them caricaturing themselves, feeding off their old roles. They have almost nothing left of that young actor we responded to— and still find ourselves responding to—except the distinctive voice and a few crotchets. There are those of us who, when we watch old movies, sit there murmuring the names as the actors appear (Florence Bates, Henry Daniell, Ernest Thesiger, Constance Collier, Edna May Oliver, Douglas Fowley), or we recognize them but can't remember their names, yet know how well we once knew them, experiencing the failure of memory as a loss of our own past until we can supply it (Maude Eburne or Porter Hall)—with great relief. After a few seconds, I can always remember them, though I cannot remember the names of my childhood companions or of the prizefighter I once dated, or even of the boy who took me to the senior prom. We are eager to hear again that line we know is coming. We hate to miss anything. Our memories are jarred by cuts. We want to see the movie to the end.

The graveyard of *Our Town* affords such a tiny perspective compared to this. Old movies on television are a gigantic, panoramic novel that we can tune in to and out of. People watch avidly for a few weeks or months or years and then give up; others tune in when they're away from home in lonely hotel rooms, or regularly, at home, a few nights a week or every night. The rest of the family may ignore the passing show, may often interrupt, because individual lines of dialogue or details of plot hardly seem to matter as they did originally. A movie on television is no longer just a drama in itself; it is part of a huge ongoing parade. To a new generation, what does it matter if a few gestures and a nuance are lost, when they know they can't watch the parade on all the channels at all hours anyway? It's like traffic on the street. The television generation knows there is no end; it all just goes on. When television watchers are surveyed and asked what kind of programming they want or how they feel television can be improved, some of them not only have no answers but can't understand the questions. What they get on their sets is television—that's it.

1967

D. J. ENRIGHT

The Marquis and the Madame

*J*USTINE: *Pamela* rewritten as pornography. *Juliette*: a female
Tom Jones reconstructed in the same spirit. *Miss Henriette Stralson*:
Clarissa minus Clarissa. *Augustine de Villeblanche*: refined smut
yielding to gross sentimentalism, very much of and for its
period. *Les 120 Journées de Sodome*: just what it says. But if there
seems little reason for literary people to concern themselves with
Sade, he has found a new lease of life among philosophers and
anthropologists. Bored and uneasy with our little lives we resort
to the greater amplitude of symbols. Bardot, Byron, Hitler,
Hemingway, Monroe, Sade: we do not require our heroes to be
subtle, just to be big. Then we can depend on someone to make
them subtle.

In her essay, 'Must we burn Sade?', Mme de Beauvoir devotes
some seventy pages of subtle explication to the plain Marquis.
Scandalized by the neglect into which he had fallen, yet
repudiating the obvious topsy-turvy whereby he has been deified, she asks that he be regarded as a man and a writer. Even so
it is not exactly as author nor as sexual pervert that he interests
her, but by his efforts to justify his perversions, to 'erect his
tastes into principles'. We are all great moralizers, especially
where our 'tastes' are concerned. 'He dreamed of an ideal society
from which his special tastes would not exclude him.' Don't we
all? And Sade's dream is clearly defined in *Les 120 Journées de
Sodome*, with its small society of libertines, protected from the
mean prejudices of the outside world, wealthy enough to pro-
cure any diversion they can think up and unlimited supplies of
every variety of human flesh. Or in Minski's castle, in *Juliette*, in
which the furniture consists of naked girls artistically arranged
and the owner lives '*selon l'état de mes couilles*,' which are kept in
prime condition by large helpings of gammon of boy accompanied by sixty bottles of Burgundy at a sitting.

Minski's castle, it would seem forgiveable to suppose, is the dream world of the impotent and Minski himself bears the hero's expected characteristics: '. . . *de dix-huit pouces de long, sur seize de circonférence, surmonté d'un champignon vermeil et large comme le cul d' un chapeau.*' The obvious thing about Sade's fiction—it was outside Mme de Beauvoir's purpose to remark the obvious—is that it is *par excellence* obscenity of the most basic sort, as it were a pattern for pornographers. It runs, quite insultingly, to the form which it lays down for itself. Incidentally, Minski's ejaculatory ability reappears, though with a nice touch of humour, in a potboiler by his admirer Apollinaire.

In *The Marquis de Sade*, a selection recently published in English by John Calder, the following passage is left in French. I will risk translating it:

One of my friends is living with the daughter he has had by his own mother; only a week ago he deflowered a boy of thirteen years, fruit of his commerce with that daughter; in a few years this same boy will marry his mother

—and, continues the narrator, the friend is still young and intends 'to enjoy still more fruits that shall be born of this wedding'. These genealogical acrobatics correspond to (and are followed by) the physical acrobatics beloved of the hack pornographer. But *honi soit* . . . and Mme de Beauvoir thinks better than I do. 'Juliette was saved and Justine lost from the beginning of time.' A less cosmogonical interpretation of the two might have it that Juliette is the bad girl who will do anything and Justine is the good girl to whom everything is done, the two stock figures of the pornographic scene.

As for Sade's philosophical disquisitions, they might serve to justify his tastes as revealed in the fiction, except that the philosophy is ludicrous—a *Modest Proposal* unironically intended— and the tastes are (I should have thought) self-evidently unamenable to justification. The philosophy is basically this: if you enjoy wickedness, it shows that Nature intended you to be wicked, and it would be wicked not to be. There seems nothing very original here. 'Thou, Nature, art my goddess': Edmund is a somewhat richer character than any of Sade's creations. The rest follows predictably from this first principle. 'There is nothing more refined than the carnal liaison of families': or, in language less gallant, incest is more Natural than non-incestuous connection. Similarly, for the pornographer any hint of tenderness is to be shunned like the plague: genuine perversion is an onanistic activity to which a second party is necessary as a tool, whereas tenderness implies the recognition of the other party as a person in his or her own right. The case against love is put neatly by Belmor in *Juliette*: it is useless in that it doesn't increase sexual enjoyment and positively pernicious in that it 'causes us to neglect our own interests for those of the thing loved' and adds this thing's pains and troubles to the sum of our own.

Notwithstanding, Mme de Beauvoir tells us that 'eroticism appears in Sade as a mode of communciation, the only valid one' between persons. Since she then admits that 'every time we side with a child whose throat has been slit by a sex-maniac, we take a stand against him', it is possible that she is using the word 'communication' in some highly special sense, paradoxical to the rest of us, reserved to philosophically-trained intellectuals. For the rest of us, Sade's message on this point might seem to come to, F—— you, Jack (or Jill), I'm all right.

Some of the ideas advanced in *Français, encore un effort* are respectable enough—his republicanism, anti-clericalism and opposition to the death penalty—but they are neither novel nor respectably argued. Elsewhere his recommendations are simple: murder, rape, torture, sodomy, cannibalism, arson, coprophagy, necrophilism, bestiality ('*le dindon est délicieux*'), etc. For Mme de Beauvoir, Sade's value, his contemporary importance, lies in the fact that 'he chose cruelty rather than indifference'. His sincerity—indifference, I suppose, cannot be sincere—encourages her to hail him as 'a great moralist'. The aptest comment on this description is M. de Bressac's casual remark to his servant, which I quote from memory: 'Now, Joseph, you b—— Justine, and then we shall feed her to the dogs.' Mme de Beauvoir shares one of her protégé's characteristics: humourlessness.

The Calder selection contains nothing from *Justine* but, considering the obvious difficulties, it conveys (with the help of short passages left in their native French) a just hint of its author's preposterousness. 'Must we burn Sade?' asks Mme de Beauvoir. Now that you mention it, why not? The world is littered with literature. And Sade teaches us little about human nature which we couldn't gather from a few minutes of honest introspection. But maybe we can learn something more useful from Mme de Beauvoir's solemn excogitations, something about our scornful reluctance to face the realities of our selves and of the selves of others, and our preferred contemplation of modish dummies, those highbrow status symbols, ourselves as heroic monsters or grand victims, our inflation or reduction of ourselves and of others to ingeniously explicated strip-cartoons, as unreal as the wicked Juliette and as empty of life as the virtuous Justine.

1953

The Savage Seventh

IT was that verse about becoming again as a little child that caused the first sharp waning of my Christian sympathies. If the Kingdom of Heaven could be entered only by those fulfilling such a condition I knew I should be unhappy there. It was not the prospect of being deprived of money, keys, wallet, letters, books, long-playing records, drinks, the opposite sex, and other solaces of adulthood that upset me (I should have been about eleven), but having to put up indefinitely with the company of other children, their noise, their nastiness, their boasting, their back-answers, their cruelty, their silliness. Until I began to meet grown-ups on more or less equal terms I fancied myself a kind of Ishmael. The realization that it was not people I disliked but children was for me one of those celebrated moments of revelation, comparable to reading Haeckel or Ingersoll in the last century. The knowledge that I should never (except by deliberate act of folly) get mixed up with them again more than compensated for having to start earning a living.

Today I am more tolerant. It's not that I loathe the little scum, as Hesketh Pearson put it; merely that 'the fact is that a child is a nuisance to a grown-up person. What is more, the nuisance becomes more and more intolerable as the grown-up person becomes more cultivated, more sensitive, and more engaged in the highest methods of adult work' (Shaw). I don't know about highest methods of adult work: what makes the contest between them so unequal is that the child is younger and so in better physical shape, life hasn't yet cut it down to size, it's not worried about anything, it hasn't been to work all today and hasn't got to go to work all tomorrow, all of which makes it quite unbearable but for none of which can it fairly be blamed. The two chief characteristics of childhood, and the two things that make it so seductive to a certain type of adult mind, are its freedom from reason and its freedom from responsibility. It is these that give it its peculiar heartless, savage strength.

These few commonplaces are intended to prepare the reader for the unflattering approach of Mr and Mrs Opie in their new book.[1] 'The worldwide fraternity of children', they quote from Douglas Newton, 'is the greatest of savage tribes, and the only one that shows no sign of dying out,' and they lose no time in implanting in their reader's mind the notion that the whole seven-million-strong community of children can be likened to a separate more primitive population suitable for frank anthropological study, like Trobrianders or the nineteenth-century poor. With this assumption, Mr and Mrs Opie suspected that such a self-contained world held a great deal of traditional lore and sayings, and hence enlisted the aid of numerous field-workers who appear to have spent eight years accumulating and reporting what they found. Since these workers included teachers at over seventy schools throughout the British Isles, the coverage was thorough, but the field-work was clearly backed up with extensive reading and private correspondence. The authors' wish, if a large body of oral material was discovered to exist, was to get it down on paper in an accurate, unidealized way. Clearly their expectations were gratified, and they have brought to the task of recording the results the blend of charm and thoroughness already evinced in their nursery rhyme collections. Their 400-page book takes the reader right into the heart of the child country. What does he find there?

Leaving aside games (to be the subject of a second volume later), the mass of sayings and customs here presented refers to amost every aspect of the unofficial social life of childhood between the ages six and fourteen. It is made up of rhymes, parodies, jokes, riddles, nicknames and repartee, together with more practical formulae of promise, barter, friendship, fortune and superstition, and a miscellaneous collection of calendar customs, pranks, and such expertise as the use of lean bacon rashers to deaden caning. The vast majority involve rhyming. Children love rhymes, however pointless, just because they are rhymes:

[1] Iona and Peter Opie, *The Lore and Language of Schoolchildren* (Oxford: Clarendon Press, 1959).

Have you seen Pa
Smoking a cigar
Riding on a bicycle?
Ha! Ha! Ha!

and a belief or prayer or promise is felt to be truer or more effective or more binding if in the form of a jingle:

Touch your head, touch your toes
Hope I never go in one of those
(*On seeing an ambulance*.)

The authors claim that this susceptibility goes deeper. 'When on their own they burst into rhyme, of no recognisable relevancy, as a cover in unexpected situations, to pass off an awkward meeting, to fill a silence, to hide a deeply-felt emotion, or in a gasp of excitement.' This does not mean that children are natural poets. The many lovers of the Opies' earlier books should be warned not to expect another harvest of ageless magical-simple ditties of cottage and countryside. The rhymes children do not let die (as opposed to those preserved for them by their elders) have no obvious qualifications for immortality:

I'm a man that came from Scotland
Shooting peas up a Nannie goat's bottom,
I'm the man that came from Scotland
Shooting peas away.

All the same, they frequently have unexpectedly long ancestries. In 1954 children were skipping in York to a rhyme the authors could trace back in an unbroken line to 1725: this is true oral tradition, exemplifying the innate conservatism of childhood in these matters that was one of the authors' chief discoveries. Norman Douglas, writing in *London Street Games* (1916), thought he was showing 'how wide-awake our youngsters are, to be able to go on inventing games out of their heads all the time'. But Douglas was wrong: the Opies report that of the 137 chants and fragments he records, 108 are still being sung today, and were presumably as traditional then as now. 'Boys continue to crack jokes that Swift collected from his friends in Queen Anne's time; they play tricks which lads used to play on each other in the heyday of Beau Brummel; they ask riddles that were posed when

Henry VIII was a boy.' A verse reported from Regency days by
Edmund Gosse's father was sent in by a twelve-year-old Spenny-
moor girl 130 years later; in 1952 Wiltshire girls were skipping
to:

> Kaiser Bill went up the hill
>   To see if the war was over;
> General French got out of his trench
>   And kicked him into Dover.
> He say if the Bone Man come
>   Stick your bayonet up his bum.

To come upon the shadowy figure of Kaiser Wilhelm II, and
the still more shadowy Napoleon Bonaparte, standing in a children's song like ghosts at midsummer noontide shows as well as
anything could the way a particular rhyme will be transmitted
unthinkingly from generation to generation until it loses all
significance. Yet, paradoxically, the child has a keen sense of the
topical. Lottie Collins becomes Diana Dors; Bonnie Prince
Charlie becomes Charlie Chaplin; Jack the Ripper becomes
Kruger and then Mickey Mouse. There are even purely modern
songs:

> Catch a Perry Como
> Wash him in some Omo
> Hang him on a line to dry.

The authors explain this paradox by insisting that 'schoolchild
chant and chatter' is made up of two very distinct streams of
oral lore: the modern mass of catch-phrases, slang, fashionable
jokes and nicknames, and the traditional inheritance of dialect
and custom governing such things as playing truant, giving
warning, sneaking, swearing, tormenting, fighting, and in general the darker and sterner side of life. This dichotomy receives
curious reinforcement from the discovery that while terms of
approval (smashing, bang on, flashy, lush, smack on, snazzy, etc)
change rapidly with the fashion, terms of disapproval (blinking
awful, bloomin' 'orrible, boring, cheesy, corny, daft, disgraceful,
flippin' awful, foul, fusty, frowsty, etc) show very little altera-
tion. But the persistence of tradition is seen even more clearly in
non-verbal ways: in calendar customs, for instance, in superstitions, in mysterious convictions connected with assembling a

million milk-bottle tops, of saying 'rabbits' on the first of the month. Many of these are strictly local. Egg-rolling at Easter, widely practised north of the Trent, is quite unknown in the Midlands and the South; Mischief Night (4 November) occupies a belt east to west across the country between, say, Derby and Saltburn. (From my observation this custom is spreading and growing more violent and disagreeable: I suggest a Herod's Eve to coincide with it, on which bands of adults might roam the streets and bash hell out of anyone under sixteen found out of doors.)

Long before the reader finishes the last chapter he will be asking himself what this tumult of rhyming, joking, riddling, jeers and epithets (the extent of which I have done no more than hint at) really amounts to in terms of knowledge about children today. Here the authors are not helpful. No doubt designedly, they have spent their space on recording the greatest possible number of jingles, nicknames, synonyms, customs and conundrums for posterity, rather than trying to draw conclusions from them. The trouble is their material is not sufficiently interesting to stand by itself. To me it demonstrated that on the whole children are quite as boring and nearly as unpleasant as I remember them. To read the chapter 'Wit and Repartee' is to live again those appalling half-hours in playground, corridor or cloakroom when the feeble backchat almost suffocated one by its staleness. And since the authors assure us that they are not concerned with the delinquent, the verses called 'Today's Menu' ('Scab and matter custard ...') must not be regarded as untypical.

Nevertheless, I can't quite subscribe to the Opies' delineation of all children as an entirely separate race of quasi-savages, or not without some reservations. All their examples are collected from non-private, non-fee-paying schools, which means in practice that, like most folklorists, they are sampling from the least literate section of the community: the title of the book should be modified mentally in consequence. Again, I cannot accept unquestioned the authors' remark that '[childhood] is as unnoticed by the sophisticated world, and quite as little affected by it, as is the culture of some dwindling aboriginal tribe living out its helpless existence in the hinterland of a native reserve.' Children copy adults ceaselessly. In fact, it might be argued that both streams of oral lore, topical and traditional, are largely cast-offs from the grown-up world. The fact that children cross their legs in examinations for luck like eighteenth-century gamblers suggests that customs and superstitions persist in childhood long after maturity has abandoned them. Already we are beginning to call Christmas 'the children's festival'. How long will 'Here's the Bible open, Here's the Bible shut, if I don't tell the truth,' etc, continue to be chanted after the present legal form of taking the oath has vanished?

Above all, though, children are linked to adults by the simple fact that they are in process of turning into them. For this they may be forgiven much. Children are bound to be inferior to adults, or there is no incentive to grow up. But there has been much agitation recently about whether grown-ups themselves are deteriorating by reason of addiction to mass media, loss of traditional self-amusements, and the like. To me (if I may quote *After Many a Summer*) 'they look as if they were having a pretty good time, in their own way of course,' but the question may be asked whether there is any evidence in this book that the hypothetical blight is spreading backwards into childhood. It is not an easy one to answer. During the time that the Opies were collecting their material, television licences increased from 800,000 to 8,000,000. It is possible, therefore, that the lore they record will soon be largely obsolete. On the other hand, we cannot be certain of this until a comparable investigation is made fifty or a hundred years hence. It is likely that the whole traditional corpus is expiring at a slower rate than we can measure, just as it has among adults, and if this is so many will regret it. But I do not think it can be said to matter seriously provided childhood retains the vitality to convert and adapt new material to its obscure and secret ends. Norman Douglas took a pessimistic view of the future: 'the standardisation of youth proceeds endlessly.' The Opies do not: 'we cannot but feel that [this] is a virile generation.' The reader is left feeling, in short, that the old rhymes are not so marvellous that it matters if children forget to sing them. The important thing is that they should sing. And there is no evidence here that they are forgetting how to do that.

1959

REYNER BANHAM

The Crisp at the Crossroads

Aᴍᴏɴɢ the triumphs of progressive technology that Luddites and Leavisites alike have lately been spared is the toad-in-the-hole-flavoured crisp with a hole in it! Scores of other equally nutty (literally so, in some cases) flavours for the familiar old potato crisp have been mooted lately. Don't imagine that cheese-and-onion or barbecue-bacon exhausts the ingenuity of the industry, now that flavours are sprinkled on as a dry powder before the crisp is cooked.

Not all the possibilities have got beyond idle brainstorming (blackberry-and-apple? smoked salmon? how about *crème de menthe*?), but the story about the party that was stoned right out the window at Redondo Beach by LSD crisps is true, apparently, and one memorable week in Montreal I breakfasted (for reasons beyond human belief) on Boursin, instant coffee and rainbow crisps. Rainbow? You bet; red, blue, green, white, natural, and all tasting like cardboard.

The potato crisp is at the crossroads, and to judge by the sundry aromas arising from the secret kitchens of R-and-D departments, the industry can't guess which way it will go. Whoever guesses right could make a real killing. The value of Britain's annual crop has doubled since 1964 and now stands around 62 million quid—crunch that! In the process, Smiths, with only 30-odd per cent of the market left, has had to concede victory to Golden Wonder, with over 45; and the old basic salted crisp has lost almost half the market to new fancy flavours.

It's been a real stir-up, and it has consequences for the arts of design, because the old basic crisp they still eat down at the Rovers' Return, even if it is doomed elsewhere, was unique among the works of man in being as neatly related to its pack as was the egg to its shell. Different kind of neat, but almost as instructive to look at.

For a start, it is an inherently unconformable shape. The cooking process that makes it crisp also crumples it into rigid but irregular corrugations. There is no way to make it pack closely with its neighbours, so that any quantity of crisps must also contain an even larger quantity of air. Bulk for bulk, as packed, crisps contain even less weight of food than cornflakes, and thus give conviction to the myth that they just *can't* be fattening.

This sense that there is no diet-busting substance in crisps is reinforced by their performance in the mouth. Apply tooth-pressure and you get deafening action; bite again and there's nothing left. It's a food that vanishes in the mouth, so, I mean, it can't be fattening, can it? It certainly isn't satisfying in any nor-mal food sense; the satisfactions of crisps, over and above the sting of flavour, are audio-masticatory—lots of response for little substance.

The pack is analogous in its performance. Keeping the crisp means keeping water-vapour away from it; and until recently the only cheap, paper-tape flexible materials that formed effective vapour-barriers were comparatively brittle and *in*flexible, and thus produced a lot of crinkling sound effects whenever they were handled. What with the crisps rattling about inside, and the pack crackling and rustling outside, you got an audio signal dis-tinctive enough to be picked up by childish ears at 200 or 300 yards.

But more than this, the traditional method of sealing off the top of the pack produced a closure that could only be opened destructively and couldn't be resealed. So eating crisps was an in-vitation to product-sadism. You tear the pack open to get at the contents, rip it further to get at the corner-lurkers in the bottom, and then crush it crackling-flat in the fist before throwing it away. It's the first and most familiar of Total-Destructo products and probably sublimates more aggression per annum than any quantity of dramaturgical catharsis.

However you look at it, or listen to it, the total relationship of crisp to package is a deafening symbiosis that comes near to per-fection. And it's the kind of perfection that not even a towering genius could have invented from scratch. The neatness of the relationship has almost a vernacular quality about it, like some survivor from a lost golden age of peasant technologies that have matured long in the wood and hand: the oar, the axe, the rolling-pin. But in the crisp's case, the golden age was recent, a threshold between two ages of industrial technology—the transitional period between the grinding poverty that nineteenth-century social moralists found so repugnant and the new affluence that twentieth-century social moralists find so repugnant.

In the history of rising genteelism, the potato crisp is a key piece of the technology that enabled a woman to go into a pub and still emerge a lady. By asking for 'a bag of crisps, instead', a lady could avoid having another drink without dropping out of a round; could participate in the social rituals of receiving goodies from the bar, or passing them on to others, without finding herself confronted with yet another jar of ultimate senselessness—and, above all, without incurring the accusations of airs, graces and 'going all la-di-dah' that would follow if she ordered Babycham.

Now that categories like 'woman' and 'lady' are no longer distinguishable, or worth distinguishing, when any bird can share a joint or a bottle of plonk with any bloke without being mistaken for what she wishes she wasn't, the ancient function of the crisp is crumbling almost as fast as the crisp itself. For, in its new functions, the crisp just does not possess the mechanical strength it needs.

Next time you go to one of these functions, and find yourself thinking that a splodge of onion dip would go nicely with the glass of foaming Silesian sherry the dean has just pressed on you, have a good look at the contents of the bowl of dip. The chances are that you will see a surface so pocked over by the shards of wrecked crisps that it looks like the Goodwin Sands during the Battle of Britain. For every Smith or Golden Wonder that actually comes up with a scoop of dip, four or five will die the death between the fingers of the would-be dipper.

Right now, the British crisp certainly isn't keeping up with technological adventures abroad—those big, white, symmetrical ones from Holland, for instance. They may look, and taste, like foamed polystyrene, but they have the structural strength to lift a lot of dip. Even so they are a poor shape for the job, compared with current models of the American 'taco chip'.

I don't know how long the taco chip has been around. I didn't really become conscious of it until I was doing my own housekeeping in Pasadena last winter. But its mastery of the problems of a savoury-dipping culture was immediately apparent. A cheerfully synthetic product, not notably derived from sliced spud, its flavour patently sprinkled on, not bred in, the taco chip comes on the equilateral-triangle format, about two inches on the side, handsomely tanned and only slightly wrinkled.

Not only is it better-looking than the pallid old spud-based product, but it also has the mechanical strength (without being inedibly tough) to take advantage of the excellent dipping performance it derives from its sharp, 60-degree corners. All in all, the taco chip is a classic US 'engineering solution', a worthy manifestation of the spirit that puts men on the moon and Mace in the campuses.

The surest indication of the crisp's escape from the pub-and-chara context of prole-cult is the fact that something like 40 per cent of the product is now bought as part of the weekly groceries in suitable family-economy packs, while licensed premises now handle only 10 per cent of the trade. Scotland, apparently, is still where the bulk of Britain's crisps are eaten—a surprising statistic since it is difficult to relate the crisp's low ratio of substance to side-effects with the Scots' alleged sensitivity to value for money.

For, if food value were the criterion for purchase, the crisp would be unsaleable. It isn't even an economical way of buying calories, compared with, say, porridge. Crisps, taco chips and their likes must be seen as ritual substitutes for solid food, the kind of token victuals that ancient peoples buried with their dead, the nutriment of angels rather than mortal flesh. In fact, the more I think about the comparison with porridge, the more worried I get. Could it be that the world's greatest anthropologue has got his polarities wrong, and should have written, say, *The Boiled and the Crisp*?

1970

Stranger in the Village

From all available evidence no black man had ever set foot in this tiny Swiss village before I came. I was told before arriving that I would probably be a 'sight' for the village; I took this to mean that people of my complexion were rarely seen in Switzerland, and also that city people are always something of a 'sight' outside of the city. It did not occur to me—possibly because I am an American—that there could be people anywhere who had never seen a Negro.

It is a fact that cannot be explained on the basis of the inaccessibility of the village. The village is very high, but it is only four hours from Milan and three hours from Lausanne. It is true that it is virtually unknown. Few people making plans for a holiday would elect to come here. On the other hand, the villagers are able, presumably, to come and go as they please— which they do: to another town at the foot of the mountain, with a population of approximately five thousand, the nearest place to see a movie or go to the bank. In the village there is no movie house, no bank, no library, no theater; very few radios, one jeep, one station wagon; and, at the moment, one typewriter, mine, an invention which the woman next door to me here had never seen. There are about six hundred people living here, all Catholic—I conclude this from the fact that the Catholic church is open all year round, whereas the Protestant chapel, set off on a hill a little removed from the village, is open only in the summertime when the tourists arrive. There are four or five hotels, all closed now, and four or five *bistros*, of which, however, only two do any business during the winter. These two do not do any great deal, for life in the village seems to end around nine or ten o'clock. There are a few stores, butcher, baker, *épicerie,* a hardware store, and a money-changer—who cannot change travelers' checks, but must send them down to the bank, an operation which takes two or three days. There is something called the *Bal-*

let Haus, closed in the winter and used for God knows what, certainly not ballet, during the summer. There seems to be only one schoolhouse in the village, and this for the quite young children; I suppose this to mean that their older brothers and sisters at some point descend from these mountains in order to complete their education—possibly, again, to the town just below. The landscape is absolutely forbidding, mountains towering on all four sides, ice and snow as far as the eye can reach. In this white wilderness, men and women and children move all day, carrying washing, wood, buckets of milk or water, sometimes skiing on Sunday afternoons. All week long boys and young men are to be seen shoveling snow off the rooftops, or dragging wood down from the forest in sleds.

The village's only real attraction, which explains the tourist season, is the hot spring water. A disquietingly high proportion of these tourists are cripples, or semicripples, who come year after year—from other parts of Switzerland, usually—to take the waters. This lends the village, at the height of the season, a rather terrifying air of sanctity, as though it were a lesser Lourdes. There is often something beautiful, there is always something awful, in the spectacle of a person who has lost one of his faculties, a faculty he never questioned until it was gone, and who struggles to recover it. Yet people remain people, on crutches or indeed on deathbeds; and wherever I passed, the first summer I was here, among the native villagers or among the lame, a wind passed with me—of astonishment, curiosity, amusement, and outrage. That first summer I stayed two weeks and never intended to return. But I did return in the winter, to work; the village offers, obviously, no distractions whatever and has the further advantage of being extremely cheap. Now it is winter again, a year later, and I am here again. Everyone in the village knows my name, though they scarcely ever use it, knows that I come from America—though, this, apparently, they will never really believe: black men come from Africa—and everyone knows that I am the friend of the son of a woman who was born here, and that I am staying in their chalet. But I remain as much a stranger today as I was the first day I arrived, and the children shout *Neger! Neger!* as I walk along the streets.

It must be admitted that in the beginning I was far too shocked to have any real reaction. In so far as I reacted at all, I reacted by trying to be pleasant—it being a great part of the American Negro's education (long before he goes to school) that he must make people 'like' him. This smile-and-the-world-smiles-with-you routine worked about as well in this situation as it had in the situation for which it was designed, which is to say that it did not work at all. No one, after all, can be liked whose human weight and complexity cannot be, or has not been, admitted. My smile was simply another unheard-of phenomenon which allowed them to see my teeth—they did not, really, see my smile and I began to think that, should I take to snarling, no one would notice any difference. All of the physical characteristics of the Negro which had caused me, in America, a very different and almost forgotten pain were nothing less than miraculous—or infernal—in the eyes of the village people. Some thought my hair was the color of tar, that it had the texture of wire, or the texture of cotton. It was jocularly suggested that I might let it all grow long and make myself a winter coat. If I sat in the sun for more than five minutes some daring creature was certain to come along and gingerly put his fingers on my hair, as though he were afraid of an electric shock, or put his hand on my hand, astonished that the color did not rub off. In all of this, in which it must be conceded there was the charm of genuine wonder and in which there was certainly no element of intentional unkindness, there was yet no suggestion that I was human: I was simply a living wonder.

I knew that they did not mean to be unkind, and I know it now; it is necessary, nevertheless, for me to repeat this to myself each time that I walk out of the chalet. The children who shout *Neger!* have no way of knowing the echoes this sound raises in me. They are brimming with good humor and the more daring swell with pride when I stop to speak with them. Just the same, there are days when I cannot pause and smile, when I have no heart to play with them; when, indeed, I mutter sourly to myself, exactly as I muttered on the streets of a city these children have never seen, when I was no bigger than these children are now: *Your* mother *was a nigger*. Joyce is right about history being a nightmare—but it may be the nightmare from which no one *can* awaken. People are trapped in history and history is trapped in them.

There is a custom in the village—I am told it is repeated

in many villages—of 'buying' African natives for the purpose of converting them to Christianity. There stands in the church all year round a small box with a slot for money, decorated with a black figurine, and into this box the villagers drop their francs. During the *carnaval* which precedes Lent, two village children have their faces blackened—out of which bloodless darkness their blue eyes shine like ice—and fantastic horsehair wigs are placed on their blond heads; thus disguised, they solicit among the villagers for money for the missionaries in Africa. Between the box in the church and the blackened children, the village 'bought' last year six or eight African natives. This was reported to me with pride by the wife of one of the *bistro* owners and I was careful to express astonishment and pleasure at the solicitude shown by the village for the souls of black folk. The *bistro* owner's wife beamed with a pleasure far more genuine than my own and seemed to feel that I might now breathe more easily concerning the souls of at least six of my kinsmen.

I tried not to think of these so lately baptized kinsmen, of the price paid for them, or the peculiar price they themselves would pay, and said nothing about my father, who having taken his own conversion too literally never, at bottom, forgave the white world (which he described as heathen) for having saddled him with a Christ in whom, to judge at least from their treatment of him, they themselves no longer believed. I thought of white men arriving for the first time in an African village, strangers there, as I am a stranger here, and tried to imagine the astounded populace touching their hair and marveling at the color of their skin. But there is a great difference between being the first white man to be seen by Africans and being the first black man to be seen by whites. The white man takes the astonishment as tribute, for he arrives to conquer and to convert the natives, whose inferiority in relation to himself is not even to be questioned; whereas I, without a thought of conquest, find myself among a people whose culture controls me, has even, in a sense, created me, people who have cost me more in anguish and rage than they will ever know, who yet do not even know of my existence. The astonishment with which I might have greeted them, should they have stumbled into my African village a few hundred years ago, might have rejoiced their hearts. But the astonishment with which they greet me today can only poison mine.

And this is so despite everything I may do to feel differently, despite my friendly conversations with the *bistro* owner's wife, despite their three-year-old son who has at last become my friend, despite the *saluts* and *bonsoirs* which I exchange with people as I walk, despite the fact that I know that no individual can be taken to task for what history is doing, or has done. I say that the culture of these people controls me—but they can scarely be held responsible for European culture. America comes out of Europe, but these people have never seen America, nor have most of them seen more of Europe than the hamlet at the foot of their mountain. Yet they move with an authority which I shall never have; and they regard me, quite rightly, not only as a stranger in their village but as a suspect latecomer, bearing no credentials, to everything they have—however unconsciously—inherited.

For this village, even were it incomparably more remote and incredibly more primitive, is the West, the West onto which I have been so strangely grafted. These people cannot be, from the point of view of power, strangers anywhere in the world; they have made the modern world, in effect, even if they do not know it. The most illiterate among them is related, in a way that I am not, to Dante, Shakespeare, Michelangelo, Aeschylus, Da Vinci, Rembrandt, and Racine; the cathedral at Chartres says something to them which it cannot say to me, as indeed would New York's Empire State Building, should anyone here ever see it. Out of their hymns and dances come Beethoven and Bach. Go back a few centuries and they are in their full glory—but I am in Africa, watching the conquerors arrive.

The rage of the disesteemed is personally fruitless, but it is also absolutely inevitable; this rage, so generally discounted, so little understood even among the people whose daily bread it is, is one of the things that makes history. Rage can only with difficulty, and never entirely, be brought under the domination of the intelligence and is therefore not susceptible to any arguments whatever. This is a fact which ordinary representatives of the *Herrenvolk*, having never felt this rage and being unable to imagine it, quite fail to understand. Also, rage cannot be hidden, it can only be dissembled. This dissembling deludes the thoughtless, and strengthens rage and adds, to rage, contempt. There are, no doubt, as many ways of coping with the resulting complex of tensions as there are black men in the world, but no black

man can hope ever to be entirely liberated from this internal warfare—rage, dissembling, and contempt having inevitably accompanied his first realization of the power of white men. What is
crucial here is that, since white men represent in the black man's
world so heavy a weight, white men have for black men a reality
which is far from being reciprocal; and hence all black men have
toward all white men an attitude which is designed, really, either
to rob the white man of the jewel of his naïveté, or else to make
it cost him dear.

The black man insists, by whatever means he finds at his disposal, that the white man cease to regard him as an exotic rarity
and recognize him as a human being. This is a very charged and
difficult moment, for there is a great deal of will power involved
in the white man's naïveté. Most people are not naturally reflective any more than they are naturally malicious, and the white
man prefers to keep the black man at a certain human remove
because it is easier for him thus to preserve his simplicity and
avoid being called to account for crimes committed by his
forefathers, or his neighbors. He is inescapably aware, nevertheless, that he is in a better position in the world than black men
are, nor can he quite put to death the suspicion that he is hated
by black men therefore. He does not wish to be hated, neither
does he wish to change places, and at this point in his uneasiness
he can scarcely avoid having recourse to those legends which
white men have created about black men, the most usual effect of
which is that the white man finds himself enmeshed, so to speak,
in his own language which describes hell, as well as the attributes
which lead one to hell, as being as black as night.

Every legend, moreover, contains its residuum of truth, and
the root function of language is to control the universe by
describing it. It is of quite considerable significance that black
men remain, in the imagination, and in overwhelming numbers
in fact, beyond the disciplines of salvation; and this despite the
fact that the West has been 'buying' African natives for centuries. There is, I should hazard, an instantaneous necessity to
be divorced from this so visibly unsaved stranger, in whose
heart, moreover, one cannot guess what dreams of vengeance are
being nourished; and, at the same time, there are few things on
earth more attractive than the idea of the unspeakable liberty
which is allowed the unredeemed. When, beneath the black

mask, a human being begins to make himself felt one cannot excape a certain awful wonder as to what kind of human being it is. What one's imagination makes of other people is dictated, of course, by the laws of one's own personality and it is one of the ironies of black–white relations that, by means of what the white man imagines the black man to be, the black man is enabled to know who the white man is.

I have said, for example, that I am as much a stranger in this village today as I was the first summer I arrived, but this is not quite true. The villagers wonder less about the texture of my hair than they did then, and wonder rather more about me. And the fact that their wonder now exists on another level is reflected in their attitudes and in their eyes. There are the children who make those delightful, hilarious, sometimes astonishingly grave overtures of friendship in the unpredictable fashion of children; other children, having been taught that the devil is a black man, scream in genuine anguish as I approach. Some of the older women never pass without a friendly greeting, never pass, indeed, if it seems that they will be able to engage me in conversation; other women look down or look away or rather contemptuously smirk. Some of the men drink with me and suggest that I learn how to ski—partly, I gather, because they cannot imagine what I would look like on skis—and want to know if I am married, and ask questions about my *métier*. But some of the men have accused *le sale nègre*—behind my back—of stealing wood and there is already in the eyes of some of them that peculiar, intent, paranoiac malevolence which one sometimes surprises in the eyes of American white men when, out walking with their Sunday girl, they see a Negro male approach.

There is a dreadful abyss between the streets of this village and the streets of the city in which I was born, between the children who shout *Neger!* today and those who shouted *Nigger!* yesterday—the abyss is experience, the American experience. The syllable hurled behind me today expresses, above all, wonder: I am a stranger here. But I am not a stranger in America and the same syllable riding on the American air expresses the war my presence has occasioned in the American soul.

For this village brings home to me this fact: that there was a day, and not really a very distant day, when Americans were scarcely Americans at all but discontented Europeans, facing a great unconquered continent and strolling, say, into a market-place and seeing black men for the first time. The shock this spectacle afforded is suggested, surely, by the promptness with which they decided that these black men were not really men but cattle. It is true that the necessity on the part of the settlers of the New World of reconciling their moral assumptions with the fact—and the necessity—of slavery enhanced immensely the charm of this idea, and it is also true that this idea expresses, with a truly American bluntness, the attitude which to varying extents all masters have had toward all slaves.

But between all former slaves and slave-owners and the drama which begins for Americans over three hundred years ago at Jamestown, there are at least two differences to be observed. The American Negro slave could not suppose, for one thing, as slaves in past epochs had supposed and often done, that he would ever be able to wrest the power from his master's hands. This was a supposition which the modern era, which was to bring about such vast changes in the aims and dimensions of power, put to death; it only begins, in unprecedented fashion, and with dreadful implications, to be resurrected today. But even had this supposition persisted with undiminished force, the American Negro slave could not have used it to lend his condition dignity, for the reason that this supposition rests on another: that the slave in exile yet remains related to his past, has some means—if only in memory—of revering and sustaining the forms of his former life, is able, in short, to maintain his identity.

This was not the case with the American Negro slave. He is unique among the black men of the world in that his past was taken from him, almost literally, at one blow. One wonders what on earth the first slave found to say to the first dark child he bore. I am told that there are Haitians able to trace their ancestry back to African kings, but any American Negro wishing to go back so far will find his journey through time abruptly arrested by the signature on the bill of sale which served as the entrance paper for his ancestor. At the time—to say nothing of the circumstances—of the enslavement of the captive black man who was to become the American Negro, there was not the remotest possibility that he would ever take power from his master's hands. There was no reason to suppose that his situation would ever change, nor was there, shortly, anything to indicate that his situation had ever been different. It was his necessity, in the words of E. Franklin Frazier, to find a 'motive for living under American culture or die.' The identity of the American Negro comes out of this extreme situation, and the evolution of this identity was a source of the most intolerable anxiety in the minds and the lives of his masters.

For the history of the American Negro is unique also in this: that the question of his humanity, and of his rights therefore as a human being, became a burning one for several generations of Americans, so burning a question that it ultimately became one of those used to divide the nation. It is out of this argument that the venom of the epithet *Nigger!* is derived. It is an argument which Europe has never had, and hence Europe quite sincerely fails to understand how or why the argument arose in the first place, why its effects are so frequently disastrous and always so unpredictable, why it refuses until today to be entirely settled. Europe's black possessions remained—and do remain—in Europe's colonies, at which remove they represented no threat whatever to European identity. If they posed any problem at all for the European conscience, it was a problem which remained comfortingly abstract: in effect, the black man, *as a man*, did not exist for Europe. But in America, even as a slave, he was an inescapable part of the general social fabric and no American could escape having an attitude toward him. Americans attempt until today to make an abstraction of the Negro, but the very nature of these abstractions reveals the tremendous effects the presence of the Negro has had on the American character.

When one considers the history of the Negro in America it is of the greatest importance to recognize that the moral beliefs of a person, or a people, are never really as tenuous as life—which is not moral—very often causes them to appear; these create for them a frame of reference and a necessary hope, the hope being that when life has done its worst they will be enabled to rise above themselves and to triumph over life. Life would scarcely be bearable if this hope did not exist. Again, even when the worst has been said, to betray a belief is not by any means to have put oneself beyond its power; the betrayal of a belief is not the same thing as ceasing to believe. If this were not so there would be no moral standards in the world at all. Yet one must also recognize that morality is based on ideas and that all ideas are danger-

ous—dangerous because ideas can only lead to action and where
the action leads no man can say. And dangerous in this respect:
that confronted with the impossibility of remaining faithful to
one's beliefs, and the equal impossibility of becoming free of
them, one can be driven to the most inhuman excesses. The ideas
on which American beliefs are based are not, though Americans
often seem to think so, ideas which originated in America. They
came out of Europe. And the establishment of democracy on the
American continent was scarcely as radical a break with the past
as was the necessity, which Americans faced, of broadening this
concept to include black men.

This was, literally, a hard necessity. It was impossible, for one
thing, for Americans to abandon their beliefs, not only because
these beliefs alone seemed able to justify the sacrifices they had
endured and the blood that they had spilled, but also because
these beliefs afforded them their only bulwark against a moral
chaos as absolute as the physical chaos of the continent it was
their destiny to conquer. But in the situation in which Americans found themselves, these beliefs threatened an idea which,
whether or not one likes to think so, is the very warp and woof
of the heritage of the West, the idea of white supremacy.

Americans have made themselves notorious by the shrillness
and the brutality with which they have insisted on this idea, but
they did not invent it; and it has escaped the world's notice that
those very excesses of which Americans have been guilty imply a
certain, unprecedented uneasiness over the idea's life and power,
if not, indeed, the idea's validity. The idea of white supremacy
rests simply on the fact that white men are the creators of
civilization (the present civilization, which is the only one that
matters; all previous civilizations are simply 'contributions' to
our own) and are therefore civilization's guardians and defenders. Thus it was impossible for Americans to accept the black
man as one of themselves, for to do so was to jeopardize their
status as white men. But not so to accept him was to deny his
human reality, his human weight and complexity, and the strain
of denying the overwhelmingly undeniable forced Americans
into rationalizations so fantastic that they approached the patho-
logical.

At the root of the American Negro problem is the necessity of
the American white man to find a way of living with the Negro

in order to be able to live with himself. And the history of this problem can be reduced to the means used by Americans—lynch law and law, segregation and legal acceptance, terrorization and concession—either to come to terms with this necessity, or to find a way around it, or (most usually) to find a way of doing both these things at once. The resulting spectacle, at once foolish and dreadful, led someone to make the quite accurate observation that 'the Negro-in-America is a form of insanity which overtakes white men.'

In this long battle, a battle by no means finished, the unforeseeable effects of which will be felt by many future generations, the white man's motive was the protection of his identity; the black man was motivated by the need to establish an identity. And despite the terrorization which the Negro in American endured and endures sporadically until today, despite the cruel and totally inescapable ambivalence of his status in his country, the battle for his identity has long ago been won. He is not a visitor to the West, but a citizen there, an American; as American as the Americans who despise him, the Americans who fear him, the Americans who love him—the Americans who became less than themselves, or rose to be greater than themselves by virtue of the fact that the challenge he represented was inescapable. He is perhaps the only black man in the world whose relationship to white men is more terrible, more subtle, and more meaningful than the relationship of bitter possessed to uncertain possessor. His survival depended, and his development depends, on his ability to turn his peculiar status in the Western world to his own advantage and, it may be, to the very great advantage of that world. It remains for him to fashion out of his experience that which will give him sustenance, and a voice.

The cathedral at Chartres, I have said, says something to the people of this village which it cannot say to me; but it is important to understand that this cathedral says something to me which it cannot say to them. Perhaps they are struck by the power of the spires, the glory of the windows; but they have known God, after all, longer than I have known him, and in a different way, and I am terrified by the slippery bottomless well to be found in the crypt, down which heretics were hurled to death, and by the obscene, inescapable gargoyles jutting out of the stone and seeming to say that God and the devil can never be divorced. I

doubt that the villagers think of the devil when they face a
cathedral because they have never been identified with the devil.
But I must accept the status which myth, if nothing else, gives
me in the West before I can hope to change the myth.

Yet, if the American Negro has arrived at his identity by
virtue of the absoluteness of his estrangement from his past,
American white men still nourish the illusion that there is some
means of recovering the European innocence, of returning to a
state in which black men do not exist. This is one of the greatest
errors Americans can make. The identity they fought so hard
to protect has, by virtue of that battle, undergone a change:
Americans are as unlike any other white people in the world as it
is possible to be. I do not think, for example, that it is too much
to suggest that the American vision of the world—which allows
so little reality, generally speaking, for any of the darker forces in
human life, which tends until today to paint moral issues in
glaring black and white—owes a great deal to the battle waged
by Americans to maintain between themselves and black men a
human separation which could not be bridged. It is only now
beginning to be borne in on us—very faintly, it must be admitted, very slowly, and very much against our will—that this
vision of the world is dangerously inaccurate, and perfectly useless. For it protects our moral high-mindedness at the terrible
expense of weakening our grasp of reality. People who shut their
eyes to reality simply invite their own destruction, and anyone
who insists on remaining in a state of innocence long after that
innocence is dead turns himself into a monster.

The time has come to realize that the interracial drama acted
out on the American continent has not only created a new black
man, it has created a new white man, too. No road whatever will
lead Americans back to the simplicity of this European village
where white men still have the luxury of looking on me as a
stranger. I am not, really, a stranger any longer for any American
alive. One of the things that distinguishes Americans from other
people is that no other people has ever been so deeply involved
in the lives of black men, and vice versa. This fact faced, with all
its implications, it can be seen that the history of the American
Negro problem is not merely shameful, it is also something of an
achievement. For even when the worst has been said, it must
also be added that the perpetual challenge posed by this problem

was always, somehow, perpetually met. It is precisely this black-white experience which may prove of indispensable value to us in the world we face today. This world is white no longer, and it will never be white again.

1953

Robert Graves and
the Twelve Caesars

TIBERIUS, Capri. Pool of water. Small children ... So far so good. One's laborious translation was making awful sense. Then ... Fish. Fish? The erotic mental image became surreal. Another victory for the Loeb Library's sly translator, J. C. Rolfe, who, correctly anticipating the pruriency of schoolboy readers, left Suetonius's gaudier passages in the hard original. One failed to crack those intriguing footnotes not because the syntax was so difficult (though it was not easy for students drilled in military rather than civilian Latin) but because the range of vice revealed was considerably beyond the imagination of even the most depraved schoolboy. There was a point at which one rejected one's own translation. Tiberius and the little fish, for instance.

Happily, we now have a full translation of the text, the work of Mr Robert Graves, who, under the spell of his Triple Goddess, has lately been retranslating the classics. One of his first tributes to her was a fine rendering of *The Golden Ass*; then Lucan's *Pharsalia*; then the *Greek Myths*, a collation aimed at rearranging the hierarchy of Olympus to afford his Goddess (the female principle) a central position at the expense of the male. (Beware Apollo's wrath, Graves: the 'godling' is more than front man for the 'Ninefold Muse-Goddess.') Now, as a diversion, Mr Graves has given us *The Twelve Caesars* of Suetonius in a good, dry, no-nonsense style; and, pleasantly enough, the Ancient Mother of Us All is remarkable only by her absence, perhaps a subtle criticism of an intensely masculine period in history.

Gaius Suetonius Tranquillus—lawyer and author of a dozen books, among them *Lives of Famous Whores* and *The Physical Defects of Mankind* (What was that about?)—worked for a time as private secretary to the Emperor Hadrian. Presumably it was during this period that he had access to the imperial archives, where he got the material for *The Twelve Caesars*, the only complete book of his to survive. Suetonius was born in AD 69, the year of the three Caesars Galba, Otho, Vitellius; and he grew up under the Flavians: Vespasian, Titus, Domitian, whom he deals with as contemporaries. He was also close enough in time to the first six Caesars to have known men who knew them intimately, at least from Tiberius on, and it is this place in time which gives such immediacy to his history.

Suetonius saw the world's history from 49 BC to AD 96 as the intimate narrative of twelve men wielding absolute power. With impressive curiosity he tracked down anecdotes, recording them dispassionately, despite a somewhat stylized reactionary bias. Like his fellow historians from Livy to the stuffy but interesting Dion Cassius, Suetonius was a political reactionary to whom the old Republic was the time of virtue and the Empire, implicitly, was not. But it is not for his political convictions that we read Suetonius. Rather, it is his gift for telling us what we want to know. I am delighted to read that Augustus was under five feet seven, blond, wore lifts in his sandals to appear taller, had seven birthmarks and weak eyes; that he softened the hairs of his legs with hot walnut shells, and liked to gamble. Or to learn that the droll Vespasian's last words were: 'Dear me, I must be turning into a god.' ('Dear me' being Graves for '*Vae*.') The stories, true or not, are entertaining, and when they deal with sex startling, even to a post-Kinseyan.

Gibbon, in his stately way, mourned that of the twelve Caesars only Claudius was sexually 'regular.' From the sexual opportunism of Julius Caesar to the sadism of Nero to the doddering pederasty of Galba, the sexual lives of the Caesars encompassed every aspect of what our post-medieval time has termed 'sexual abnormality.' It would be wrong, however, to dismiss, as so many commentators have, the wide variety of Caesarean sensuality as simply the viciousness of twelve abnormal men. They were, after all, a fairly representative lot. They differed from us—and their contemporaries—only in the fact of power, which made it possible for each to act out his most recondite sexual fantasies. This is the psychological fascination of Suetonius. What will men so placed do? The answer, apparently, is anything and everything. Alfred Whitehead once remarked that one got the essence of a culture not by those things which were said at the time but by those things which were *not* said, the underlying assumptions of the society, too obvious to be stated. Now it is an underlying assumption of twentieth-century America that human beings are either heterosexual or, through some arresting of normal psychic growth, homosexual, with very little traffic back and forth. To us, the norm is heterosexual; the family is central; all else is deviation, pleasing or not depending on one's own tastes and moral preoccupations. Suetonius reveals a very different world. His underlying assumption is that man is bisexual and that given complete freedom to love—or, perhaps more to the point in the case of the Caesars, to violate—others, he will do so, going blithely from male to female as fancy dictates. Nor is Suetonius alone in this assumption of man's variousness. From Plato to the rise of Pauline Christianity, which tried to put the lid on sex, it is explicit in classical writing. Yet to this day Christian, Freudian and Marxian commentators have all decreed or ignored this fact of nature in the interest each of a patented approach to the Kingdom of Heaven. It is an odd experience for a contemporary to read of Nero's simultaneous passion for both a man and a woman. Something seems wrong. It must be one or the other, not both. And yet this sexual eclecticism recurs again and again. And though some of the Caesars quite obviously preferred women to men (Augustus had a particular penchant for Nabokovian nymphets), their sexual crisscrossing is extraordinary in its lack of pattern. And one suspects that despite the stern moral legislation of our own time human beings are no different. If nothing else, Dr Kinsey revealed in his dogged, arithmetical way that we are all a good deal less predictable and bland than anyone had suspected.

One of the few engaging aspects of the Julio-Claudians was authorship. They all wrote; some wrote well. Julius Caesar, in addition to his account of that famed crusade in Gaul, wrote an *Oedipus*. Augustus wrote an *Ajax*, with some difficulty. When asked by a friend what his *Ajax* had been up to lately, Augustus sighed: 'He has fallen not on his sword, but wiped himself out on my sponge.' Tiberius wrote an *Elegy on the Death of Julius Caesar*. The scatterbrained Claudius, a charmingly dim prince, was a devoted pedant who tried to reform the alphabet. He was also among the first to have a serious go at Etruscan history. Nero of course is remembered as a poet. Julius Caesar and Augustus were distinguished prose writers; each preferred plain old-fashioned Latin. Augustus particularly disliked what he called the 'Asiatic' style, favored by, among others, his rival Marc Antony, whose speeches he found imprecise and 'stinking of far-fetched phrases.'

Other than the fact of power, the twelve Caesars as men had little in common with one another. But that little was significant: a fear of the knife in the dark. Of the twelve, eight (perhaps nine) were murdered. As Domitian remarked not long before he himself was struck down: 'Emperors are necessarily wretched men since only their assassination can convince the public that the conspiracies against their lives are real.' In an understandable attempt to outguess destiny, they studied omens, cast horoscopes, and analyzed dreams (they were ingenious symbolists, anticipating Dr Freud, himself a Roman buff). The view of life from Palatine Hill was not comforting, and though none of the Caesars was religious in our sense of the word, all inclined to the Stoic. It was Tiberius, with characteristic bleakness, who underscored their dangerous estate when he declared that it was Fate, not the gods, which ordered the lives of men.

Yet what, finally, was the effect of absolute power on twelve representative men? Suetonius makes it quite plain: disastrous. Caligula was certifiably mad. Nero, who started well, became progressively irrational. Even the stern Tiberius's character weakened. In fact, Tacitus, in covering the same period as Suetonius, observes: 'Even after his enormous experience of public affairs, Tiberius was ruined and transformed by the violent influence of absolute power.' Caligula gave the game away when he told a critic, 'Bear in mind that I can treat anyone exactly as I please.' And that cruelty which is innate in human beings, now given the opportunity to use others as toys, flowered monstrously in the Caesars. Suetonius's case history (and it is precisely that) of Domitian is particularly fascinating. An intelligent man of some charm, trained to govern, Domitian when he first succeeded to the Principate contented himself with tearing the wings off flies, an infantile pastime which gradually palled until, inevitably, for flies he substituted men. His favorite game was to talk gently of mercy to a nervous victim; then, once all fears had

been allayed, execute him. Nor were the Caesars entirely unob-
jective about their bizarre position. There is an oddly revealing
letter of Tiberius to a Senate which had offered to ensure in
advance approbation of all his future deeds. Tiberius declined
the offer: 'So long as my wits do not fail me, you can count on
the consistency of my behavior; but I should not like you to
set the precedent of binding yourselves to approve a man's every
action; for what if something happened to alter that man's
character?' In terror of their lives, haunted by dreams and
omens, giddy with dominion, it is no wonder that actual insanity
was often the Caesarean refuge from a reality so intoxicating.

The unifying *Leitmotiv* in these lives is Alexander the Great.
The Caesars were fascinated by him. He was their touchstone of
greatness. The young Julius Caesar sighed enviously at his tomb.
Augustus had the tomb opened and stared long at the conqueror's face. Caligula stole the breastplate from the corpse and
wore it. Nero called his guard the 'Phalanx of Alexander the
Great.' And the significance of this fascination? Power for the
sake of power. Conquest for the sake of conquest. Earthly
dominion as an end in itself: no Utopian vision, no dissembling,
no hypocrisy. I knock you down; now *I* am king of the castle.
Why should young Julius Caesar be envious of Alexander? It
does not occur to Suetonius to explain. He assumes that *any*
young man would like to conquer the world. And why did
Julius Caesar, a man of first-rate mind, want the world? Simply,
to have it. Even the resulting Pax Romana was not a calculated
policy but a fortunate accident. Caesar and Augustus, the makers
of the Principate, represent the naked will to power for its own
sake. And though our own society has much changed from the
Roman (we may point with somber pride to Hitler and Stalin,
who lent a real Neronian hell to our days), we have, nevertheless, got so into the habit of dissembling motives, of denying
certain dark constants of human behavior, that it is difficult to
find a reputable American historian who will acknowledge the
crude fact that a Franklin Roosevelt, say, wanted to be President
merely to wield power, to be famed and to be feared. To learn
this simple fact one must wade through a sea of evasions: history
as sociology, leaders as teachers, bland benevolence as a motive
force, when, finally, power *is* an end to itself, and the instinctive
urge to prevail the most important single human trait, the

necessary force without which no city was built, no city destroyed. Yet many contemporary sociologists and religionists turned historians will propose, quite seriously: If there had not been a Julius Caesar then the *Zeitgeist* would have provided another like him, even though it is quite evident that had this particular Caesar not existed no one would have dared invent him. World events are the work of individuals whose motives are often frivolous, even casual. Had Claudius not wanted an easy conquest so that he might celebrate a triumph at Rome, Britain would not have been conquered in AD 44. If Britain had not been colonized in the first century ... the chain of causality is plain.

One understands of course why the role of the individual in history is instinctively played down by a would-be egalitarian society. We are, quite naturally, afraid of being victimized by reckless adventurers. To avoid this we have created the myth of the ineluctable mass ('other-directedness') which governs all. Science, we are told, is not a matter of individual inquiry but of collective effort. Even the surface storminess of our elections disguises a fundamental indifference to human personality: if not this man, then that one; it's all the same, life will go on. Up to a point there is some virtue in this; and though none can deny that there is a prevailing grayness in our placid land, it is certainly better to be non-ruled by mediocrities than enslaved by Caesars. But to deny the dark nature of human personality is not only fatuous but dangerous. For in our insistence on the surrender of private will ('inner-directedness') to a conception of the human race as some teeming bacteria in the stream of time, unaffected by individual deeds, we have been made vulnerable not only to boredom, to that sense of meaninglessness which more than anything else is characteristic of our age, but vulnerable to the first messiah who offers the young and bored some splendid prospect, some Caesarean certainty. That is the political danger, and it is a real one.

Most of the world today is governed by Caesars. Men are more and more treated as things. Torture is ubiquitous. And, as Sartre wrote in his preface to Henri Alleg's chilling book about Algeria, 'Anybody, at any time, may equally find himself victim or executioner.' Suetonius, in holding up a mirror to those

Caesars of diverting legend, reflects not only them but ourselves: half-tamed creatures, whose great moral task it is to hold in balance the angel and the monster within—for we are both, and to ignore this duality is to invite disaster.

1959

La Paz

SOUTHWARDS from the glistening steel-blue Titicaca runs the highway through the Bolivian altiplano, leaving the Peruvian highlands behind. To the east stand the splendours of the Andean cordillera, rank upon rank of noble snow-peaks, but the road passes through a landscape more lunar than celestial, an arid, drear, friendless kind of country, fourteen thousand feet above the sea. It is littered with the poor mud huts of the Aymara Indians, and the piles of stones they have scraped and scrabbled from their miserable soil, and sometimes you meet a peasant with his donkeys or his llamas, and sometimes you set the dust flying in an adobe village, and sometimes you see far away across the wilderness some solitary Indian woman, like a huddled witch on a moor, hastening bent-back across the rubble.

For sixty miles the road plods on through this monotony, and then it falls over a precipice. Suddenly it crosses the lip of the high plateau and tumbles helter-skelter, lickety-split into a chasm: and as you slither down the horse-shoe bends you see in the ravine below you, secreted in a fold of the massif, the city of La Paz. Its red roofs and mud huts pile up against the canyon walls and spill away into the river valley below. All around it is the immensity of the altiplano, and high above it to the south meditates the lovely white mountain called Illimani, where the royal condor of Inca legend folded its great wings in sleep. La Paz is the highest of the world's big cities, at twelve thousand feet. It is a tumultuous, feverish, often maddening, generally harum-scarum kind of place: but nobody with an eye to country or a taste for drama could fail to respond to its excitements, or resist the superb improbability of its situation.

After such an approach, in such an environment, you might reasonably expect to find, like the old voyagers, men with three eyes, or heads slung beneath their shoulders. Well, La Paz does its best. Consider a few simple facts about the city. Its atmosphere is so rarefied that virtually the only purpose of the single municipal fire engine is to squirt indelibly coloured water at political demonstrators. One of its liveliest institutions is a smugglers' trade union, the Syndicate of Frontier Merchants, and by far its best shopping centre is the *mercado negro*, a vast open-air emporium of illegally imported goods in which I once ran into a very respectable Customs official happily buying himself some illicit gramophone records. Half the women of this city wear bowler hats, reverently removing them when they enter a church, and among the old-fashioned cottage remedies readily available are foetus of llama, skin of cat, and horn of armadillo. La Paz has known 179 coups and revolutions since Bolivia became independent in 1825, and its currency is such that when I emptied my pockets there one day I found myself in possession of 683,700 bolivianos (I needed a million odd to pay my hotel bill, plus a few thousand, of course, for the bell-boy).

There, I am laughing at it, but only with wry affection, for I have seldom found a city more enthralling. It is anything but comic, beneath the veneer. It is pathetic, tragic, stimulating and menacing, and it still retains some of the savage glare and glitter that the Spaniards brought when they founded it four centuries ago. It is not in itself a beautiful place. Its few old buildings are swamped in half-hearted modernism, and all around it in the bowl of its canyon the Indians have built their terraced streets of mud and corrugated iron; but it possesses nevertheless, to an almost eccentric degree, the quality of individualism. It is a brittle metropolis. There is nowhere else much like it on the face of the earth, but if I had to find an analogy I would suggest some quivering desert city, Amman, say, or Kairouan, miraculously transplanted to a declivity in the Tibetan plateau.

It is a city of the Andes, and it is the swarming Andean Indians who nowadays set its style. The men are sometimes striking enough, with their ear-flapped woollen hats and Inca faces; but the women are fascinating beyond description. With their rakishly cocked bowler hats, their blinding blouses and skirts, their foaming flounces of petticoats, the babies like infant potentates upon their backs and the sandals made of old tyres

upon their feet—gorgeously accoutred and endlessly industrious,
plumed often with a handsome dignity and assurance, they give
to La Paz a flavour part gipsy, part coster, and all pungency.
There are, I swear it, no more magnificent ladies in the world
than the market-women of La Paz. Bowlers cockily atilt, like
bookies', they sit high on trestle tables in the covered market,
their bosoms grandly heaving beneath white overalls, their faces
at once lofty, cunning, all-observant and condescending; and
they are invested so closely by all their wares, so heaped about
with pineapples and bananas, so wallowing in papayas, mandarins, nuts and flowers, that they put old Marvell quite in the
shade, in the luscious sensuality of the lives they lead.

It is an Indian, highland turbulence that keeps this city tense and
wary, and makes the midnight curfew more the rule than the
exception. In the halls of Congress, beneath the painted scrutiny of Bolívar and Murillo, they are mostly Spanish faces, de-
claiming Latin polemics; but high in the balcony above the
debate, peering silently over the railing, are the dark, attentive,
enigmatic eyes of the Aymaras. In La Paz you feel everywhere
the rising awareness of the Indian people, together with the
smouldering of latent violence. It is a city of rumours and
echoes. Sometimes the tin miners of Catavi are about to march
upon the capital, dragging their hostages behind them. Sometimes, before daybreak, you may hear the tread of marching feet
and the singing of slogans outside your window. Sometimes
masked carabinieri, slung about with tommy-guns, ransack your
car for arms, and sometimes you find a chain slung across the
city gate on the hilltop, and a civilian with a rifle vigilantly
beside it. Fifteen years ago the mob of La Paz hung the mutilated
body of their President from a lamp-post in the Plaza Murillo,
and today the old square is stiff with soldiers, in German steel
helmets and thick high-collared jackets, self-consciously ceremonial on little platforms outside the Presidency, unobtrusively
watchful upon the roof of the Cathedral.

    All this passion, all this energy, thumps through the city night
and day, sharpened into something knife-like and tremulous by
the breathless clarity of the altitude. You can feel it on the promenade of the Prado at weekends, when the wide-eyed girls and
men with small moustaches chatter with a gay intensity at the

tables of the Copacabana. You can hear it in the conversations of the place, dark with plots but humorous with tall stories, cynical but often secretive. You can see it in the slogans daubed on almost every wall, with their baffling permutations of political initials and the paint that drips down in frenzied blobs from their exhortations. You can even see it reflected in the smiling, bustling and wagging of the city's enchanting Carpaccio dogs. The marvellous glacial air of La Paz, which sends the tourists puffing and dizzy to their beds, makes for fizz, bounce, and heady enthusiasm, and the isolation of this queer city, mountain metropolis of a land-locked State, gives it a sense of introvert obsession.

And most of all you will know the pressure of La Paz if you visit the high Indian quarters after dark. They tumble and straggle dustily upon the hillside, dim-lit and padlocked, but at night they are tumultuous with activity. It is not a noisy sort of energy—it has a padded, hushed insinuation to it—but it is tremendously purposeful and intent. Crouching along every alley are the indefatigable street sellers, huddled about some hissing brazier, or sprawling, a confusion of skirts, shawls and babies, behind their stalls of mandarins. Hundreds of candles illuminate the pavement counters; beneath a multitude of canvas awnings, like the market restaurants of Singapore, the Indians eat their thick stews or sip their coca tea; outside each dark and balconied courtyard, the caravanserai of La Paz, the lorries are preparing for the dawn journey—down to the steaming Yungas for tropical fruits and jungle vegetables, across the altiplano for the fabulous rainbow trout of Titicaca.

The scene is shadowy and cluttered, and you cannot always make out the detail as you push through the crowd; but the impression it leaves is one of ceaseless, tireless energy, a blur of strange faces and sinewy limbs, a haze of ill-understood intentions, a laugh from a small Mongol in dungarees, a sudden stink from an open drain, a cavalcade of tilted bowlers in the candlelight—and above it all, so clear, so close that you confuse the galaxies with the street lamps, the wide blue bowl of the Bolivian sky and the brilliant, cloudless stars of the south.

But here's an odd thing. When you come to La Paz from the north, over the escarpment, it seems a very prodigy among cities; but if you drive away from it towards Illimani and the south, looking back over your shoulder as you cross the last ridge, why, all the magic has drained from it, all the colour has faded, all that taut neurosis seems an illusion, and it looks like some drab old mining camp, sluttish among the tailings.

1963

A Visit from Royalty

THE royal visit was the most ballyhooed event that I can remember in South Africa. The royal family was dinned into us from every newspaper, every cinema, every wireless broadcast, every shop window, every decoration hung across every street. The royal family was here; the royal family was there; the royal family did this; the royal family did that. They had been in South Africa for weeks before they arrived in Johannesburg, and by that time hysteria was inescapable. A female announcer of the South African Broadcasting Corporation burst into tears over the air when the royal family came round her corner; a reporter on one of the dailies claimed that he had been stopped by 'an ordinary man in the street' in one of the Reef mining towns, who had exclaimed: 'What a golden eagle among men is the King!'

And at last, one rather cloudy day, the royal family came to beflagged, ecstatic Johannesburg. I saw them in the morning, rushing up Eloff Street in an open car, with outriders on motor cycles, and a ripple of applause coming from the people, fading before it had begun; the car was gone so quickly. The royal car was followed by a succession of big American cars with nameless people in them, all moving at a breathtaking pace. The policemen relaxed, an officer took his hand away from his cap, and the people turned to one another with reluctant, drawn faces, like sleepers awakened from a dream. People began picking up their folding chairs, children ran across the street where the cars had passed, the crowds on both sides of the street broke up, wavered, walked towards the station or the tram termini, carrying the little flags they had hardly had time to wave. I do not know what the people had been expecting, for I had not been among them before the convoy of cars had come past, and had, indeed, been taken by surprise by the tired, known faces rushing past, and the quick, too-late applause. The people dispersed with no exaltation or disappointment: they were strange to see at that moment, as though one were in a thousand bedrooms as day returned and the sleepers reluctantly admitted the light between their lids.

In the evening the emotions were different. With night, with darkness, with the thousands of coloured lights, the crowds were awake and wild. All over Johannesburg there were huge throngs of people, walking, yelling; the bars were full and noisy; and, as one does so often in Johannesburg, one caught the feel or violence in the dark streets with their buildings towering on either side. There were no Africans about; for their own safety, perhaps, they had kept away. The liveliness of the streets that are usually empty of pedestrians after nightfall had something terrifying about it: the city was alive, bristling like an animal. And the passion that filled the people, that drove them to walk up and down the pavements, and in and out of bars, that made them wait on street corners, and change their places repeatedly on the stands, was elemental and powerful. It was curiosity.

I have never known anything like it. There was a huge animal passion of curiosity among the people, that was like a hunger, and was later to become a rage. They walked and waited and talked, with an anticipation so intense one might have thought something without which they could not live was about to be shown to them. It seemed to be some final, lasting knowledge that they were seeking; a spectacle which would satisfy them forever. And all the night was tedious and tense, until that moment would come. Then they, who lived so far from Europe, from England, from Buckingham Palace, would at last *see*.

We waited. The policemen forbade people to cross Commissioner Street, so we settled down hopefully on the stands; then became restive again. Someone threw orange peel at a policeman, who fell into a rage, and drew his baton. He said he'd kill the person who did it. But the crowd told him to shut up. They called him Major, and Colonel, and, in an even wilder flight of fancy, Field-Marshal Smuts. So the young constable put his baton away, muttering to himself. Then a new sport began. People started slipping across the road, and the policemen tried to stop them. A man would wait until all the policemen on a particular stretch of road were busy chasing someone else, and then he would dash across, a small hurrying figure running across the dark tar, with the policeman after him. If he did manage to get across, a cheer went up from the crowd; if he didn't, a groan of commiseration. People called to the police, distracting their attention to help others. It was all quite good-humoured, but eventually one of the policemen hit one of the people he had caught with his baton, and the game ended in anger.

But we soon forgot the man the police had dragged away with blood coming from his forehead. We were waiting for the two princesses to go to a ball; and now young couples who had been invited to the ball were walking down the middle of Commissioner Street, the men wearing evening suits and the girls in long dresses. So we cheered them, mockingly and enviously; for white South Africans are democrats among themselves, and do not readily admit anyone else's right to be cheered just like that, unless he is a politician or a rugby player. The people we cheered were also white South Africans, and so were embarrassed by the cheers; when we saw that we cheered even more loudly, of course; and made rude remarks about the girls. '*Sis!*' a woman next to me exclaimed, in protest against the behaviour of the crowd, 'These people have got no respect.' She must have been one of nature's Englishwomen, for the rest of us had no respect at all, and no shame at not having any.

But all this, we knew, was preparatory, and everyone was relieved when the last of the couples had gone, the street was cleared, and the policemen came to attention. 'When they coming, General Smuts?' someone asked the policeman nearest to us. He said: 'Two minutes' time,' and we settled down in silence. We hunched, waiting for their coming. Then—a bright glow of car headlights, and a shout from the people farther down the road, the shout coming nearer, not yet really loud, and then it was upon us—a glimpse, a vision of pale glittering faces in a black car that was past us, again, before we could really shout, before we could really do anything. And now it was gone. There was nothing now, except for empty Commissioner Street, and the receding tail light of a motor-car and some motor cycles.

Nothing had been given us. As in the morning, there was a momentary silence, a kind of numbness. Then the animal awoke—not begrudgingly, as in the morning, but with a full throat. A roar went up from the crowd, a huge animal yell that rang in the streets. All along the road people were shouting, in a great, cheated roar. No answer had been given to them. And the yell died into silence as suddenly as it started.

A moment later the mob broke and began running down the road, past the Kensington tram terminus and towards the City Hall. People screamed and ran, from both sides of Commissioner Street. The police were unable to stop them. Jackets and dresses were flying loose, hundreds of feet were beating on the tar, hundreds of voices were screaming at the night, at nothing. A woman fell, and people jumped over her, or side-stepped to get away from what was just an obstruction in their path, and not a crying woman on the pavement. But no sooner had she been helped to her feet than she jerked away from her helpers and ran on screaming like all the others.

The princesses apparently had entered the City Hall through the door facing the Cenotaph, for our mob ran straight into another huge crowd gathered there. In the blaze of floodlights, people were pushing and screaming, and waving their hands though there was nothing to wave at, for the princesses had already gone inside. The crowd was possessed; in a rage, a frenzy, its passion unabated. Something had to be given to them—glimpses of two shining girls could not slake this thirst. So their passion focused itself on the nearest thing to hand: the car the princesses had arrived in. The car became their target—to see the car, to touch it, to hold it, to destroy it perhaps. But no, they did not want to destroy it. They just had to touch the car. They pushed and fought with one another, driving forward in surges. A woman next to me was carrying a baby in her arms, but she too was pushing, the child's face smothered in her sleeve. She screamed at me in Afrikaans, '*Eina!* You're pushing like a Kaffir!' and for a moment I remembered reading in one of the papers about the almost miraculous spirit of good will between the races that had been spread throughout the country by the royal visit. Miraculous, apparently, was the word. But that was lost, the woman, her words, the baby, as the crowd again gave a heave and we were all carried forward, this time right against the backs of the policemen who, with linked arms, were shoving us away from the car as determinedly as we were shoving towards it. The night was pandemonium; and all in a blaze of light that made every white face shine as though transfigured, that illu-minated every open mouth and gleaming eye. And the police shoved the crowds back, shoved them back, until a passage was cleared and the car drove away, though a thousand voices called after it in a gasp, 'Ah!' and again, as the car turned a corner, 'Ah!' from the back of a thousand throats.

With some pushing I managed to make my way through the bodies and feet, hands and handbags, and finally get out of the pressure of the crowd, to the side of the City Hall. Few people seemed to be leaving: most of the crowd was still heaving about immediately around where the princesses' car had been. The last thing I remember before I left was a small, English-speaking South African, in a neat brown suit and shirtcuffs neat at his wrists, speaking to himself, or possibly to others, in the hope of whipping them into action that he himself was afraid to take. He was pointing at a group of Indian youths on the outskirts of the crowd, and his face bore that pale, fanatical look, self-absorbed, as though listening to God within himself, that white South Africans often wear when they are working up to violence on those with darker skins than their own. 'Look at them,' he was saying. 'Look at them. Filthy f—— coolies, coming to look at the King and Queen, as if they're white men. Look at them, f—— cheeky coolies. Let's do something.' His lips were trembling; the tremor spread to his hands. He stared at the Indians: he also was committing himself to a passion, perhaps one related to that of the crowds who, as the next morning's papers put it, had gathered to show their love for the princesses.

1953

Is It Alas, Yorick?

R ECENTLY I have started doing sums in my head: how old was my father when I was my sons' age, and how did he deal with me? I am always startled. What—that fifty-year-old figure of authority that I remember? (No, not authority but confidence, completedness, a man who had solved his problems and now stood on top of the hill looking calmly back and calmly on.) Am I really his age now, who am not like that at all, neither calm nor complete, and unlikely to be so?

I try to console myself with the thought that I probably appear to my sons as he did to me (though I doubt this) and tell myself, more confidently, that he was not really like that either.

I have been reading some early letters of his, written to his distant parents, explaining why he had given up his medical studies and had decided, against their wishes, to get married. I recognized the tone at once, slightly blustering, self-exculpatory; I had used it myself, to him. One of his excuses for his lack of industry is so far-fetched as to be possibly true. Overcome by thirst in the laboratory he had swigged a beaker of clear fluid which turned out to contain some deadly acid. This affected his work and, he tells them, permanently damaged his health. As he often boasted to me that he had never had a day's illness in his life, which in my experience was true, he had either conveniently forgotten this or, to his parents, was pulling the longbow as far as his bulging cheek would allow. But there is definitely a young man's rather whingeing note in the letter; slight, but sad. So when did he become the confident figure I remember? No, that's not right, he wasn't confident, he was diffident, but I derived confidence from him. . . . So do I go on puzzling, nuzzling at his shade.

If you push into a thicket you sometimes come across a green-ing sheep-skull. Outside the thicket, plump, living sheep tear confidently at the grass, undismayed. We are like that, and should be. We miss our dead—sometimes we even grieve. It is hard to imagine that one day we shall be among them. But meanwhile it is well to remember that through all sorts of cycles of change they nourish us, and continue to set us problems that will be with us until we die. So, having done my sums, I try to remember how he, fifty, dealt with me at fourteen—but I can't. Perhaps he didn't deal with me at all. But maybe it is not too late; he was, after all, a Roman Catholic, as in essence I am, so perhaps I should pray to him and ask his advice? But I can't do that either, can only imagine him embarrassed, evasive of my intensities, as he was in life; properly so, as I now think. Thus he eludes me still, which is perhaps why I think of him so much.

But it is also because I feel myself in a special position because of him. He was not, I feel pig-headedly certain, like any other father. He didn't even look like any other father. He wore black, broad-brimmed hats for a start. Not very broad-brimmed, not sombreros, but not neat trilbys either. He was red-faced and red-haired, balding, with a huge dome, and was stout, to the extent of appearing almost square when I knew him, balanced on tiny feet. Someone painted a portrait of him and friends complained the painting made him look like a butcher. He could have looked like a butcher but he did not. Whether he was dis-tinguished-looking I have no idea but he was certainly distinct-ive. I have never seen anyone who reminded me of him, even remotely. With his neat red moustache he might have been a bank manager, I suppose, or a retired army officer—but it would have been impossible to imagine which bank, or which army. A conundrum, you just could not place him; New Zealander-Irish, he was as near as possible classless. When he briefly had a large desk in an office of his own making, even when he sat behind it, he gave the impression he was just passing through, was about to reach for his hat and go out into his beloved London—if indeed he wasn't already wearing the hat, which was usually the case.

He had no job, like other fathers, not a real one. He began adult life as a remittance man and when the remittances stopped he remained impoverished for a time, apparently not noticing. Then he suddenly began writing sketches for comedians and these, as the years passed, turned into radio shows which culminated in ITMA, the wartime programme that made him famous and, briefly, in funds; both of which he enjoyed. My point is this: he was, and is, my exemplar of order, an order I would like to pass on to my sons: but how can I when my own father spent his life, earned his living, presenting a sort of inspired, zany *dis*order as a source of true heart's ease? And every kind of authority as ludicrous? If he invented a mayor that mayor was amiably bent; if he wrote of a doctor that doctor was the source of every possible confusion to his patients. Each generation tries to be less pompous than its predecessor. I have watched my friends, the interesting ones, define themselves by reacting against their parents—against a too-limiting sense of class, or convention, or morality. (I have also sometimes seen them look aghast at the disordered world they have created, and attempt, too late, to swing back to the values of their parents.) But my father did not impose himself like that. Not that it was Liberty Hall. I once turned him purple with rage (an almost unique occurrence) because I used a fairly mild swear-word. But I can hardly define myself by going around cussing all the time.

His working methods were cottage-industry and bordered on the chaotic. He would leave it till the deadline and then get up in the middle of the night and sit at a tiny table in front of the electric fire, Parker pen in his stubby fingers, filling the ash-tray and sheets of lined foolscap which he dropped on the floor. By the time I surfaced he had covered the carpet with paper and was wandering in to the kitchen, reading bits out to my mother to see if she laughed, laughing himself anyway.

Of course, it (and he) was not all good. That period of unemployed impoverishment showed irresponsibility—he had a family. Later he often drank too much—in company, never at home. Indeed I suspected that, among friends, he hardly noticed that he was drinking at all, just emptied what came to hand, as he had in the laboratory. In company the puns flowed easily without malice. They no longer work, they were born of the moment, but I remember one occasion when a friend called Watt warned him about the excessive drinking of one of my father's business associates named Blatt: 'But surely,' said my father, 'that's a case of the Watt calling the kettle Blatt?' and what could have been an embarrassing, possiby unpleasant, moment went up in a shout of laughter. I remember admiring him for that, for the quickness of it, the geniality.

The image of the sheep-skull keeps returning to me.... The most famous skull is Yorick's, and he was a jester too. I have sometimes suspected that Yorick is the secret hero of *Hamlet*. Certainly he is a hero of Hamlet's: does he not, at the intensest moments of his confusion and grief, express himself in wild whirling word-plays?

My father didn't go in for giving advice (though he did once solemnly recommend that I keep a bottle of Vichy water by my bedside). He didn't himself—he drank enormous quantities of lime-juice instead and I certainly never saw him with a hangover however much he had deserved one—but he saw no self-contradiction in that. He told me that I should never go bald if I massaged my scalp in a certain way, as he always had. He placed his square finger-tips on his shining dome in order to show me how. However, when I was hit by a grief, he did venture an oblique suggestion during the course of a shy tête-à-tête lunch. 'Now that something terrible has happened to you,' he said, 'perhaps you'll write comedy.' Coming from him that is not quite the show-must-go-on, laugh, clown, laugh cliché it might otherwise sound. For what was there in my father's life? There was God, there was fellowship, and there were jokes. It is not a bad recipe. At times I have detested jokes—a son must react against his father in some fashion. I have seen jokes for the evasions they are, what Edward Thomas called the 'monkey, humour', praising Richard Jefferies for his lack of it. If you sit in a room with a television comedy going on next door and hear the automaton-like bursts of hilarity it is possible to hate laughter itself.

But my pendulum swings. Sometimes I think jokes are the only truly serious response to our absurd fates. Who can match the desperate humourlessness of the adolescent who thinks he is the first to discover seriousness? (I was probably like that, which is why my father ducked.) For after all the show *must* go on. The alternative is not a joke.

Maybe what my father meant, but was too gentle to say, was that now something terrible had happened to me perhaps I might grow up. I would like to ask him about that now because I suspect he never quite did so himself and this has impeded my own growing-up, for which I bless him, however tiresome I may be to others. For I am never at ease with those who have come too surely to terms with life. I would also like to ask Yorick, that fellow of infinite jest, what he would have said if he had heard Hamlet say 'Alas'. Something to the point, surely, but not portentous.

So I go on puzzling, nuzzling the green grass outside the thicket.

1983

V. S. NAIPAUL

Columbus and Crusoe

THE adventure of Columbus is like *Robinson Crusoe*. No one can imaginatively possess the whole; everything beyond the legend is tedious and complicating. It is so even in Björn Landström's book, *Columbus*, which makes the difficult adventure as accessible as it can be made. The text itself is a retelling from the usual sources. The maps and illustrations are more important. The maps make medieval ideas of geography clear. The illustrations, a true labour of love, are numerous and exact: ships, the islands, the people, the weather, the vegetation, and even the Flemish hawk's bell which delighted the natives until it became a measure of the gold dust the discoverer required them to collect.

In the legend Columbus is persecuted by many enemies; he goes back to Spain white-haired, in chains, and he dies in poverty and disgrace. It is Columbus's own picture: he had a feeling for theatre. His concern for gold exceeded his sovereign's: he expected to get a tenth of all that was found. The chains were not necessary; he was begged to take them off. He wore them for effect, just as, after the previous disaster, he had returned in the Franciscan habit. That disaster had its profitable side. He had sent back slaves, as he had always intended. He claimed, or his son claimed for him, that he had got rid of two-thirds of the natives of Hispaniola in two years; the remainder had been set to gathering gold dust. (This was an exaggeration: he had only got rid of a third.) Even after his disgrace he fussed about his coat-of-arms, appropriating a red field for the castle of Castile, as on the royal coat-of-arms. He complained to the end about his poverty, but one of his personal gold shipments, again after his disgrace, amounted to 405 pounds. His father was a weaver; his sister married a cheesemonger; his son married a lady of royal blood. And at his death Spain hadn't gained very much. Mexico was thirteen years away; and the Indies, the source of his gold, where he thought he had discovered the Terrestrial Paradise, had become, largely through his example, *anus mundi*.

It is a story of extended horror. But it isn't only the horror that numbs response. Nor is it that the discoverer deteriorates so steadily after the discovery. It is the banality of the man. He was looking less for America or Asia than for gold; and the banality of expectation matches a continuing banality of perception. At the heart of the seamanship, the toughness, the avarice, the vindictiveness and the brutality, there is only this:

16 September. Here the Admiral says that on that day and all succeeding days they met with very mild breezes, and the mornings were very sweet, with naught lacking save the song of the nightingale. He adds: 'And the weather was like April in Andalusia.'
29 September. The air was very sweet and refreshing, so that the only thing lacking was the song of the nightingale; the sea was as calm as a river.

This is from *The Book of the First Voyage*, when he was at his most alert. The concrete details are deceptive. The sea and its life are observed, but mainly for signs of the nearness of land; just as, at the moment of discovery, the natives are studied, but only by a man 'vigilant'—his own word—for gold. 'Their hair is not curly ... they are not at all black.' Not an anthropological interest, not the response of wonder—disappointment rather: Columbus believed that where Negroes were, there was gold. Beyond this vigilance the words and the perceptions fail. The nightingale, April in Andalusia: the props of a banal poetry are used again and again until they are without meaning. They are at an even lower level than the recent astronaut's 'Wow'—there is nothing like this pure cry of delight in Columbus. After the discovery, his gold-seeking seaman's banalities become repetitive, destroying romance and making the great adventure trivial. A book about Columbus needs to have pictures, and this is why Mr Landström's book is so valuable.

The medieval mind? But Queen Isabella wrote during the second voyage to find out what the climate was like. April in Andalusia wasn't enough: she wanted pictures, and the romance. Marco Polo, whom Columbus had read, dealt in romance; and Amerigo Vespucci, after whom the continent is not unfairly

named. Vespucci thought it worth mentioning that the natives of
the islands and the Main pissed casually into the hot sand during conversation, without turning aside; that the women were
wanton and used a certain animal poison, sometimes lastingly
fatal to virility, to increase the size of the male member. Perhaps
he made this up; but though he too was vigilant and his own
voyage ended in profitable slave-trading, he sought in the tradition of travel-romance to awaken wonder at the fact of the
New World.

The facts about Columbus have always been known. In his
own writings and in all his actions his egoism is like an exposed
deformity; he condemns himself. But the heroic gloss, which is
not even his own, has come down through the centuries. When
the flagship ran aground at Haiti on the first voyage, the Indians
were more than helpful: they wept to show their sympathy.
Columbus was vigilant: he noted that it would be easy to subdue
this 'cowardly' unarmed race. This was what he presently did.
Mr Landström suggests that it was unfortunate and not really
meant: it is the traditional gloss. On the third voyage Columbus
thought he had discovered the Terrestrial Paradise. Mr Landström, again following the gloss, says that Columbus wasn't very
well at the time. But it was just this sort of geography that had
made him attempt the Ocean Sea.

In this adventure, as in today's adventures in space, the
romance is something we ourselves have to supply. The discovery needs a hero; the contempt settles on the country that, in
the legend, betrays the hero. The discovery—and it would have
come without Columbus—could not but be horrible. Primitive
people, once exposed, have to be subdued and utilized and somehow put down, in the Indies, Australasia, the United States,
Southern Africa; even India has its aboriginal problems. Four
hundred years after the great Spanish debate, convened by the
Emperor, on the treatment of primitive people, Rhodesia is
an imperial issue. The parallel is there; only the contemporary
debate, conducted before a mass-electorate on one side and a
dispossessed but indifferent primitive people on the other, is
necessarily more debased.

There is no Australian or American black legend; there is at
the most a romantic, self-flattering guilt. But the black legend of
Spain will persist, as will the heroic legend of Columbus. The

dream of the untouched, complete world, the thing for ourselves alone, the dream of Shangri-la, is an enduring human fantasy. It fell to the Spaniards to have the unique experience. Generosity and romance, then, to the discoverer; but the Spaniards will never be forgiven. And even in the violated New World the Spaniards themselves remained subject to the fantasy. The quest for El Dorado became like a recapitulation of the whole New World adventure, a wish to have it all over again; more men and money were expended on this in twenty expeditions than on the conquest of Mexico, Peru and New Granada.

Robinson Crusoe, in its essential myth-making middle part, is an aspect of the same fantasy. It is a monologue; it is all in the mind. It is the dream of being the first man in the world, of watching the first crop grow. Not only a dream of innocence: it is the dream of being suddenly, just as one is, in unquestionable control of the physical world, of possessing 'the first gun that had been fired there since the creation of the world'. It is the dream of total power. 'First, I made him know his name should be Friday, which was the day I saved his life. I called him so for the memory of the time. I likewise taught him to say master, and then let him know that was to be my name.' Friday is awkward about religion; Crusoe cannot answer. Power brings problems. Crusoe sees some cannibals about to kill and eat a man. He runs to liberate. But then he stops. What is his right to interfere? Is it just the gun? Some Spaniards are to be rescued. How will his freedom and power continue? How will they obey? Where do sanctions start in the empty world? They must sign a contract. But there is no pen, no paper: a difficulty as particular and irrational as in a nightmare. It is from more than a desert island that he is rescued. The issues can never be resolved.

Later Crusoe makes good, in that very New World, but in the settled, beaten-down slave society of Brazil. The horror of the discovery, of being the first totally powerful man in the world: that happened a long time before.

1967

JOHN UPDIKE

The Bankrupt Man

THE bankrupt man dances. Perhaps, on other occasions, he sings. Certainly he spends money in restaurants and tips generously. In what sense, then, is he bankrupt?

He has been declared so. He has declared himself so. He returns from the city agitated and pale, complaining of hours spent with the lawyers. Then he pours himself a drink. How does he pay for the liquor inside the drink, if he is bankrupt?

One is too shy to ask. Bankruptcy is a sacred state, a condition beyond conditions, as theologians might say, and attempts to investigate it are necessarily obscene, like spiritualism. One knows only that he has passed into it and lives beyond us, in a condition not ours.

He is dancing at the Chilblains Relief Association Fund Ball. His heels kick high. The mauve spotlight caresses his shoulders, then the gold. His wife's hair glistens like a beehive of tinsel above her bare shoulders and dulcet neck. Where does she get the money, to pay the hairdresser to tease and singe and set her so dazzlingly? We are afraid to ask but cannot tear our eyes from the dancing couple.

The bankrupt man buys himself a motorcycle. He is going to hotdog it all the way to Santa Barbara and back. He has a bankrupt sister in Santa Barbara. Also, there are business details to be cleared up along the way, in Pittsburgh, South Bend, Dodge City, Santa Fe, and Palm Springs. Being bankrupt is an expansionist process; it generates ever new horizons.

We all want to dance with the bankrupt man's wife. Sexual health swirls from her like meadow mist, she sparkles head to toe, her feet are shod in slippers of crystal with caracul liners. 'How do you manage to keep up ap——?' We drown our presumptuous question murmurously in her corsage; her breasts billow, violet and gold, about our necktie.

The bankrupt man is elected to high civic office and declines, due to press of business. He can be seen on the streets, rushing everywhere, important-looking papers flying from his hands. He is being sued for astronomical amounts. He wears now only the trendiest clothes—unisex jumpsuits, detachable porcelain collars, coat sleeves that really unbutton. He goes to the same hairdresser as his wife. His children are all fat.

Why do we envy him, the bankrupt man? He has discovered something about America that we should have known all along. He has found the premise that has eluded us. At our interview, his answers are laconic, assured, delivered with a twinkle and well-spaced, conspiratorial, delicious lowerings of his fine baritone.

Q: When did you first know that you were bankrupt?

A: I think from birth I intuited I was headed that way. I didn't cry, like other infants.

Q: Do you see any possibility for yourself of ever being non-bankrupt?

A: The instant bankruptcy is declared, laws on the federal, state, and local levels work in harmony to erode the condition. Some assets are exempted, others are sheltered. In order to maintain bankruptcy, fresh investments must be undertaken, and opportunities seized as they arise. A sharp eye on economic indicators must be kept lest the whole package slip back into the black. Being bankrupt is not a lazy man's game.

Q: Have you any word of advice for those of us who are not bankrupt?

A [*with that twinkle*]: Eat your hearts out.

The interview is concluded. Other appointments press. He and his family must put in a splendid appearance at the Meter Readers' Benefit Picnic. They feed grapes to one another, laughing. The children tumble in the tall grass, in their private-school uniforms. The bankrupt man's wife is beginning to look fat, sunlight dappling her shoulders. Only he maintains a hard edge, a look of bronze. He wins the quoit toss and captains the winning tug-of-war team; the other side, all solvent small-business men in gray suits, falls into the ditch. Magnanimously, he holds down to them a huge helping hand. By acclamation, he is elected to the

vestry of all the local Protestant churches and eats the first piece
of the Meter Readers' Bicentennial Chocolate Layer Cake.

This galls us. We wish to destroy him, this clown of legerity,
who bounces higher and higher off the net of laws that would
enmesh us, who weightlessly spiders up the rigging to the
dizzying spotlit tip of the tent-space and stands there in a glittering trapeze suit, all white, like the chalk-daubed clown who
among the Australian aborigines moves in and out of the sacred
ceremonial, mocking it. We spread ugly rumors, we mutter that
he is not bankrupt at all, that he is as sound as the pound, as the
dollar, that his bankruptcy is a sham. He hears of the rumor and
in a note on one-hundred-percent-rag stationery, with embossed
letterhead, he challenges us to meet him on West Main Street, by
the corner of the Corn Exchange, under the iron statue of Cyrus
Shenanigan, the great Civil War profiteer. We accept the challenge. We experience butterflies in the stomach. We go look at
our face in the mirror. It is craven and shrivelled, embittered by
ungenerous thoughts.

Comes the dawn. Without parked cars, West Main Street
seems immensely wide. The bankrupt man's shoulders eclipse
the sun. He takes his paces, turns, swiftly reaches down and pulls
out the lining of both pants pockets. Verily, they are empty. We
fumble at our own, and the rattle of silver is drowned in the triumphant roar of the witnessing mob. We would have been torn
limb from limb had not the bankrupt man with characteristic
magnanimity extended to us a protective embrace, redolent of
cologne and smoking turf and wood violets.

In the locker room, we hear the bankrupt man singing. His
baritone strips the tiles from the walls like cascading dominoes.
He has just shot a minus sixty-seven, turning the old course
record inside out.

He ascends because he transcends. He deals from the bottom
of the deck. He builds castles in air. He makes America grow.
His interests ramify. He is in close touch with Arabian oil. With
Jamaican bauxite. With antarctic refrigeration. He creates employment for squads of lawyers. He gets on his motorcycle. He
tugs a thousand creditors in his wake, taking them over horizons
they had never dreamt of hitherto.

He proves there is an afterlife.

1983

At the Dam

Since the afternoon in 1967 when I first saw Hoover Dam, its image has never been entirely absent from my inner eye. I will be talking to someone in Los Angeles, say, or New York, and suddenly the dam will materialize, its pristine concave face gleaming white against the harsh rusts and taupes and mauves of that rock canyon hundreds or thousands of miles from where I am. I will be driving down Sunset Boulevard, or about to enter a freeway, and abruptly those power transmission towers will appear before me, canted vertiginously over the tailrace. Sometimes I am confronted by the intakes and sometimes by the shadow of the heavy cable that spans the canyon and sometimes by the ominous outlets to unused spillways, black in the lunar clarity of the desert light. Quite often I hear the turbines. Frequently I wonder what is happening at the dam this instant, at this precise intersection of time and space, how much water is being released to fill downstream orders and what lights are flashing and which generators are in full use and which just spinning free.

I used to wonder what it was about the dam that made me think of it at times and in places where I once thought of the Mindanao Trench, or of the stars wheeling in their courses, or of the words *As it was in the beginning, is now and ever shall be, world without end, amen.* Dams, after all, are commonplace: we have all seen one. This particular dam had existed as an idea in the world's mind for almost forty years before I saw it. Hoover Dam, showpiece of the Boulder Canyon project, the several million tons of concrete that made the Southwest plausible, the *fait accompli* that was to convey, in the innocent time of its construction, the notion that mankind's brightest promise lay in American engineering.

Of course the dam derives some of its emotional effect from precisely that aspect, that sense of being a monument to a faith since misplaced. 'They died to make the desert bloom,' reads a plaque dedicated to the ninety-six men who died building this first of the great high dams, and in context the worn phrase touches, suggests all of that trust in harnessing resources, in the meliorative power of the dynamo, so central to the early Thirties. Boulder City, built in 1931 as the construction town for the dam, retains the ambience of a model city, a new town, a toy triangular grid of green lawns and trim bungalows, all fanning out from the Reclamation building. The bronze sculptures at the dam itself evoke muscular citizens of a tomorrow that never came, sheaves of wheat clutched heavenward, thunderbolts defied. Winged Victories guard the flagpole. The flag whips in the canyon wind. An empty Pepsi-Cola can clatters across the terrazzo. The place is perfectly frozen in time.

But history does not explain it all, does not entirely suggest what makes that dam so affecting. Nor, even, does energy, the massive involvement with power and pressure and the transparent sexual overtones to that involvement. Once when I revisited the dam I walked through it with a man from the Bureau of Reclamation. For a while we trailed behind a guided tour, and then we went on, went into parts of the dam where visitors do not generally go. Once in a while he would explain something, usually in that recondite language having to do with 'peaking power', with 'outages' and 'dewatering', but on the whole we spent the afternoon in a world so alien, so complete and so beautiful unto itself that it was scarcely necessary to speak at all. We saw almost no one. Cranes moved above us as if under their own volition. Generators roared. Transformers hummed. The gratings on which we stood vibrated. We watched a hundred-ton steel shaft plunging down to that place where the water was. And finally we got down to that place where the water was, where the water sucked out of Lake Mead roared through thirty-foot penstocks and then into thirteen-foot penstocks and finally into the turbines themselves. 'Touch it,' the Reclamation man said, and I did, and for a long time I just stood there with my hands on the turbine. It was a peculiar moment, but so explicit as to suggest nothing beyond itself.

There was something beyond all that, something beyond energy, beyond history, something I could not fix in my mind. When I came up from the dam that day the wind was blowing harder, through the canyon and all across the Mojave. Later, toward Henderson and Las Vegas, there would be dust blowing, blowing past the Country-Western Casino FRI & SAT NITES and blowing past the Shrine of Our Lady of Safe Journey STOP & PRAY, but out at the dam there was no dust, only the rock and the dam and a little greasewood and a few garbage cans, their tops chained, banging against a fence. I walked across the marble star map that traces a sidereal revolution of the equinox and fixes forever, the Reclamation man had told me, for all time and for all people who can read the stars, the date the dam was dedicated. The star map was, he had said, for when we were all gone and the dam was left. I had not thought much of it when he said it, but I thought of it then, with the wind whining and the sun dropping behind a mesa with the finality of a sunset in space. Of course that was the image I had seen always, seen it without quite realizing what I saw, a dynamo finally free of man, splendid at last in its absolute isolation, transmitting power and releasing water to a world where no one is.

1970

About Face*

'At fifty,' wrote Orwell, 'everyone has the face he deserves.'
I believe this and repeat it with confidence, being myself forty-
six and hopeful that for me there is still time. I hope, that is, that
within the next four years I shall be able to develop a noble
brow, a strong chin, a deep and penetrating gaze, a nose that
doesn't disappoint. This may take some doing, for I have been
told by different people at different times that I resemble the
following odd cast of characters: the actors Sal Mineo, Russ
Tamblyn, and Ken Berry, the scholar Walter Kaufmann, the
assassin Lee Harvey Oswald, and a now-deceased Yorkshire ter-
rier named Max. Despite this, and even though no one has ever
noted a resemblance in me to Alexander the Great or Lord
Byron, I tend to think of myself, as I expect most men do, as
a nice-enough looking chap. Beyond that I am not prepared to
go, for I have long appreciated the fact that the limits of self-
knowledge begin at one's own kisser. To have stared at the
damned thing so long and yet still not to know what it reveals is
a true tribute to the difficulties of self-analysis. So while I tend to
believe, with Orwell, that everyone has the face he deserves, I
gaze into the mirror and cannot tell whether justice has been
done.

The notion that the face is a text to be read for clues to human
character is one with a long history. It goes back at least as far as
Aristotle, among whose works is that entitled *History of Animals
and A Treatise on Physiognomy*. Almost all work in physiognomy,
the putative science dealing with the connection between facial
features and psychological characteristics, has been disqualified,
and the *Encyclopaedia Britannica*, in a brief article on the subject,
notes: 'Since many efforts to specify such relationships [between
facial features and personal character] have been discredited, the
term physiognomy commonly connotes pseudoscience or charlat-

* This is an abridged version of the published essay.

anry (see Fortunetelling; Palmistry).' Which makes very good sense, except that I cannot bring myself altogether to believe it. On the subject of physiognomy, I find myself in the condition of a man I once heard about who, at the end of a career of thirty-odd years working for the Anti-Defamation League, remarked that, after fighting all that time against every racial and religious stereotype, he had come to believe that perhaps there was more to these stereotypes than he had thought when he had started on the job. Rather like that man, I fear that, while I believe physiognomy to have been largely discredited, there may be more to it than an intelligent person is supposed to allow.

But let me take a paragraph to hedge, qualify, and tone down what I have just written. I do not, for example, believe that a large head implies great intelligence, or even that a high forehead implies ample intellectual capacity, though apparently Shakespeare, himself well-endowed in this respect, did. Nor do I believe that a strong jaw inevitably translates into a character of great determination. I do believe, with the poet, that the eyes are the windows of the soul; yet I do not go so far as to say that Elizabeth Taylor, who has the most beautiful public eyes of our day, therefore has the most beautiful soul. I do not believe bad teeth or bad skin symbolic of a grave flaw in character.

The mystery of personality is written in the human face—this I do believe. But, as with all truly intricate mysteries, this one must be read subtly, patiently, penetratingly.

I have always had an intense interest in faces and from as early as I can remember have watched them the way bird-watchers do birds. One of the pleasures that living in a large city provides is the delight of viewing a large human aviary. Can there be any doubt that the human face, even though it is of a very long run, is still the best game in town? Consider: we are all playing with essentially the same cards—eyes, a nose, a chin, a mouth, cheeks, eyebrows, hair, ears, a forehead—dealt out on the cloth of skin over the front of our skulls. But how inexhaustibly interestingly these cards have been dealt. Noses retroussé or Gogolian, lips sensuous or forbidding, eyebrows wispy or bushy, cheeks puffy or gaunt, chins prognathous or nonexistent, eyes though available in a limited number of colors nonetheless of limitless expressive possibilities—what variety, what modalities within the variety, what variegation within the modalities!

The given in the human face is, of course, heredity. Yet I wonder if heredity—providing skin and eye and hair color, bone structure, et cetera—really furnishes anything more than the broad canvas on which the more delicate and interesting strokes are painted by time and personal fate. What usually makes a face interesting—a priggish nose, quizzical eyebrows, sarcastic lips, lines and wrinkles oddly placed—is there as a result not of heredity but of experience. What time does to a face is most fascinating of all, and I sometimes think that no face, unless it be one of rare beauty or especial hideousness, is of great interest—rather like wine that hasn't had time to age properly—much before thirty.

Perhaps it is impossible to predict the way a face will age. Most people of a physiognomic bent tend to work backward, which is to say from hindsight. Thus, to cite an example, Richard Perceval Graves, the recent biographer of A. E. Housman, writes of Housman's father: 'Photographs of Edward [Housman] reinforce the impression of a man who has inherited some of his father's intelligence, but more of his determination than of his judgment. The mouth and jaw are firm, even obstinate, but the eyes are weak and uncertain.' But this reading is entirely ex post facto; Mr Graves already knows that A. E. Housman's father, though in some ways determined, even obstinate, was a man of poor judgment, uncertainty, and weakness. What he first found in the man's life he afterward discovered in his face. It is the way most of us work.

Yet read faces we must, for however unreliable a method it may be, none other exists for taking at least a rough measure of others. The face, the seat of four of the five human senses, is also the meter of the emotions. The art of the actor is based on this fact. Feelings veiled in fleshy shadows, secret enmities that must not be misread, insincerities that the voice and even the mouth may be able to disguise but not the eyes—all these are to be found in the face. Goodwill and admiration, possibly even love, are writ in the disposition of facial features, and these, too, must be correctly gauged. The significance of a tic could be decisive to one's fate.

Another question is why some faces are photogenic and others are not. It may be that good bones render one more photogenic, but good bones do not necessarily make for a good face.

Photographs, like statistics, often lie. Except in the hands of a photographer who is himself an artist, the camera generally misses what is most interesting in the human face. The reason is that faces are almost always most striking in animation. Some people, on the other hand, seem almost too pliantly camera-ready. Truman Capote, for instance, has for me the look of someone who has been photographed much too often, the equivalent of a woman who has slept with too many men.

Does that last sentence strike you as goofy? Does it ring sexist, mystical, a mite mad? In his novel *Mr Sammler's Planet,* Saul Bellow has a woman character whom he describes as showing, through her eyes, evidence of having slept with too many different men. Do such things show in the eyes? John Brophy, in his fine book, *The Human Face Reconsidered,* writes of the eyes, 'Although the eyes can thus make vivid communications, their power of expression is restricted: they can plead but not argue; they can state but not analyze; they can declare effects but are helpless to explain causes.' Still, to plead, state, or declare effects is to do a very great deal. The eyes are generally conceded to be the most expressive part of the face, though some say that the mouth can be equally expressive. But in this matter I go with the Polish proverb that runs, 'Watch closely the eyes of him who bows the lowest.'

I know I need to look at, if not deeply into, the eyes of someone with whom I am talking. I find myself slightly resentful—perhaps irritated comes closer to it—at having to talk to someone wearing sunglasses. Worst of all are those mirrored-lens sunglasses that, when you look into them, throw back two slightly distorted pictures of yourself, rather like old-time funhouse mirrors. I like eyes not only to be up front, where God put them, but out front, where I can see them.

What goes for eyes goes for other facial features. The ears are said to be the least expressive parts of the face—some talented people can twitch theirs while the ears of others redden when they lie or are under stress—but, in men at any rate, I prefer not to shoot conversationally till I see the lobes of their ears, a thing not always possible under the dispensation of recent masculine hairdos. Charles de Gaulle had big ears; John O'Hara had ears that stuck out from his head; and so do my own, though I do not own up to this fact easily. None of us, I suspect, easily owns up to his own irregularities. I was recently to be met at an airport by someone I had never met before. When I asked him what he looked like, so that I might recognize him upon arrival, he said he was blond, had a mustache, and would be wearing a blue suit. All of which turned out to be quite true, except that he neglected to mention that he also weighed around three hundred pounds.

In my neighborhood there walks a man who—through a war injury? a fire? an industrial accident?—has had the left side of his face blown away. Where features once were, a drape of flesh has been drawn. He is small, tidy, wears a cap, and through his walk and general demeanor gives an impression of thoughtfulness. The effect upon first seeing him is jolting. Life must be hard for him, and one wonders if he has ever grown inured to watching strangers recoil upon initial sight of him. But why is one jolted, why does one recoil? As much as from anything, I think it has to do with one's inability to read his face. One cannot sense his mood or know what he is (even roughly) thinking—and the result is disconcerting in the extreme.

Reading Faces by Leopold Bellak, MD, and Samm Sinclair Baker not only maintains that the project of reading faces is a sensible one but offers a method for doing so. This method is called the Zone System, and the way it works is to divide the human face vertically down the center and horizontally under the eyes. It operates on the correct assumption that the face is asymmetrical. It speculates on the possibility that the division of the brain into left and right functions may have effects on the left and right sides of the face. One cannot say of this book, as Gibbon said of some *Lives* by Jerome, that 'the only defect in these pleasing compositions is the want of truth and common sense.' But as a self-help book it is, I think, helpful only in a very limited way. For example, by dividing a face horizontally one can sometimes determine that, though its mouth is smiling, its eyes are cold and scrutinizing. It is also interesting to note that, divided vertically, one side of a person's face can seem cheery, while the other seems wary. One might go from there to say that a face so divided may bespeak a person riven in some fundamental way.

But whenever *Reading Faces* goes much beyond this it becomes slightly suspect. Sensibly enough, its authors write, 'What one reads in the face are *potentialities,* from which further inferences can be drawn—from conversation, observation, and experience with the person over a period of time.' The problem is, though, that most of the faces submitted for study are those of well-known people from politics, sports, and show business, and the analyses offered of their faces by the authors are more than a touch commonplace. In some cases, they show a political bias in favor of old-style New Deal Democrats. Of Eleanor Roosevelt they write, 'It is a most unusual face about which one can only say good things.' Having been brought up in a home in which Franklin and Eleanor Roosevelt were well regarded, I tend to go along with this reading. But where our authors find such traits in Mrs Roosevelt's face as intelligence, compassion, and optimism, an old-line Taft Republican could as easily find naiveté, smugness, and self-righteousness.

One serious question about faces is whether one can find beautiful or even agreeable-looking someone whom one despises. Moral judgments, as Santayana noted, take precedence over aesthetic ones, or at least do so for most of us. So when confronted with a person one detests, perhaps the best one can say is that he or she is very good-looking—yet one is likely to add, 'at least to the superficial observer.' What makes this observer superficial, of course, is that he is not privy to the real lowdown about the despicable character in question. Yet how much easier it is to read backward, through hindsight, from behavior to evidence of behavior in the face. As John Brophy reminds us, during Hitler's rise and early years in power, no one detected the insanity we now see so clearly in his face. The aged, puffy, baby face of Winston Churchill, a cigar clamped in its mouth, might appear, to someone who has no knowledge of what Churchill accomplished, as a perfect subject for an anti-smoking poster.

The genius of the unknown sculptor is to have created what sometimes seems a rather limited number of human facial types yet, within this limited number of types, an infinite variety. With only rare exceptions, almost every face one sees one has seen before, if not in life, then in the work of the great painters. Walking the streets one sees here a pair of kindly Holbeinesque lips; there the porcelain cheeks of a Botticelli; elsewhere the rubicund coloring of one of Bruegel's peasants; and sometimes a face taken over from Rembrandt entire. If flesh and bone be the material of the face, time supplies its varnish. And what extraordinary things time does, leaving this face unmarked, that one looking as if it were a salmon mousse left out in the rain. To read the effects of time on a face requires, as the New Critics used to call it, close reading. 'For in order to understand how beautiful an elderly lady can once have been,' Proust wrote, 'one must not only study but interpret every line of her face.'

Nothing so improves the appearance as a high opinion of oneself. Let this stand as the first in a paragraph riddled with risky generalizations. Love of one's work tends to make one's face interesting. Artists have animated faces, and performing musicians the most animated of all. Suffering, too, confers interest on a face, but only suffering that, if not necessarily understood, has been thought about at length. Uninterested people have uninteresting faces. In ways blatant or subtle, personality sets its seal on every face. Some people have historical seals set on their faces as well; thus some men and women walk the streets today with Romanesque, Elizabethan, or Victorian faces. Intelligence is more readily gauged in a face than is stupidity. As a final generalization, let me say that the more precisely one thinks of the relation of face to character, and the more carefully one attempts to formulate the connection between the two, the madder the entire business begins to seem.

Yet what choice have we but to continue reading faces as best we can, bringing to the job all that we have in the way of intuition, experience, intelligence? We read most subtly of course those people we know most closely: our friends, our known enemies, our families. In the faces of such people we can recognize shifting moods, hurt and pride, all the delicate shades of feeling. But of that person we supposedly know most intimately, ourself, the project remains hopeless. Study photographs of ourselves though we may, stare at ourselves in mirrors though we do, our self-scrutiny generally comes to naught. If you don't believe me, stop a moment and attempt to describe yourself to someone who has never seen you. The best I can do is the following: 'I look a bit like Lee Harvey Oswald and I also rather resemble my dog, though I seem more dilapidated. You can't miss me.'

1983

A Blizzard of Tiny Kisses

Princess Daisy by Judith Krantz

To be a really lousy writer takes energy. The average novelist remains unread not because he is bad but because he is flat. On the evidence of *Princess Daisy*, Judith Krantz deserves her high place in the best-seller lists. This is the second time she has been up there. The first time was for a book called *Scruples*, which I will probably never get around to reading. But I don't begrudge the time I have put into reading *Princess Daisy*. As a work of art it has the same status as a long conversation between two not very bright drunks, but as best-sellers go it argues for a reassuringly robust connection between fiction and the reading public. If cheap dreams get no worse than this, there will not be much for the cultural analyst to complain about. *Princess Daisy* is a terrible book only in the sense that it is almost totally inept. Frightening it isn't.

In fact, it wouldn't even be particularly boring if only Mrs Krantz could quell her artistic urge. 'Above all,' said Conrad, 'to make you see.' Mrs Krantz strains every nerve to make you see. She pops her valves in the unrelenting effort to bring it all alive. Unfortunately she has the opposite of a pictorial talent. The more detail she piles on, the less clear things become. Take the meeting of Stash and Francesca. Mrs Krantz defines Prince Alexander Vassilivitch Valensky, alias Stash, as 'the great war hero and incomparable polo-player'. Stash is Daisy's father. Francesca Vernon, the film star, is her mother. Francesca possesses 'a combination of tranquillity and pure sensuality in the composition of the essential triangle of eyes and mouth'. Not just essential but well-nigh indispensable, one would have thought. Or perhaps that's what she means.

This, however, is to quibble, because before Stash and

Francesca can generate Daisy they first have to meet, and theirs is a meeting of transfigurative force, as of Apollo catching up with Daphne. The scene is Deauville, 1952. Francesca the film star, she of the pure sensuality, is a reluctant spectator at a polo game—reluctant, that is, until she claps eyes on Stash. Here is a description of her eyes, together with the remaining component of the essential triangle, namely her mouth. 'Her black eyes were long and widely spaced, her mouth, even in repose, was made meaningful by the grace of its shape: the gentle arc of her upper lip dipped in the centre to meet the lovely pillow of her lower lip in a line that had the power of an embrace.'

And this is Stash, the great war hero and incomparable polo-player: 'Valensky had the physical presence of a great athlete who has punished his body without pity throughout his life and the watchful, fighting eyes of a natural predator. His glance was bold and his thick brows were many shades darker than his blonde hair, cropped short and as coarse as the coat of a hastily brushed dog.... His nose, broken many times, gave him the air of a roughneck.... Not only did Valensky never employ un-necessary force on the bit and reins but he had been born, as some men are, with an instinct for establishing a communication between himself and his pony which made it seem as if the ani-mal was merely an extension of his mind, rather than a beast with a will of its own.'

Dog-haired, horse-brained and with a bashed conk, Stash is too much for Francesca's equilibrium. Her hat flies off.

Oh no!' she exclaimed in dismay, but as she spoke, Stash Valensky leaned down from his pony and scooped her up in one arm. Holding her easily, across his chest, he urged his mount after the wayward hat. It had come to rest two hundred yards away, and Valensky, leaving Francesca mounted, jumped down from his saddle, picked the hat up by its ribbons and carefully replaced it on her head. The stands rang with laughter and applause.

Francesca heard nothing of the noise the spectators made. Time, as she knew it, had stopped. By instinct, she remained silent and waiting, passive against Stash's soaking-wet polo shirt. She could smell his sweat and it confounded her with desire. Her mouth filled with saliva. She wanted to sink her teeth into his tan neck, to bite him until she would taste his blood, to lick up the rivulets of sweat which ran down to his open collar. She wanted him to fall to the ground with her in his arms, just as he was, flushed, steaming, still breathing heavily from the game, and grind himself into her.

But this is the first of many points at which Mrs Krantz's minus capability for evocation leaves you puzzled. How did Stash get the hat back on Francesca's head? Did he remount, or is he just very tall? If he did remount, couldn't that have been specified? Mrs Krantz gives you all the details you don't need to form a mental picture, while carefully withholding those you do. Half the trick of pictorial writing is to give only the indispensable points and let the reader's imagination do the rest. Writers who not only give the indispensable points but supply all the concrete details as well can leave you feeling bored with their brilliance—Wyndham Lewis is an outstanding example. But a writer who supplies the concrete details and leaves out the indispensable points can only exhaust you. Mrs Krantz is right to pride herself on the accuracy of her research into every department of the high life. What she says is rarely inaccurate, as far as I can tell. It is, however, almost invariably irrelevant.

Anyway, the book starts with a picture of Daisy ('Her dark eyes, not quite black, but the colour of the innermost heart of a giant purple pansy, caught the late afternoon light and held it fast . . .') and then goes on to describe the meeting of her parents. It then goes on to tell you a lot about what her parents got up to before they met. Then it goes on to tell you about *their* parents. The book is continually going backwards instead of forwards, a canny insurance against the reader's impulse to skip. At one stage I tried skipping a chapter and missed out on about a century. From the upper West Side of New York I was suddenly in the Russian Revolution. That's where Stash gets his fiery temperament from—Russia.

'At Chez Mahu they found that they were able only to talk of unimportant things. Stash tried to explain polo to Francesca but she scarcely listened, mesmerised as she was with the abrupt movements of his tanned hands on which light blonde hair grew, the hands of a great male animal.' A bison? Typically, Mrs Krantz has failed to be specific at the exact moment when specificity would be a virtue. Perhaps Stash is like a horse not just in brain but in body. This would account for his tendency to view Francesca as a creature of equine provenance. 'Francesca

listened to Valensky's low voice, which had traces of an English
accent, a brutal man's voice which seemed to vibrate with
an underlying tenderness, as if he were talking to a newborn
foal...'

There is a lot more about Stash and Francesca before the
reader can get to Daisy. Indeed, the writer herself might never
have got to Daisy if she (i.e. Mrs Krantz) had not first wiped out
Stash and Francesca. But before they can be killed, Mrs Krantz
must expend about a hundred and fifty pages on various desperate attempts to bring them alive. In World War Two the incomparable polo-player becomes the great war hero. Those keen to
see Stash crash, however, are doomed to disappointment, since
before Stash can win medals in his Hurricane we must hear
about his first love affair. Stash is fourteen years old and the
Marquise Clair de Champery is a sex-pot of a certain age. 'She
felt the congestion of blood rushing between her primly pressed
together thighs, proof positive that she had been right to provoke the boy.' Stash, meanwhile, shows his customary tendency
to metamorphose into an indeterminate life-form. 'He took her
hand and put it on his penis. The hot sticky organ was already
beginning to rise and fill. It moved under her touch like an animal.' A field mouse? A boa constrictor?

Receiving the benefit of Stash's extensive sexual education,
Francesca conceives twins. One of the twins turns out to be
Daisy and the other her retarded sister, Danielle. But first Stash
has to get to the clinic. 'As soon as the doctor telephoned, Stash
raced to the clinic at 95 miles an hour.' Miserly as always with
essentials, Mrs Krantz trusts the reader to supply the information
that Stash is attaining this speed by some form of motorised
transport.

Stash rejects Danielle, Francesca flees with Danielle and Daisy.
Stash consoles himself with his collection of jet aircraft. Mrs
Krantz has done a lot of research in this area but it is trans-
parently research, which is not the same thing as knowledge.
Calling a Junkers 88 a Junker 88 might be a misprint, but her
rhapsody about Stash's prize purchase of 1953 is a dead giveaway. 'He tracked down and bought the most recent model
available of the Lockheed XP-80, known as the Shooting Star, a
jet which for many years could out-manoeuvre and out-perform
almost every other aircraft in the world.' USAF fighter aircraft

carried 'X' numbers only before being accepted for service. By 1953 the Shooting Star was known as the F-80, had been in service for years, and was practically the slowest thing of its type in the sky. But Mrs Krantz is too fascinated by that 'X' to let it go. She deserves marks, however, for her determination to catch up on the arcane nomenclature of boys' toys.

Stash finally buys a farm during a flying display in 1967. An old Spitfire packs up on him. 'The undercarriage of the 27-year-old plane stuck and the landing gear could not be released.' Undercarriage and landing gear are the same thing—her vocabularies have collided over the Atlantic. Also an airworthy 27-year-old Spitfire in 1967 would have been a very rare bird indeed: no wonder the undercarriage got in the road of the landing gear. But Mrs Krantz goes some way towards capturing the excitement of machines and should not be mocked for her efforts. Francesca, incidentally, dies in a car crash, with the make of car unspecified.

One trusts that Mrs Krantz's documentation of less particularly masculine activities is as meticulous as it is undoubtedly exhaustive, although even in such straightforward matters as food and drink she can sometimes be caught making the elementary mistake of piling on the fatal few details too many. Before Stash gets killed he takes Daisy to lunch every Sunday at the Connaught. After he gets killed he is forced to give up this practice, although there is no real reason why he should not have continued, since he is no more animated before his prang than after. Mrs Krantz has researched the Connaught so heavily that she must have made herself part of the furniture. It is duly noted that the menu has a brown and gold border. It is unduly noted that the menu has the date printed at the bottom. Admittedly such a thing would not happen at the nearest branch of the Golden Egg, but it is not necessarily the mark of a great restaurant. Mrs Krantz would probably hate to hear it said, but she gives the impression of having been included late amongst the exclusiveness she so admires. There is nothing wrong with gusto, but when easy familiarity is what you are trying to convey, gush is to be avoided.

Full of grand meals served and consumed at chapter length, *Princess Daisy* reads like *Buddenbrooks* without the talent. Food is important to Mrs Krantz: so important that her characters keep turning into it, when they are not turning into animals. Daisy has a half-brother called Ram, who rapes her, arouses her sexually, beats her up, rapes her again, and does his best to wreck her life because she rejects his love. His passion is understandable, when you consider Daisy's high nutritional value. 'He gave up the struggle and devoured her lips with his own, kissing her as if he were dying of thirst and her mouth were a moist fruit.' A mango? Daisy fears Ram but goes for what he dishes out. 'Deep within her something sounded, as if the string of a great cello had been plucked, a note of remote, mysterious but unmistakable warning.' Boing.

Daisy heeds the warning and lights out for the USA, where she becomes a producer of television commercials in order to pay Danielle's hospital bills. She pals up with a patrician girl called Kiki, whose breasts quiver in indignation—the first breasts to have done that for a long, long time. At such moments one is reminded of Mrs Krantz's true literary ancestry, which stretches all the way back to Elinor Glyn, E. M. Hull and Gertrude Atherton. She is wasting a lot of her time and too much of ours trying to be John O'Hara. At the slightest surge of congested blood between her primly pressed together thighs, all Mrs Krantz's carefully garnered social detail gives way to eyes like twin dark stars, mouths like moist fruit and breasts quivering with indignation.

There is also the warm curve of Daisy's neck where the jaw joins the throat. Inheriting this topographical feature from her mother, Daisy carries it around throughout the novel waiting for the right man to kiss it *tutto tremante*. Ram will definitely not do. A disconsolate rapist, he searches hopelessly among the eligible young English ladies—Jane Bonham-Carter and Sabrina Guinness are both considered—before choosing the almost inconceivably well-connected Sarah Fane. Having violated Sarah in his by now standard manner, Ram is left with nothing to do except blow Daisy's secret and commit suicide. As Ram bites the dust, the world learns that the famous Princess Daisy, star of a multi-million-dollar perfume promotion, has a retarded sister. Will this put the kibosh on the promotion, not to mention Daisy's love for the man in charge, the wheeler-dealer head of Supracorp, Pat Shannon ('larky bandit', 'freebooter' etc.)?

Daisy's libido, dimmed at first by Ram's rape, has already been reawakened by the director of her commercials, a ruthless but prodigiously creative character referred to as North. Yet North finally lacks what it takes to reach the warm curve of Daisy's neck. Success in that area is reserved for Shannon. He it is who undoes all the damage and fully arouses her hot blood. 'It seemed a long time before Shannon began to imprint a blizzard of tiny kisses at the point where Daisy's jaw joined her throat, that particularly warm curve, spendthrift with beauty, that he had not allowed himself to realise had haunted him for weeks. Daisy felt fragile and warm to Shannon, as if he'd trapped a young unicorn [horses again—C.J.], some strange, mythological creature. Her hair was the most intense source of light in the room, since it reflected the moonlight creeping through the windows, and by its light he saw her eyes, open, rapt and glowing; twin dark stars.'

Shannon might think he's got hold of some kind of horse, but as far as Daisy's concerned she's a species of cetacean. 'It was she who guided his hands down the length of her body, she who touched him wherever she could reach, as playfully as a dolphin, until he realised that her fragility was strength, and that she wanted him without reserve.'

Daisy is so moved by this belated but shatteringly complete experience that she can be forgiven for what she does next. 'Afterward, as they lay together, half asleep, but unwilling to drift apart into unconsciousness, Daisy farted, in a tiny series of absolutely irrepressible little pops that seemed to her to go on for a minute.' It takes bad art to teach us how good art gets done. Knowing that the dithyrambs have gone on long enough, Mrs Krantz has tried to undercut them with something earthy. Her tone goes wrong, but her intention is worthy of respect. It is like one of those clumsy attempts at naturalism in a late-medieval painting—less pathetic than portentous, since it adumbrates the great age to come. Mrs Krantz will never be much of an artist but she has more than a touch of the artist's ambition.

Princess Daisy is not to be despised. Nor should it be deplored for its concern with aristocracy, glamour, status, success and things like that. On the evidence of her prose, Mrs Krantz has not enough humour to write tongue-in-cheek, but other people are perfectly capable of reading that way. People don't get their morality from their reading matter: they bring their morality to it. The assumption that ordinary people's lives could be controlled and limited by what entertained them was always too condescending to be anything but fatuous.

Mrs Krantz, having dined at Mark's Club, insists that it is exclusive. There would not have been much point to her dining there if she did not think that. An even bigger snob than she is might point out that the best reason for not dining at Mark's Club is the chance of finding Mrs Krantz there. It takes only common sense, though, to tell you that on those terms exclusiveness is not just chimerical but plain tedious. You would keep better company eating Kentucky Fried Chicken in a launderette. But if some of this book's readers find themselves daydreaming of the high life, let us be grateful that Mrs Krantz exists to help give their vague aspirations a local habitation and a name. They would dream anyway, and without Mrs Krantz they would dream unaided.

To pour abuse on a book like this makes no more sense than to kick a powder-puff. *Princess Daisy* is not even reprehensible for the three million dollars its author was paid for it in advance. It would probably have made most of the money back without a dime spent on publicity. The only bad thing is the effect on Mrs Krantz's personality. Until lately she was a nice Jewish lady harbouring the usual bourgeois fancies about the aristocracy. But now she gives interviews extolling her own hard head. 'Like so many of us,' she told the *Daily Mail* on 28 April, 'I happen to believe that being young, beautiful and rich is more desirable than being old, ugly and destitute.' Mrs Krantz is fifty years old, but to judge from the photograph on the back of the book she is engaged in a series of hard-fought delaying actions against time. This, I believe, is one dream that intelligent people ought not to connive at, since the inevitable result of any attempt to prolong youth is a graceless old age.

1980